T0135337

Time-dependent Problems in Imaging and Parameter Identification

Barbara Kaltenbacher • Thomas Schuster
Anne Wald

Editors

Time-dependent Problems in Imaging and Parameter Identification

 Springer

Editors
Barbara Kaltenbacher
Department of Mathematics
Alpen-Adria-Universität Klagenfurt
Klagenfurt, Austria

Thomas Schuster
Department of Mathematics
Saarland University
Saarbrücken, Saarland, Germany

Anne Wald
Department of Mathematics
Saarland University
Saarbrücken, Saarland, Germany

ISBN 978-3-030-57786-5   ISBN 978-3-030-57784-1   (eBook)
https://doi.org/10.1007/978-3-030-57784-1

This Springer imprint is published by the registered company Springer Nature Switzerland AG
The registered company address is: Gewerbestrasse 11, 6330 Cham, Switzerland

Dedicated to our beloved families:

Manfred, Robert, Cornelia, and Katharina
Petra, Adrian, and Selina
Philip and Kira

Preface

Inverse problems deal with the reconstruction of unknown quantities from indirect observations and have a large number of applications ranging from medical imaging via nondestructive testing to geophysical prospection, to name just a few exemplary areas. These problems are by their nature ill-posed in the sense that small perturbations in the given data, e.g., due to measurement noise, can lead to large deviations in the reconstructions, unless appropriate mathematical tools—regularization methods—are developed and used. In addition, the question of uniqueness, i.e., whether the given observations unambiguously determine the searched-for quantity, is essential and often challenging to answer. Another important aspect in inverse problems is mathematical modeling, i.e., the correct description of the underlying relations and processes governing the connection between the parameters to be reconstructed on the one hand and the observed data on the other hand.

While investigations on these questions have so far mainly been focused on static problems, the dependence of parameters and/or states not only on space but also on time plays an increasingly important role in inverse problems applications: Time-resolved observations on the one hand allow to image evolutionary phenomena such as blood flow or a motion of the object, for example, a beating heart. On the other hand, they enable super-resolution in microscopy. Correspondingly, the underlying physical mechanisms are instationary and thus require modeling with time-dependent partial differential equations (PDEs)—typically wave, diffusion, or transport equations. The aim of this book is to collect some novel contributions on time-dependent parameter identification and imaging problems, therewith providing an overview on recent developments as well as a stimulus on further research in this area.

A key step for working with novel time-resolved imaging techniques is their proper **modeling** based on the underlying physics. Consequently, a considerable emphasis of the contributions in chapters "Joint Phase Reconstruction and Magnitude Segmentation from Velocity-Encoded MRI Data", "Quantitative OCT Reconstructions for Dispersive Media", "Inverse Problems of Single Molecule Localization Microscopy", and "Parameter Identification for the Landau–Lifshitz–Gilbert

Equation in Magnetic Particle Imaging" is on describing or actually even deriving such models.

Chapter "Joint Phase Reconstruction and Magnitude Segmentation from Velocity-Encoded MRI Data" provides a complete description of the inverse problem of velocity-encoded MRI from the acquisition process to the spin proton density estimation and devises a joint variational model to simultaneously estimate phase and magnitude reconstruction and its segmentation. The resulting non-convex and nonlinear optimization problem is then solved by a Bregman iteration.

In chapter "Quantitative OCT Reconstructions for Dispersive Media", optical coherence tomography (OCT) for layered media is described as an inverse problem for a one-dimensional wave equation, which can be explicitly solved in the nondispersive case, thus leading to a layer stripping type reconstruction algorithm. This approach is then extended to the practically relevant dispersive setting and the case of an absorbing medium is addressed as well.

Also, chapter "Parameter Identification for the Landau–Lifshitz–Gilbert Equation in Magnetic Particle Imaging" contains a modeling section, describing the physics of magnetic particle imaging (MPI) and yielding the Landau–Lifshitz–Gilbert (LLG) equation as a mathematical model. The task of calibration thus becomes a parameter identification problem for the LLG equation and is considered from two different inverse problem perspectives: A classical reduced approach relying on a parameter-to-state map and an all-at-once approach considering model and observations as a simultaneous system of operator equations for parameter and state.

Finally, chapter "Inverse Problems of Single Molecule Localization Microscopy" is entirely devoted to the derivation of a model for a novel super-resolution microscopy technique, namely single-molecule localization microscopy (SMLM). Here the resolution limit determined by Abbe's limit of diffraction is overcome by replacing the simultaneous acquisition of a microscopy image with a sequence of frames recording single blinking events. The chapter provides a complete mathematical description of the measurement setup by Maxwell's equations and adaptations thereof, thus finally establishing SMLM as an inverse problem for a coupled system of PDEs.

One of the key issues in the context of time-dependent imaging is **motion** of the target. This may be of periodic nature or irregular, and the actual motion may be the object of interest or an unwanted side effect—think of tracking flow in a pipe on the one hand and the movement of a patient during a recording on the other hand. Thus, detection and estimation of motion, as well as its compensation for proper imaging are key topics in chapters "Motion Compensation Strategies in Tomography", "Microlocal Properties of Dynamic Fourier Integral Operators", "Joint Motion Estimation and Source Identification Using Convective Regularisation with an Application to the Analysis of Laser Nanoablations", and "Tomographic Reconstruction for Single Conjugate Adaptive Optics".

Chapter "Joint Motion Estimation and Source Identification Using Convective Regularisation with an Application to the Analysis of Laser Nanoablations" considers spatially one-dimensional imaging of a certain morphogenesis process and proposes a variational method for joint motion estimation and source identification in the underlying one-dimensional mechanical model of tissue formation. Here the data misfit term in the variational model is defined as the L^2 residual of the nonhomogeneous continuity equation and regularization is based on the convective derivative of the source.

Chapter "Tomographic Reconstruction for Single Conjugate Adaptive Optics" deals with the reconstruction of atmospheric turbulence profiles from wavefront measurements. Here the wind speeds, which cause the motion of the different turbulent layers, play a crucial role for uniqueness: It is proven that mutually different velocities lead to unique identifiability of the turbulence profiles. This result is extended to the practically relevant setting of finite telescope aperture based on Fourier expansions.

Chapter "Motion Compensation Strategies in Tomography" provides a framework for compensating known motion in tomographic imaging in order to avoid artifacts in the reconstruction. The motion is described as a sequence of diffeomorphisms, corresponding to deformation vector fields. Moderate deformations, where the dynamic forward operator results as an appropriate concatenation of the static forward operator with a (multiplicative and superposition type) function of a diffeomorphism, allow for an explicit reversal of motion which, combined with a static reconstruction, yields a correct image from dynamic data. In the complementary case of strong deformations, the method of the approximate inverse is shown to provide efficient motion compensation.

In chapter "Microlocal Properties of Dynamic Fourier Integral Operators", methods from microlocal analysis are used to answer the question which features (in the sense of singularities) of the imaged object can be recovered from the given data and in which locations motion artifacts can arise. The analysis is based on the theory of Fourier integral operators, which covers many relevant dynamic tomographic methods based on integral transforms, such as the classical Radon transform in X-ray CT or the circular Radon transform in a particular setting of photoacoustic tomography but is probably also applicable to PDE-based models.

Reconstruction schemes are often built on existing paradigms known from static inverse and imaging problems and need to be adapted to the time-dependent setting in order to work efficiently. In view of practical applications, the solution of time-dependent problems often involves elaborate computations, which demands fast reconstruction schemes. In the case of ill-posed inverse problems, they also need to incorporate regularization. Iterative reconstruction schemes, as considered, e.g., in chapters "Joint Phase Reconstruction and Magnitude Segmentation from Velocity-Encoded MRI Data", "Dynamic Inverse Problems for the Acoustic Wave Equation", "Sequential Subspace Optimization for Recovering Stored Energy Functions in Hyperelastic Materials from Time-Dependent Data", and "Parameter Identification for the Landau–Lifshitz–Gilbert Equation in Magnetic Particle Imaging", generate successive approximations of the searched-for quantity and achieve regularization by a well-chosen stopping criterion. To guarantee convergence of these methods, conditions on the forward problem need to be verified, such as differentiability of the forward operator and certain restrictions on its nonlinearity such as the so-called tangential cone condition.

The latter is investigated in chapter "The Tangential Cone Condition for Some Coefficient Identification Model Problems in Parabolic PDEs", first of all in an abstract setting of evolution models and then for several concrete inverse source or coefficient problems for linear and nonlinear parabolic PDEs, contrasting the reduced and the all-at-once approach of formulating the inverse problem.

For the identification of space- and time-dependent coefficients in an acoustic wave equation, chapter "Dynamic Inverse Problems for the Acoustic Wave Equation" provides results on the derivative of the forward operator in appropriate function spaces, as well as its adjoint, as required for iterative regularization. Likewise, this is done in chapter "Parameter Identification for the Landau–Lifshitz–Gilbert Equation in Magnetic Particle Imaging" for magnetic particle imaging and in chapter "Sequential Subspace Optimization for Recovering Stored Energy Functions in Hyperelastic Materials from Time-Dependent Data" for structural health monitoring with hyperelastic waves.

The latter, i.e., chapter "Sequential Subspace Optimization for Recovering Stored Energy Functions in Hyperelastic Materials from Time-Dependent Data", also devises an accelerated iteration scheme, namely sequential subspace optimization, which uses, instead of only one gradient step, a linear combination of directions with optimized step sizes. This method is applied to a dictionary discretization of the stored energy function characterizing the hyperelastic material and thus, via its spatial variability, allowing to detect damage in the inspected medium.

Whenever a quantity is to be determined from indirect measurements, the question arises whether these data uniquely determine the hidden object. The more complex the underlying model that connects the searched-for parameters with the given observations, the more challenging is the answer to this question. **Uniqueness** proves are provided in chapters "Holmgren-John Unique Continuation Theorem for Viscoelastic Systems" and "An Inverse Source Problem Related to Acoustic Nonlinearity Parameter Imaging".

Chapter "Holmgren-John Unique Continuation Theorem for Viscoelastic Systems" establishes the unique continuation property for a viscoelastic system, i.e., a hyperbolic PDE with a memory term, based on the Holmgren-John concept of proving uniqueness for problems with analytic coefficients and noncharacteristic boundaries. This is essential for solving inverse problems such as the detection of obstacles or inclusions.

For a nonlinear acoustic wave equation (the Moore–Gibson–Thompson equation), the uniqueness of an inverse source problem as well as its conditional stability is established in chapter "An Inverse Source Problem Related to Acoustic Nonlinearity Parameter Imaging", as a first step into the mathematical analysis of ultrasonic nonlinearity imaging.

Segmentation and **Registration**, i.e., the decomposition of image regions according to their different intensities, as well as the alignment of different frames or modalities to a common coordinate system, are imaging tasks that play a role in many applications and require special adaptation in the time-dependent context.

Here similarity measures are crucial in order to identify objects of the same type on the one hand and to distinguish between different ones on the other hand. Chapter "Review of Image Similarity Measures for Joint Image Reconstruction from Multiple Measurements" provides a broad overview on similarity measures with a particular view on joint multi-modality imaging, as often also encountered in time-dependent settings.

In the context of imaging flow motion by velocity-encoded MRI, segmentation is implemented in chapter "Joint Phase Reconstruction and Magnitude Segmentation from Velocity-Encoded MRI Data" by means of a spatially resolved penalty term in a variational model.

Future research on time-dependent parameter identification and imaging problems will on the one hand be certainly directed into complementation and unification of the theory, as it has to a certain extent already been accomplished for static inverse problems. On the other hand, many new tailored methods for specific dynamic inverse problems as well as innovative general solution paradigms to tackle time-dependence are just about to emerge. In particular, questions concerning the optimal sampling of data or the promotion of special features of solutions such as spatial and/or temporal sparsity will be further investigated. Last but not least, in the exploding field of machine learning, evolutionary models turn out to play a crucial role.

1. Joint Phase Reconstruction and Magnitude Segmentation from Velocity-Encoded MRI Data (Veronica Corona, Martin Benning, Lynn F. Gladden, Andi Reci, Andrew J. Sederman, and Carola-Bibiane Schönlieb)
2. Dynamic Inverse Problems for the Acoustic Wave Equation (Thies Gerken)
3. Motion Compensation Strategies in Tomography (Bernadette N. Hahn)
4. Microlocal Properties of Dynamic Fourier Integral Operators (Bernadette N. Hahn, Melina-L. Kienle Garrido, and Eric Todd Quinto)
5. The Tangential Cone Condition for Some Coefficient Identification Model Problems in Parabolic PDEs (Barbara Kaltenbacher, Tram Thi Ngoc Nguyen, and Otmar Scherzer)
6. Sequential Subspace Optimization for Recovering Stored Energy Functions in Hyperelastic Materials from Time-Dependent Data (Rebecca Klein, Thomas Schuster, and Anne Wald)
7. Joint Motion Estimation and Source Identification Using Convective Regularisation with an Application to the Analysis of Laser Nanoablations (Lukas F. Lang, Nilankur Dutta, Elena Scarpa, Bénédicte Sanson, Carola-Bibiane Schönlieb, and Jocelyn Étienne)
8. Quantitative OCT Reconstructions for Dispersive Media (Peter Elbau, Leonidas Mindrinos, and Leopold Veselka)
9. Review on Image Similarity Measures for Joint Multi-modality Image Reconstruction (Ming Jiang)
10. Holmgren-John Unique Continuation Theorem for Viscoelastic Systems (Maarten V. de Hoop, Ching-Lung Lin, and Gen Nakamura)

Contents

Joint Phase Reconstruction and Magnitude Segmentation from Velocity-Encoded MRI Data

Veronica Corona, Martin Benning, Lynn F. Gladden, Andi Reci, Andrew J. Sederman, and Carola-Bibiane Schönliebs

Abstract Velocity-encoded MRI is an imaging technique used in different areas to assess flow motion. Some applications include medical imaging such as cardiovascular blood flow studies, and industrial settings in the areas of rheology, pipe flows, and reactor hydrodynamics, where the goal is to characterise dynamic components of some quantity of interest. The problem of estimating velocities from such measurements is a nonlinear dynamic inverse problem. To retrieve time-dependent velocity information, careful mathematical modelling and appropriate regularisation is required. In this work, we use an optimisation algorithm based on non-convex Bregman iteration to jointly estimate velocity-, magnitude- and segmentation-information for the application of bubbly flow imaging. Furthermore, we demonstrate through numerical experiments on synthetic and real data that the joint model improves velocity, magnitude and segmentation over a classical sequential approach.

1 Introduction

Magnetic resonance imaging (MRI) is an imaging technique that allows to visualise the chemical composition of patients or materials in a non-invasive fashion. Besides resolving in great detail the morphology of the object under consideration, MRI

V. Corona · C.-B. Schönlieb
Department of Applied Mathematics and Theoretical Physics, University of Cambridge, Cambridge, UK
e-mail: vc324@cam.ac.uk; cbs31@cam.ac.uk

M. Benning (✉)
School of Mathematical Sciences, Queen Mary University of London, London, UK
e-mail: m.benning@qmul.ac.uk

L. F. Gladden · A. Reci · A. J. Sederman
Department of Chemical Engineering and Biotechnology, University of Cambridge, Cambridge, UK
e-mail: lfg1@cam.ac.uk; ar622@cam.ac.uk; ajs40@cam.ac.uk

© Springer Nature Switzerland AG 2021
B. Kaltenbacher et al. (eds.), *Time-dependent Problems in Imaging and Parameter Identification*, https://doi.org/10.1007/978-3-030-57784-1_1

is intrinsically sensitive to motion, flow and diffusion [1, 2]. This means that in a single experiment, MRI can produce both structural and functional information. By designing the acquisition protocol appropriately, MRI can provide flow and motion estimation. This technique is known as MR velocimetry or phase-encoded MR velocity imaging [3–6]. In this work, we will focus on the inverse problem involved in recovering velocities from this kind of data.

In many MRI applications, the goal is not only to extract the structure of the object of interest, but also to estimate some functional features. An example is flow imaging, in which the aim is to reconstruct the velocity of the fluid that is moving in some structure. In order to acquire the velocity information and assess flow motion, phase-encoded MR velocity imaging is widely used in different areas. In medical imaging, this is used for example in cardiovascular blood flow studies to assess the distribution and variation in flow in blood vessels around the heart [7]. Other industrial applications include the study of the rheology of complex fluids [8], liquids and gases flowing through packed beds [9–11], granular flows [4, 12] and multiphase turbulence [13].

MRI scanners use strong magnetic fields and radio waves to excite subatomic particles (like protons) that subsequently emit radio frequency signals which can be measured by the radio frequency coils. Because the local magnetisation of the spins is a vector quantity, it is possible to derive both magnitude and phase images from the signal. Furthermore, for appropriately designed experiments, the velocity information can be estimated from the phase image. The problem of retrieving magnitude and phase (and therefore velocities) from such measurements is non-linear. Many standard approaches reduce this inverse problem to a complex but linear inverse problem, where magnitude and phase are estimated subsequently. With this strategy, however, it is impossible to impose regularity on the velocity information. In this work, we therefore propose a joint framework to simultaneously estimate magnitude and phase from undersampled velocity-encoded MRI. Based on [14], we additionally introduce a third task, that is the segmentation on the magnitude, to improve the overall reconstruction quality. The main motivation is that by estimating edges simultaneously from the data, both magnitude and segmentation are reconstructed more accurately. By enhancing the magnitude reconstruction, we expect in turn to improve the corresponding phase image and therefore the final velocity estimation.

Contributions

In this work we consider the problem of estimating flow, magnitude and segmentation of regions of interest from undersampled velocity-encoded MRI data. The problem is of great interest in different areas including cardiovascular blood flow analysis in medical imaging and rheology of complex fluids in industrial applications. To this end, we propose a joint variational model for undersampled velocity-encoded MRI. The significance of our approach is that by tackling the phase and magnitude reconstruction *jointly*, we can exploit their strong correlation and finally impose regularity on the velocity component. This is further assisted by the introduction of a segmentation term as additional prior to enhance edges of the regions of interest. Our main contributions are

- A description of the forward and inverse problem of velocity-encoded MRI in the setting of bubbly flow estimation.
- A joint variational framework for the approximation of the non-linear inverse problem of velocity estimation. We show that by exploiting the strong correlation in the data, our joint method yields an accurate estimation of the underlying flow, alongside a magnitude reconstruction that preserves and enhances intrinsic structures and edges, due to a joint segmentation approach. Moreover, we achieve an accurate segmentation to discern between different areas of interest, e.g. fluid and air.
- An alternating Bregman iteration method for non-convex optimisation problems.
- Numerical experiments on synthetic and real data in which we demonstrate the suitability and potential of our approach and provide a comparison with sequential approach.

Organisation of the Paper
This paper is organised as follows. In Sect. 2 we describe the derivation of the inverse problem of velocity-encoded MRI from the acquisition process to the spin proton density estimation. In Sect. 3 we present our joint variational model to jointly estimate phase and magnitude reconstruction and its segmentation. In Sect. 4 we propose an optimisation scheme to solve the non-convex and non-linear problem using Bregman iteration. To conclude, in Sect. 5 we demonstrate the performance of our proposed joint method in comparison with a sequential approach for synthetic and real MRI data.

2 Velocity-Encoded MRI

In the following we will briefly describe the mathematics of the acquisition process involved in MRI velocimetry. Subsequently we are going to see that finding the unknown spin proton density basically leads to solving the inverse problem of the Fourier transform.

2.1 *From the Bloch Equations to the Inverse Problem*

The magnetisation of a so-called spin isochromat can be described by the Bloch equations

$$\frac{d}{dt}\begin{pmatrix} M_x(t) \\ M_y(t) \\ M_z(t) \end{pmatrix} = \begin{pmatrix} -\frac{1}{T_2} & \gamma B_z(t) & -\gamma B_y(t) \\ -\gamma B_z(t) & -\frac{1}{T_2} & \gamma B_x(t) \\ \gamma B_y(t) & -\gamma B_x(t) & -\frac{1}{T_1} \end{pmatrix}\begin{pmatrix} M_x(t) \\ M_y(t) \\ M_z(t) \end{pmatrix} + \begin{pmatrix} 0 \\ 0 \\ \frac{M_0}{T_1} \end{pmatrix}.$$

(1)

Here $M(t) = (M_x(t), M_y(t), M_z(t))$ is the nuclear magnetisation (of the spin isochromat), γ is the gyromagnetic ratio, $B(t) = (B_x(t), B_y(t), B_z(t))$ denotes the magnetic field experienced by the nuclei, T_1 is the longitudinal and T_2 the transverse relaxation time and M_0 the magnetisation in thermal equilibrium. If we define $M_{xy}(t) = M_x(t) + i M_y(t)$ and $B_{xy}(t) = B_x(t) + i B_y(t)$, with $i^2 = -1$, we can rewrite (1) to

$$\frac{d}{dt}M_{xy}(t) = -i\gamma \left(M_{xy}(t)B_z(t) - M_z(t)B_{xy}(t) \right) - \frac{M_{xy}(t)}{T_2}$$

(2a)

$$\frac{d}{dt}M_z(t) = i\frac{\gamma}{2} \left(M_{xy}(t)\overline{B_{xy}}(t) - \overline{M_{xy}}(t)B_{xy}(t) \right) - \frac{M_z(t) - M_0}{T_1}$$

(2b)

with $\overline{\cdot}$ denoting the complex conjugate of $\cdot$.

If we assume for instance that $B = (0, 0, B_0)$ is just a constant magnetic field in z-direction, (2) reduces to the decoupled equations

$$\frac{d}{dt}M_{xy}(t) = -i\gamma B_0 M_{xy}(t) - \frac{M_{xy}(t)}{T_2},$$

(3a)

$$\frac{d}{dt}M_z(t) = -\frac{M_z(t) - M_0}{T_1}.$$

(3b)

It is easy to see that this system of Eqs. (3) has the unique solution

$$M_{xy}(t) = e^{-t(i\omega_0 + 1/T_2)}M_{xy}(t_j)$$

(4a)

$$M_z(t) = M_z(t_j)e^{-\frac{t}{T_1}} + M_0\left(1 - e^{-\frac{t}{T_1}}\right)$$

(4b)

for $\omega_0 := \gamma B_0$ denoting the Larmor frequency, and $M_{xy}(t_j)$, $M_z(t_j)$ being the initial magnetisations at time $t = t_j \geq 0$.

2.2 Signal Recovery

The key idea to enable spatially resolved nuclear magnetic resonance spectrometry is to add a magnetic field $\hat{B}(t)$ to the constant magnetic field B_0 in z-direction that varies spatially over time. Then, (3a) changes to

$$\frac{d}{dt}M_{xy}(t) = -i\gamma(B_0 + \hat{B}(t))M_{xy}(t) - \frac{M_{xy}(t)}{T_2},$$

which, for initial value $M_{xy}(t_j)$, has the unique solution

$$M_{xy}(t) = e^{-i\gamma\left(B_0 t + \int_{t_j}^t \hat{B}(\tau)\,d\tau\right)} e^{-\frac{t}{T_2}} M_{xy}(t_j), \tag{5}$$

if we ensure $\hat{B}(t_j) = 0$. If now $x(t)$ denotes the spatial location of a considered spin isochromat at time t, we can write $\hat{B}(t)$ as $\hat{B}(t) = x(t) \cdot g(t)$, with a vectorial function $g : [0, \infty) \to \mathbb{R}^3$ that describes the influence of the magnetic field gradient over time.

If a radio-frequency (RF) pulse that has been used to induce magnetisation in the x-y-plane is subsequently turned off at some time $t = t_*$ and thus, $B_x(t) = 0$ and $B_y(t) = 0$ for $t > t_* > t_j$, the same coils that have been used to induce the RF pulse can be used to measure the x-y magnetisation. Using (4a) and assuming $t_* < t \ll T_2$ for all $x \in \mathbb{R}^3$, this gives rise to the following model-equation:

$$M_{xy}(t) = e^{-i\gamma\left(B_0 t + \int_{t_j}^t x(\tau)\cdot g(\tau)\,d\tau\right)} M_{xy}(t_j). \tag{6}$$

In the following we assume that $x(t)$ can be approximated reasonably well via its Taylor approximation around a time $t = t_j$, i.e.

$$x(t) = \sum_{n=0}^{\infty} \frac{x^{(n)}(t_j)}{n!}\left(t - t_j\right)^n,$$

which yields

$$\int_{t_j}^t x(s) \cdot g(s)\,ds = \sum_{n=0}^{\infty}\left[\frac{x^{(n)}(t_j)}{n!} \cdot \int_{t_j}^t g(s)\left(s - t_j\right)^n ds\right]$$

$$= \sum_{n=0}^{\infty}\left[\frac{x^{(n)}(t_j)}{n!} \cdot \int_0^{t-t_j} g(\tau + t_j)\,\tau^n\,d\tau\right], \tag{7}$$

for $t \geq t_j$. It is well-known that appropriate application of gradients (i.e. appropriate design of g) enables the approximation of individual moments of (7). If we further assume that the system to be observed does only contain zero- and first-order moments, we can assume

$$\int_{t_j}^t x(s) \cdot g(s)\,ds = x_j \cdot \int_{t_j}^t g(\tau + t_j)\,d\tau + \varphi_j \cdot \int_{t_j}^t g(\tau + t_j)\,\tau\,d\tau, \tag{8}$$

where x_j is now short for $x(t_j)$ and $\varphi_j := x'(t_j)$ is the corresponding velocity information.

Equation (8) allows us to turn (6) into a useful mathematical model as we can encode velocity information and remove the temporal dependency of x. For notational convenience, we denote

$$\xi_j(t) := \int_{t_j}^{t} g(\tau + t_j) \, d\tau \quad \text{and} \quad \zeta_j(t) := \int_{t_j}^{t} g(\tau + t_j) \tau \, d\tau .$$

Note that throughout this work we will refer to ζ_j as the velocity-encoding gradients.

Since the RF-coils measure a volume of the whole x-y net-magnetisation, the acquired signal then equals

$$f_j(t) = \int_{\mathbb{R}^3} u(x_j) \, e^{-i\gamma \left(B_0(x_j)t + \varphi_j \cdot \zeta_j(t) \right)} \, e^{-i\gamma x_j \cdot \xi_j(t)} \, dx_j . \tag{9}$$

with $u(x_j)$ denoting the spin-proton density $M_{xy}(t_j)$ at a specific spatial coordinate $x_j \in \mathbb{R}^3$. Note that for $r(x_j) := u(x_j) \, e^{-i\gamma \left(B_0(x_j)t + \varphi_j \cdot \zeta_j(t) \right)}$ we observe that f is just the Fourier transform of the complex signal $r(x_j)$ with magnitude $u(x_j)$ and phase $-\gamma(B_0(x_j)t + \varphi_j \cdot \zeta_j)$.

2.3 Removal of Background Magnetic Field

Our goal is to recover the velocity information φ from f. Assuming that we do not know B_0, we can alternatively conduct two experiments, where the setup is identical apart from the velocity-encoding gradients having opposite polarities, i.e. we take two measurements with velocity-encoding gradients ζ_j^+ and ζ_j^- that satisfy $\zeta_j^- = -\zeta_j^+$. Dropping the $\pm$-notation for ζ_j and replacing it with a single ζ_j, we measure

$$f_j^+(t) = \int_{\mathbb{R}^3} u(x_j) \, e^{-i\gamma \left(B_0(x_j)t + \varphi(x_j) \cdot \zeta_j(t) \right)} \, e^{-i\gamma x_j \cdot \xi_j(t)} \, dx_j , \tag{10a}$$

$$f_j^-(t) = \int_{\mathbb{R}^3} u(x_j) \, e^{-i\gamma \left(B_0(x_j)t - \varphi(x_j) \cdot \zeta_j(t) \right)} \, e^{-i\gamma_j x \cdot \xi_j(t)} \, dx_j . \tag{10b}$$

If we define $\varphi_j^+(x_j, t) := B_0(x_j)t + \varphi(x_j) \cdot \zeta_j(t)$ and $\varphi_j^-(x_j, t) := B_0(x_j)t - \varphi(x_j) \cdot \zeta_j(t)$, we immediately observe

$$\varphi(x_j) \cdot \zeta_j(t) = \frac{1}{2} \left(\varphi_j^+(x_j, t) - \varphi_j^-(x_j, t) \right) .$$

The inverse problem of (10) is to recover $u(x_j)$ and φ_j from f_j^+ and f_j^-.

2.4 Zero-Flow Experiment

As we are going to focus on applications in chemical engineering, we have the ability to manipulate our sample and perform additional experiments to correct for certain artefacts in the acquisition. In particular, a zero-flow experiment is conducted. This experiment is to account for imperfections in the measurement system which causes an added signal between the positive and negative ζ experiments even in the absence of flow, and enables a correction that allows direct quantification of flow and tissue motion. We refer to this technique as flow compensation, which consists of acquiring a reference scan, with any flow switched-off, with vanishing zero and first gradient moments, before the actual velocity encoding scan with added bipolar gradients is performed. In this way, we obtain background phase images from the reference scan, and velocity sensitivity with the second flow-sensitive scan. In practice, this means that in addition to (10), the following two measurements are taken:

$$f_j^{\text{noflow}^+}(t) = \int_{\mathbb{R}^3} u(\boldsymbol{x}_j)\, e^{-i\gamma\varphi_j^{\text{noflow}^+}(\boldsymbol{x}_j,t)}\, e^{-i\gamma\boldsymbol{x}_j\cdot\xi_j(t)}\, d\boldsymbol{x}_j\,, \tag{11a}$$

$$f_j^{\text{noflow}^-}(t) = \int_{\mathbb{R}^3} u(\boldsymbol{x}_j)\, e^{-i\gamma\varphi_j^{\text{noflow}^-}(\boldsymbol{x}_j,t)}\, e^{-i\gamma\boldsymbol{x}_j\cdot\xi_j(t)}\, d\boldsymbol{x}_j\,, \tag{11b}$$

so that the actual velocity information can be recovered via

$$\varphi(\boldsymbol{x}_j)\cdot\zeta_j(t) = \frac{1}{2}\left(\left(\varphi_j^+(\boldsymbol{x}_j,t) - \varphi_j^-(\boldsymbol{x}_j,t)\right) - \left(\varphi_j^{\text{noflow}^+}(\boldsymbol{x}_j,t) - \varphi_j^{\text{noflow}^-}(\boldsymbol{x}_j,t)\right)\right). \tag{12}$$

The inverse problem is to recover u and φ from (10) and (11) via (12). More details on phase-encoded MR velocity imaging can be found in [15].

In other words, for a given direction of the velocity to be measured (x, y or z), the corresponding component velocity map (φ_x, φ_y or φ_z) is acquired by applying repeatedly a pulse sequence with the velocity-encoding gradient in the respective direction (x, y or z) and with alternating polarity between consecutive pulse sequences. The difference between the phase of the MRI image reconstructed from the acquired measurements of consecutive pulse sequences, and the reference to a zero flow experiment, yields the component velocity map.

2.5 Sampling

The discrete MRI signal is acquired by sampling the continuous signals of f_j^+, f_j^-, $f_j^{\text{noflow}^+}$ and $f_j^{\text{noflow}^-}$ at m discrete points in time. Based on the previous model assumptions, the acquisition of an individual sample reads

$$(f_j)_l := \int_{t_j}^{t_{j+1}} \Psi(t, t_l^j) \left[\int_{\mathbb{R}^3} u(\mathbf{x}_j) e^{-i\gamma \varphi_j(\mathbf{x}_j, t)} e^{-i\gamma \mathbf{x}_j \cdot \xi_j(t)} d\mathbf{x}_j \right] dt , \qquad (13)$$

for $t_j \leq t_1^j < t_2^j < \ldots < t_m^j \leq t_{j+1}$, between two consecutive times t_j and t_{j+1} with $0 < t_j < t_{j+1}$, $f_j \in \{f_j^+, f_j^-, f^{\text{noflow}_j^+}, f^{\text{noflow}_j^-}\}$ and $\varphi_j \in \{\varphi_j^+, \varphi_j^-, \varphi_j^{\text{noflow}^+}, \varphi_j^{\text{noflow}^-}\}$. Here Ψ denotes the sampling function or distribution. If we for example assume $\Psi(t, t_l^j) = \delta(t - t_l^j)$, where δ denotes the Dirac delta distribution, then (13) simplifies to

$$(f_j)_l = \int_{\mathbb{R}^3} u(\mathbf{x}_j) e^{-i\gamma (\varphi_j(\mathbf{x}_j))_l} e^{-i\gamma \mathbf{x}_j \cdot (\xi_j)_l} d\mathbf{x}_j , \qquad l \in \{1, \ldots, m\}, \qquad (14)$$

for

$$(\xi_j)_l := \int_{t_j}^{t_l^j} g(\tau + t_j) d\tau \quad \text{and} \quad (\varphi_j(\mathbf{x}_j))_l := \varphi_j(\mathbf{x}_j, t_l^j) .$$

This, together with the relation

$$\varphi(\mathbf{x}_j) \cdot (\zeta_j)_l = \frac{1}{2} \left(\left((\varphi_j^+(\mathbf{x}_j))_l - (\varphi_j^-(\mathbf{x}_j))_l \right) - \left((\varphi_j^{\text{noflow}^+}(\mathbf{x}_j))_l - (\varphi_j^{\text{noflow}^-}(\mathbf{x}_j))_l \right) \right)$$

for

$$(\zeta_j)_l := \int_{t_j}^{t_l^j} g(\tau + t_j) \tau \, d\tau$$

is our final acquisition model. We record and store all m measurements in vectors $f_j \in \mathbb{C}^m$. For the remainder of this work, we rewrite (14) (in vectorial form) as

$$f_j = SF \left(u_j e^{i\varphi_j} \right) , \qquad (15)$$

where SF denotes the (sub-sampled) Fourier transform, $f_j \in \mathbb{C}^m$ denotes the vector of Fourier-samples, and the space of Fourier samples is commonly referred to as k-space. Please note that we have fixed $\gamma = 1$ to simplify notation, but different choices of γ can certainly be used. Sampling strategies are very important to reduce the acquisition times and therefore to be able to image dynamic systems using velocity-encoded MRI through fast imaging techniques. The main idea is to exploit redundancy in some specific domain of the measured data. This approach is strongly related to the theory of compressed sensing (CS) [16–18] and many image reconstruction techniques have been proposed [11, 13, 19–22].

Depending on whether $\gamma\xi$ is sampled on a uniform or non-uniform grid, SF can be realised via the Fast Fourier Transform (FFT) [23] or via a non-uniform Fourier Transform such as NUFFT [24].

2.6 Dynamic Inverse Problem

We want to highlight that the index j in (15) suggests that the spin-proton density u and the velocity φ in (15) can be studied over time. To do so, one could take a sequence of s measurements each, at (initial) times $\{t_j\}_{j=0}^{s-1}$, for $0 < t_1 < t_2 < \ldots < t_s$, so that we have a sequence of measurements $\{f_j\}_{j=0}^{s-1}$. This way, we would easily introduce a discrete temporal dimension to our inverse problem that potentially allows us to exploit any temporal correlation between frames $\{u_j\}_{j=0}^{s-1}$ and $\{\varphi_j\}_{j=0}^{s-1}$. In this work, however, we will only consider the reconstruction of individual frames for reasons that we are going to address later.

In the following, we will only consider an individual frame of the dynamic inverse problem for velocity-encoded MRI in the discrete setting and under the presence of noise, making use of the notation of the discrete Fourier transform operator SF.

3 Mathematical Modelling

In this section we first present the velocity-encoded MRI reconstruction inverse problem in the presence of noise and discuss a sequential variational regularisation scheme to approximate the solution. Secondly, we introduce our joint reconstruction and segmentation approach in a Bregman iteration framework to jointly estimate phase, magnitude and segmentation.

3.1 Indirect Phase-Encoded MR Velocity Imaging

The velocity-encoded MRI image reconstruction problem is described as follows. Let $u, \varphi \in \mathbb{R}^n$ be the proton density or magnitude image and correspondent phase image, respectively, in a discretised image domain $\Omega := \{1, \ldots, n_1\} \times \{1, \ldots, n_2\}$, with $n = n_1 n_2$. The vector $f = (f_l)_{l=1}^m \in \mathbb{C}^m$ with $m \ll n$ are the measured Fourier coefficients obtained from (15). Based on (15) the forward model for noisy data is given by

$$f = SF\left(ue^{i\varphi}\right) + \eta, \tag{16}$$

where η is Gaussian noise with zero mean and standard deviation σ. For brevity we will follow the notation $A = SF$. As explained in the previous section, velocity information is encoded in the phase image. However, during the acquisition the phase is perturbed by an error due to field inhomogeneity and chemical shift. To account for this error, usually different measurements corresponding to different polarities of encoding flow gradients are acquired. Then the velocity (in one direction) at one particular time will be estimated as in (12), where ζ is a constant known from the acquisition setting.

Given the presence of noise and partial observation of the data due to undersampling, the problem described in (16) is ill-posed. A simple strategy to obtain an approximated solution is to replace with zero the missing Fourier coefficients and compute the so-called zero-filling solution

$$r_z = A^* f \tag{17}$$

where $r = ue^{i\varphi}$. However, these reconstructed images will contain aliasing artefacts because of the undersampling. A classical approach to solve this problem is to compute approximate solutions of (16) using a variational regularisation approach. We consider a Tikhonov-type regularisation approach that reads

$$r_j \in \arg\min_r \left\{ \frac{1}{2} \|A_j r - f_j\|_2^2 + \alpha J(r) \right\}, \tag{18}$$

for $j \in \{1, \ldots, 4\}$ being the different measurements, where the first term is the data fidelity that imposes consistency between the reconstruction and the given measurements f, the second term is the regularisation, which incorporates some prior knowledge of the solution. The parameter $\alpha > 0$ is a regularisation parameter that balances the two terms in the variational scheme. In this setting, the survey proposed in [25] describes different choices for the regularisation functional J, including wavelets and higher-order total variation (TV) schemes. Subsequently, the phases can be extracted from these complex images $r_j = u_j e^{i\varphi_j}$ as

$$\varphi_j = \arg(r_j). \tag{19}$$

More recently, other reconstruction approaches have been proposed to regularise the phase of the image [26–30]. All these methods rely on modelling separately prior knowledge on the magnitude and on phase images and differ on the optimisation schemes involved in the non-convex and non-linear problem. However, while it is possible to exploit information about the velocity from fluid mechanics, it is in general hard to assume specific knowledge on the individual phases. As explained in the previous section and described in (12), velocities are computed as phase differences of different MR measurements and therefore the regularisation needs to be imposed on the phase difference rather than individual phases. In this work, we step away from the approach of only regularising individual phases and propose instead to regularise the velocity as difference of phases. In the following we describe our choice of regularisation and algorithmic framework for velocity-encoded MRI.

3.2 Joint Variational Model

In many industrial applications, velocity-encoded MRI is used to estimate flow of different chemical species in different physical status, such as gas-liquid systems [31]. In this case, one aims at recovering a piecewise constant image or an image with sharp edges to facilitate further analysis such as identification of regions of interest. It was proposed in [14] to use a segmentation task as additional regularisation on the reconstruction to impose regularity in terms of sharp edges. It was shown there that this is highly beneficial for very low undersampling rates in MRI. In this work, we expand this idea to the phase-encoded MR velocity imaging data, where the idea is to jointly solve for magnitude, segmentation and phase improving performances on the three tasks.

Following the work in [14], we are interested in the *joint model* to recover magnitude u_j and velocity φ components through the measured phases φ_j from undersampled MRI data f_j and to estimate a segmentation v_j on the magnitude images. As described in the previous section, we are dealing with four MRI measurements to obtain one component velocity image. Defining the shorthand notations $u := \{u_j\}_{j=1}^4$, $v := \{v_j\}_{j=1}^4$ and $\varphi := \{\varphi_j\}_{j=1}^4$, this joint model reads as

$$
E(u, v, \varphi) = \sum_{j=1}^4 \left\{ \underbrace{\frac{1}{2} \|A(u_j e^{i\varphi_j}) - f_j\|_2^2}_{\text{reconstruction}} \right.
$$

$$
\left. + \delta \underbrace{\sum_n v_{nj}(c_1 - u_{nj})^2 + (1 - v_{nj})(c_2 - u_{nj})^2}_{\text{segmentation}} \right\}. \tag{20}
$$

The first term in (20) describes the reconstruction fidelity term for the magnitudes u and phases φ for the given data $f := \{f_j\}_{j=1}^4$. Note that we now write $1, \ldots, 4$, instead of $+$, $-$, noflow$+$, noflow$-$ for simplicity in the notation. The second term represents the segmentation problem to find partitions v, with $v_{nj} \in [0, 1]$, of the images u in two disjoint regions that have mean intensity values close to the constants c_1 and c_2 [32, 33]. Note that in our experiments we solve the problems for fixed constants leveraging on the prior knowledge that air appears dark and water bright. However, it is also possible to minimise over c_1 and c_2, fixing the other variables, yielding to a simple update as the average intensity of the image inside the segmented region. The parameter δ weighs the effect of the segmentation onto the reconstruction. The underlying idea is to exploit structure and redundancy in the data, estimating edges simultaneously from the data, ultimately improving the reconstruction. By incorporating prior knowledge of the regions of interest we impose additional regularity of the solution.

The joint cost function (20) is non-convex. While sub-problems in u and v (leaving the other parameters fixed) are convex, the sub-problems in φ are non-linear and non-convex. In the next section we present a unified framework based on non-convex Bregman iterations to solve the joint model.

4 Optimisation

There are many ways of minimising (20). We want to pursue a strategy that guarantees smooth velocity-components, piecewise-constant segmentations and magnitude images with sharp transitions in an inverse scale-space fashion. In order to achieve those features, we aim to approximate minimisers of (20) via an alternating Bregman proximal method or Bregman iteration of the form

$$u_l^{k+1} \in \arg \min_u \left\{ E(u_1^{k+1}, \ldots, u_{l-1}^{k+1}, u, u_{l+1}^k, \ldots, u_d^k, v^k, \varphi^k) + D_{J_u}^{p_l^k}(u, u_l^k) \right\}, \tag{21a}$$

$$p_l^{k+1} = p_l^k - \frac{\partial}{\partial u_l} E(u_1^{k+1}, \ldots, u_{l-1}^{k+1}, u_l^{k+1}, u_{l+1}^k, \ldots, u_d^k, v^k, \varphi^k), \tag{21b}$$

$$v_l^{k+1} \in \arg \min_v \left\{ E(u^{k+1}, v_1^{k+1}, \ldots, v_{l-1}^{k+1}, v, v_{l+1}^k, \ldots, v_d^k, \varphi^k) + D_{J_v}^{q_l^k}(v, v_l^k) \right\}, \tag{21c}$$

$$q_l^{k+1} = q_l^k - \frac{\partial}{\partial v_l} E(u^{k+1}, v_1^{k+1}, \ldots, v_{l-1}^{k+1}, v_l^{k+1}, v_{l+1}^k, \ldots, v_d^k, \varphi^k), \tag{21d}$$

$$\varphi^{k+1} \in \arg \min_\varphi \left\{ \langle \partial_\varphi E(u^{k+1}, v^{k+1}, \varphi^k), \varphi \rangle + D_{J_\varphi}^{w^k}(\varphi, \varphi^k) \right\}, \tag{21e}$$

$$w^{k+1} = w^k - \frac{\partial}{\partial \varphi} E(u^{k+1}, v^{k+1}, \varphi^{k+1}). \tag{21f}$$

for $l = 1, \ldots, d := 4$, $u := (u_l)_{l=1}^d$, $v := (v_l)_{l=1}^d$ and $\varphi := (\varphi_l)_{l=1}^d$. Here J_u, J_v and J_φ are proper, lower semi-continuous and convex functions and $D_{J_u}^{p_l^k}(u, u_l^k)$, $D_{J_v}^{q_l^k}(v, v_l^k)$ and $D_{J_\varphi}^{w^k}(\varphi, \varphi^k)$ are the corresponding generalised Bregman distances [34, 35] with arguments and corresponding subgradients p_l^k, q_l^k and w^k. A generalised Bregman distance is the distance between a function J evaluated at argument u and its linearisation around argument v, i.e.

$$D_J^q(u, v) = J(u) - J(v) - \langle q, u - v \rangle,$$

for a subgradient $q \in \partial J(v)$. Note that algorithm (21) has update rules for the subgradients, as J_u, J_v and J_φ are allowed to be non-smooth, which makes the selection of particular subgradients necessary. Note that the chosen update formulas guarantee that p_l^{k+1}, q_l^{k+1} and w^{k+1} are valid subgradients again.

The algorithm is a hybrid of the algorithms proposed in [36] and [14]. For both algorithms global convergence results, motivated by Xu and Yin [37] and Bolte et al. [38], have been established. Convergence for (21) is out of the scope of this paper as we focus on the application of velocity-encoded MRI, and thus has not been proven here. However, we are confident that convergence can be derived under suitable assumptions following the analysis in [36] and [14].

Since we deal with imperfect data potentially corrupted by measurement noise and numerical errors, we will use (21) in combination with an early-stopping criterion in order not to converge to a minimiser of (20) but to approximate the solution of (16) via iterative regularisation.

The crucial part for the application of (21) are the choices of the underlying functions J_u, J_v and J_φ of the corresponding Bregman distances. We want both the magnitude images and the segmentations to maintain sharp discontinuities and therefore want to penalise their discretised, isotropic, total variation. On the other hand, we want to guarantee smooth components of our velocity field, which is why we penalise them with the two-norm of a discretised gradient. In particular, we choose

$$J_u(u) = \alpha\, \mathrm{TV}(u) := \alpha \|\,|\nabla u|\,\|_1\,, \qquad J_v(v) := \beta\, \mathrm{TV}(v), \tag{22}$$

to be the isotropic total variation with weights $\alpha > 0$ and $\beta > 0$, where $\nabla : \mathbb{R}^n \to \mathbb{R}^{2n}$ denotes a forward finite-difference approximation of the gradient, $|\cdot|$ the Euclidean vector norm and $\|\cdot\|_1$ the pixel-wise one-norm. Further, we choose J_φ in a way that allows to enable an H^1-norm-type smoothing on the difference of the phases, i.e.

$$J_\varphi k + 1(\varphi) = \frac{1}{2\tau}\left(\eta\|\,|\nabla\left((\varphi_1 - \varphi_2) - (\varphi_3 - \varphi_4)\right)|\,\|^2 + \sum_{l=1}^{d}\|\varphi_l\|^2\right),$$

where $\eta > 0$ and $\tau > 0$ denote some weights. In the first term, we penalise the difference of the 4 measurements introduced in Sects. 2.3 and 2.4 to impose that our actual velocity component is smooth. The additional 2-norm penalisation is introduced to ensure that problem (21e) is coercive and therefore a solution to (21e) exists. Note that all convex sub-optimisation-problems in (21) are solved numerically with a primal-dual hybrid gradient (PDHG) method [39–42]. Once we have approximated the magnitudes, labels and phases with this iterative regularisation strategy, we can compute the velocity components via (12).

5   Numerical Results

In this section we present numerical results of our method for the specific application of bubble burst hydrodynamics using MR velocimetry. The hydrodynamics of bursting bubbles is important in many different areas such as geophysical processes and bioreactor design. We refer to [43] for an overview on the field and the description of results on the first experimental measurement of the liquid velocity field map during the burst of a bubble at the liquid surface interface.

5.1   Case-Study on Simulated Dataset

To quantitatively evaluate our method, we consider the simulated k-space data of a rising spherical bubble in an infinite fluid during Stokes flow regime. The simulated data consists of 32 time frames, but for the sake of compactness we will show some visual outputs for one time step $t = 19$.

We assess the performance of our approach for velocity and magnitude estimation by comparing our solutions with respect to the groundtruth and using the mean squared error (MSE) defined as $\|x^{\text{groundtruth}} - x\|_2^2/n$, where n is the number of pixels in the image.

We also present a comparison with a sequential approach, where the magnitude is obtained with a classic CS TV-regularised approach and the phase is subsequently estimated using the method proposed in [36] and presented in [43] for the evaluation of bubbly flow estimation.

In Fig. 1 we can see the results for the sequential approach compared to the joint approach when sampling only 11% of the k-space data. Although visually there is not significant change, the MSE shows a big improvement for the joint approach. This confirms that using our joint model is relevant for the problem of velocity-encoded MRI. For the 32 frames, we report the average MSE for magnitude and phase in Table 1 where can see a drastic improvement compared to the sequential approach.

The parameters used in these experiments were set as follows: $\alpha = 1$, $\beta = 0.1$, $\delta = 1$, $c_1 = 0.9$, $c_2 = 0.1$, $\tau = 1.9$, and $\eta = 0.5$.

5.2   Real Dataset

In this section we present our model performance on real data acquired with the following protocol described in [43] and briefly reported here.

Acquisition Protocol
The experiments were conducted on an AV-400 Bruker magnet, operating at a resonant frequency of 400.25 MHz for ^{1}H observation with an RF coil of 25 mm

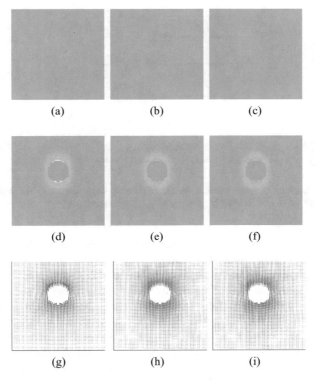

Fig. 1 Phase reconstructions for the sequential approach and our joint approach compared to the ground truth. Top row: x direction, middle row: z direction, bottom row: velocity plots. We sampled 11% of the k-space data. (**a**) Groundtruth. (**b**) Sequential MSE = 0.0030. (**c**) Joint MSE = 0.0020. (**d**) Groundtruth. (**e**) Sequential MSE = 0.0046. (**f**) Joint MSE = 0.0035. (**g**) Groundtruth. (**h**) Sequential. (**i**) Joint

Table 1 MSE for phase (φ_1 and φ_2) and magnitude (u_1 and u_2) images for the sequential and joint approaches. The error is significantly decreased using our proposed joint approach

	u_1	u_2	φ_1	φ_2
Sequential	0.0019	0.0028	0.0032	0.0059
Joint	**0.0011**	**0.0012**	**0.0018**	**0.0051**

diameter. The maximum magnetic field gradient amplitude available in each spatial direction is $146\,\text{Gcm}^{-1}$. The velocity images were acquired with a 2D MR spiral imaging technique developed and published in [44]. Images were acquired with 64×64 pixels over a field of view of 17×17 mm resulting in an image resolution of 265×265 mm, over a slice thickness of $150\,\mu\text{m}$. Data in k-space were acquired along a spiral trajectory at a sampling rate corresponding to 25% of full Nyquist sampling over a time of 2.05 ms for the entire image.

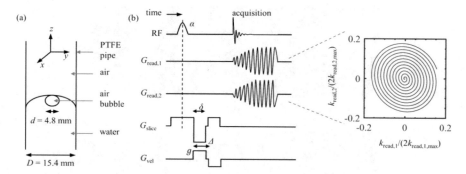

Fig. 2 (**a**) Schematic of experimental setup. (**b**) Pulse sequence used for MR velocimetry acquisitions and the corresponding k-space traversal. Taken from [43]

We acquire the three velocity components for a transverse slice (perpendicular to the axis of the pipe) and a longitudinal slice (parallel to the axis of the pipe), cutting through approximately the centre of the bubble. For a given slice direction (transverse or longitudinal) and a given direction of the velocity, four measurements corresponding to the application of the velocity-encoding gradient with alternating polarity and to the flow compensation, are taken, as discussed in Sect. 2 (see Fig. 2b). The final velocity for each component is then obtained as the difference between the phase of the MRI images reconstructed from the acquired k-space data of consecutive pulse sequences with flow on, and the reference to the zero flow experiment (see Sects. 2.3 and 2.4, respectively).

Experimental Results on Real Data

We now present the results for our joint model in comparison with the zero-filling solution and the corresponding sequential approach for real data acquired with the protocol described above. In Fig. 3 we show the result for a specific time frame for a bubble in a transversal and longitudinal view. At this specific time, the bubble is bursting which corresponds to an upward jet being ejected. As we can see, the zero-filling solution gives an indication of the flow velocity but it is very noisy and imprecise. In contrast, the joint approach removes noise and successfully estimates the velocity flow. The sequential approach on the other hand, although it produces a smoother reconstruction, results in small errors (see e.g. Fig. 3e on the left). In Fig. 6 we observe similar results for a different time frame. We refer to the Appendix for the full dynamic sequence result.

We also present the results for the magnitude and segmentation for the zero-filling solution, sequential approach and joint approach. We can see in Figs. 4 and 5 that the joint approach exploits the structure in the data and presents more accurate magnitude reconstructions and segmentations. It is clear that, even in this rather simple segmentation problem, the joint approach is able to improve the results of

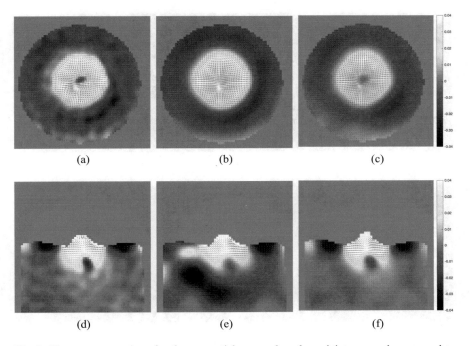

(a)   (b)   (c)

(d)   (e)   (f)

Fig. 3 Phase reconstructions for the sequential approach and our joint approach compared to the zero-filling solution. Results for a bursting bubble from a transversal view (top row) and longitudinal view (bottom row). (**a**) Zero-filled. (**b**) Sequential. (**c**) Joint. (**d**) Zero-filled. (**e**) Sequential. (**f**) Joint

both tasks. This gain is significant in Fig. 5f. Additionally, the joint magnitudes present very sharp edges distinguishing air and fluid thanks to the segmentation coupling term in the model, which acts as additional prior to reconstruct images exploiting prior knowledge on the region of interest. The parameters used in these experiments were set as follows: $\alpha = 0.08$, $\beta = 0.1$, $\delta = 3$, $c_1 = 0.9$, $c_2 = 0.1$, $\tau = 1.9$, $\eta = 1$ (Fig. 6).

6 Conclusion and Outlook

In this work we have presented a joint framework for flow estimation, magnitude reconstruction and segmentation from undersampled velocity-encoded MRI data. After having described the corresponding dynamic inverse problem, we have presented a joint variational model based on a non-convex Bregman iteration. We

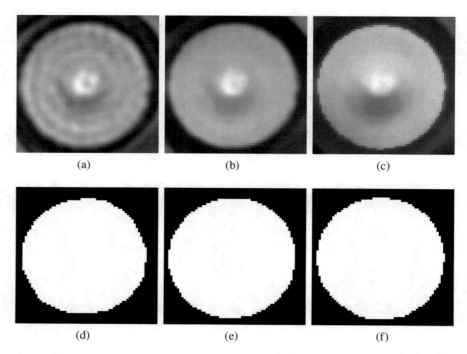

Fig. 4 Magnitude reconstructions (top row) and corresponding segmentations (bottom row) for the sequential approach and our joint approach compared to the zero-filling solution. Transversal view. (**a**) Zero-filled. (**b**) Sequential. (**c**) Joint. (**d**) Zero-filled. (**e**) Sequential. (**f**) Joint

have demonstrated that by imposing regularity on the individual components (in contrast to the sequential approach), our joint method achieves accurate estimations of the velocities, as well as an enhanced magnitude reconstruction with sharp edges, thanks to the joint segmentation. Furthermore, we assessed the performance of our joint approach on synthetic and real data. In this context, we have shown that the joint model improves the performances of the different imaging tasks compared to the classical sequential approaches.

Future work includes the investigation of the full joint temporal and spatial optimisation. By extending the model to the full 4D setting, we believe the performance will be enhanced further, as temporal correlation e.g. in the segmentation can be exploited. The current limitation is the lack of such 4D dataset. Indeed, as described in the acquisition protocol, the velocity data was acquired separately for each spatial component to speed up the acquisition.

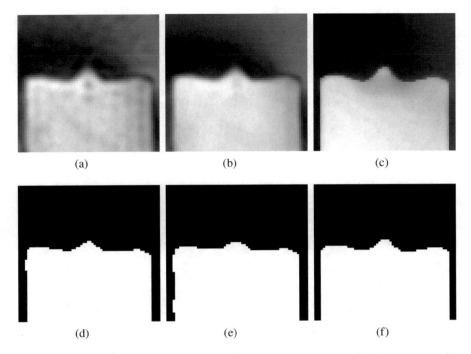

Fig. 5 Magnitude reconstructions (top row) and corresponding segmentations (bottom row) for the sequential approach and our joint approach compared to the zero-filling solution. Longitudinal view. (**a**) Zero-filled. (**b**) Sequential. (**c**) Joint. (**d**) Zero-filled. (**e**) Sequential. (**f**) Joint

Acknowledgments VC acknowledges the financial support of the Cambridge Cancer Centre and Cancer Research UK. MB acknowledges the Leverhulme Trust Early Career Fellowship ECF-2016–611, "Learning from mistakes: a supervised feedback-loop for imaging applications". CBS acknowledge support from Leverhulme Trust project "Breaking the non-convexity barrier", EPSRC grant "EP/M00483X/1", EPSRC centre "EP/N014588/1", the Cantab Capital Institute for the Mathematics of Information, and from CHiPS and NoMADS (Horizon 2020 RISE project grant). Moreover, CBS is thankful for support by the Alan Turing Institute.

Appendix 1: Details of Our Algorithm for Velocity-Encoded MRI

In this section, we write our proposed general algorithm (21) for the specific problem of joint velocity-encoded MRI reconstruction and magnitude segmentation. The algorithm for one measurement reads as

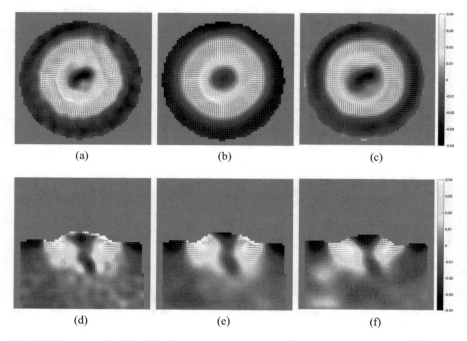

Fig. 6 Phase reconstructions for the sequential approach and our joint approach compared to the zero-filling solution. Top row: transversal view. Bottom row: longitudinal view. (**a**) Zero-filled. (**b**) Sequential. (**c**) Joint. (**d**) Zero-filled. (**e**) Sequential. (**f**) Joint

$$
u^{k+1} = \arg\min_u \left\{ \frac{1}{2} \|A(ue^{i\varphi^k}) - f\|_2^2 + D_{J_u}^{p^k}(u, u^k) \right.
$$

$$
\left. + \delta \sum_n v_n^k(c_1 - u_n)^2 + (1 - v_n^k)(c_2 - u_n)^2 \right\}
$$

$$
p^{k+1} = p^k - \left(\mathrm{Re}\left(e^{i\varphi^k}(A^*(A(u^{k+1}e^{i\varphi^k}) - f)) \right) \right.
$$

$$
\left. + 2\delta \left(v_n^k(u_n^{k+1} - c_1) + (1 - v_n^k)(u_n^{k+1} - c_2) \right) \right)
$$

$$
v^{k+1} = \arg\min_v \left\{ \delta \sum_n v_n(c_1 - u_n^{k+1})^2 + (1 - v_n)(c_2 - u_n^{k+1})^2 + D_{J_v}^{q^k}(v, v^k) \right\}
$$

$$
q^{k+1} = q^k - \delta \left((c_1 - u_n^{k+1})^2 - (c_2 - u_n^{k+1})^2 \right)
$$

$$
\varphi^{k+1} = \arg\min_\varphi \left\{ \tau \left\langle \varphi - \varphi^k, \mathrm{Re}\left(e^{i\varphi^k} A^*(A(u^{k+1}e^{i\varphi^k}) - f) \right) \right\rangle + D_{J_\varphi}^{r^k}(\varphi, \varphi^k) \right\}
$$

$$
r^{k+1} = r^k - \mathrm{Re}\left(e^{i\varphi^{k+1}} A^*(A(u^{k+1}e^{i\varphi^{k+1}}) - f) \right)
$$

The individual subproblems are solved in an alternating fashion by fixing the remaining variables. The optimisation for subproblems in u and in v is described in [14], with the minor modification of the phase φ, which remains fixed. The subproblem in φ is solved using the code in [45], again with minor modifications to include our regularisation functions and 4 raw measurements instead of 2.

Appendix 2: Further Numerical Results

In this section we show the full dynamic sequence of a bubble burst event. At time $t = 1$ the bubble resting at the air-liquid interface. When the thin liquid film breaks, the bubble burst, causing the formation of an upward and downward jet. The upward jet moves in the empty space left by the bubble and reached its maximum at $t = 4$. After that, the jet falls down into the liquid pool, causing a downward jet and some oscillation. At around $t = 8$ the liquid motion stops (Figs. 7 and 8).

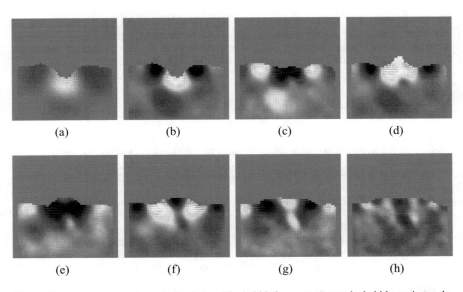

(a)   (b)   (c)   (d)

(e)   (f)   (g)   (h)

Fig. 7 Full time sequence. Longitudinal view. The bubble burst event sees the bubble resting at the interface between liquid and air, before this film is finally broken. The bursting causes an upward jet that moves the liquid at its highest position at $t = 4$. Subsequently, the jet drops into a downward jet, causing oscillation in the liquid, until it finally dies out at $t = 8$. (**a**) $t = 1$. (**b**) $t = 2$. (**c**) $t = 3$. (**d**) $t = 4$. (**e**) $t = 5$. (**f**) $t = 6$. (**g**) $t = 7$. (**h**) $t = 8$

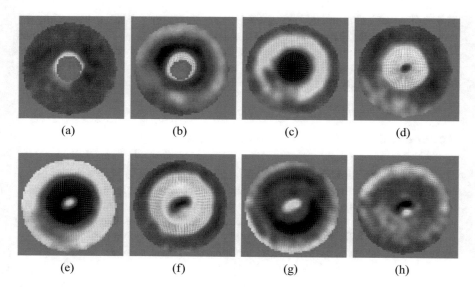

Fig. 8 Full time sequence. Transversal view through the middle of the bubble. We can see the bubble burst event and the upward/inward jet caused by the empty space left by the bubble. Subsequently, the jet falls down into the liquid pool causing a downward/outward jet, until it dies out at $t = 8$. (**a**) $t = 1$. (**b**) $t = 2$. (**c**) $t = 3$. (**d**) $t = 4$. (**e**) $t = 5$. (**f**) $t = 6$. (**g**) $t = 7$. (**h**) $t = 8$

References

1. C.T. Burt, NMR measurements and flow. J. Nucl. Med. **23**(11), 1044–1045 (1982)
2. L Axel, Blood flow effects in magnetic resonance imaging. Am. J. Roentgenol. **143**(6), 1157–1166 (1984)
3. P.T. Callaghan, *Translational Dynamics and Magnetic Resonance* (Oxford University Press, 2011)
4. E. Fukushima, Nuclear magnetic resonance as a tool to study flow. Ann. Rev. Fluid Mech. **31**(1), 95–123 (1999)
5. C.J. Elkins, M.T. Alley, Magnetic resonance velocimetry: applications of magnetic resonance imaging in the measurement of fluid motion. Exp. Fluids **43**(6), 823–858 (2007)
6. L.F. Gladden, A.J. Sederman, Recent advances in flow MRI. J. Magn. Reson. **229**, 2–11 (2013)
7. P.D. Gatehouse, J. Keegan, L.A. Crowe, S. Masood, R.H. Mohiaddin, K.-F. Kreitner, D.N. Firmin, Applications of phase-contrast flow and velocity imaging in cardiovascular MRI. Eur. Radiol. **15**(10), 2172–2184 (2005)
8. P.T. Callaghan, Rheo-NMR: nuclear magnetic resonance and the rheology of complex fluids. Rep. Progr. Phys. **62**(4), 599 (1999)
9. A.J. Sederman, M.L. Johns, P. Alexander, L.F. Gladden, Structure-flow correlations in packed beds. Chem. Eng. Sci. **53**(12), 2117–2128 (1998)
10. M.H. Sankey, D.J. Holland, A.J. Sederman, L.F. Gladden, Magnetic resonance velocity imaging of liquid and gas two-phase flow in packed beds. J. Magn. Reson. **196**(2), 142–148 (2009)
11. D.J. Holland, D.M. Malioutov, A. Blake, A.J. Sederman, L.F. Gladden, Reducing data acquisition times in phase-encoded velocity imaging using compressed sensing. J. Magn. Reson. **203**(2), 236–246 (2010)

12. D.J. Holland, C.R. Müller, J.S. Dennis, L.F. Gladden, A.J. Sederman, Spatially resolved measurement of anisotropic granular temperature in gas-fluidized beds. Powder Technol. **182**(2), 171–181 (2008)
13. A.B. Tayler, D.J. Holland, A.J. Sederman, L.F. Gladden, Exploring the origins of turbulence in multiphase flow using compressed sensing MRI. Phys. Rev. Lett. **108**(26), 264505 (2012)
14. V. Corona, M. Benning, M.J. Ehrhardt, L.F. Gladden, R. Mair, A. Reci, A.J. Sederman, S. Reichelt, C.-B. Schönlieb, Enhancing joint reconstruction and segmentation with non-convex Bregman iteration. Inverse Prob. **5**(5), 055001 (2019)
15. M. Markl, *Velocity Encoding and Flow Imaging*. University Hospital Freiburg, Dept. of Diagnostic Radiology, Germany, 2006
16. E. Candes, J. Romberg, T. Tao, Robust uncertainty principles: exact signal reconstruction from highly incomplete frequency information. IEEE Trans. Inf. Theory **52**(2)489–509 (2006)
17. D.L. Donoho, Compressed sensing. IEEE Trans. Inf. Theory **52**(4), 1289–1306 (2006)
18. M. Lustig, D. Donoho, J.M. Pauly, Sparse MRI: the application of compressed sensing for rapid MR imaging. Magn. Reson. Med. **58**(6), 1182–1195 (2007)
19. B. Jung, M. Honal, P. Ullmann, J. Hennig, M. Markl, Highly k-t-space-accelerated phase-contrast MRI. Magn. Reson. Med **60**(5):1169–1177 (2008)
20. E. Paciok, B. Blümich, Ultrafast microscopy of microfluidics: compressed sensing and remote detection. Angew. Chem. Int. Ed. **50**(23), 5258–5260 (2011)
21. J. Paulsen, V.S. Bajaj, A. Pines, Compressed sensing of remotely detected MRI velocimetry in microfluidics. J. Magn. Reson. **205**(2), 196–201 (2010)
22. P. Parasoglou, D. Malioutov, A.J. Sederman, J. Rasburn, H. Powell, L.F. Gladden, A. Blake, M.L. Johns, Quantitative single point imaging with compressed sensing. J. Magn. Reson. **201**(1), 72–80 (2009)
23. J.W. Cooley, J.W. Tukey, An algorithm for the machine calculation of complex Fourier series. Math. Comput. **19**(90), 297–301 (1965)
24. J.A. Fessler, B.P. Sutton, Nonuniform fast Fourier transforms using min-max interpolation. IEEE Trans. Signal Process. **51**(2), 560–574 (2003)
25. M. Benning, L. Gladden, D. Holland, C.-B. Schönlieb, T. Valkonen, Phase reconstruction from velocity-encoded MRI measurements—a survey of sparsity-promoting variational approaches. J. Magn. Reson. **238**, 26–43 (2014)
26. J.A. Fessler, D.C. Noll, Iterative image reconstruction in MRI with separate magnitude and phase regularization, in *Proceedings of the Second IEEE International Symposium on Biomedical Imaging: Nano to Macro (IEEE Cat No. 04EX821)*, 2004, pp. 209–212
27. M.V.W. Zibetti, A.R. De Pierro, Separate magnitude and phase regularization in MRI with incomplete data: preliminary results, in *Proceedings of the IEEE International Symposium on Biomedical Imaging: From Nano to Macro*, 2010, pp. 736–739
28. F. Zhao, D.C. Noll, J. Nielsen, J.A. Fessler, Separate magnitude and phase regularization via compressed sensing. IEEE Trans. Med. Imaging **31**(9), 1713–1723 (2012)
29. T. Valkonen, A primal–dual hybrid gradient method for nonlinear operators with applications to MRI. Inverse Prob. **30**(5), 055012 (2014)
30. M.V. Zibetti, A.R. Pierro, Improving compressive sensing in MRI with separate magnitude and phase priors. Multidim. Syst. Sign. Process. **28**(4), 1109–1131 (2017)
31. L.F. Gladden, A.J. Sederman, Magnetic resonance imaging and velocity mapping in chemical engineering applications. Ann. Rev. Chem. Biomol. Eng. **8**(1), 227–247 (2017). PMID: 28592175
32. T.F. Chan, L.A. Vese, Active contours without edges. IEEE Trans. Image Process. **10**(2), 266–277 (2001)
33. T.F. Chan, S. Esedoglu, M. Nikolova, Algorithms for finding global minimizers of image segmentation and denoising models. SIAM J. Appl. Math. **66**(5), 1632–1648 (2006)
34. L.M. Bregman, The relaxation method of finding the common point of convex sets and its application to the solution of problems in convex programming. USSR Comput. Math. Math. Phys. **7**(3), 200–217 (1967)

35. K.C. Kiwiel, Proximal minimization methods with generalized Bregman functions. SIAM J. Control Optim. **35**(4), 1142–1168 (1997)
36. M. Benning, M.M. Betcke, M.J. Ehrhardt, C.-B. Schönlieb, Choose your path wisely: gradient descent in a Bregman distance framework. arXiv preprint arXiv:1712.04045 (2017)
37. Y. Xu, W. Yin, A block coordinate descent method for regularized multiconvex optimization with applications to nonnegative tensor factorization and completion. SIAM J. Imaging Sci. **6**(3), 1758–1789 (2013)
38. J. Bolte, S. Sabach, M. Teboulle, Proximal alternating linearized minimization for nonconvex and nonsmooth problems. Math. Program. **146**(1–2), 459–494 (2014)
39. M. Zhu, T. Chan, An efficient primal-dual hybrid gradient algorithm for total variation image restoration. UCLA CAM Report, 34, 2008
40. T. Pock, D. Cremers, H. Bischof, A. Chambolle, An algorithm for minimizing the Mumford-Shah functional, in *2009 IEEE 12th International Conference on Computer Vision* (IEEE, Piscataway, 2009), pp. 1133–1140
41. E. Esser, X. Zhang, T.F. Chan, A general framework for a class of first order primal-dual algorithms for convex optimization in imaging science. SIAM J. Imaging Sci. **3**(4), 1015–1046 (2010)
42. A. Chambolle, T. Pock, A first-order primal-dual algorithm for convex problems with applications to imaging. J. Math. Imaging Vis. **40**(1), 120–145 (2011)
43. A. Reci, *Signal Sampling and Processing in Magnetic Resonance Applications*. Ph.D. thesis, University of Cambridge, 2019
44. A.B. Tayler, D.J. Holland, A.J. Sederman, L.F. Gladden, Exploring the origins of turbulence in multiphase flow using compressed sensing MRI. Phys. Rev. Lett. **108**(26), 264505 (2012)
45. M. Benning, Research data supporting "Gradient descent in a generalised Bregman distance framework". (2016) https://www.repository.cam.ac.uk/handle/1810/261518

Dynamic Inverse Problems for the Acoustic Wave Equation

Thies Gerken

Abstract We consider the identification of a time- and space dependent wave speed and mass density based on the knowledge of the wave field. The wave propagation is modeled by the acoustic wave equation. By making use of an abstract framework for parameter reconstruction in hyperbolic partial differential equations, we are able to obtain a well-defined forward operator. Furthermore, we prove the Fréchet-differentiability of this forward operator and the local ill-posedness of the inverse problems. In order to facilitate the application of regularization schemes, we also calculate the necessary adjoint operators. The theoretical considerations are complemented by a numerical demonstration of the inversion using the regularization method CG-REGINN in two- and three-dimensional settings. There we present the numerically obtained convergence rates and show that even in this time-dependent setting one can obtain good reconstructions with reasonable computational effort.

1 Introduction

In this article we will consider dynamic inverse-coefficient problems for the acoustic wave equation. For a *wave speed c* and *mass density ρ* that do not depend on time, a variant of this equation reads

$$\frac{1}{c(x)^2 \rho(x)} u''(t, x) - \operatorname{div} \frac{\nabla u(t, x)}{\rho(x)} = f(t, x) \tag{1}$$

and serves (together with suitable initial—and boundary conditions) as a simple model for the propagation of acoustic waves in fluids [4, 7–9] and also seismic waves [10, 13]. In these cases the unknown u in the equation above is called the *acoustic pressure*.

T. Gerken (✉)
University of Bremen, Bremen, Germany
e-mail: tgerken@math.uni-bremen.de

© Springer Nature Switzerland AG 2021
B. Kaltenbacher et al. (eds.), *Time-dependent Problems in Imaging and Parameter Identification*, https://doi.org/10.1007/978-3-030-57784-1_2

The identification of time- and space dependent coefficients in hyperbolic equations is a new topic of research. It poses several new difficulties compared to the static case, which can be analyzed using the frameworks [3, 11]. On the one hand, the dynamic case involves a more complicated theory to even obtain a well-defined forward model. On the other hand, one has to deal with a lot more degrees of freedom in numerical simulations. From a modeling point of view it is not even clear, whether it is advisable to also employ the partial differential equation (1) for dynamic wave speed and mass density. The position of static parameters c, ρ in relation to the time derivatives can be changed without changing the equation, but for time-dependent parameters this is obviously not the case. The variant we chose for this work is the equation

$$\frac{1}{\rho(t, x)} \frac{\mathrm{d}}{\mathrm{d}t} \left(\frac{1}{c(t, x)^2} u'(t, x) \right) - \operatorname{div} \frac{\nabla u(t, x)}{\rho(t, x)} = f(t, x). \tag{2}$$

which gives rise to the dynamic inverse problems of finding the time- and space-dependent coefficients c or ρ from the solution of this equation. Note that other positions of c and ρ in the equation in relation to the time derivatives yield almost the same theoretical results. The same holds if one wants to include additional zero- and first-order terms qu, vu' with functions q, v that are also to be identified.

Our theoretical analysis is based on the framework presented in [5], which deals with the inverse problem of finding the operators A, B and C from the solution u of the evolution equation

$$\frac{\mathrm{d}}{\mathrm{d}t} C(t)u'(t) + B(t)u'(t) + A(t)u(t) = f(t),$$

to be solved for almost all t that belong to some bounded time domain I. In particular, it deals with the well-posedness and Fréchet-differentiability of the forward operator that maps these operators onto u. Note that we do not assume that the reader is familiar with the abstract theory of [5]. We will briefly present the required results (omitting the proofs) before using them.

In order to apply the abstract framework we first have to restate the acoustic wave equation as an evolution equation, which we carry out in Sect. 2 and also have to set up a *value operator* that maps the two unknown parameters onto the operators A, B and C. This operator can then be composed with the forward operator from the general theory. This also allows showing Fréchet-differentiability of the parameter-to-solution map in the subsequent Sect. 3, where we further give a characterization of the adjoint of this derivative. We conclude the theoretical analysis by proving the ill-posedness of the corresponding inverse problems in Sect. 4. The theoretical considerations will then be put to practical use in Sect. 5, where we discretize (2) and tackle the numerical reconstruction of the parameters c and ρ using a Newton-approach.

2   Construction of the Forward Operator

Let $n \in \mathbb{N}$, $\Omega \subset \mathbb{R}^n$ a bounded domain and $I = [0, T]$ for some $T > 0$. We concentrate on the initial boundary value problem

$$\frac{1}{\rho}\left(\frac{1}{c^2}u'\right)' - \operatorname{div}\frac{\nabla u}{\rho} = f \qquad \text{in } I \times \Omega, \tag{3a}$$

$$u(0) = (c^{-2}\rho^{-1}u')(0) = 0 \qquad \text{in } \Omega, \tag{3b}$$

$$u = 0 \qquad \text{on } I \times \partial\Omega. \tag{3c}$$

The wave field u, the right-hand side f and the two unknown coefficients c and ρ are assumed to be real functions on $I \times \Omega$. The formulation of the second initial condition in (3b) keeps in mind that the weak solution to (3) might not yield a well-defined $u'(0)$. However, if the solution u and the coefficients are regular enough, then the solution will of course equivalently satisfy $u'(0) = 0$.

Due to the presence of ρ^{-1} outside of the time derivatives in the leading term this problem does not immediately yield an evolution equation of the form $(Cu')' + \mathcal{B}u' + \mathcal{A}u = f$. We can reformulate the equation by observing that

$$\frac{1}{\rho}\left(\frac{1}{c^2}u'\right)' = \left(\frac{1}{\rho c^2}u'\right)' - \frac{\rho'}{\rho^2 c^2}u',$$

at least as long as $c^{-2}u'$ is weakly differentiable in time. Hence, we simply replace (3) by

$$\left(\frac{1}{\rho c^2}u'\right)' + \left(\frac{\rho'}{\rho^2 c^2}\right)u' - \operatorname{div}\frac{\nabla u}{\rho} = f \qquad \text{in } I \times \Omega, \tag{4a}$$

$$u(0) = (c^{-2}\rho^{-1}u')(0) = 0 \qquad \text{in } \Omega, \tag{4b}$$

$$u = 0 \qquad \text{on } I \times \partial\Omega \tag{4c}$$

and base our analysis on this restated problem. The regularity results of [6] can then be used a-posteriori to conclude that a weak solution of (4) also weakly solves the original problem (3a).

Due to the boundary conditions, the suitable function space for $u(t)$ is $H_0^1(\Omega)$. By identifying the dual space of $L^2(\Omega)$ with itself, we obtain the Gelfand triple

$$H_0^1(\Omega) \subset L^2(\Omega) \subset H^{-1}(\Omega).$$

This relation implies that we regard every function belonging to $H_0^1(\Omega)$ as an element of $H^{-1}(\Omega)$ through the inner product of $L^2(\Omega)$. To ease notation, we will sometimes omit the "(Ω)" part for Lebesgue- or Sobolev spaces connected to the domain Ω whenever the expressions tend to become unwieldy, for example when they appear in Bochner spaces.

The weak formulation of (4) is immediately obtained by integrating over the domain and then formally integrating by parts and reads

$$\frac{d}{dt}\big(C_{c,\rho}(t)u'(t), \varphi\big) + \big(B_{c,\rho}(t)u'(t), \varphi\big) + \big\langle A_\rho(t)u(t), \varphi\big\rangle = \langle f(t), \varphi\rangle \qquad (5)$$

for all $\varphi \in H_0^1(\Omega)$, which in turn should be fulfilled for almost all $t \in I$. We seek a solution $u \in L^2(I; H_0^1(\Omega)) \cap H^1(I; L^2(\Omega))$ that additionally adheres to the initial conditions $u(0) = 0$ (as an equality in $L^2(\Omega)$) and $\big(C_{c,\rho}u'\big)(0) = 0$ (holding in $H^{-1}(\Omega)$). We denote with $(\cdot, \cdot)$ the inner product of $L^2(\Omega)$ and with $\langle \cdot, \cdot \rangle$ the dual pairing of $H^{-1}(\Omega)$ and $H_0^1(\Omega)$. The operators that appear in the weak formulation are defined for $t \in I$, $\varphi, \psi \in H_0^1(\Omega)$ and $v \in L^2(\Omega)$ by

$$\big\langle A_\rho(t)\psi, \varphi\big\rangle := \int_\Omega \frac{\nabla\psi(x) \cdot \nabla\varphi(x)}{\rho(t, x)}\, dx, \qquad C_{c,\rho}(t)v := \frac{1}{\rho(t)c(t)^2}\, v,$$

$$\text{and } B_{c,\rho}(t)v := \frac{\rho'(t)}{\rho(t)^2 c(t)^2}v.$$

With "$\cdot$" we denote the usual inner product of $\mathbb{R}^n$. By making use of distributional derivatives we can also write $A_\rho(t)\psi = -\operatorname{div}\big(\rho(t)^{-1}\nabla\psi\big) \in H_0^1(\Omega)$, which avoids the appearance of the test function φ. Furthermore, we make use of calligraphic font for the pointwise application of the operators to a time-dependent function, e.g. $\mathcal{A}_\rho$ is defined as

$$\mathcal{A}_\rho \in \mathcal{L}\big(L^2(I; H_0^1(\Omega)), L^\infty(I; H^{-1}(\Omega))\big), \qquad (\mathcal{A}_\rho v)(t) := A_\rho(t)v(t).$$

With this notation, the weak formulation (5), joined with the homogeneous initial conditions (4b), is equivalent to the evolution problem

$$(C_{c,\rho}u')' + B_{c,\rho}u' + \mathcal{A}_\rho u = f \quad \text{in } L^2(I; H^{-1}(\Omega)), \qquad (6a)$$

$$u(0) = 0 \text{ and } (C_{c,\rho}u')(0) = 0. \qquad (6b)$$

The forward operator to our inverse problems reads

$$F(c, \rho) := u,$$

where u solves (6). It can be decomposed into $F = S \circ P$, with the "value operator"

$$P(c, \rho) := (A_\rho, B_{c,\rho}, C_{c,\rho}), \qquad (7)$$

which maps the two parameters to the operators that appear in (6), and the solution operator S, which then maps these operators to the solution u of (6).

The operator S has already been thoroughly analyzed in [5]. We will now briefly present the relevant well-posedness result, adapted to the setting at hand. Unsurprisingly, the operators A and C need to be self-adjoint and coercive. These restrictions are captured in the sets

$$\mathcal{L}^{\mathrm{sa}}(Z, Z^*) := \left\{ G \in \mathcal{L}(Z, Z^*) \mid G^* = G \right\},$$

$$\mathcal{L}_{\alpha}^{\mathrm{sa}}(Z, Z^*) := \left\{ G \in \mathcal{L}^{\mathrm{sa}}(Z, Z^*) \mid \langle Gz, z \rangle \geq \alpha \|z\|_Z^2 \text{ for all } z \in Z \right\},$$

which make sense for $\alpha \geq 0$ and any Hilbert space Z. In these definitions we identify Z^{**} with Z, i.e. both G and G^* belong to $\mathcal{L}(Z, Z^*)$. Because we also identify the dual space of $L^2(\Omega)$ with the space itself, this also gives rise to $\mathcal{L}^{\mathrm{sa}}(L^2(\Omega))$.

For every "degree of regularity" $k \in \mathbb{N}_0 := \mathbb{N} \cup \{0\}$, we would like to regard S as an operator between the two Banach spaces

$$X^{(k)} := W^{k+1,\infty}\big(I;\, \mathcal{L}^{\mathrm{sa}}\big(H_0^1(\Omega),\, H^{-1}(\Omega)\big)\big) \times W^{k_1,\infty}\big(I;\, \mathcal{L}\big(L^2(\Omega)\big)\big)$$

$$\times W^{k+1,\infty}\big(I;\, \mathcal{L}^{\mathrm{sa}}\big(L^2(\Omega)\big)\big)$$

$$\text{and } Y^{(k)} := W^{k,\infty}\big(I;\, H_0^1(\Omega)\big) \cap W^{k+1,\infty}\big(I;\, L^2(\Omega)\big)$$

with $k_1 := \max\{k, 1\}$. In order to ensure the coercivity, we define the domain of definition of S to be

$$D(S) := \Big\{ (A, B, C) \in X^{(0)} \ \Big| \ A(t) \in \mathcal{L}_{A_0+\varepsilon}^{\mathrm{sa}}\big(H_0^1(\Omega),\, H^{-1}(\Omega)\big) \text{ and}$$

$$C(t) \in \mathcal{L}_{C_0+\varepsilon}^{\mathrm{sa}}\big(L^2(\Omega)\big) \text{ for almost all } t \in I \text{ for some } \varepsilon > 0 \Big\}.$$

Indeed, if the right-hand side f belongs to

$$\mathcal{F}^{(k)} := \Big\{ f \in H^k(I;\, L^2(\Omega)) \cup H^{k+1}(I;\, H^{-1}(\Omega)) \ \Big| \ f^{(k-1)}(0) \in L^2(\Omega) \text{ if } k \geq 1$$

$$\text{and } f^{(j)}(0) = 0 \text{ for all } j = 0, \ldots, k-2 \text{ if } k \geq 2 \Big\},$$

then the abstract framework proves that

$$S \colon D(S) \cap X^{(k)} \to Y^{(k)}, \quad (A, B, C) \mapsto u$$

is well-defined. Note that $D(S) \cap X^{(k)}$ is an open subset of $X^{(k)}$.

In order to formally set up our forward operator F, we are left to find a domain of definition for P as defined in (7) such that its image is a subset of $D(S) \cap X^{(k)}$. Clearly, differentiability of the parameters (c, ρ) directly translates to differentiability of $P(c, \rho)$ with respect to time. To ensure that $A_\rho(t)$ and $C_{c,\rho}(t)$ are well-defined we assume that $\rho(t, x) \geq \rho_0 > 0$ and $c(t, x) \geq c_0 > 0$ hold for almost all $(t, x) \in I \times \Omega$. Both operators are self-adjoint; regarding their coercivity we observe that if $\rho(t, x) \leq \rho_1 < \infty$ and $c(t, x) \leq c_1 < \infty$ for almost all $(t, x) \in I \times \Omega$, then

$$\left(C_{c,\rho}(t)v, v\right) = \int_\Omega \frac{v(x)^2}{\rho(t, x)c(t, x)^2} \, dx \geq \rho_1^{-1} c_1^{-2} \|v\|^2_{L^2(\Omega)}$$

holds for all $v \in L^2(\Omega)$. Since Ω is bounded, it provides a Poincaré inequality of the form $\|\psi\|_{H_0^1(\Omega)} \leq C_p \|\nabla \psi\|_{L^2(\Omega;\mathbb{R}^n)}$ with a constant $C_p > 0$. Hence,

$$\left\langle A_\rho(t)\psi, \psi\right\rangle = \int_\Omega \frac{|\nabla \psi(x)|^2}{\rho(t, x)} \, dx \geq \rho_1^{-1} C_p^{-2} \|\psi\|^2_{H_0^1(\Omega)}$$

is valid for all $\psi \in H_0^1(\Omega)$. Let the constants ρ_0, ρ_1, c_0 and c_1 be fixed in the sequel. The previous considerations motivate the definitions

$$W^{(k)} := W^{k+1,\infty}\left(I; L^\infty(\Omega)\right) \times W^{k+1,\infty}\left(I; L^\infty(\Omega)\right) \tag{8}$$

$$D(P) := \left\{ (c, \rho) \in W^{(0)} \;\middle|\; \rho_0 + \varepsilon \leq \rho \leq \rho_1 - \varepsilon \text{ and} \right.$$

$$\left. c_0 + \varepsilon \leq c \leq c_1 - \varepsilon \text{ a.e. in } I \times \Omega \text{ for a } \varepsilon > 0 \right\}, \tag{9}$$

because in this way we obtain a well-defined

$$P: D(P) \cap W^{(k)} \to D(S) \cap X^{(k)}, \qquad (c, \rho) \mapsto (A_\rho, B_{c,\rho}, C_{c,\rho}).$$

For this we set the constants in $D(S)$ to be $C_0 := \rho_1^{-1} c_1^{-2}$ and $A_0 := \rho_1^{-1} C_p^{-2}$. Moreover, $D(P) \cap W^{(k)}$ forms an open subset of $W^{(k)}$ for all $k \in \mathbb{N}_0$.

The forward operator $F = S \circ P$ can then be viewed as

$$F: D(F) \cap W^{(k)} \to Y^{(k)},$$

$$(c, \rho) \mapsto u$$

with $k \in \mathbb{N}_0$, $D(F) := D(P)$, and the weak solution u of the "re-stated" problem (4).

3 Fréchet-Differentiability

We depend on differentiability of F for its numerical inversion using Newton-based methods. The forward operator is comprised of the operators P and S, and the differentiability of S has already been discussed in [5]. Thus, we mainly need to establish differentiability of the value operator P.

The parameters c and ρ enter into P only by their reciprocal values. Knowing that these are also differentiable is enough to define P, but to prove that P is Fréchet-differentiable (or just continuous) with respect to the norm of $X^{(k)}$, we depend on norm estimates for derivatives of such reciprocal functions. Estimates for derivatives of arbitrary order are given by the following formula.

Lemma 1 Let $m \in \mathbb{N}$, $g_0 > 0$ and $g \in W^{m,\infty}(I)$ with $g(t) \geq g_0 > 0$ almost everywhere. Then

$$\left\| \frac{1}{g} \right\|_{W^{m,\infty}(I)} \leq M \left(1 + g_0^{-1} \right)^{m+1} \left(1 + \|g\|_{W^{m,\infty}(I)} \right)^m$$

holds, where $M > 0$ is a constant that only depends on m.

Keeping this lemma in mind, we can turn to proving differentiability of P.

Theorem 1 Let $k \in \mathbb{N}_0$. The map $P \colon D(P) \cap W^{(k)} \to X^{(k)}$ is Fréchet-differentiable, and its derivative $\partial P \colon D(P) \cap W^{(k)} \to \mathcal{L}\left(W^{(k)}, X^{(k)} \right)$, evaluated at $(c, \rho) \in D(P) \cap W^{(k)}$ and $(\bar{c}, \bar{\rho}) \in W^{(k)}$ is given by

$$\partial P(c, \rho)[\bar{c}, \bar{\rho}] = \begin{pmatrix} \partial A_\rho[\bar{\rho}] \\ \partial B_{c,\rho}[\bar{c}, \bar{\rho}] \\ \partial C_{c,\rho}[\bar{c}, \bar{\rho}] \end{pmatrix}$$

$$= t \mapsto \begin{pmatrix} \varphi \in H_0^1 \mapsto \operatorname{div}\left(\dfrac{\bar{\rho}(t)}{\rho(t)^2} \nabla \varphi \right) \\ v \in L^2 \mapsto \left(\dfrac{\bar{\rho}'(t)}{\rho(t)^2 c(t)^2} - \dfrac{2\rho'(t)\bar{\rho}(t)}{\rho(t)^3 c(t)^2} - \dfrac{2\rho'(t)\bar{c}(t)}{\rho(t)^2 c(t)^3} \right) v \\ v \in L^2 \mapsto -\left(\dfrac{\bar{\rho}(t)}{\rho(t)^2 c(t)^2} + \dfrac{2\bar{c}(t)}{\rho(t)c(t)^3} \right) v \end{pmatrix}.$$

Moreover, $\partial P \colon D(P) \cap W^{(k)} \to \mathcal{L}\left(W^{(k)}, X^{(k)} \right)$ is continuous.

Proof The image of P consists of finitely many components, therefore it is enough to look at each component on its own. The proposed candidates for the derivatives of each of the operators A, B, and C with respect to the parameters can be obtained by formally treating them as if they were ordinary rational functions with scalar arguments c and ρ; for B we only have to note that differentiation in time is a linear operator. The linearity of the resulting operators is obvious, and their boundedness is also easy to see: For instance, time derivatives of $\partial A_\rho[\bar{\rho}]$ of order $i = 0, \ldots, k+1$ can be bounded by

$$\left|\left\langle \left(\partial A_\rho[\bar\rho]\right)^{(i)}(t)\varphi, \psi\right\rangle\right| \le \int_\Omega \left|\left(\bar\rho/\rho^2\right)^{(i)}(t)\right| |\nabla\varphi \cdot \nabla\psi|\, dx$$

$$\le \left\|\left(\bar\rho/\rho^2\right)^{(i)}(t)\right\|_{L^\infty(\Omega)} \|\varphi\|_{H_0^1(\Omega)} \|\psi\|_{H_0^1(\Omega)}$$

$$\le 2^i \|\bar\rho\|_{W^{i,\infty}(I, L^\infty(\Omega))} \left\|\rho^{-2}\right\|_{W^{i,\infty}(I, L^\infty(\Omega))} \|\varphi\|_{H_0^1(\Omega)} \|\psi\|_{H_0^1(\Omega)}$$

by making use of the Leibniz rule. To conclude boundedness of the linear map

$$\partial A_\rho: W^{k+1,\infty}(I; L^\infty(\Omega)) \to W^{k+1,\infty}\left(I; \mathcal{L}(H_0^1(\Omega), H^{-1}(\Omega))\right)$$

we can leave the norm of ρ^{-2} as-is, because we only need it to be finite. However, applying Lemma 1 to it yields the continuity of ∂A_ρ in ρ, and therefore also in the tuple (c, ρ) in the norm of $W^{(k)}$.

We demonstrate the estimation of the linearization error in the context of the third component of ∂P, the operator C. For this we need to assume that $\bar c$ and $\bar\rho$ are small enough such that $(c + \bar c, \rho + \bar\rho)$ belongs to the open set $D(P) \cap X^{(k)}$. Then we calculate

$$e := \left\|C_{c+\bar c, \rho+\bar\rho} - C_{c,\rho} - \partial C_{c,\rho}[\bar c, \bar\rho]\right\|_{W^{k+1,\infty}(I; \mathcal{L}(L^2(\Omega)))}$$

$$= \left\|\frac{1}{(\rho + \bar\rho)(c + \bar c)^2} - \frac{1}{\rho c^2} + \frac{\bar\rho}{\rho^2 c^2} + \frac{2\bar c}{\rho c^3}\right\|_{W^{k+1,\infty}(I; L^\infty(\Omega))}$$

$$= \left\|\frac{\bar\rho^2}{c^2\rho^2(\rho + \bar\rho)} + \frac{3\bar c^2 c + 2\bar c^3}{c^3(c + \bar c)^2(\rho + \bar\rho)} + \frac{2\bar c\bar\rho}{c^3\rho(\rho + \bar\rho)}\right\|_{W^{k+1,\infty}(I; L^\infty(\Omega))}.$$

Again, we are required to bound not only this difference with respect to $L^\infty(I; L^\infty(\Omega))$, but also its time derivatives. On each fraction we can invoke the product rule, and Lemma 1 shows the norms of the denominators to remain bounded when $(\bar c, \bar\rho) \to 0$. We observe that the $W^{k+1,\infty}(I; L^\infty(\Omega))$-norms of the numerators are of order $O\left(\|(\bar c, \bar\rho)\|^2\right)$, thus the linearization error e has to be as well. □

Obtaining differentiability of F is now simply a matter of applying the chain rule.

Theorem 2 *Let $k \ge 2$ and $f \in \mathcal{F}^{(k)}$. Then $F: D(P) \cap W^{(k)} \to Y^{(k-2)}$ is Fréchet-differentiable. For every $x = (c, \rho) \in D(P) \cap W^{(k)}$ and $h = (\bar c, \bar\rho) \in W^{(k)}$, the value $\partial F(x)[h]$ is the unique weak solution $u_h \in Y^{(k-1)}$ of the partial differential equation*

$$\frac{1}{\rho}\left(\frac{u_h'}{c^2}\right)' - \mathrm{div}\left(\frac{\nabla u_h}{\rho}\right) = \frac{\bar\rho}{\rho^2}\left(\frac{u'}{c^2}\right)' - \mathrm{div}\left(\frac{\bar\rho}{\rho^2}\nabla u\right) + \frac{2}{\rho}\left(\frac{\bar c}{c^3}u'\right)' \tag{10}$$

together with homogeneous initial values $u_h(0) = u'_h(0) = 0$. As always, $u = F(x)$ denotes the solution of the forward problem.

Proof Since $F = S \circ P$, we have

$$\partial F(x) = \partial S(P(x)) \circ \partial P(x),$$

and from [5] we know that $\partial S(A, B, C)[\bar{A}, \bar{B}, \bar{C}] = u_h \in Y^{(k-1)}$, where u_h has homogeneous initial values and solves

$$(Cu'_h)' + \mathcal{B}u'_h + \mathcal{A}u_h = g_A(u)[\bar{A}] + g_B(u)[\bar{B}] + g_C(u)[\bar{C}]$$

$$= -(\bar{A}(\cdot))[u(\cdot)] - \bar{B}(\cdot)[u'(\cdot)] - (\bar{C}(\cdot)[u'(\cdot)])'.$$

To evaluate $\partial F(x)$ at $h \in W^{(k)}$, we simply need to substitute $(A, B, C) = P(x)$ and $(\bar{A}, \bar{B}, \bar{C}) = \partial P(x)[h]$, which results in the PDE

$$\left(\frac{u'_h}{\rho c^2}\right)' + \frac{\rho'}{\rho^2 c^2} u'_h - \text{div}\left(\frac{\nabla u_h}{\rho}\right) + q u_h = -\text{div}\left(\frac{\bar{\rho}}{\rho^2}\nabla u\right) + \left(\left(\frac{\bar{\rho}}{\rho^2 c^2} + \frac{2\bar{c}}{\rho c^3}\right)u'\right)'$$

$$- \left(\frac{\bar{\rho}'}{\rho^2 c^2} - \frac{2\rho'\bar{\rho}}{\rho^3 c^2} - \frac{2\rho'\bar{c}}{\rho^2 c^3}\right)u'.$$

Both sides of this equation can be simplified since both u and u_h belong to $Y^{(1)} = W^{2,\infty}(I; L^2(\Omega)) \cap W^{1,\infty}(I; H^1_0(\Omega)) \subset C^1(I; L^2(\Omega))$. Through the product rule we obtain the PDE in the assertion. The same holds for the homogeneous initial values of u_h: Because u'_h is continuous (with values in $L^2(\Omega)$), the second initial condition $(u'_h/(\rho c^2))(0) = 0$ is equivalent to $u'_h(0) = 0$.   □

We would like to remark that the PDE for ∂F is exactly what one would expect, and that it can also be deduced through formal linearization of the wave equation (4). The significance of our result consists of showing that (especially in the context of time-dependent parameters) this linearized equation is well-posed and that it actually is the Fréchet-derivative of the operator F, at least for $k \geq 2$.

3.1   Adjoint of the Derivative

Most regularization methods will need access to the adjoint of $\partial F(x)$ to solve the inverse problem, and our approach in Sect. 5 (based on conjugate gradients) will be no exception. However, since $Y^{(k)}$ is too regular to contain measurement noise, we will replace the image space with $L^2(I; L^2(\Omega))$. This means that we should look at the adjoint of $\partial F(x) \in \mathcal{L}(W^{(k)}, L^2(I; L^2(\Omega)))$. Naturally, since $F = S \circ P$, we have

$$\partial F(x)^* = \partial P(x)^* \circ \partial S(P(x))^* \in \mathcal{L}\big(L^2(I; H), \big(W^{(k)}\big)^*\big). \tag{11}$$

The adjoint of $\partial S(P(x))$ can be characterized as follows.

Theorem 3 *Let* $k \geq 2$, $p = (A, B, C) \in D(S) \cap X^{(k)}$, $f \in \mathcal{F}^{(k)}$ *and* $u = S(p)$. *For* $v \in L^2(I; L^2(\Omega))$, *there exists a unique solution* $w_v \in Y^{(0)}$ *of*

$$(Cw_v')' - \mathcal{B}^* w_v' + (\mathcal{A} - (\mathcal{B}^*)')w_v = v \text{ in } L^2(I; H^{-1}(\Omega)) \tag{12}$$

with homogeneous end conditions $w_v(T) = (Cw_v')(T) = 0$. *Furthermore, the adjoint*

$$(\partial S(p))^* \in \mathcal{L}\big(L^2(I; H), \big(X^{(k)}\big)^*\big) \quad of \quad \partial S(p) \in \mathcal{L}\big(X^{(k)}, L^2(I; H)\big)$$

can be evaluated at $v \in L^2(I; H)$, $h = (\bar{A}, \bar{B}, \bar{C}) \in X^{(k)}$ *using*

$$\big\langle (\partial S(p))^*[v], h \big\rangle_{(X^{(k)})^* \times X^{(k)}} = \int_0^T \big(\bar{C}(t)u'(t), w_v'(t)\big) - \big(\bar{B}(t)u'(t), w_v(t)\big)$$
$$- \big\langle \bar{A}(t)u(t), w_v(t)\big\rangle \, dt.$$

Proof See [5].   □

For the derivative itself we have analyzed ∂S and ∂P independently of each other, and only then used the chain rule to combine the two. However, even with the simple structure of P, a characterization of $\partial P(x)^* \in \mathcal{L}\big(\big(X^{(k)}\big)^*, \big(W^{(k)}\big)^*\big)$ is not possible because of insufficient knowledge about the dual space of $X^{(k)}$, i.e. how a general $v \in (X^{(k)})^*$ could act on $\partial P(x)[h]$. Fortunately, as seen on (11), we do not need to evaluate $P^*(z)$ for arbitrary z, but only for $z \in \mathcal{R}(\partial S(P(x))^*)$. This way, we can directly obtain a characterization of the adjoint of ∂F.

Theorem 4 *Let* $k \geq 2$, $f \in \mathcal{F}^{(k)}$ *and* $x = (c, \rho) \in D(P) \cap W^{(k)}$. *The application of the adjoint of* $\partial F(x) \in \mathcal{L}(W^{(k)}, L^2(I; L^2(\Omega)))$ *on* $v \in L^2(I; L^2(\Omega))$ *can be written as*

$$\partial F(x)^*[v] = \begin{pmatrix} \dfrac{2u'}{c^3}\left(\dfrac{w_v}{\rho}\right)' \\ \dfrac{\nabla u \cdot \nabla w_v}{\rho^2} + \dfrac{w_v}{\rho^2}\left(\dfrac{u'}{c^2}\right)' \end{pmatrix} \in L^\infty(I; L^1(\Omega))^2, \tag{13}$$

where the embedding of $L^\infty(I; L^1(\Omega))$ *into* $\big(W^{(k)}\big)^*$ *has to be understood using the inner product of* $L^2(I; L^2(\Omega))$, $u = F(x) \in Y^{(2)}$ *and* $w_v \in Y$ *denotes the solution of the adjoint equation*

$$\left(\dfrac{1}{c^2}\left(\dfrac{w_v}{\rho}\right)'\right)' - \text{div}\left(\dfrac{\nabla w_v}{\rho}\right) = v \tag{14a}$$

in $L^2(I; H^{-1}(\Omega))$, together with homogeneous end conditions

$$w_v(T) = \left(\frac{w_v}{\rho c^2}\right)'(T) = 0. \tag{14b}$$

Proof Let $h = (\bar{c}, \bar{\rho}) \in W^{(k)}$. To calculate $\langle \partial F(x)^*[v], h \rangle$, we substitute ∂P using Theorem 2 and $\partial S(P(x))^*$ using Theorem 3 and obtain

$$\langle \partial F(x)^*[v], h \rangle_{(W^{(k)})^* \times W^{(k)}} = \langle \partial S(P(x))^*[v], \partial P(x)[h] \rangle_{(X^{(k)})^* \times X^{(k)}}$$

$$= \int_0^T \left(\partial C_{c,\rho}[\bar{c}, \bar{\rho}](t) u'(t), w_v'(t) \right) - \left(\partial B_{c,\rho}[\bar{c}, \bar{\rho}](t) u'(t), w_v(t) \right)$$

$$- \langle \partial A_\rho[\bar{\rho}](t) u(t), w_v(t) \rangle \, dt$$

$$= \int_0^T \left(\left(\frac{\bar{\rho}(t)}{\rho(t)^2 c(t)^2} + \frac{2\bar{c}(t)}{\rho(t) c(t)^3} \right) u'(t), w_v'(t) \right)$$

$$- \left(\left(\frac{\bar{\rho}'(t)}{\rho(t)^2 c(t)^2} - \frac{2\rho'(t)\bar{\rho}(t)}{\rho(t)^3 c(t)^2} - \frac{2\rho'(t)\bar{c}(t)}{\rho(t)^2 c(t)^3} \right) u'(t), w_v(t) \right)$$

$$- \left\langle \text{div} \left(\frac{\bar{\rho}(t)}{\rho(t)^2} \nabla u(t) \right), w_v(t) \right\rangle dt.$$

Here, w_v denotes the solution of the adjoint Eq. (12). Our goal is to reshape this expression into some kind of dual product that has h on one side. Inside all of the $L^2(\Omega)$ inner products we can shift from one side to the other as we wish, as long as both sides of the inner product belong to $L^2(\Omega)$. In general, neither the product $u(t)w_v(t)$ of two $H_0^1(\Omega)$-functions, nor the multiplication of $u'(t) \in H_0^1(\Omega)$ and $w_v'(t) \in L^2(\Omega)$ will belong to $L^2(\Omega)$. However, they do lie in $L^1(\Omega)$. Thus, we resort to regarding the resulting integrals as dual products between $L^\infty(\Omega)$ and $L^1(\Omega)$. This and some further reorganizing yields

$$\langle \partial F(x)^*[v], h \rangle = \int_0^T \left\langle \frac{2u'(t)}{c(t)^3} \frac{d}{dt}\left(\frac{w_v(t)}{\rho(t)}\right), \bar{c}(t) \right\rangle_{L^\infty \times L^1}$$

$$+ \left\langle \frac{\nabla u(t) \cdot \nabla w_v(t)}{\rho(t)^2} + \frac{w_v(t)}{\rho(t)^2} \frac{d}{dt}\left(\frac{u'(t)}{c(t)^2}\right), \bar{\rho}(t) \right\rangle_{L^\infty \times L^1} dt.$$

This proves that the adjoint has the asserted form, but we still need to show that w_v solves (14). Since $B_{c,\rho}$ is self-adjoint, Eq. (12) reads

$$(C_{c,\rho} w_v')' - B_{c,\rho} w_v' + (\mathcal{A}_\rho - \mathcal{B}_{c,\rho}') w_v = v, \tag{15}$$

to be solved in the $L^2(I; H^{-1}(\Omega))$-sense and is joined by the end conditions $w_v(T) = (C_{c,\rho}w'_v)(T) = 0$, which directly translate to those given in the assertion. Substituting the operators with their definitions, Eq. (15) becomes

$$\left(\frac{w'_v}{\rho c^2}\right)' - \frac{\rho'}{\rho^2 c^2}w'_v - \left(\frac{\rho'}{\rho^2 c^2}\right)' w_v - \operatorname{div}\left(\frac{\nabla w_v}{\rho}\right) = v.$$

Two applications of the product rule prove that this equation is equivalent to the more compact differential equation (14a).   □

Note that $\partial F(x)^*[v]$ (as in (13)) will not only belong to $L^\infty(I; L^1(\Omega))^2$: For $k \geq 2$ we have $u \in W^{2,\infty}(I; H^1_0(\Omega))$, $w_v \in L^\infty(I; H^1_0(\Omega))$ and $w'_v \in L^\infty(I; L^2(\Omega))$, and the embedding theorems for Sobolev spaces (cf. [1]) yield $H^1_0(\Omega) \subset L^p(\Omega)$ for $p > 2$, but this depends on the space dimension n. Combining this with the Hölder-inequality, we see that in the case of a one- two- or three-dimensional problem the following holds:

- If $n = 1$, then $\left(\partial F(x)^*[v]\right)(t) \in L^2(\Omega) \times L^1(\Omega)$.
- If $n = 2$, then $\left(\partial F(x)^*[v]\right)(t) \in L^q(\Omega) \times L^1(\Omega)$ for all $1 \leq q < 2$.
- If $n = 3$, then $\left(\partial F(x)^*[v]\right)(t) \in L^q(\Omega) \times L^1(\Omega)$ for all $1 \leq q < 3/2$.

Note that we cannot use regularity results for w_v because this would require more than just $v \in L^2(I; L^2(\Omega))$ which means we would have to change the space of the measurements. We see that for the parameter c we can obtain a setting for the adjoint that only involves reflexive Banach spaces. However, for the second component of $\partial F(x)^*[v]$ (which contains $\nabla u \cdot \nabla w_v$) to belong to something else than $L^1(\Omega)$ we would first need to prove better *spatial* regularity results for u or w_v.

4 Ill-posedness

The operator S is locally ill-posed, but this does not automatically imply that $F = S \circ P$ is ill-posed as well. For example, all sequences in $D(S) \cap X^{(k)}$ that do not converge, but have convergent images under S, might not belong to the range of P. However, in [5] it was analyzed under which circumstances sequences of images under S converge. We can then construct sequences for P that fulfill these properties. Precisely, the result we need is the following.

Theorem 5 *Let $k \in \mathbb{N}_0$ and $f \in \mathcal{F}^{(k)}$. Further, let $p = (A, B, C) \in D(S) \cap X^{(k)}$ and $u = S(p)$.*

  (i) *If $(R_j)_{j \in \mathbb{N}} \subset W^{k_1,\infty}(I; \mathcal{L}(L^2))$ satisfies $\|R_j\| \leq \Gamma$ and $\mathcal{R}_j v' \to 0$ in $H^k(I; L^2)$ for all $v \in Y^{(k)}$, then $S(A, B + R_j, C) \to u$ in $Y^{(k)}$ when $j \to \infty$.*
  (ii) *Let $k > 0$ and $(R_j)_{j \in \mathbb{N}} \subset W^{k+1,\infty}(I; \mathcal{L}(H^1_0, H^{-1}))$ with $\|R_j\| \leq \Gamma$, with Γ small enough to guarantee $(A + R_j, B, C) \in D(S)$ for all j, and $\mathcal{R}_j v \to 0$*

in $H^k(I; H^{-1})$ for all $v \in Y^{(k)}$. Then $S(A + R_j, B, C) \to u$ in $Y^{(k-1)}$ when $j \to \infty$.

(iii) Let $k > 0$ and $(R_j)_{j \in \mathbb{N}} \subset W^{k+1,\infty}(I; \mathcal{L}(L^2))$ with $\|R_j\| \leq \Gamma$, with Γ small enough to guarantee $(A, B, C + R_j) \in D(S)$ for all j, and $(R_j v')' \to 0$ in $H^{k-1}(I; L^2)$ for all $v \in Y^{(k)}$. Then $S(A, B, C + R_j) \to u$ in $Y^{(k-1)}$ when $j \to \infty$.

In each case the convergence is uniform in (A, B, C) on every bounded subset of $D(S) \cap X^{(k)}$.

Our main focus is the time-dependence of the parameters c and ρ, hence we will also make use of time-dependent disturbances R_j. Working in the time variable is more difficult, because the parameters have to be differentiable. The following lemma provides suitable smooth auxiliary functions.

Lemma 2 Let $r \in \mathbb{N}_0$. There exists $(\alpha_j)_{j \in \mathbb{N}} \subset C_c^\infty(I)$ which satisfies

$$0 < \gamma \leq \|\alpha_j\|_{W^{r,\infty}(I)} \leq 1 \quad \text{for all } j \in \mathbb{N}$$

and $\alpha_j \varphi \to 0$ in $H^m(I)$ as $j \to \infty$ for all fixed $\varphi \in H^m(I)$ with $m = 0, \ldots, r$. Moreover, if $r > 0$ then $\|\alpha_j\|_{W^{r-1,\infty}(I)} \to 0$ when $j \to \infty$.

Proof These are the same sequences that were used in [5] to show ill-posedness in a setting based on the elastic wave equation. □

Finally, we can employ these sequences to show the local ill-posedness of F.

Theorem 6 Let $k \in \mathbb{N}$, $f \in \mathcal{F}^{(k)}$ and $p = (c, \rho) \in D(P) \cap W^{(k)}$. The tasks of finding c or ρ such that $F(p) = y \in Y^{(k-1)}$ holds are locally ill-posed.

Proof Let $p = (c, \rho) \in D(P) \cap W^{(k)}$ and $(A, B, C) := P(p)$. Since $D(P) \cap W^{(k)}$ forms an open subset of $W^{(k)}$ there exists $\delta_0 > 0$ such that $B(p, \delta_0) \subset D(P) \cap W^{(k)}$. Let $0 < \delta \leq \delta_0$.

(i) Reconstruction of ρ: The parameter ρ is involved in three operators, making this part of the proof more complicated. Moreover, we have to take care that the perturbations still satisfy the coercivity constraints. Let $(\alpha_j)_{j \in \mathbb{N}}$ denote the sequence from Lemma 2 for $r = k + 1$. We define

$$\rho_j(t, x) := \frac{1}{\varepsilon \alpha_j(t) + \rho(t, x)^{-1}}$$

and $p_j := (c, \rho_j)$, where we would like to set $\varepsilon > 0$ in such a way that $\rho_j \in B(\rho, \delta)$ for all $j \in \mathbb{N}$. To make ρ_j well-defined, we require $\varepsilon < 1/\|\rho\|_{L^\infty(I; L^\infty(\Omega))}$ and to secure some wiggle room we further restrict this to $\varepsilon \leq 1/(2\rho_1)$. This implies that $\varepsilon \alpha_j + \rho^{-1} \geq 1/(2\rho_1)$. Denoting with M the positive constant from Lemma 1 for $m = k + 1$, we see that

$$\|\rho_j - \rho\|_{W^{k+1,\infty}(I;L^\infty(\Omega))} = \left\|\frac{\varepsilon\alpha_j}{\varepsilon\alpha_j + \rho^{-1}}\right\|_{W^{k+1,\infty}(I;L^\infty(\Omega))}$$

$$\leq 2^{k+1}\varepsilon\|\alpha_j\|_{W^{k+1,\infty}(I;L^\infty)}\left\|(\varepsilon\alpha_j + \rho^{-1})^{-1}\right\|_{W^{k+1,\infty}(I;L^\infty)}$$

$$\leq 2^{k+1}\varepsilon M(1 + 2\rho_1)^{k+2}\left(1 + \left\|\varepsilon\alpha_j + \rho^{-1}\right\|_{W^{k+1,\infty}(I;L^\infty)}\right)^{k+1}$$

$$\leq 2^{k+1}\varepsilon M(1 + 2\rho_1)^{k+2}\left(1 + (2\rho_1)^{-1} + \left\|\rho^{-1}\right\|_{W^{k+1,\infty}(I;L^\infty)}\right)^{k+1} =: \varepsilon\Lambda,$$

with a constant Λ that only depends on ρ and k. Thus, by $\varepsilon :=$ $\min\{\delta/\Lambda, 1/(2\rho_1)\}$ we obtain $\rho_j \in B(\rho, \delta)$ for all $j \in \mathbb{N}$ and thus also $p_j \in B(p, \delta) \subset D(P)$. We will now verify that $\rho_j \nrightarrow \rho$ in $W^{k+1,\infty}(I; L^\infty(\Omega))$ by showing that the derivatives of order $k + 1$ do not converge with respect to $L^\infty(I; L^\infty(\Omega))$. Expanding $(\rho_j - \rho)^{(k+1)}$ using the product rule leads to

$$(\rho_j - \rho)^{(k+1)} = \left(\frac{\varepsilon\alpha_j}{\varepsilon\alpha_j + \rho^{-1}}\right)^{(k+1)} = \varepsilon\sum_{i=0}^{k+1}\binom{k+1}{i}\alpha_j^{(i)}\left(\frac{1}{\varepsilon\alpha_j + \rho^{-1}}\right)^{(k+1-i)}.$$

Derivatives of $1/(\varepsilon\alpha_j + \rho^{-1})$ remain bounded when $j \to \infty$, and $\alpha_j^{(i)} \to 0$ in $L^\infty(I)$ for all $i = 0, \ldots, k$. Hence all except for the last summand converge to zero. It is therefore enough to show that the last summand does not converge to zero in order to conclude this for the whole sum. Indeed, $\alpha_j^{(k+1)}$ does not vanish in the limit, and we observe $1/(\varepsilon\alpha_j + \rho^{-1}) > 1/(\varepsilon + \rho_0^{-1})$ almost everywhere. The only thing left to show is the convergence $F(p_j) \to F(p)$ in $Y^{(k-1)}$. We see that $P(p_j) = (A + R_j^A, B + R_j^B, C + R_j^C)$ with

$$R_j^A(t)v = \text{div}\left(\left(\frac{1}{\rho_j(t)} - \frac{1}{\rho(t)}\right)\nabla v\right) = \text{div}\left(\varepsilon\alpha_j(t)\nabla v\right)$$

$$R_j^B(t)u = \left(\frac{\rho_j'(t)}{\rho_j(t)^2 c(t)^2} - \frac{\rho'(t)}{\rho(t)^2 c(t)^2}\right)u$$

$$= \frac{d}{dt}\left(\frac{1}{\rho(t)} - \frac{1}{\rho_j(t)}\right)\frac{1}{c(t)^2}u = -\frac{\varepsilon\alpha'(t)}{c(t)^2}u$$

$$R_j^C(t)u = \left(\frac{1}{\rho_j(t)c(t)^2} - \frac{1}{\rho(t)c(t)^2}\right)u = \frac{\varepsilon\alpha_j(t)}{c(t)^2}u$$

for all $v \in H_0^1(\Omega)$, $u \in L^2(\Omega)$ and almost all $t \in I$. Since P is continuous, the norms of R_j^A, R_j^B and R_j^C have to remain bounded when $j \to \infty$. Moreover, for every $u \in H^{k+1}(I; L^2(\Omega))$ we see

$$\left\| \left(\mathcal{R}_j^C u' \right)' \right\|_{H^{k-1}(I;L^2)} \leq \left\| \mathcal{R}_j^C u' \right\|_{H^k(I;L^2)} = \varepsilon \left\| \alpha_j u'/c^2 \right\|_{H^k(I;L^2)} \to 0$$

as $j \to \infty$ due to the design of the α_j. Likewise $\|\mathcal{R}_j^A v\|_{H^k(I;H^{-1})} = \varepsilon \|\alpha_j \Delta v\|_{H^k(I;H^{-1})}$ has to vanish in the limit, and for all fixed $u \in H^k(I; L^2)$ holds

$$\left\| \mathcal{R}_j^B u' \right\|_{H^{k-1}(I;L^2)} = \varepsilon \left\| \alpha_j' u'/c^2 \right\|_{H^{k-1}(I;L^2)} \leq \varepsilon \left\| \alpha_j u'/c^2 \right\|_{H^k(I;L^2)} \to 0.$$

Due to the uniform convergences in Theorem 5, we can apply it simultaneously to multiple components of S and thus finally conclude this part of the proof with

$$S(A + R_j^A, B + R_j^B, C + R_j^C) \to S(A, B, C) = F(p),$$

holding in $Y^{(k-1)}$ as $j \to \infty$.

(ii) Reconstruction of c: Similar to ρ, but with squared reciprocals: We define for $(t, x) \in I \times \Omega$

$$c_j(t, x) := \left(\varepsilon \alpha_j(t) + c(t, x)^{-2} \right)^{-1/2}$$

and again see that $\varepsilon > 0$ can be chosen in such a way that $c_j \in B(c, \delta)$ for all $j \in \mathbb{N}$. Moreover, $c_j \not\to c$ in $W^{k+1,\infty}(I; L^\infty(\Omega))$ and $P(c_j, \rho) = (A, B + R_j^B, C + R_j^C)$ with

$$R_j^B(t)u = \left(\frac{\rho'(t)}{\rho(t)^2 c_j(t)^2} - \frac{\rho'(t)}{\rho(t)^2 c(t)^2} \right) u = \frac{\varepsilon \alpha(t)\rho'(t)}{\rho(t)} u$$

$$R_j^C(t)u = \left(\frac{1}{\rho(t)c_j(t)^2} - \frac{1}{\rho(t)c(t)^2} \right) u = \frac{\varepsilon \alpha_j(t)}{\rho(t)} u.$$

By design, these operators possess the same properties as those that appeared in the proof for ρ.

$\square$

A direct consequence of the theorem is the ill-posedness of $F(p) = y \in Y^{(k-1)}$ in all c or ρ such that $p = (c, \rho) \in D(P) \cap W^{(k)}$ with $k \geq 1$. Note that this is a stronger result than the local ill-posedness of $F(p) = y \in Y^{(k-1)}$ in all tuples $p \in D(P) \cap W^{(k)}$, because the latter problem would already be ill-posed if finding either one of the parameters yielded an ill-posed problem.

When applying a Newton solver to the nonlinear inverse problem, knowing about the ill-posedness of ∂F is even more crucial than the ill-posedness of F.

Corollary 1 *Let $k \geq 2$ and $f \in \mathcal{F}^{(k)}$. We consider $F: D(P) \cap W^{(k)} \to Z$ with $Z = W^{j,p}(I; H)$ or $Z = C^j(I; H)$ for $0 \leq j \leq k$ and $1 \leq p < \infty$. For every $p = (c, \rho) \in D(P) \cap W^{(k)}$ its linearization $\partial F(p) \in \mathcal{L}(W^{(k)}, Z)$ is a compact operator.*

Proof $\partial F(p) = \partial S(P(p)) \circ \partial P(p)$ with linear and continuous $\partial P(p)$, and it was shown in [5] that $\partial S(P(p))$ is compact for the image spaces as in the assertion. □

It could be that $\partial F(p)$ is compact because it has finite dimensional range, which would make the resulting problems well-posed in the sense of linear inverse problems (ill-posed in the sense of Hadamard, but with a continuous generalized inverse). Like in the abstract framework for S, this is not the case here.

Lemma 3 *Let $k \geq 2$ and $f \in \mathcal{F}^{(k)} \setminus \{0\}$. The ranges of*

$$\partial_c F(x) \in \mathcal{L}\left(W^{k+1,\infty}\left(I; L^\infty(\Omega)\right), Y^{(k-1)}\right), \text{ and}$$

$$\partial_\rho F(x) \in \mathcal{L}\left(W^{k+1,\infty}\left(I; L^\infty(\Omega)\right), Y^{(k-1)}\right)$$

are infinite-dimensional at every $x = (c, \rho) \in D(P) \cap W^{(k)}$.

Proof If for example $\partial_c F(x)$ had finite-dimensional range, i.e. the set of all $u_h = \partial_c F(x)[h]$ with $h \in W^{(k)}$ was finite-dimensional, then this would imply that the set of all right-hand sides to the linearized equation (10) was finite-dimensional as well. We prove that the latter is not the case for both of our parameters, starting with ρ.

(i) From $f \neq 0$ follows $u := F(x) \neq 0$. Let $(\beta_i)_{i \in \mathbb{N}} \subset C^\infty(I)$. We define $h_i(t, x) := \beta_i(t)$ for $(t, x) \in I \times \Omega$, and in doing so obtain $h_i \in W^{k+1,\infty}(I; L^\infty(\Omega))$. The right-hand side of the partial differential equation that is solved by $\partial_\rho F(c, \rho)[h_i]$ reads

$$\beta_i \left(\frac{1}{\rho^2} \left(\frac{u'}{c^2} \right)' - \operatorname{div}\left(\frac{\nabla u}{\rho^2} \right) \right) =: \beta_i w.$$

Clearly, $w \in C^1(I; H^{-1}(\Omega))$ because $k \geq 2$. Moreover, if $w = 0$ then u would solve the wave equation

$$\frac{1}{\rho^2}\left(\frac{u'}{c^2} \right)' - \operatorname{div}\left(\frac{\nabla u}{\rho^2} \right) = 0$$

with homogeneous initial- and boundary conditions, which would contradict $u \neq 0$. Since w is continuous in time and does not vanish everywhere we can choose the β_i in such a way that $\{\beta_i w\}_{i \in \mathbb{N}}$ is linearly independent.

(ii) In the derivative with respect to c we choose $\beta_i \in C^\infty(I)$ such that $\{\beta_i u'\}_{i \in \mathbb{N}}$ is linearly independent. Then the right-hand sides $(\beta_i u'/c^3)' = \beta_i' u'/c^3 + \beta_i (u'/c^3)'$ have pointwise disjoint supports inside $\operatorname{spt} u'$ (which is the same as $\operatorname{spt} u'/c^3$), thus they must be linearly independent as long as they do not vanish everywhere. However, it could be that $(\beta_i u'/c^3)' = 0$ for some $i \in \mathbb{N}$. We show how this can be remedied. Suppose this is the case and fix some $t_0 \in I$ and $\varepsilon > 0$ such that $\beta_i(t_0) \neq 0$, $\beta_i'(t_0) \neq 0$ and $\operatorname{spt} \beta_i \subset (t_0 - \varepsilon, t_0 + \varepsilon) \subset I$. We replace β_i with its mirrored version

$$\bar{\beta}_i(t) := \begin{cases} \beta_i(2t_0 - t), & \text{if } t \in (t_0 - \varepsilon, t_0 + \varepsilon), \\ 0, & \text{otherwise} \end{cases}$$

and conclude the proof by observing that

$$\left(\bar{\beta}_i \frac{u'}{c^3} \right)'(t_0) = -\beta_i'(t_0) \frac{u'(t_0)}{c(t_0)^3} + \beta_i(t_0) \left(\frac{u'}{c^3} \right)'(t_0)$$

$$= \left(\beta_i \frac{u'}{c^3} \right)'(t_0) - 2\beta_i'(t_0) \frac{u'(t_0)}{c(t_0)^3} = -2\beta_i'(t_0) \frac{u'(t_0)}{c(t_0)^3} \neq 0.$$

$\square$

5 Numerical Experiments

Before we turn to the numerical treatment of the inverse problems we give a brief overview of the discretization. The acoustic wave equation (2) is discretized in time using Crank-Nicholson, and then in space using piecewise linear finite elements on a rectangular mesh. For this we made use of the C++ library *deal.II* [2]. The parameters c and ρ are discretized on the same grid. To ease the presentation we use the hypercube $\Omega := [-1, 1]^n$ as the space domain. The time interval is given by $I := [0, 2\pi]$, and partitioned into N equally sized sub-intervals.

We will only consider the reconstruction of either c or ρ, i.e. assume the other parameter is known. Furthermore, for the numerical experiments it is more convenient to write the searched-for parameter as the sum of a known, smooth "background" function $c_b, \rho_b \colon I \times \Omega \to \mathbb{R}$ (most likely a constant function) and a perturbation, and only reconstruct this perturbation. For the inversion, this has the same effect as using the background parameter as the initial guess, but it has the advantage that reconstruction errors will not be tainted by the a-priori information about the background medium.

Obviously we can write $D(F)$ as the Cartesian product $D(F) = D(F_c) \times D(F_\rho)$ that separates the constraints on c and ρ into the two sets

$$D(F_c) := \left\{ c \in W^{1,\infty}(I; L^\infty(\Omega)) \, \middle| \, c_0 + \delta \leq c_b + c \leq c_1 - \delta \right.$$

$$\left. \text{a.e. in } I \times \Omega \text{ for a } \delta > 0 \right\},$$

$$D(F_\rho) := \left\{ \rho \in W^{1,\infty}(I; L^\infty(\Omega)) \, \middle| \, \rho_0 + \delta \leq \rho_b + \rho \leq \rho_1 - \delta \right.$$

$$\left. \text{a.e. in } I \times \Omega \text{ for a } \delta > 0 \right\}.$$

This allows to define the forward operators

$$F_c : D(F_c) \cap W^{3,\infty}(I; L^\infty(\Omega)) \to L^2(I; L^2(\Omega)), \qquad F_c := F(c_b + \cdot, \rho_b),$$

$$F_\rho : D(F_\rho) \cap W^{3,\infty}(I; L^\infty(\Omega)) \to L^2(I; L^2(\Omega)), \qquad F_\rho := F(c_b, \rho_b + \cdot)$$

that are well-defined and differentiable as long as $\rho_b, c_b \in W^{3,\infty}(I; L^\infty(\Omega))$. In our experiments we employ $\rho_b(t, x) := 1$ and $c_b(t, x) := 0.3$ for all $(t, x) \in I \times \Omega$.

We will now present the function that we wish to use to build exact parameters for the two inverse problems. We know from our theoretical results that we must use essentially bounded functions, that also have to possess some smoothness in time. To ensure that this also holds after discretization we will use a product ansatz. In space, we want the function to consist of both discontinuities and smooth parts. Let $\Omega_L(\alpha, \beta) \subset \Omega$ denote an L-shape with distance $\alpha > 0$ from the coordinate axes and width $\beta > 0$. While for $n = 2$ this description should be sufficient, it is not clear at all what it means in three dimensions. In an attempt to make the area/volume of these sets approximately equal, we define it to be the union of three thin plates, each aligned with one coordinate plane. In both cases, it can be described by

$$\Omega_L(\alpha, \beta) = \left\{ x \in \Omega \, \middle| \, \text{there is } i_0 \in \{1, \ldots, n\} \text{ s.t. } x_{i_0} \in [\alpha - 1, \alpha + \beta - 1] \right.$$

$$\left. \text{and } x_i \in [\alpha - 1, 1 - \alpha] \text{ for } i \neq i_0 \right\}.$$

For the continuous part we define for $\omega \in [-1, 1]$

$$\lambda_{\omega,r}(x) := \begin{cases} 1 - r^{-2} \|x - (\omega, \ldots, \omega)\|^2 & \text{if } \|x - (\omega, \ldots, \omega)\| \leq r, \\ 0 & \text{otherwise.} \end{cases}$$

We combine Ω_L and λ by defining the test parameter to be

$$\Lambda_{\text{LDot}}(t, x) := \left(0.2 + 0.8 \cdot \sin(t)^2 \right) \left(\chi_{\Omega_L(0.2, 0.3)}(x) - \lambda_{0.35, 0.45}(x) \right).$$

Its discretization is depicted in Fig. 1. Note that the 3D presentation only shows isosurfaces at half the maximal- and half the minimal values of the function. The meshes we employed here are the same meshes we will be using for the inverse

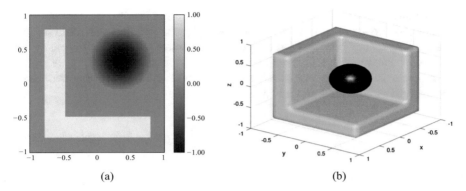

(a)   (b)

Fig. 1 Test parameter Λ_{LDot}, evaluated at $t = \pi/2$. (**a**) $n = 2$. (**b**) $n = 3$

problems. They consist of $N = 256$ discretization points in time and $M = 4225$ (if $n = 2$) or $M = 35937$ (if $n = 3$) nodes in the finite element space.

We continue by discussing our regularization approach. Conforming to the theoretical setting of F_c and F_ρ is very difficult because they are defined on L^∞-type spaces (which are not reflexive), thus even Banach-space reconstruction methods are not applicable. Furthermore, due to the large number of degrees of freedom (over one million in 2D and about ten million in 3D) we have to pick a fast reconstruction method. Therefore we choose to stay in the classical Hilbert space framework by treating F_c, F_ρ as being well-defined on open subsets of $H^1(I; L^2(\Omega))$. However, this L^2-approach will not enforce the boundedness constraints on c and ρ. We remedy this by composing the forward operators with the pointwise application of the transform $\gamma: \mathbb{R} \to (a, b)$ given for $s \in \mathbb{R}$ by

$$\gamma(s) := \frac{a+b}{2} + \frac{a-b}{2} \tanh(s)$$

for $s \in \mathbb{R}$. The values for a and b depend on the problem at hand; for ρ we use $(a, b) := (-0.1, 100)$, whereas for c we set $(a, b) := (-0.27, 30)$. This modification causes $\rho + \rho_b$ and $c + c_b$ to remain uniformly positive throughout the inversion.

Note that the data, which we denote by u^ε, is generated using the same discretized operators and subsequently perturbed using uniformly distributed white noise that is scaled to the appropriate relative noise level ε. To make sure that the data actually contains all of the information about the unknown parameters, we set the right-hand side f to be active almost everywhere by $f(t, x) := \cos(2t)$. Note that this is not required for the successful termination of the reconstruction. However, in order to showcase the regularization properties of our implementation we want the error to be able to converge to zero, which simply might not be possible if we use a very localized right-hand side.

The inversion of the transformed forward operators is handled by the regularization method CG-REGINN [12]. It is started using zero as an initial guess and stopped via the discrepancy principle, i.e. as soon as the iterates produce a discrepancy that is smaller than $\tau \varepsilon \|u^\varepsilon\|$ with $\tau := 2$.

In order to keep the presentation brief we will show results for two cases: reconstruction of c in 2D and ρ in 3D.

5.1 Reconstruction of the Wave Speed in 2D

We start with the reconstruction of c in a two-dimensional setting and $0.15\Lambda_{\text{LDot}}$ as the ground truth. For a noise level of 1% we obtain a reconstruction c^ε as shown in Fig. 2. For the presentation we chose to evaluate the reconstructed parameter at two time instances: $t = \frac{1}{2}\pi$ is near the start of the simulation, and $t = \frac{4}{3}\pi$ is closer to $T = 2\pi$. Although both images are blurry, we can clearly recognize the resemblance to the ground truth. We also note that there is a small "shadow" around the L-shape in both pictures, and that the approximation near the ball is better for the lower value of t. The fact that the reconstruction quality degrades in t makes sense, since changing the parameter near $t = T$ has almost no effect on the data.

If we go to a very low noise level of $\varepsilon = 10^{-4}$, as seen in Fig. 3, then the shadow around the L is gone. It is noteworthy how well the L-shape is approximated; in its vicinity the reconstruction seems to be piecewise constant, although we are using L^2-based norms that do not enforce sparsity of ∇c at all. However, we should also note that the fact that the edges of the ground truth are perfectly aligned with the mesh, which can be seen as a-priori information. Again, the approximation of the ball-shaped part of Λ_{LDot} seems to be better for $t = \frac{1}{2}\pi$, because for $t = \frac{4}{3}\pi$ some artifacts begin to develop.

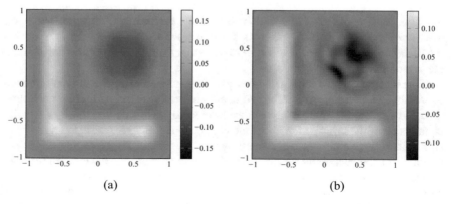

(a)   (b)

Fig. 2 Reconstruction of $c = 0.15\Lambda_{\text{LDot}}$ in 2D with $\varepsilon = 10^{-2}$. (a) Evaluated at $t = \frac{1}{2}\pi$. (b) Evaluated at $t = \frac{4}{3}\pi$

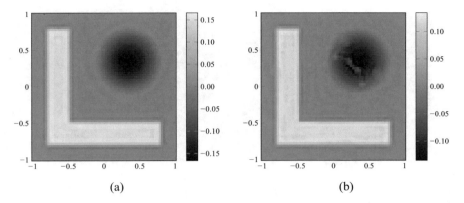

Fig. 3 Reconstruction of $c = 0.15\Lambda_{\text{LDot}}$ in 2D with $\varepsilon = 10^{-4}$. (**a**) Evaluated at $t = \frac{1}{2}\pi$. (**b**) Evaluated at $t = \frac{4}{3}\pi$

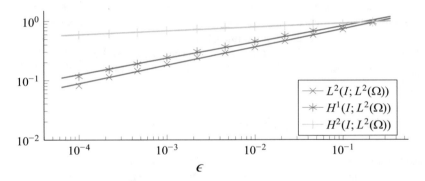

Fig. 4 Errors for $c = 0.15\Lambda_{\text{LDot}}$ in 2D depending on ε

The relative $H^1(I; L^2)$-errors for $\varepsilon = 10^{-2}$ and $\varepsilon = 10^{-4}$ are 46% and 12%, respectively. These values are also depicted in Fig. 4, along with the errors for nine other values for ε. In the figure we have also included the corresponding $L^2(I; L^2)$—and $H^2(I; L^2)$ errors. Regardless of the norm we use to gauge the error, on a logarithmic scale they describe straight lines. If we assume this behavior to hold for $\varepsilon \to 0$, then we are able to conclude the rates

$$\left\| c - c^\varepsilon \right\|_{L^2(I;L^2)} \approx O(\varepsilon^{0.31}), \quad \left\| c - c^\varepsilon \right\|_{H^1(I;L^2)} \approx O(\varepsilon^{0.27}),$$
$$\text{and } \left\| c - c^\varepsilon \right\|_{H^2(I;L^2)} \approx O(\varepsilon^{0.06}).$$

The slope for the H^2-error is so small that we cannot really speak of "convergence"; the error seems to be almost unaffected by the noise level. The most important rate is the one for the H^1-norm, because this is the norm we used to obtain c^ε. Convergence

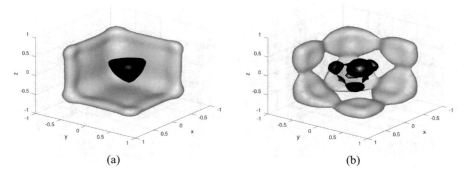

(a)   (b)

Fig. 5 Reconstruction of $\rho = 0.75\Lambda_{\mathrm{LDot}}$ in 3D with $\varepsilon = 4.6 \cdot 10^{-2}$. (a) Evaluated at $t = \frac{1}{2}\pi$. (b) Evaluated at $t = \frac{4}{3}\pi$

rates provided by the theory for REGINN of course depend on source conditions for the ground truth, which we are not able to verify in practice. However, considering that the highest possible guaranteed rate of convergence for REGINN is 0.5, a rate of 0.27 seems reasonable. The fact that the convergence rate in the $L^2(I; L^2)$-norm is only slightly higher implies that the H^1-error is not dominated by $\left\|c' - (c^\varepsilon)'\right\|$.

5.2   Reconstruction of the Mass Density in 3D

Due to the increased computation times in 3D we present results for slightly higher noise levels, the lowest being $\varepsilon = 10^{-3}$. We scale Λ_{LDot} by 0.75 and use it as the ground truth for ρ. Isosurfaces of the reconstruction for $\varepsilon = 4.6 \cdot 10^{-2}$ can be found in Fig. 5. Even for this relatively high level of noise the qualitative behavior of the exact parameter is already apparent in the reconstruction. As in the 2D case, the reconstruction is better for smaller $t \in I$. The relative $H^1(I; L^2(\Omega))$-error of this reconstructed ρ^ε to the ground truth is about 70%, which is reduced to 40% for $\varepsilon = 10^{-3}$. Plots of the latter reconstruction are presented as Fig. 6, and exhibit much sharper edges in the L-shape than for the high ε. Further, for $t = \pi/2$ the isosurfaces close to the ball-shaped part of Λ_{LDot} have become smoother.

Reconstruction errors for more values for ε are shown in Fig. 7. We see the same behavior as for c in the 2D setting, albeit with higher reconstruction errors. By examining the respective slopes, we are led to the rates

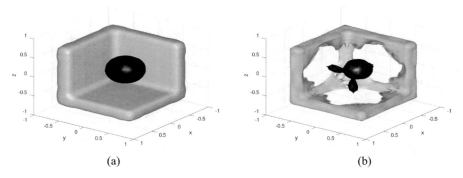

(a)   (b)

Fig. 6 Reconstruction of $\rho = 0.75\Lambda_{\text{LDot}}$ in 3D with $\varepsilon = 10^{-3}$. (a) Evaluated at $t = \frac{1}{2}\pi$. (b) Evaluated at $t = \frac{4}{3}\pi$

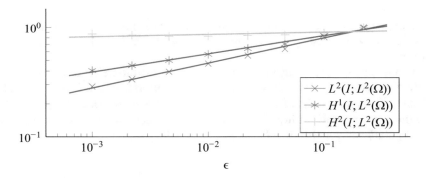

Fig. 7 Errors for $\rho = 0.75\Lambda_{\text{LDot}}$ in 3D depending on ε

$$\left\|\rho - \rho^{\varepsilon}\right\|_{L^2(I;L^2)} \approx O(\varepsilon^{0.23}), \quad \left\|\rho - \rho^{\varepsilon}\right\|_{H^1(I;L^2)} \approx O(\varepsilon^{0.16}),$$
$$\text{and } \left\|\rho - \rho^{\varepsilon}\right\|_{H^2(I;L^2)} \approx O(\varepsilon^{0.02}).$$

Again, the rates for H^2 do not decrease, but we also do not expect them to. The H^1- and L^2-rates are smaller than the their counterparts for c, which might indicate that Λ_{LDot} satisfies a worse source condition with respect to F_{ρ} than it does for F_c.

5.3 Computational Effort

We would like to conclude the numerical examples with an analysis of how the required computational effort increases as $\varepsilon \to 0$. Obviously, the time that is needed to run REGINN until the discrepancy principe is satisfied is highly dependent on the

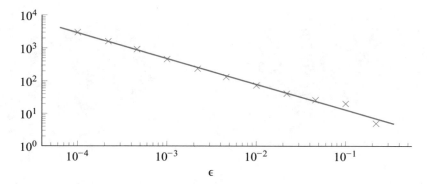

Fig. 8 PDE solutions required for $c = 0.15 \Lambda_{\text{LDot}}$ in 2D depending on ε

system and the implementation. Instead, we will measure the effort by how many solutions of the acoustic wave equation were needed in the whole reconstruction process.

Figure 8 shows the behavior of this quantity for the different noise levels in the setting we used for c. From it we deduce that this effort is of order $O\left(\varepsilon^{-0.8}\right)$. This behavior is similar to the reconstruction of ρ, where we observed the rate $\varepsilon^{-0.9}$ in the three-dimensional setting. These rates are consistent with the expected effort of CG-regularization applied to a *linear* inverse problem.

References

1. R.A. Adams, J.J.F. Fournier, *Sobolev Spaces*, 2nd edn. Pure and Applied Mathematics. (Academic Press, London, 2003)
2. G. Alzetta et al., The deal. II library, version 9.0. J. Numer. Math. **26**(4), 173–183 (2018)
3. K.D. Blazek, C. Stolk, W.W. Symes, A mathematical framework for inverse wave problems in heterogeneous media. Inverse Prob. **29**(6), 065001 (2013)
4. D. Colton, R. Kress, *Inverse Acoustic and Electromagnetic Scattering Theory*, 3rd edn. Applied Mathematical Sciences (Springer, New York, 2013)
5. T. Gerken, Dynamic inverse wave problems—part II: operator identification and applications. Inverse Prob. **36**(2), 024005 (2020)
6. T. Gerken, S. Grützner, Dynamic inverse wave problems—part I: regularity for the direct problem. Inverse Prob. **36**(2), 024004 (2020)
7. M. Ikawa, *Hyperbolic Partial Differential Equations and Wave Phenomena*, 1st edn. Transla-tions of Mathematical Monographs (American Mathematical Society, Providence, 2000)
8. F.B. Jensen, W.A. Kuperman, M.B. Porter, H. Schmidt, *Computational Ocean Acoustics*, 2nd edn. Modern Acoustics and Signal Processing (Springer, New York, 2011)
9. A. Kirsch, *An Introduction to the Mathematical Theory of Inverse Problems*, 2nd edn. Applied Mathematical Sciences (Springer, New York, 2011)
10. A. Kirsch, A. Rieder, On the linearization of operators related to the full waveform inversion in seismology. Math. Methods Appl. Sci. **37**(18), 2995–3007 (2014)

11. A. Kirsch, A. Rieder, Inverse problems for abstract evolution equations with applications in electrodynamics and elasticity. Inverse Prob. **32**(8), 085001 (2016)
12. A. Rieder, Inexact Newton regularization using conjugate gradients as inner iteration. SIAM J. Numer. Anal. **43**(2), 604–622 (2005)
13. W.W. Symes, The seismic reflection inverse problem. Inverse Prob. **25**(12), 123008 (2009)

Motion Compensation Strategies in Tomography

Bernadette N. Hahn

Abstract Imaging modalities have been developed and established as important and powerful tools to recover characteristics of the interior structure of a studied specimen from induced measurements. The reconstruction process constitutes a well-known application of the theory of inverse problems and is well understood if the investigated object is stationary.

However, in many medical and industrial applications, the studied quantity shows a time-dependency, for instance due to patient or organ motion. Most imaging modalities record the data sequentially, i.e. temporal changes of the object during the measuring process lead to inconsistent data sets. Therefore, standard reconstruction techniques which solve the underlying inverse problem in the static case lead to motion artefacts in the computed image and hence to a degraded image quality.

Consequently, suitable models and algorithms with a specific treatment of the dynamics have to be developed in order to solve such time-dependent imaging problems. This article provides a respective theoretical framework as well as numerical results from different imaging applications, including a study of 3D cone-beam CT.

1 Motivation and State-of-the-Art

Over the past decades, tomographic techniques have been developed and established as powerful and important tools for non-invasive imaging with various applications from clinical diagnosis to non-destructive testing. Exploiting the properties of an imaging agent, e.g. propagation of electromagnetic waves, the induced response from a studied medium is measured. The reconstruction of the searched-for function, characteristic of the medium, from the collected data thus matches with solving an associated inverse problem. If the object under investigation is stationary

B. N. Hahn (✉)
Department of Mathematics, University of Stuttgart, Stuttgart, Germany
e-mail: bernadette.hahn@imng.uni-stuttgart.de

© Springer Nature Switzerland AG 2021

B. Kaltenbacher et al. (eds.), *Time-dependent Problems in Imaging and Parameter Identification*, https://doi.org/10.1007/978-3-030-57784-1_3

51

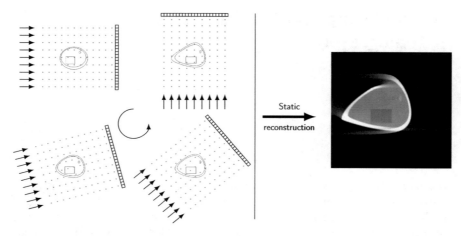

Fig. 1 Temporal changes of a specimen during the data acquisition in computerized tomography (left) and standard reconstruction applied to dynamic data (right)

during the time-dependent scanning, the reconstruction process is well known for most of the imaging systems, see [37].

However, the stationary-assumption is often not satisfied. Prominent examples arise in medical imaging due to respiratory and cardiac motion, gastrointestinal motility, blood flow or body movement of Parkinson patients or infants. Besides clinical applications, investigating dynamic objects arouses the interest in non-destructive testing such as imaging driven liquid fronts for oil recovery studies [2], performing elasticity experiments during the scan to determine material parameters [22], or imaging objects in working stage, e.g. aircraft engines [5].

The dynamic behaviour of the investigated object during the data collection leads to an inconsistent data set. Therefore, standard reconstruction techniques which solve the underlying inverse problem in the static case lead to motion artefacts in the computed image (e.g. blurring, ghosting, distortions) which can significantly degrade the image quality and hence misleads the diagnosis [12, 27, 47], see also Fig. 1. For hybrid imaging methods, these artefacts lead to spatial misalignments of the reconstructions which significantly reduce the diagnostic accuracy and hence affect the success of the treatment [36].

Dynamic Inverse Problems
Following [44], we refer to an inverse problem, where the investigated object is allowed to change during the measuring process, as *dynamic inverse problem*.

1.1 Hardware-Based Artefact Reduction Strategies

In medical imaging, the periodic nature of physiologic motion can be exploited to reduce motion artefacts by hardware-based gating methods. External devices, e.g. electrocardiographs and thoracic belts, detect respiratory expansion and/or cardiovascular motion, and are then used to collect and assort the measured data to specific phases in the motion cycle [9, 13]. A main drawback of the described artefact reduction procedures is their restriction to periodic (patient) motion and hence, it cannot be extended e.g. to applications in non-destructive testing.

Another, intuitive approach is to reduce the required data acquisition time for individual imaging modalities by faster scanners or reduced sampling in data space. In [42], and recently in [38], a multi-source computerized tomography set-up is proposed to avoid the time-consuming rotation of a single radiation source. However, this decreases the signal-to-noise-ratio and hence the quality of the reconstructed image.

1.2 Reconstruction Techniques for Motion Compensation

A more general approach is provided by *motion compensation* methods, where the dynamical information is incorporated in the reconstruction step.

For individual imaging modalities like CT, MRI or PET, several methods of this type have been proposed in the literature, see below for an overview.

Gating methods in general neglect the strong temporal correlation between the single phases. By taking temporal redundancies into account, the reconstruction step can be formulated as a variational problem [10, 39]. If explicit deformation models are incorporated, e.g. in terms of an optimal flow constraint or shape information, this approach leads to non-convex optimization problems [3–5, 28, 29].

For special deformations which preserve the underlying data acquisition geometry, exact analytic reconstruction methods have been derived, especially in computerized tomography, where this type of motion includes affine deformations, [7, 8, 14, 43]. In this case, techniques for rebinning the measured data to make them feasible for standard reconstruction methods are proposed as well, [5, 34]. Besides iterative methods, e.g. [1, 21], approximate inversion formulas have been derived in computerized tomography to compensate for general, non-affine deformations [23, 24].

So far, only a few regularization techniques have been developed in the general context of dynamic linear inverse problems [25, 44, 45], which have been applied in computerized and impedance tomography, respectively. The more recent article [6] proposes a computational method in a Bayesian framework along with an approach to quantify uncertainties of the obtained solution. However, especially the method in [44, 45], suffers from high computational costs and the motion artefacts are not entirely eliminated.

1.3   Outline of the Article

This article is devoted to the study of regularization methods for dynamic inverse problems, summarizing the theoretical framework provided in [14–16] and presenting novel numerical results from various imaging applications. More precisely, we study the application of our theory in the context of photoacoustic tomography and 3D cone-beam CT, whereas the mentioned previous articles evaluated the respective theory at the example of 2D computerized tomography with parallel scanning geometry.

In Sect. 2, we incorporate the time-dependency of the investigated object in the inverse problem associated to the static case by means of diffeomorphic motion models. We then provide an overview of strategies to estimate the motion information from the measured data, which allows to assume the motion to be known prior to the actual reconstruction step.

The resulting mathematical model of dynamic inverse problems gives then rise to a classification scheme distinguishing two cases depending on the object's motion. Section 3 summarizes a general regularization theory for the first category of *moderate deformations*, a subclass of affine deformations, which was developed in detail in [16]. The theoretical results are evaluated at an example from photoacoustic tomography.

For the more general second category of *strong deformations*, a regularization strategy is developed in Sect. 4 by extending the *method of the approximate inverse* to the time-dependent setting as initially proposed in [15]. The design of efficient algorithms is discussed and evaluated at the example of 3D cone-beam computerized tomography.

2   The Mathematical Model of Dynamic Inverse Problems

This section is devoted to the derivation of suitable mathematical models for dynamic inverse problems with a specific treatment of the dynamics.

First, we derive a motion model based on the physical observation that the particles forming the material body change their position in space over time. An object which is changing in time is described by a sequence of functions $f_t : \mathbb{R}^n \to \mathbb{R}$, $t \in [0, T] \subset \mathbb{R}$, representing the different configurations over time. Thus, the motion can be described by a sequence of displacements which correlate the different states of the body to one reference configuration. In particular, this motion model corresponds to the *Lagrangian description* which gives the trajectory of each material particle starting from the initial position [48].

Finally, the model describing the dynamic inverse problem is obtained by combining the motion model with the forward operator from the underlying static scenario.

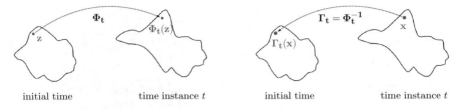

Fig. 2 Illustration of the motion model in terms of Φ_t (left) and Γ_t (right)

2.1 Diffeomorphic Motion Models

Throughout the article, let $[0, T] \subset \mathbb{R}$ denote an interval covering the time period required for the measurement process. Without loss of generality, we consider the initial state of the object, denoted by f_0, as the reference configuration represented in the cartesian coordinate system of $\mathbb{R}^n$. The motion of the particles can then be expressed by a sequence of mappings $\Phi_t : \mathbb{R}^n \to \mathbb{R}^n$, $t \in [0, T]$ with $\Phi_0(x) = x$. Considering the particle initially located at position $x \in \mathbb{R}^n$, the vector $\Phi_t(x)$ denotes its position at time t, see Fig. 2 (left).

Motivated by medical applications and elastic deformations in non-destructive testing, Φ_t is assumed to be a diffeomorphism for all $t \in [0, T]$ and we denote $\Gamma_t := \Phi_t^{-1}$. The descriptive interpretation of the mapping Γ_t is the following: The particle located at x at time t was at the initial time at position $\Gamma_t x$, see Fig. 2 (right).

Using the *motion functions* Γ_t, $t \in [0, T]$ and the initial state function f_0, we find the state of the object at time instance t to be

$$f(t, x) = f_0(\Gamma_t(x)). \tag{1}$$

To simplify the notation, we write $\Gamma_t x$ instead of $\Gamma_t(x)$.

Remark 1 This motion model is intensity preserving, i.e. each particle keeps its initial intensity over time. Analogously, a mass preserving model of type

$$f(t, x) = f_0(\Gamma_t x) \, |\det D\Gamma_t x|$$

could be considered. Please note that this simply results in different weights within the mathematical model of dynamic inverse problems. In particular, this does not alter the nature of our reconstruction algorithms, as explained in [18].

Support Condition
In applications, the studied specimen, more precisely f_0 and all its transformed versions f_t, $t \in [0, T]$ typically have compact support. In particular,

(continued)

we assume

$$\text{supp}(f_0(\Gamma_t \cdot)) \subset \Omega_X \quad \text{for all } t \in [0, T] \tag{2}$$

with a bounded subset $\Omega_X \subset \mathbb{R}^n$. Further, throughout the article, we make use of the continuous extension $f_0(x) := 0$ for $x \notin \Omega_X$.

We next address how such motion information can be extracted from measured data.

Extraction of Motion Information

In applications, the exact motion, i.e. the motion functions $\Gamma_t : \mathbb{R}^n \to \mathbb{R}^n$, $t \in [0, T]$ are in general unknown. If modelled by suitable basis functions b_k, e.g. B-splines [50] with coefficients $w_k(t) \in \mathbb{R}$,

$$\Gamma_t(x) = \sum_{k=1}^{N} w_k(t) b_k(t, x),$$

this requires to estimate the paramters $w_k(t)$ prior to or within the reconstruction step.

Recovering both the unknown parameters and the reference image of the object simultaneously leads to non-convex optimization problems of extremely large size, [3]. This complexity however can be reduced by decoupling the two tasks.

For instance, the calibration of the deformation parameters is proposed to be performed via additional measurements with external devices or via additional images, eventually obtained from another imaging modality [1, 7, 35, 40, 41]. In [34], linear scaling and translation parameters are estimated directly from dynamic, two-dimensional CT-data without any prior knowledge about the object or any additional measurement. This approach is extended in [17] to general parametrized deformation maps. The authors in [24] propose an iterative procedure: If edges look cluttered in an initial reconstruction, the reconstruction step is repeated with an updated motion model.

In the following, we want to focus on the aspect of motion compensation. Therefore, throughout the article, we assume the deformation maps Γ_t, $t \in [0, T]$, to be known.

2.2   Model Operators for Dynamic Linear Inverse Problems

We now turn to the derivation of forward operators modelling dynamic inverse problems. To this end, we combine our motion model with the mathematical model that characterizes the underlying static case.

Many imaging modalities can be modelled mathematically by a linear integral operator represented by a kernel $k : [0, T] \times \mathbb{R}^m \times \mathbb{R}^n \to \mathbb{R}$ (or $\mathbb{C}$) via

$$\mathcal{A} : L_2(\Omega_X) \longrightarrow L_2([0, T] \times \Omega_Y)$$

$$\mathcal{A}h(t, y) = \int_{\mathbb{R}^n} h(x)\, k(t, y, x)\, \mathrm{d}x, \tag{3}$$

where Ω_X and Ω_Y denote bounded subsets of $\mathbb{R}^n$ and $\mathbb{R}^m$, respectively. In this model, the codomain of $\mathcal{A}$ is already given in the time-resolved form (i.e. the time instance t arises explicitly as one of the data variables) accounting for a time-dependent data acquisition. However, the investigated object described by h is assumed to be static. Therefore, we refer to the problem

$$\text{" Find } h \text{ from } \mathcal{A}h(t, y) = g(t, y), \quad t \in [0, T], y \in \Omega_Y\text{"} \tag{4}$$

as *static inverse problem*.

Example (Static CT)

The mathematical model for 2D computerized tomography (CT) is given by the 2D Radon transform

$$\mathcal{R} : L_2(V_1(0)) \longrightarrow L_2([0, 2\pi] \times \mathbb{R})$$

$$\mathcal{R}h(\varphi, s) = \int_{\mathbb{R}^2} h(x)\, \delta(s - x^T \theta(\varphi))\, \mathrm{d}x$$

with $\theta(\varphi) = (\cos(\varphi), \sin(\varphi))^T$, the delta-distribution δ and the unit circle $V_1(0)$. This model corresponds to the integration of the searched-for static quantity h, which is compactly supported in $V_1(0)$, along the straight lines

$$L(\varphi, s) := \{x \in \mathbb{R}^2 : x^T \theta(\varphi) = s\}. \tag{5}$$

In modern CT scanners, all detector points record simultaneously. Thus, the time consuming step of the data acquisition protocol is the rotation of the radiation source around the specimen. Since the source position is

(continued)

characterized by the angle φ, this is the data variable that can be uniquely identified by a time instance t and vice versa. Thus, the mapping

$$\mathcal{R} : L_2(V_1(0)) \longrightarrow L_2([0, 2\pi] \times \mathbb{R})$$

$$\mathcal{R}h(t, s) = \int_{\mathbb{R}^2} h(x)\, \delta(s - x^T \theta(t))\, \mathrm{d}x$$

matches the time-resolved representation (3).

We now derive the mathematical model for the associated time-dependent inverse problem. Let the sequence of functions $(f_t)_{t \in [0,T]}$, $f_t : \mathbb{R}^n \to \mathbb{R}$, characterize the time-dependent object with compact support in $\Omega_X \subset \mathbb{R}^n$. Then, at time instance t, the measurement $g(t, \cdot)$ encodes the state f_t. Thus, the associated dynamic inverse problem is given by

$$\mathcal{A}^{dyn} f(t, y) = g(t, y) \tag{6}$$

with the dynamic operator

$$\mathcal{A}^{dyn} f(t, y) := \mathcal{A} f_t(t, y)$$

and $f \in L_2([0, T] \times \Omega_X)$, $f(t, x) := f_t(x)$. Thus, $\mathcal{A}^{dyn}$ can be considered as mapping from $L_2([0, T] \times \Omega_X)$ into $L_2([0, T] \times \Omega_Y)$. If the static operator $\mathcal{A}$ is of type (3), then

$$\mathcal{A}^{dyn} f(t, y) = \int_{\mathbb{R}^n} f(t, x)\, k(t, y, x)\, \mathrm{d}x.$$

From this representation, it becomes clear that additional information are required in order to extract the time-dependent quantity f from the dynamic data $g = \mathcal{A}^{dyn} f$.

Additional Information Required

The added time dimension regarding the searched-for quantity results in an incomplete data problem: In the static case, all measured data, i.e. $g(t, \cdot) \,\forall t \in [0, T]$, encode the information about one single object state. For instance in CT, this corresponds to recovering a function from all its line integrals. In contrast, in the dynamic scenario, only a very small portion of the data, namely $g(t, \cdot)$ for one single time instance t, encode each individual state. In

(continued)

CT, this corresponds to the task of recovering each state f_t from a subset of its line integrals (namely only from line integrals in direction $\theta(t)^{\perp}$).

Thus, solving dynamic inverse problems typically requires the incorporation of some additional information. Hence, we now incorporate temporal correlations of the individual object states in terms of a motion model as described in Sect. 2.1.

Incorporating correlation (1), i.e. $f(t, x) = f_0(\Gamma_t x)$, in the definition of the dynamic forward operator $\mathcal{A}^{dyn}$, we obtain the following operator $\mathcal{A}_\Gamma$ for the initial state function

$$\mathcal{A}_\Gamma f_0(t, y) := \int_{\mathbb{R}^n} f_0(\Gamma_t x) \, k(t, y, x) \, dx,$$

which depends on the motion functions $\Gamma_t, t \in [0, T]$. In particular, the substitution $x \mapsto \Gamma_t x$ yields the equivalent representation

$$\mathcal{A}_\Gamma f_0(t, y) = \int_{\mathbb{R}^n} |\det D\Gamma_t^{-1}(x)| \, f_0(x) \, k(t, y, \Gamma_t^{-1} x) \, dx. \tag{7}$$

The support condition (2) ensures that the range $R(\mathcal{A}_\Gamma)$ is a subset of $L_2([0, T] \times \Omega_Y)$. Thus, $\mathcal{A}_\Gamma$ can be considered as mapping from $L_2(\Omega_X) \to L_2([0, T] \times \Omega_Y)$.

If the deformation fields Γ_t are known, the dynamic inverse problem (6) reduces to determining f_0 from the equation

$$\mathcal{A}_\Gamma f_0 = g. \tag{8}$$

Example (Dynamic CT)
In dynamic 2D CT, the inverse problem

$$\mathcal{R}_\Gamma f_0 = g$$

has to be solved with the dynamic forward operator

$$\mathcal{R}_\Gamma f_0(t, s) = \int |\det D\Gamma_t^{-1} x| \, f_0(x) \, \delta(s - (\Gamma_t^{-1} x)^T \theta(t)) \, dx.$$

This operator integrates a weighted version of the reference state f_0 along the curved lines

(continued)

Fig. 3 Integration curves in
the static case (left) and in
case of a non-affine
deformation (right)

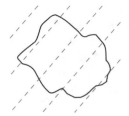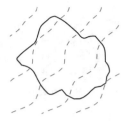

$$C_\Gamma(t, s) = \{x \in \mathbb{R}^2 \mid (\Gamma_t^{-1} x)^T \theta(t) = s\},$$

see also Fig. 3.

If the dynamic behaviour is described by affine deformations, i.e. $\Gamma_t x :=$
$A_t x + b_t$ with $A_t \in \mathbb{R}^{2 \times 2}$ and $b_t \in \mathbb{R}^2$ for all $t \in [0, T]$, then the integration
curves simplify to

$$C_{A_t, b_t}(t, s) = \{x \in \mathbb{R}^2 \mid x^T (A_t^T \theta(t)) = s + A_t^T b_t\}.$$

Thus, in this particular case, they correspond to shifted and rotated versions of
the original straight lines $L(t, s)$ from the static case, see (5), and the dynamic
operator $\mathcal{R}_\Gamma$ can be related to the underlying static operator $\mathcal{R}$ by a change of
coordinates in data space. This means

$$\mathcal{R}_\Gamma = \mathcal{V} \mathcal{R}$$

with suitable transformation $\mathcal{V}$.

In general, however, it is not possible to express a curved line as rigid
transformation of a straight line. In this case, the dynamic model $\mathcal{R}_\Gamma$ cannot
be related to $\mathcal{R}$ by modifying the data acquisition scheme, so we can say *they
differ strongly*.

This observation from dynamic CT motivates the following classification scheme
for dynamic inverse problems.

Classification Scheme [16]

Let $\mathcal{A}$ be a static operator and let $(\Gamma_t)_{t \in [0, T]}$ be a motion model. If there
exists a diffeomorphism $M : [0, T] \times \mathbb{R}^m \to [0, T] \times \mathbb{R}^m$ and a continuous
function $\alpha : [0, T] \times \mathbb{R}^m \to [0, T] \times \mathbb{R}^m \setminus \{0\}$ such that

(continued)

$$\mathcal{A}_\Gamma = \mathcal{V}\mathcal{A}$$

with an operator

$$\mathcal{V}g(y) = \alpha(t, y)\, g(M(t, y)), \tag{9}$$

then the motion model $(\Gamma_t)_{t \in [0,T]}$ is called *moderate* with respect to $\mathcal{A}$. Otherwise, we speak of a *strong* motion model with respect to $\mathcal{A}$.

The operator $\mathcal{V}$ as in (9) is studied in more detail in Theorem 1.

Inverse problems, including (6) and (8), are in general ill-posed and thus, a regularization method is required to solve these problems. In the following sections, we address the derivation of suitable dynamic regularizations for both types of deformations.

We conclude this section by stating the representation of the adjoint operators $\mathcal{A}_\Gamma^*$ and $\mathcal{A}^{dyn^*}$, since they play an important role throughout the article. For the time-resolved operator $\mathcal{A}^{dyn^*}$, we calculate

$$\mathcal{A}^{dyn^*}g(t, x) = \int_{\Omega_Y} \overline{k(t, y, x)}\, g(t, y)\, \mathrm{d}y.$$

A change of coordinates in the integral $\langle \mathcal{A}_\Gamma f, g \rangle_{L_2}$ leads to the representation

$$\mathcal{A}_\Gamma^* g(x) = \int_{[0,T] \times \Omega_Y} \overline{k(t, y, \Gamma_t^{-1} x)}\, g(t, y)\, \mathrm{d}t\, \mathrm{d}y. \tag{10}$$

If we denote

$$\mathcal{A}_t : L_2(\Omega_X) \longrightarrow L_2(\mathbb{R}^m)$$
$$f \mapsto \mathcal{A}_t f(y) := \mathcal{A}f(t, y)$$

for fixed $t \in [0, T]$, then, with

$$\mathcal{A}^* g(x) = \int_{[0,T]} \mathcal{A}_t^* g_t(x)\, \mathrm{d}t,$$

it holds

$$\mathcal{A}^{dyn^*}g(x) = \mathcal{A}_t^* g_t(x), \tag{11}$$

and

$$\mathcal{A}_\Gamma^* g(x) = \int_{[0,T]} |\det D\Gamma_t^{-1} x| \, \mathcal{A}_t^* g_t(\Gamma_t^{-1} x) \, dt \qquad (12)$$

for $g \in L_2([0, T] \times \mathbb{R}^m)$ with $g_t(y) := g(t, y)$.

With the mathematical model at hand, we now develop suitable regularization methods within the subsequent sections.

3   Compensating Moderate Deformations

In this section, we study regularization strategies for dynamic inverse problems with moderate motion. To this purpose, we consider the more general setting of $\mathcal{A}$ being a mapping into a weighted L_2-space, i.e.

$$\mathcal{A} : L_2(\Omega_X) \to L_2([0, T] \times \Omega_Y, w)$$

with a measurable weight w. Considering such weighted L_2-spaces has several advantages, for instance with respect to mapping properties or the derivation of a singular value decomposition. Regarding the Radon transform $\mathcal{R}$ for instance, the singular value decomposition is known if $\mathcal{R}$ is considered as mapping $L_2(V_1(0)) \to L_2([0, 2\pi] \times [-1, 1], w)$ with weight $w(s) := (1 - s^2)^{-1/2}$.

In case of a moderate deformation, the dynamic forward operator $\mathcal{A}_\Gamma$ is given by $\mathcal{A}_\Gamma = \mathcal{V}\mathcal{A}$ with an operator $\mathcal{V}$ as stated in (9). We start by summarizing properties of this mapping $\mathcal{V}$ from [16].

Theorem 1 *The operator*

$$\mathcal{V} : L_2([0, T] \times \Omega_Y, w) \longrightarrow L_2(M([0, T] \times \mathbb{R}^m), w_\Gamma)$$

$$\mathcal{V}g(t, y) = \alpha(t, y) \, g(M(t, y))$$

with weight $w_\Gamma(t, y) = |\det DM(t, y)| \, \alpha(t, y)^{-2} \, w(M(t, y))$ is linear and bijective with inverse

$$\mathcal{V}^{-1} = \mathcal{V}^*.$$

Proof According to its Definition, $\mathcal{V}$ is linear. We first compute its adjoint

$$\mathcal{V}^* : L_2(M([0, T] \times \Omega_Y), w_\Gamma) \longrightarrow L_2([0, T] \times \Omega_Y, w).$$

It holds

$$\langle \mathcal{V}g, h \rangle_{L_2([0,T] \times \Omega_Y, w_\Gamma)} = \int_{[0,T] \times \Omega_Y} \mathcal{V}g(t, y) \, h(t, y) \, w_\Gamma(t, y) \, dt dy$$

$$= \int_{[0,T]\times\Omega_Y} \alpha(t,y)^{-1} g(M(t,y)) h(t,y) \, |\det DM(t,y)| \, w(M(t,y)) \, dt \, dy$$

$$= \int_{M([0,T]\times\Omega_Y)} g(t,y) \, m\left(M^{-1}(t,y)\right)^{-1} h(M^{-1}(t,y)) \, w(t,y) \, dt \, dy$$

$$= \langle g, \mathcal{V}^* h \rangle_{L_2(M([0,T]\times\Omega_Y),w)}$$

with $\mathcal{V}^* h(t,y) = m\left(M^{-1}(t,y)\right)^{-1} h(M^{-1}(t,y))$. For $g \in L_2([0,T] \times \Omega_Y, w)$, we further obtain

$$\mathcal{V}^* \mathcal{V} g(t,y) = m\left(M^{-1}(t,y)\right)^{-1} \mathcal{V} g(M^{-1}(t,y))$$

$$= m\left(M^{-1}(t,y)\right)^{-1} m\left(M^{-1}(t,y)\right) g(M(M^{-1}(t,y)))$$

$$= g(t,y),$$

and respectively for $g \in L_2(M([0,T] \times \Omega_Y), w_\Gamma)$

$$\mathcal{V}\mathcal{V}^* g(t,y) = g(t,y),$$

i.e. $\mathcal{V}^{-1} = \mathcal{V}^*$. □

Due to the properties of $\mathcal{V}$ verified in Theorem 1, many properties of the static operator $\mathcal{A}$ transfer directly to its dynamic counterpart $\mathcal{A}_\Gamma$. A detailed overview is given in [16]. The following Lemma states some of these properties which are relevant regarding the formulation of suitable dynamic regularization methods.

Lemma 1

(i) *If $\mathcal{A} : L_2(\Omega_X) \to L_2([0,T] \times \Omega_Y, w)$ is continuous, then*

$$\mathcal{A}_\Gamma : L_2(\Omega_X) \to L_2(M([0,T] \times \Omega_Y), w_\Gamma)$$

is continuous.

(ii) *Regarding the nullspace, noted N, it holds*

$$N(\mathcal{A}_\Gamma) = N(\mathcal{A}). \tag{13}$$

(iii) *Let $\mathcal{A}^\dagger$ be the generalized inverse of $\mathcal{A}$. Then, the pseudoinverse of $\mathcal{A}_\Gamma$ is given by*

$$\mathcal{A}_\Gamma^\dagger = \mathcal{A}^\dagger \mathcal{V}^{-1}.$$

Proof

(i) Since $\mathcal{V}$ is a unitary transformation, it holds

$$\|\mathcal{A}_\Gamma\|_{L_2(M([0,T]\times\Omega_Y), w_\Gamma)} = \|\mathcal{A}\|_{L_2([0,T]\times\Omega_Y, w)}.$$

(ii) The nullspace property follows from the bijectivity of $\mathcal{V}$.
(iii) Let $f = \mathcal{A}_\Gamma^\dagger g$, i.e. $f \in N(\mathcal{A}_\Gamma)^\perp$ and $\mathcal{A}_\Gamma^*\mathcal{A}_\Gamma f = \mathcal{A}_\Gamma^* g$. Since $\mathcal{V}$ is a unitary operator, it holds

$$\mathcal{A}_\Gamma^*\mathcal{A}_\Gamma = \mathcal{A}^*\mathcal{V}^*\mathcal{V}\mathcal{A} = \mathcal{A}^*\mathcal{A},$$

and further

$$\mathcal{A}^*\mathcal{A}f = \mathcal{A}^*\mathcal{V}^{-1}g.$$

Due to the nullspace property (13), $f \in N(\mathcal{A}_\Gamma)^\perp$ implies $f \in N(\mathcal{A})^\perp$. Thus, $f = \mathcal{A}^\dagger\mathcal{V}^{-1}g$.

$\square$

From the proof of Lemma 1 iii), it follows directly for the domain $\mathcal{D}(\mathcal{A}_\Gamma^\dagger) = R(\mathcal{A}_\Gamma) \oplus R(\mathcal{A}_\Gamma)^\perp$, where $R(\mathcal{A}_\Gamma)$ denotes the range of $\mathcal{A}_\Gamma$:

Corollary 1 *For $g \in \mathcal{D}(\mathcal{A}_\Gamma^\dagger)$, it holds $\mathcal{V}^{-1}g \in \mathcal{D}(\mathcal{A}^\dagger)$.*

With these properties, we can show the following regularization property.

Theorem 2 *Let the family $(\mathcal{T}_\gamma)_{\gamma\in(0,\infty)}$ be a regularization for $\mathcal{A}^\dagger$. Then, the family $(\mathcal{S}_\gamma)_{\gamma\in(0,\infty)}$ with*

$$\mathcal{S}_\gamma := \mathcal{T}_\gamma\mathcal{V}^{-1}$$

is a regularization for $\mathcal{A}_\Gamma^\dagger$.

Proof Let $g \in \mathcal{D}(\mathcal{A}_\Gamma^\dagger)$ and $\|g - g^\epsilon\| \leq \epsilon$. With Corollary 1, it follows $\mathcal{V}^{-1}g \in \mathcal{D}(\mathcal{A}^\dagger)$ and due to the unitary property of $\mathcal{V}$, it holds $\|\mathcal{V}^{-1}g - \mathcal{V}^{-1}g^\epsilon\| = \|g - g^\epsilon\| \leq \epsilon$. Since $(\mathcal{T}_\gamma)_{\gamma\in(0,\infty)}$ is a regularization for $\mathcal{A}^\dagger$, we obtain with the parameter choice rule $\gamma = \gamma(\epsilon, g^\epsilon)$ and the regularizing property of $(\mathcal{T}_\gamma)_{\gamma\in(0,\infty)}$

$$\lim_{\substack{\epsilon\to 0 \\ g^\epsilon\to g}} \mathcal{S}_{\gamma(\epsilon,g^\epsilon)}g^\epsilon = \lim_{\substack{\epsilon\to 0 \\ g^\epsilon\to g}} \mathcal{T}_{\gamma(\epsilon,g^\epsilon)}\mathcal{V}^{-1}g^\epsilon = \mathcal{A}^\dagger\mathcal{V}^{-1}g = \mathcal{A}_\Gamma^\dagger g.$$

This concludes the proof.

$\square$

Thus, for moderate deformations, we obtain a dynamic regularization method for solving $\mathcal{A}_\Gamma f_0 = g$ by adapting any static regularization for $\mathcal{A}$ according to the transform $\mathcal{V}$.

Further properties, including a singular value decomposition and a characterization of the ill-posedness of the dynamic forward operator $\mathcal{A}_\Gamma$ under moderate deformation Γ can be found in [16].

Example: Photoacoustic Tomography

To illustrate the theoretical results of this section, we consider the static inverse problem $\mathcal{A}f = g$ with the circular Radon transform

$$\mathcal{A}f(\theta(t), r) = \frac{1}{2\pi r} \int_{V_1(0)} f(x)\,\delta(r - \|\theta(t) - x\|)\,dx, \tag{14}$$

which integrates a measurable function f supported inside the unit disk $V_1(0) \subset \mathbb{R}^2$ along circles

$$C(t, r) = \{x \in \mathbb{R}^2 : \|x - \theta(t)\| = r\}$$

with $\theta(t) = (\cos(t), \sin(t))^T$ and $(t, r) \in [0, 2\pi] \times (0, \infty)$. This operator represents for instance a simplified mathematical model in 2D photoacoustic tomography (PAT), see for instance [26].

Theorem 3 *Let $(\Gamma_t)_{t \in [0,T]}$ describe a rotational movement of the initial state f_0, i.e. $\Gamma_t x := A_t x$ with unitary matrix*

$$A_t = \begin{pmatrix} \cos(\omega_t) & -\sin(\omega_t) \\ \sin(\omega_t) & \cos(\omega_t) \end{pmatrix} \in \mathbb{R}^{2 \times 2}$$

for all $t \in [0, 2\pi]$ with $\omega_t \in \mathbb{R}$ such that $\{\theta(t + \omega_t) : t \in [0, 2\pi]\} = S^1$. Then, the dynamic operator $\mathcal{A}_\Gamma$ is related to the static transform $\mathcal{A}$ via

$$\mathcal{A}_\Gamma = \mathcal{V}\mathcal{A}$$

with $\mathcal{V}g(\theta_t, r) = g(\theta_{\omega_t + t}, r)$.

Thus, rotations as stated in the Theorem are moderate deformations with respect to the spherical Radon transform.

Proof According to (7), the dynamic operator $\mathcal{A}_\Gamma$ with the stated motion model is given by

$$\mathcal{A}_\Gamma f(t, r) = \frac{1}{2\pi r} \int_{V_1(0)} |\det A_t^{-1}|\,f(x)\,\delta(r - \|\theta(t) - A_t^{-1}x\|)\,dx.$$

(continued)

For each $t \in [0, 2\pi]$, A_t represents a rotation with angle ω_t, i.e. it holds $|\det A_t^{-1}| = 1$, and we further obtain

$$\mathcal{A}_\Gamma f(\theta(t), r) = \frac{1}{2\pi r} \int_{V_1(0)} f(x)\, \delta(r - \|A_t\theta(t) - x\|)\, dx$$
$$= \mathcal{V}\mathcal{A}_\Gamma(\theta(t), r)$$

with $\mathcal{V}g(\theta(t), r) = g(\theta(\omega_t + t), r)$.   □

Remark 2

(i) Please note that the property $\{\theta(t + \omega_t) : t \in [0, 2\pi]\} = S^1$ guarantees the required diffeomorphism property of the transform $T : S^1 \times (0, \infty) \longrightarrow S^1 \times (0, \infty)$. Descriptively, this condition ensures that all information about the object f are actually encoded in the dynamic data $g = \mathcal{A}_\Gamma f$. This is studied in more detail in the subsequent book chapter *Microlocal properties of dynamic Fourier integral operators*.

(ii) Theorem 3 states, that in the presence of an object rotation, the dynamic operator $\mathcal{A}_\Gamma$ still integrates along circles. The additional constraint on the rotation sequence $(A_t)_{t \in [0, 2\pi]}$ ensures, that all these modified circles cover the complete unit disk (i.e. the support of the object).

Theorem 4 *A suitable reconstruction method* $\mathcal{S}^{\mathrm{DFBP}} : L_2(S^1 \times (0, 2)) \to L_2(V_1(0))$ *for dynamic photoacoustic tomography with rotational movement as stated above is given by*

$$\mathcal{S}^{\mathrm{DFBP}}g(x) = \frac{1}{2\pi} \int_0^{2\pi} \int_0^2 (\partial_r r \partial_r g)(\theta(t), r) \, \log \left| r^2 - \|x - \Gamma_t\theta(t)\|^2 \right| \, dr\, dt.$$

Proof Since rotational deformations as stated above are moderate deformations with respect to the circular Radon transform, we obtain a suitable dynamic reconstruction method by adapting an established regularization strategy from the static case. The circular Radon transform as given in (14) is well known in the literature and various inversion formulae were worked out, for instance in the 2D case as

$$f(x) = \frac{1}{2\pi} \int_0^{2\pi} \int_0^2 (\partial_r r \partial_r \mathcal{A}f)(\theta(t), r) \, \log \left| r^2 - \|x - \theta(t)\|^2 \right| \, dr\, dt$$
$$=: \mathcal{T}^{FBP} f(x),$$

(continued)

see [11], providing a static reconstruction method denoted $\mathcal{T}^{FBP}$. Thus, $\mathcal{S}^{DFBP} := \mathcal{T}^{FBP}\mathcal{V}^{-1}$ is a dynamic reconstruction method according to Theorem 2 and with the representation of $\mathcal{V}$, it holds

$$\mathcal{S}^{DFBP}g(x) = \frac{1}{2\pi} \int_0^{2\pi} \int_0^2 (\partial_r r \partial_r \mathcal{V}^{-1}g)(\theta(t), r) \log \left| r^2 - \|x - \theta(t)\|^2 \right| \, dr \, dt$$

$$= \frac{1}{2\pi} \int_0^{2\pi} \int_0^2 \mathcal{V}^{-1}(\partial_r r \partial_r g)(\theta(t), r) \log \left| r^2 - \|x - \theta(t)\|^2 \right| \, dr \, dt$$

$$= \frac{1}{2\pi} \int_0^{2\pi} \int_0^2 (\partial_r r \partial_r g)(\theta(t), r) \log \left| r^2 - \|x - A_t \theta(t)\|^2 \right| \, dr \, dt.$$

This concludes the proof. ◻

For the numerical evaluation, we consider the phantom depicted in Fig. 4 (left). In this example, the phantom performs on the time interval $[0, \pi]$ a rotational movement given by the angles $\omega_t = t/10$, $t \in [0, \pi]$ and during $[\pi, 2\pi]$ returns to its initial state. The state of the object at the end of the scanning is shown in Fig. 4 (right).

The respective PAT data are simulated by discretizing the forward operator $\mathcal{A}_\Gamma$ with the trapezoidal rule with 1400 samples. More precisely, we hereby obtain the discrete data

$$g_{j,k} := (\mathcal{A}_{\Gamma_{t_j}} f)(t_j, r_k), \quad j = 1, \dots, N \text{ and } k = 1, \dots, M,$$

where t_j are uniformly distributed angles in $[0, 2\pi)$, r_k uniformly distributed in $[0, 2]$ with $N = 300$, $M = 300$. Furthermore, in order to test stability, we add a sample of White Noise to the data set, corresponding to a noise level of 2.5%.

The result of the above stated reconstruction method is illustrated in Fig. 5 (left), which shows the reconstructed initial state of the object on a 512×512 grid. Figure 5 (right) illustrates the result of the static filtered backprojection algorithm applied to the dynamic data. The comparison with the exact initial state shows that the dynamic reconstruction technique in fact compensates for the motion while the static algorithm causes strong distortion artefacts.

Further examples, including a detailed evaluation regarding computerized tomography can be found in [16].

Fig. 4 Phantom at the initial time $t = 0$ (left) and at half time of the scanning, i.e. at time $t = \pi$ (right)

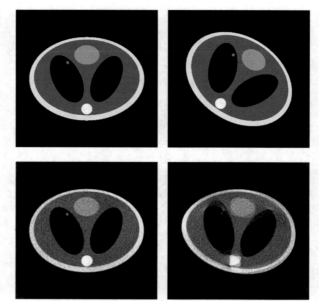

Fig. 5 Dynamic reconstruction at the initial time (left) and static reconstruction (right) from noisy data

4 Compensating General Deformations via the Method of the Approximate Inverse

After working out a regularization theory for moderate deformations, we now turn towards the more general scenario of strong deformations. To this purpose, we focus on (8) and apply the method of the approximate inverse which calculates linear functionals of the sought-for solution, see [30, 32]. To simplify the notation, we consider in the following

$$\mathcal{A} : L_2(\Omega_X) \to L_2([0, T] \times \Omega_Y),$$

i.e. as mapping between classical L_2-spaces. Nevertheless, the presented theory can be easily extended to weighted L_2-spaces as well.

4.1 The Method of the Approximate Inverse

In order to obtain a stable approximation of the solution f_0, we calculate the smoothed version f_0^γ,

$$f_0(x) \approx f_0^\gamma(x) = \langle f_0, \delta_x^\gamma \rangle_{L_2(\Omega_X)}$$

with a prescribed *mollifier* δ_x^γ. The precise definition of a mollifier is given in the following, see also [46].

Definition 1 For all $x \in \Omega_X$, let $\delta_x^\gamma \in L_2(\Omega_X)$ with

$$\int_{\Omega_X} \delta_x^\gamma(z)\, \mathrm{d}z = 1, \quad \gamma > 0.$$

Let further

$$f^\gamma(x) = \int_{\Omega_X} f(z)\, \delta_x^\gamma(z)\, \mathrm{d}z$$

converge to f in $L_2(\Omega_X)$ as $\gamma \to 0$. Then, δ_x^γ is called a mollifier.

A mollified version f_0^γ can be reconstructed by evaluating linear functionals on the measured dynamic data $g = \mathcal{A}_\Gamma f_0$.

Theorem 5 *Let $\delta_x^\gamma \in L_2(\Omega_X)$ be a mollifier and let ψ_x^γ be the solution of*

$$\mathcal{A}_\Gamma^* \psi_x^\gamma = \delta_x^\gamma. \tag{15}$$

Then,

$$f_0^\gamma(x) = \langle g, \psi_x^\gamma \rangle_{L_2([0,T]\times\Omega_Y)}.$$

Equation (15) is called *auxiliary problem*, its solution ψ_x^γ called *reconstruction kernel*. Since $\mathcal{A}_\Gamma$ depends on the dynamic behavior, we speak of Eq. (15) as *dynamic auxiliary problem*, and of ψ_x^γ as *dynamic reconstruction kernel*.

As a further specification, we call ψ_x^γ a *special reconstruction kernel* since it depends on the specific reconstruction point x.

The Approximate Inverse

Theorem 5 introduces an operator $S^\gamma : L_2([0, T] \times \Omega_Y) \to L_2(\Omega_X)$ with $S^\gamma g(x) = \langle g, \psi_x^\gamma \rangle_{L_2}$, which is called *approximate inverse* of $\mathcal{A}_\Gamma$. The regularization property of the method is ensured by imposing conditions on the mollifier and on the choice of parameter γ [30, 31]. The effect of the dynamic behavior on the smoothing properties of the forward operator is analyzed in the chapter *Microlocal properties of dynamic Fourier integral operators* [19].

Since the auxiliary problem (15) is independent of the data, the reconstruction kernel ψ_x^γ can be precomputed. In principle, mollifiers for different reconstruction points x can be chosen independently. In this case, however, the auxiliary problem

(15) has to be solved for distinct right-hand sides leading to high computational costs and storage needs. This effort can be reduced by considering invariances of $\mathcal{A}_\Gamma^*$.

Theorem 6 *Let*

$$T_1^x : L_2(\Omega_X) \to L_2(\mathbb{R}^n), \quad T_2^x : L_2([0, T] \times \Omega_Y) \to L_2([0, T] \times \mathbb{R}^m)$$

be linear operators with

$$T_1^x \mathcal{A}_\Gamma^* = \mathcal{A}_\Gamma^* T_2^x, \tag{16}$$

and let ψ^γ be a solution of the auxiliary problem

$$\mathcal{A}_\Gamma^* \psi^\gamma = \delta^\gamma \tag{17}$$

with $\delta^\gamma \in L_2(\Omega_X)$. Then a solution of

$$\mathcal{A}_\Gamma^* \psi_x^\gamma = \delta_x^\gamma$$

with the special mollifier

$$\delta_x^\gamma = T_1^x \delta^\gamma \tag{18}$$

is given by

$$\psi_x^\gamma = T_2^x \psi^\gamma.$$

Proof According to the relations (16), (17), and (18), it holds

$$\mathcal{A}_\Gamma^* T_2^x \psi^\gamma = T_1^x \mathcal{A}_\Gamma^* \psi^\gamma = T_1^x \delta^\gamma = \delta_x^\gamma,$$

and thus, $T_2^x \psi_\gamma$ solves the auxiliary problem $\mathcal{A}_\Gamma^* \psi_x^\gamma = \delta_x^\gamma$.   □

Consequently, only a single auxiliary problem has to be solved while the special mollifiers and corresponding reconstruction kernels are generated by applying the operators T_1^x and T_2^x, respectively.

Remark 3 The method of the approximate inverse can be extended to enable the so-called *feature reconstruction*, where a feature $\mathcal{L}f_0$ with a linear feature operator $\mathcal{L}$ is determined directly from the measured data, see [20, 31]. In this case, the respective reconstruction kernel can be computed by solving the auxiliary problem

$$\mathcal{A}_\Gamma^* \psi_x^\gamma = \mathcal{L}^* \delta_x^\gamma,$$

and efficient algorithms are obtained by considering linear invariance properties for $\mathcal{A}_\Gamma^*$ as well as $\mathcal{L}^*$.

4.2 Computing the Dynamic Reconstruction Kernel

We now address the solution of the auxiliary problem (15). In static CT, for instance, an explicit representation of the kernel ψ^γ can be derived using the inversion formula for the Radon transform [31]. For dynamic forward operators $\mathcal{A}_\Gamma$, no general inversion formula is known so far. Thus, we present an alternative strategy to compute suitable dynamic reconstruction kernels. The idea consists in exploiting the relation with the time-resolved forward operator $\mathcal{A}^{dyn}$ and its adjoint operator.

$$\mathcal{A}^{dyn^*} g(t, x) = \int_{\mathbb{R}^m} \overline{k(x, t, y)}\, g(t, y)\, \mathrm{d}y.$$

Theorem 7 *Let δ_x^γ be a mollifier for the initial state function f_0 and denote*

$$e_{0,x}^\gamma(t, z) = \left(\int_{[0,T]} |\det D\Gamma_v^{-1}(\Gamma_t z)|\, \mathrm{d}v \right)^{-1} \delta_x^\gamma(\Gamma_t z). \qquad (19)$$

Further assume there exists $\psi_{0,x}^\gamma$ with

$$\mathcal{A}^* \psi_{0,x}^\gamma = e_{0,x}^\gamma. \qquad (20)$$

Then, it holds

(i)

$$\langle f, e_{0,x}^\gamma \rangle_{L_2([0,T] \times \Omega_X)} = \langle f_0, \delta_{0,x}^\gamma \rangle_{L_2(\Omega_X)},$$

in particular, $e_{0,x}^\gamma$ is a time-dependent mollifier incorporating the motion information,

(ii)

$$\mathcal{A}^{dyn^*}_\Gamma \psi_{0,x}^\gamma = \delta_x^\gamma,$$

i.e. $\psi_{0,x}^\gamma$ is our searched-for reconstruction kernel.

Proof

(i) From the definition of $e_{0,x}^\gamma$, it follows

$$\int_{[0,T]} |\det D\Gamma_t^{-1}z|\, e_{0,x}^{\gamma}(t, \Gamma_t^{-1}z)\, dt$$

$$= \int_{[0,T]} |\det D\Gamma_t^{-1}z|\, \delta_x^{\gamma}(z) \left(\int_{[0,T]} |\det D\Gamma_v^{-1}(z)|\, dv \right)^{-1} dt = \delta_x^{\gamma}(z).$$

Together with the temporal correlation (1), namely $f(t, x) = f_0(\Gamma_t x)$, the support property (2) and the substitution $\overline{z} := \Gamma_t(z)$, we then obtain

$$\langle f, e_{0,x}^{\gamma} \rangle_{L_2([0,T] \times \Omega_X)} = \int_{[0,T] \times \Omega_X} f(t, z)\, e_{0,x}^{\gamma}(t, z)\, dt\, dz$$

$$= \int_{[0,T] \times \mathbb{R}^n} f_0(\Gamma_t z)\, e_{0,x}^{\gamma}(t, z)\, dt\, dz$$

$$= \int_{[0,T] \times \mathbb{R}^n} f_0(\overline{z})\, |\det D\Gamma_t^{-1}\overline{z}|\, e_{0,x}^{\gamma}(t, \Gamma_t^{-1}\overline{z})\, dt\, d\overline{z}$$

$$= \int_{\mathbb{R}^n} f_0(z)\, \delta_x^{\gamma}(\overline{z})\, dt\, d\overline{z}$$

$$= \langle f_0, \delta_x^{\gamma} \rangle_{L_2(\Omega_X)}.$$

A simple calculation further shows

$$\int_{[0,T] \times \mathbb{R}^n} e_{0,x}^{\gamma}(v, z)\, dv\, dz = \int_{\mathbb{R}^n} \delta_x^{\gamma}(z)\, dz = 1,$$

i.e. $e_{0,x}^{\gamma}$ is in fact a time-dependent mollifier for $f(0, x)$ according to Definition 1.

(ii) The correlation between δ_x^{γ} and $e_{0,x}^{\gamma}$ from the proof of i) along with the equation $\mathcal{A}^{dyn*} \psi_{0,x}^{\gamma} = e_{0,x}^{\gamma}$ and the representations of $\mathcal{A}^{dyn*}$ and $\mathcal{A}_{\Gamma}^{*}$, see (11) and (12), yields

$$\delta_x^{\gamma}(z) = \int_{[0,T]} |\det D\Gamma_t^{-1}z|\, e_{0,x}^{\gamma}(t, \Gamma_t^{-1}z)\, dt$$

$$= \int_{[0,T]} |\det D\Gamma_t^{-1}z|\, \mathcal{A}^{dyn*} \psi_{0,x}^{\gamma}(t, \Gamma_t^{-1}z)\, dt$$

$$= \int_{[0,T]} |\det D\Gamma_t^{-1}z|\, \mathcal{A}_t^{*} \psi_{0,x}^{\gamma}(t, \Gamma_t^{-1}z)\, dt$$

$$= \mathcal{A}_{\Gamma}^{*} \psi_{0,x}^{\gamma}(z).$$

Thus, $\psi_{0,x}^{\gamma}$ is the searched-for dynamic reconstruction kernel.

$\square$

Exploiting invariances, it is sufficient to solve the auxiliary problem for $x = 0$, i.e. $\mathcal{A}^{dyn*}\psi^\gamma = e^\gamma$ with $e^\gamma := e^\gamma_{0,0}$.

What if e^γ Is Not in the Range of $\mathcal{A}^*$?

If e^γ is not in the range of $\mathcal{A}^*$, then the auxiliary problem $\mathcal{A}^*\psi^\gamma = e^\gamma$ has no solution in the classical sense and instead, the generalized solution via the Moore-Penrose inverse has to be computed. However, an analysis provided in [15] turns out, that in the static setting, the generalized solution of (20) does not represent an adequate approximation to the exact kernel. Thus, [15] proposed instead to approximate ψ^γ by minimizing the penalized defect

$$\left\| \mathcal{A}^{dyn*}\psi^\gamma - e^\gamma \right\|^2 + \alpha \left\| \psi^\gamma - \psi^{\gamma,stat} \right\|^2, \quad \alpha > 0,$$

or equivalently by solving the normal equation

$$(\mathcal{A}^{dyn}\mathcal{A}^{dyn*} + \alpha I)\psi^\gamma = \mathcal{A}^{dyn}e^\gamma + \alpha\psi^{\gamma,stat}$$

with the identity operator I, incorporating the exact static reconstruction kernel in the penalty term. The numerical examples in [15] as well as our results in Sect. 4.3 will illustrate that reconstruction kernels of this kind provide in fact a good motion compensation. Besides, the normal equation is an integral equation of the second kind, so it can be solved numerically without the severe problems arising for equations of the first kind.

We now address suitable invariance operators for the dynamic scenario. This is studied in detail in [15].

For affine deformations, we can adapt invariances holding in the static case to invariance properties in the dynamic case.

Theorem 8 *Let $T_1^x : L_2(\Omega_X) \longrightarrow L_2(\mathbb{R}^n)$ and $T_2^{x,t} : L_2(\Omega_Y) \longrightarrow L_2(\mathbb{R}^m)$ be invariance operators for the static problem with fixed time instance t, i.e.*

$$T_1^x \mathcal{A}_t^* = \mathcal{A}_t^* T_2^{x,t} \quad \forall \, x, t.$$

Then, for affine motion functions Γ_t, $t \in [0, T]$, it holds

$$T_1^x \mathcal{A}_\Gamma^* = \mathcal{A}_\Gamma^* T_2^{dyn}$$

with

$$T_2^{dyn} : L_2([0, T] \times \mathbb{R}^m) \longrightarrow L_2([0, T] \times \mathbb{R}^m)$$

$$T_2^{dyn} g(t, y) := T_2^{\Gamma_t^{-1} x - \Gamma_t^{-1} 0, t} g_t(y).$$

Proof Since Γ_t is an affine mapping, it holds in particular

$$\Gamma_t^{-1}(z - x) = \Gamma_t^{-1} z - (\Gamma_t^{-1} x - \Gamma_t^{-1}(0)).$$

With the definition of the involved operators, we obtain

$$
\begin{aligned}
\mathcal{A}_\Gamma^* T_2^{dyn} g(z) &= \int_{[0,T]} |\det D\Gamma_t^{-1} z| \, \mathcal{A}_t^* T_2^{\Gamma_t^{-1} x - \Gamma_t^{-1} 0, t} g_t(\Gamma_t^{-1} x) \, dt \\
&= \int_{[0,T]} |\det D\Gamma_t^{-1} z| \, T_1^{\Gamma_t^{-1} x - \Gamma_t^{-1} 0} \mathcal{A}_t^* g_t(\Gamma_t^{-1} z) \, dt \\
&= \int_{[0,T]} |\det D\Gamma_t^{-1} z| \, \mathcal{A}_t^* g_t(\Gamma_t^{-1} z - (\Gamma_t^{-1} x - \Gamma_t^{-1} 0)) \, dt \\
&= \int_{[0,T]} |\det D\Gamma_t^{-1} z| \, \mathcal{A}_t^* g_t(\Gamma_t^{-1}(z - x)) \, dt \\
&= \mathcal{A}_\Gamma^* g(z - x) \\
&= T_1^x \mathcal{A}_\Gamma^* g(z).
\end{aligned}
$$

This concludes the proof. □

Remark 4 As discussed in [15], deriving linear invariances in the presence of non-linear object motion might in general not be possible. Hence, the use of approximate invariances is suggested instead and an error analysis has been provided. For our numerical examples, we are going to use approximate invariance which are exact for affine deformations, namely by using the operators T_1^x and T_2^{dyn} as defined above.

4.3 Applications

We want to illustrate our general dynamic reconstruction technique at the example of 3D X-ray tomography. An evaluation regarding 2D computerized tomography with parallel scanning geometry can be found in [15].

Example: 3D X-Ray Tomography
We consider an X-ray source emitting a cone of X-rays through the studied specimen to a 2D detector. The movement of the combination source-detector determines different geometries. Let $M \subset \mathbb{R}^3$ describe the curve of the X-ray

(continued)

source. Then, the mathematical model of 3D-CT for a static object h is given by the cone-beam transform

$$\mathcal{D}h(a, \theta) = \int_0^\infty h(a + \beta\theta)\, d\beta$$

with $a \in M \subset \mathbb{R}^3$ denoting the position of the source and $\theta \in S^2$ characterizing the direction of the ray.

One simple realization consists in rotating the radiation source on a circle around the specimen with radius $R > 0$, i.e. $M = \{R(\cos(\varphi), \sin(\varphi), 0)^T \mid \varphi \in [0, 2\pi]\}$. Despite some drawbacks from a mathematical point of view (for instance the Tuy–Kirillov condition is not satisfied resulting in incomplete data), this geometry is used in many real-world applications. Thus, we consider this setting in the following.

As in the 2D case, see the example of the Radon transform on page 7, the rotation of the radiation source represents the time-dependent step of the data acquisition, i.e. we identify the angle φ which characterizes the current source position as time variable. Thus, we obtain the dynamic operator

$$\mathcal{D}^{dyn} f(t, \theta) = \int_0^\infty f(a(t) + \beta\theta, t)\, d\beta$$

for a time-dependent function $f \in L_2([0, T] \times \Omega_X)$. If we further incorporate the motion information, we obtain

$$\mathcal{D}_\Gamma f_0(t, \theta) = \int_0^\infty f_0(\Gamma_t(a(t) + \beta\theta))\, d\beta$$

as dynamic operator for the initial state f_0, respectively.

In order to derive a reconstruction algorithm which compensates for the motion, we apply the method proposed in Sect. 4.2. Following Theorem 7, we determine the reconstruction kernel ψ^γ by considering the auxiliary problem

$$\mathcal{D}^{dyn*} \psi^\gamma = e^\gamma, \tag{21}$$

with the time-dependent mollifier e^γ (19) stemming from the static mollifier δ^γ.

Lemma 2 *The adjoint operator $\mathcal{D}^{dyn*}$ as mapping from $L_2([0, 2\pi] \times S^2)$ to $L_2([0, 2\pi] \times \mathbb{R}^3)$ is given by*

$$\mathcal{D}^{dyn*} g(t, x) = \|x - a(t)\|^{-2}\, g\left(a(t), \frac{x - a(t)}{\|x - a(t)\|}\right).$$

(continued)

Proof Since the investigated object has compact support (and is surrounded by source and detector), there is a minimal radius L such that $\operatorname{supp} f \subset V_L(0)$ and $L < R$. Therefore, in the definition of the cone beam transform, we can restrict ourselves to the integration over a compact interval $[L_1, L_2] \subset \mathbb{R}$, where $0 < L_1 < R - L$ and $L_2 > L + R$. This results in the following representation for the dynamic operator

$$\mathcal{D}^{dyn} f(t, \theta) = \int_{L_1}^{L_2} f(a(t) + \beta\theta, t) \, d\beta.$$

With the substitution $x := a(t) + \beta\theta$, we obtain

$$\langle \mathcal{D}^{dyn} f, g \rangle_{L_2([0,2\pi] \times S^2)} = \int_{[0,2\pi]} \int_{S^2} \int_{L_1}^{L_2} f(a(t) + \beta\theta, t) \, g(t, \theta) \, d\beta \, d\theta \, dt$$

$$= \int_{[0,2\pi]} \int_{V_L(0)} f(x, t) \, \|x - a(t)\|^{-2}$$

$$g\left(t, \frac{x - a(t)}{\|x - a(t)\|}\right) \, dx \, dt,$$

and thus the stated representation for $\mathcal{D}^{dyn*}$. □

A generalized solution of (21) is computed via the penalized normal equation

$$(\mathcal{D}^{dyn} \mathcal{D}^{dyn*} + \alpha I) \psi^\gamma = \mathcal{D}^{dyn} e^\gamma + \alpha \psi^{\gamma, stat}.$$

Due to the property

$$\mathcal{D}^{dyn} \mathcal{D}^{dyn*} g(t, \theta) = \int_{L_1}^{L_2} \mathcal{D}^{dyn*} g(a(t) + \beta\theta, t) \, d\beta$$

$$= \int_{L_1}^{L_2} \|\beta\theta\|^{-2} g\left(t, \frac{\beta\theta}{\|\beta\theta\|}\right) \, d\beta$$

$$= \left(\frac{1}{L_1} - \frac{1}{L_2}\right) g(t, \theta),$$

with appropriately selected $L_1, L_2 \in \mathbb{R}$ (see proof of Lemma 2), we obtain

$$\psi^\gamma = \left(\frac{1}{L_1} - \frac{1}{L_2} + \alpha\right)^{-1} \left(\mathcal{D}^{dyn} e^\gamma + \alpha \psi^{\gamma, stat}\right).$$

(continued)

Thus, the reconstruction kernel for dynamic cone-beam tomography results from averaging the generalized solution of the dynamic auxiliary problem and the static reconstruction kernel. In the static case, suitable reconstruction kernels have been derived for the circular cone beam transform. For instance, the static reconstruction kernel associated to the Gaussian mollifier

$$\delta^{\gamma}(z) = (2\pi)^{-3/2} \frac{1}{\gamma^3} e^{-\frac{\|z\|^2}{2\gamma^2}}$$

has been computed by Weber in his PhD-thesis [49] and in [33]. Further, he and his co-authors derived the special reconstruction kernels $\psi_x^{\gamma,stat}$ by

$$\psi_x^{\gamma,stat}(a(t), \theta) = T_2^{x,t} \psi^{\gamma}(a(t), \theta)$$

with (approximate) invariance operator

$$T_2^{x,t} \psi(a(t), \theta) = \frac{R^2}{\|a - x\|^2} \psi(a, U_x^T \theta),$$

where U_x^T corresponds to the unitary matrix that rotates $\frac{a-x}{\|a-x\|}$ onto a/R, i.e.

$$U_x^T \frac{a-x}{\|a-x\|} = \frac{a}{R}.$$

We adapt this invariance operator according to Theorem 8 and Remark 4 to the dynamic setting with motion model Γ.

Numerical Results

The algorithm is tested for the three-dimensional phantom with compact support in $V_1(0)$ whose initial state is shown in Fig. 6 (first row) for three different cross sections throughout the object. The dynamic behavior is described by the nonlinear scaling

$$\Gamma_t x = \begin{pmatrix} \frac{(s_1(t)x_1+1)^5-1}{5s_1(t)} \\ \frac{(s_2(t)x_2+1)^5-1}{5s_2(t)} \\ \frac{x_3}{s_3(t)} \end{pmatrix}$$

with

(continued)

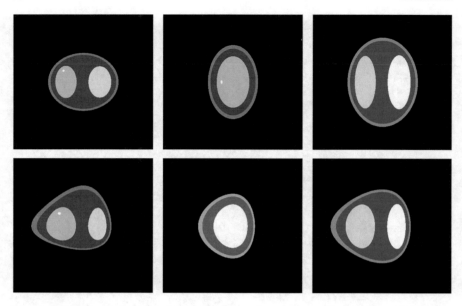

Fig. 6 Initial state (first row) and final state (second row) of our 3D phantom with nonaffine deformation. The three columns correspond to different crossections of the phantom (first column: fixed component $x_3 = 0$, second column: fixed component $x_2 = -0.27$, third column: fixed component $x_1 = 0$)

$$s_1(t) = \sqrt[4]{\sin(0.0375 \cdot t/\pi)}, \quad s_2(t) = \sqrt[4]{\sin(0.045 \cdot t/\pi)},$$

$$s_3(t) = 1 + \frac{25}{128}(s_1(t) + s_2(t)), \quad t \in [0, 2\pi].$$

To illustrate this dynamic behavior, the final state of the three cross sections is shown in Fig. 6 (last row). The respective dynamic cone-beam data are simulated for 360 source positions rotating on a circle with radius $R = 8$ and 801×801 planar detector points. In order to account for the statistical nature of photon emission, we further add noise to the simulated data characterized by the Poisson distribution resulting in an overall peak-signal-to-noise-ratio of 16 dB (corresponding to a noise level of approximately 6%).

We then apply the proposed dynamic algorithm with regularization parameters $\gamma = 0.0025$ and $\alpha = 1$. We further choose $L_1 = \frac{R-1}{2}$ and $L_2 = 2R$.

(continued)

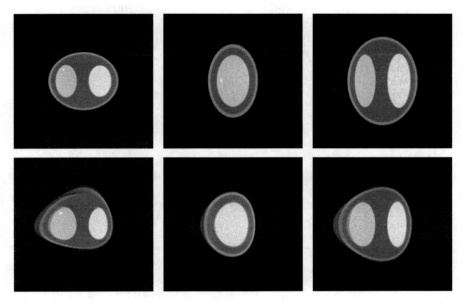

Fig. 7 Dynamic reconstruction (first row) and static reconstruction (second row) from noisy dynamic data. The three columns correspond to different crossections of the phantom (first column: fixed component $x_3 = 0$, second column: fixed component $x_2 = -0.27$, third column: fixed component $x_1 = 0$)

The respective reconstruction result is shown in Fig. 7 (first row) for the three cross sections of the object. As a comparison, the second row of Fig. 7 depicts the respective result when the algorithm with the static filter from [33] with regularization parameter $\gamma = 0.0025$ is applied to the dynamic data. Comparing the results highlights the motion compensation property of the proposed dynamic reconstruction approach. Despite the severe non-affine displacement during the data collection, the initial state is reconstructed without distortions or motion artefacts. With a static algorithm however, severe distortions arise. In particular, the small inclusion in the right ellipse (see first and second column) is not visible in the static reconstruction for $x_2 = -0.27$ (since it moved out from this cross section in the course of the data acquisition). In practical applications, the motion parameters have to be extracted beforehand, see our discussion in Sect. 2.1. Thus, we further want to evaluate how

(continued)

our dynamic reconstruction strategy performs in combination with a potential motion estimation procedure. For this purpose, we apply the dynamic reconstruction algorithm with approximate motion parameters, which are obtained by adding noise samples uniformly distributed in $[-0.09, 0.09]$ to the exact parameters. These noise samples correspond to a relative estimation error of 12, 5%. Figure 8 provides a visual comparison between the exact motion parameter s_1 and the noisy version used for the reconstruction step.

The result of the dynamic algorithm with approximate motion parameters is displayed in Fig. 9. This experiment shows that the dynamic regularization technique compensates well for the motion even if its parameters are not exactly known.

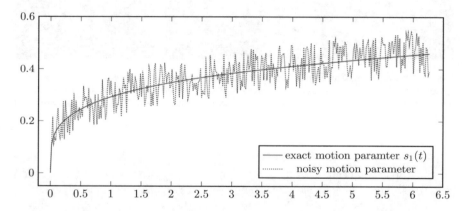

Fig. 8 Motion parameter $s_1(t)$ for $t \in [0, 2\pi]$ (solid line) and its noisy version (dashed line)

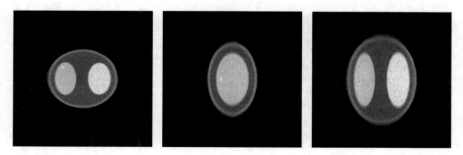

Fig. 9 Dynamic reconstruction with noisy motion parameters with nonaffine deformation from noisy dynamic data. The three columns correspond to different crossections of the phantom (first column: fixed component $x_3 = 0$, second column: fixed component $x_2 = -0.27$, third column: fixed component $x_1 = 0$)

5 Conclusion

In this chapter, we presented regularization strategies to solve general linear dynamic inverse problems with known object motion. In particular, our method based on the approximate inverse is not restricted to affine deformations. The numerical results from 3D cone-beam tomography illustrate its capability to compensate for strong, non-affine motion. The subsequent chapter provides a complementary study on the effect of the motion on the overall information content in dynamic data.

Acknowledgments The work of the author was supported by the Deutsche Forschungsgemeinschaft under grant HA 8176/1-1. Further, the author thanks Katharina Bernard for assisting with the implementation regarding dynamic 3D Cone-beam-CT and Gael Rigaud for fruitful discussions.

References

1. C. Blondel, R. Vaillant, G. Malandain, N. Ayache, 3D tomographic reconstruction of coronary arteries using a precomputed 4D motion field. Phys. Med. Biol. **49**, 2197–2208 (2004)
2. R. Boutchko, V.L. Rayz, N.T. Vandehey, J.P. O'Neil, T.F. Budinger, P.S. Nico, W.W. Moses, Imaging and modeling of flow in porous media using clinical nuclear emission tomography systems and computational fluid dynamics. J. Appl. Geochem. **76**, 74–81 (2012)
3. M. Burger, J. Modersitzki, S. Suhr, A nonlinear variational approach to motion-corrected reconstruction of density images. arXiv (2015)
4. C. Chen, B. Gris, O. Öktem, A new variational model for joint image reconstruction and motion estimation in spatiotemporal imaging. SIAM J. Imaging Sci. **12**, 1686–1719 (2019)
5. D. Chen, H. Li, Q. Wang, P. Zhang, Y. Zhu, Computed tomography for high-speed rotation object. Opt. Express **23**, 13423–13442 (2015)
6. J. Chung, A.K. Saibaba, M. Brown, E. Westman, Efficient generalized Golub–Kahan based methods for dynamic inverse problems. Inverse Prob. **34**, 024005 (2018)
7. C.R. Crawford, K.F. King, C.J. Ritchie, J.D. Godwin, Respiratory compensation in projection imaging using a magnification and displacement model. IEEE Trans. Med. Imaging **15**, 327–332 (1996)
8. L. Desbat, S. Roux, P. Grangeat, Compensation of some time dependent deformations in tomography. IEEE Trans. Med. Imaging **26**, 261–269 (2007)
9. R.L. Ehman, M.T. McNamara, M. Pallack, H. Hricak, C.B. Higgins, Magnetic resonance imaging with respiratory gating: techniques and advantages. Am. J. Roentgenol. **143**, 1175–1182 (1984)
10. L. Feng, M.B. Srichai, R.P. Lim, A. Harrison, W. King, G. Adluru, E.V. Dibella, D.K. Sodickson, R. Otazo, D. Kim, Highly accelerated real-time cardiac cine MRI using k-t SPARSE-SENSE. Magn. Reson. Med. **70**, 64–74 (2013)
11. D. Finch, M. Haltmeier, Rakesh, Inversion of spherical means and the wave equation in even dimensions. SIAM J. Appl. Math. **68**, 392–412 (2007)
12. J. Fitzgerald, P.G. Danias, Effect of motion on cardiac SPECT imaging: recognition and motion correction. J. Nucl. Cardiol. **8**, 701–706 (2001)
13. E. Gravier, Y. Yang, M. Jin, Tomographic reconstruction of dynamic cardiac image sequences. IEEE Trans. Image Process. **16**, 932–942 (2007)
14. B. Hahn, Reconstruction of dynamic objects with affine deformations in dynamic computerized tomography. J. Inverse Ill-Posed Prob. **22**, 323–339 (2014)

15. B.N. Hahn, Efficient algorithms for linear dynamic inverse problems with known motion. Inverse Prob. **30**, 035008 (2014)
16. B.N. Hahn, Dynamic linear inverse problems with moderate movements of the object: ill-posedness and regularization. Inverse Prob. Imaging **9**, 395–413 (2015)
17. B.N. Hahn, Motion estimation and compensation strategies in dynamic computerized tomography. Sensing Imaging **18** 1–20 (2017)
18. B. N. Hahn, E.T. Quinto, Detectable singularities from dynamic Radon data. SIAM J. Imaging Sci. **9**, 1195–1225 (2016)
19. B.N. Hahn, M.L. Kienle-Garrido, E.T. Quinto, Microlocal properties of dynamic Fourier integral operators, in *Time-Dependent Problems in Imaging and Parameter Identification*, ed. by B. Kaltenbacher, T. Schuster, A. Wald (Springer, Cham, 2021), pp. 85–120
20. B.N. Hahn, A.K. Louis, M. Maisl, C. Schorr, Combined reconstruction and edge detection in dimensioning. Meas. Sci. Technol. **24**, 125601 (2013)
21. A.A. Isola, A. Ziegler, T. Koehler, W.J. Niessen, M. Grass, Motion-compensated iterative cone-beam CT image reconstruction with adapted blobs as basis functions. Phys. Med. Biol. **53**, 6777–6797 (2008)
22. J. Kastner, B. Plank, C. Heinzl, Advanced X-ray computed tomography methods: high resolution CT, phase contrast CT, quantitative CT and 4DCT, in *Proceedings: Digital Industrial Radiology and Computed Tomography (DIR 2015), Ghent* (2015)
23. A. Katsevich, An accurate approximate algorithm for motion compensation in two-dimensional tomography. Inverse Prob. **26**, 065007 (2010)
24. A. Katsevich, M. Silver, A. Zamayatin, Local tomography and the motion estimation problem. SIAM J. Imaging Sci. **4**, 200–219 (2011)
25. S. Kindermann, A. Leitão, On regularization methods for inverse problems of dynamic type. Numer. Funct. Anal. Optim. **27**, 139–160 (2006)
26. P. Kuchment, L. Kunyansky, Mathematics of photoacoustic and thermoacoustic tomography, in *Handbook of Mathematical Methods in Imaging*, ed. by O. Scherzer (Springer, New York, 2015)
27. D. Le Bihan, C. Poupon, A. Amadon, F. Lethimonnier, Artifacts and pitfalls in diffusion MRI. J. Magn. Reson. Imaging **24**, 478–488 (2006)
28. S. Lingala, E. DiBella, M. Jacob, Deformation corrected compressed sensing (DC-CS): a novel framework for accelerated dynamic MRI. IEEE Trans. Med. Imaging **34**, 72–85 (2015)
29. J. Liu, X. Zhang, X. Zhang, H. Zhao, Y. Gao, D. Thomas, D.A. Low, H. Gao, 5D respiratory motion model based image reconstruction algorithm for 4D cone-beam computed tomography. Inverse Prob. **31**, 115007 (2015)
30. A.K. Louis, Approximate inverse for linear and some nonlinear problems. Inverse Prob. **12**, 175–190 (1996)
31. A.K. Louis, Feature reconstruction in inverse problems. Inverse Prob. **27**, 065010 (2011)
32. A.K. Louis, P. Maass, A mollifier method for linear operator equations of the first kind. Inverse Prob. **6**, 427–440 (1990)
33. A.K. Louis, T. Weber, D. Theis, Computing reconstruction kernels for circular 3D cone beam tomography. IEEE Trans. Med. Imaging **27**, 880–886 (2008). Special Issue on Fully 3D Image Reconstruction
34. W. Lu, T.R. Mackie, Tomographic motion detection and correction directly in sinogram space. Phys. Med. Biol. **47**, 1267–84 (2002)
35. D. Manke, K. Nehrke, P. Börnert, Novel prospective respiratory motion correction approach for free-breathing coronary MR angiography using a patient-adapted affine motion model. Magn. Reson. Med. **50**, 122–131 (2003)
36. S.J. McQuaid, B.F. Hutton, Sources of attenuation-correction artefacts in cardiac PET/CT and SPECT/CT. Eur. J. Nucl. Med. Mol. Imaging **35**, 1117–1123 (2008)
37. F. Natterer, F. Wübbeling, *Mathematical Methods in Image Reconstruction* (SIAM, Philadelphia, 2011)
38. E. Niemi, M. Lassas, A. Kallonen, L. Harhanen, K. Hämäläinen, S. Siltanen, Dynamic multi-source X-ray tomography using a spacetime level set method. J. Comput. Phys. **291**, 218–237 (2015)

39. R. Otazo, E. Candès, D.K. Sodickson, Low-rank plus sparse matrix decomposition for accelerated dynamic MRI with separation of background and dynamic components. Magn. Reson. Med. **73**, 1125–1136 (2015)
40. M. Reyes, G. Malandain, P.M. Koulibaly, M.A. González-Ballester, J. Darcourt, Model-based respiratory motion compensation for emission tomography image reconstruction. Phys. Med. Biol. **52**, 3579–3600 (2007)
41. S. Rit, D. Sarrut, L. Desbat, Comparison of analytic and algebraic methods for motion-compensated cone-beam CT reconstruction of the thorax. IEEE Trans. Med. Imaging **28**, 1513–1525 (2009)
42. E.L. Ritman, J.H. Kinsey, R.A. Robb, L.D. Harris, B.K. Gilbert, Physical and technical considerations in the design of the DSR: A high temporal resolution volume scanner. Am. J. Roentgenol. **134**, 369–374 (1980)
43. S. Roux, L. Desbat, A. Koenig, P. Grangeat, Exact reconstruction in 2D dynamic CT: compensation of time-dependent affine deformations. Phys. Med. Biol. **49**, 2169–2182 (2004)
44. U. Schmitt, A.K. Louis, Efficient algorithms for the regularization of dynamic inverse problems: I. Theory. Inverse Prob. **18**, 645–658 (2002)
45. U. Schmitt, A.K. Louis, C. Wolters, M. Vauhkonen, Efficient algorithms for the regularization of dynamic inverse problems: II. Applications. Inverse Prob. **18**, 659–676 (2002)
46. T. Schuster, *The Method of Approximate Inverse: Theory and Applications* (Springer, Berlin, 2007)
47. L. Shepp, S. Hilal, R. Schulz, The tuning fork artifact in computerized tomography. Comput. Graphics Image Process. **10**, 246–255 (1979)
48. R.M. Temam, A.M. Miranville, *Mathematical Modeling in Continuum Mechanics* 2nd edn. (Cambridge University Press, New York, 2005)
49. T. Weber, Schnelle Rekonstruktionskernberechnung in der 3D-Computertomographie. Ph.D. thesis, Saarland University (2008)
50. R. Zeng, J.A. Fessler, J.M. Balter, Respiratory motion estimation from slowly rotating x-ray projections: theory and simulation. Med. Phys. **32**, 984–991 (2005)

Microlocal Properties of Dynamic Fourier Integral Operators

Bernadette N. Hahn, Melina-L. Kienle Garrido, and Eric Todd Quinto

Abstract Following from the previous chapter *Motion compensation strategies in tomography*, this article provides a complementary study on the overall information content in dynamic tomographic data using the framework of microlocal analysis and Fourier integral operators. Based on this study, we further analyze which characteristic features of the studied specimen can be reliably reconstructed from dynamic tomographic data and which additional artifacts have to be expected in a dynamic image reconstruction. Our theoretical results, in particular the affect of the dynamic behavior on the measured data and the reconstruction result, is then illustrated in detail at various numerical examples from dynamic photoacoustic tomography.

1 On Singularities and Artifacts

In the previous chapter *Motion compensation strategies in tomography* [16], we studied regularization strategies for solving time-dependent inverse problems in tomography, which arise when the investigated specimen changes during the data acquisition process. In this article, we now provide a complementary study on the overall information content of such dynamic tomography data. In particular, we show how the respective information content affects the reconstruction quality.

Typically, the searched-for quantity f in tomographic applications can be considered as a piecewise constant function, where each value represents a specific material (e.g. bone, brain, air, etc.). In this case, the gradient ∇f–or more precisely the *singularities* of f–contain much of the information about f. A rigorous

B. N. Hahn (✉) · M.-L. Kienle Garrido
Department of Mathematics, University of Stuttgart, Stuttgart, Germany
e-mail: bernadette.hahn@imng.uni-stuttgart.de;
Melina-Loren.Kienle-Garrido@imng.uni-stuttgart.de

E. T. Quinto
Department of Mathematics, Tufts University, Medford, MA, USA
e-mail: todd.quinto@tufts.edu

© Springer Nature Switzerland AG 2021   85
B. Kaltenbacher et al. (eds.), *Time-dependent Problems in Imaging and Parameter Identification*, https://doi.org/10.1007/978-3-030-57784-1_4

mathematical definition of singularity is given in Sect. 2 along with an intuitive example.

The task "finding f from measured data $g = \mathcal{A}f$" then corresponds to "extracting the singular features of f from g". Thus, a thorough analysis on how an operator $\mathcal{A}$ encodes singular features has to be developed in order to fully understand the reconstruction process. This in turn can provide important insights regarding the design of reconstruction operators in order to avoid the formation of unwanted artifacts in the resulting reconstruction.

The most prominent example is limited-angle computerized tomography. In various applications, the radiation source cannot perform a complete 180- or 360° rotation around the specimen, such as for instance in dental diagnostics. If data are only measured for a subinterval of this angular range, the standard CT-reconstruction algorithm causes additional features, namely streak artifacts, to appear in the reconstruction results, see Figs. 2 and 3. Furthermore, certain singular features are missing in the reconstructed image.

An analysis of singularities and artifacts requires deep mathematics, namely the theory of *microlocal analysis* which goes back to techniques developed by Hörmander and others based on Fourier transforms. Over the last decades, microlocal analysis has been employed to understand image formation in static tomographic problems such as classical X-ray CT [9, 27, 31], seismics [5, 11, 29], sonar [10, 25], radar [1, 6, 28, 36], electron microscope tomography [32], Compton CT [34, 39], and geodesic transforms [8, 19].

In this article, we extend these classic results to dynamic tomography problems. In particular, we tackle the following questions:

- How does the dynamic behavior of the object affect the information content of the data g?
- Which singular features can be reliably reconstructed from dynamic tomography data?
- Which additional artifacts have to be expected in a dynamic image reconstruction?

Such a rigorous mathematical characterization can have great benefits in applications. For instance, it allows radiologists to determine whether a singularity in the reconstructed image belongs to the object or represents an artifacts, thereby making more reliable medical diagnoses. It could further serve as a basis for developing an adaptive data sampling protocol depending on the motion of the patient so that the measurements encode all relevant information. The analysis based on the model operator $\mathcal{A}$ could also be combined with data driven methods for image reconstruction or image post-processing in order to guarantee reliable results.

Microlocal analysis has begun to be used in motion-compensated CT [18, 22, 23] with extensions to generalized dynamic Radon transforms [17, 33]. The aim of this article is to provide a general framework for dynamic Fourier integral operators along with a characterization of visible singularities and added artifacts.

With this aim in mind, the article is organized as follows. In Sect. 2, we provide the basic concepts from microlocal analysis, including the concepts of singularities,

Fourier integral operators and artifacts. Next, in Sect. 3, we derive the concept of dynamic Fourier integral operators based on an underlying motion model, and we study how these operators encode the singularities of the searched-for quantity in the measured data. Due to their practical relevance in tomography, we provide, in particular, a detailed analysis for the special case of generalized dynamic Radon transforms. Section 4 addresses the reconstruction problem assuming the motion is known exactly. In particular, we characterize visible and added singularities in dynamic reconstructions using methods of filtered backprojection type. Our theoretical results are then illustrated in Sect. 5 for various numerical examples from dynamic photoacoustic tomography (PAT).

2   Basic Concepts of Microlocal Analysis

In this section we will outline the basic microlocal principles used in the article. We refer to [20, 21, 24, 37, 38] for more details.

First, we introduce some basic notation. Let $x = (x_1, x_2)$ be in $\mathbb{R}^2$ and let h be a real-valued function of variables including x. Let $G = (g_1, g_2)^T$ be an $\mathbb{R}^2$-valued function of variables including x. Then we define

$$D_x h = \left(\frac{\partial h}{\partial x_1}, \frac{\partial h}{\partial x_2} \right), \qquad D_x G = \begin{pmatrix} \frac{\partial g_1}{\partial x_1} & \frac{\partial g_1}{\partial x_2} \\ \frac{\partial g_2}{\partial x_1} & \frac{\partial g_2}{\partial x_2} \end{pmatrix} = \begin{pmatrix} D_x g_1 \\ D_x g_2 \end{pmatrix}$$

and other derivatives are defined in a similar way; for example, if h depends on t, then we define $D_t h = \frac{\partial h}{\partial t}$.

We now introduce notation for higher derivatives. Let $n \in \mathbb{N}$, then the point $\overline{\alpha} = (\alpha_1, \alpha_2, \ldots, \alpha_n) \in \{0, 1, 2, \ldots\}^n$ is called a *multi-index*. Let Ω be an open subset of $\mathbb{R}^n$ and let $h : \Omega \to \mathbb{R}$ be smooth. Then we define

$$D^{\overline{\alpha}} h = \frac{\partial^{\alpha_1}}{\partial x_1^{\alpha_1}} \frac{\partial^{\alpha_2}}{\partial x_2^{\alpha_2}} \cdots \frac{\partial^{\alpha_n}}{\partial x_n^{\alpha_n}} h \text{ and}$$

$$|\overline{\alpha}| = \alpha_1 + \alpha_2 + \cdots \alpha_n. \tag{1}$$

Now, we introduce some basic function classes. The set $\mathcal{D}(\mathbb{R}^n)$ consists of all C^∞ smooth functions of compact support in $\mathbb{R}^n$ and $f_k \to f$ in $\mathcal{D}(\mathbb{R}^n)$ if for some fixed compact set K, all f_k are supported in K and $f_k \to f$ uniformly along with all derivatives. The set $\mathcal{E}(\mathbb{R}^n)$ is the set of all C^∞ smooth functions on $\mathbb{R}^n$ with convergence in $\mathcal{E}$ being uniform convergence on compact sets along with all derivatives.

The dual space to $\mathcal{D}(\mathbb{R}^n)$ is denoted $\mathcal{D}'(\mathbb{R}^n)$ and called the space of distributions. Its topology is defined by weak convergence (i.e., $u_k \to u$ in $\mathcal{D}'(\mathbb{R}^n)$ if for every $f \in \mathcal{D}(\mathbb{R}^n)$, $u_k(f) \to u(f)$). The dual space to $\mathcal{E}(\mathbb{R}^n)$ is the set $\mathcal{E}'(\mathbb{R}^n)$ of all distributions of compact support with the topology defined by weak convergence on functions in $\mathcal{E}(\mathbb{R}^n)$. More details on these function spaces can be found, e.g., in [35].

2.1 A Rigorous Theory of Singularities

Wavefront sets are a precise classification of singularities of functions and the key to understanding them is the relation between smoothness of f and rapid decay at infinity of its Fourier transform, $\mathcal{F}f(y) = \frac{1}{2\pi} \int_{x \in \mathbb{R}^2} e^{-iy \cdot x} f(x) \mathrm{d}x$.

Smoothness and Rapid Decay

A distribution $f \in \mathcal{E}'(\mathbb{R}^n)$ is smooth if and only if $\mathcal{F}f$ is rapidly decaying at infinity (i.e., $\mathcal{F}f(\xi)$ decays at infinity faster than any power of $1/\|\xi\|$).

The proof of this statement uses the Fourier inversion formula [35], boundedness of $\mathcal{F} \colon L^1(\mathbb{R}^n) \to L^\infty(\mathbb{R}^n)$, and that, under the Fourier transform, a derivative of f becomes the product of a polynomial with $\mathcal{F}f$.

Definition 1 Let $u \in \mathcal{D}'(\mathbb{R}^n)$ and let $(x_0, \xi_0) \in \mathbb{R}^n \times (\mathbb{R}^n \setminus \{0\})$. Then u is smooth at x_0 in direction ξ_0 if there is a smooth cutoff function at x_0, $\psi \in \mathcal{D}(\mathbb{R}^n)$ (i.e., $\psi(x_0) \neq 0$) and an open cone V containing ξ_0 such that $\mathcal{F}(\psi u)(\xi)$ is rapidly decaying at infinity for all $\xi \in V$.

On the other hand, if u is not smooth at x_0 in direction ξ_0, then $(x_0, \xi_0) \in \mathrm{WF}(u)$, the C^∞ wavefront set of u.

This definition generalizes the relation between rapid decay of $\mathcal{F}f$ and smoothness of f by considering decay near individual directions rather than in all directions. Generally, the wavefront set is defined as a subset of a cotangent bundle, but we will not use that abstraction since there is a natural identification of $\mathbb{R}^n \times (\mathbb{R}^n \setminus \{0\})$ with $T^*(\mathbb{R}^n) \setminus \{0\}$.

In particular, according to its definition, the vectors $(x_0, \xi_0) \in \mathrm{WF}(u)$ characterize simultaneously the location, $x_0 \in \mathbb{R}^n$, and the direction, $\xi_0 \in \mathbb{R}^n \setminus \{0\}$ of singularities of f.

Example

The wavefront sets of characteristic functions can be understood intuitively.

First, let D be the unit disk in $\mathbb{R}^2$ and let χ_D, be its characteristic function. Note that χ_D is smooth (either identically zero or identically one) away from the boundary of D, namely the unit sphere S^1. Therefore, the wavefront set $\mathrm{WF}(\chi_D)$ should involve only points x in this boundary. In fact, $\mathrm{WF}(\chi_D)$ is the set of normals to the boundary of the disk,

$$\left\{ (x, tx) \,\middle|\, x \in S^1, t \neq 0 \right\}$$

(continued)

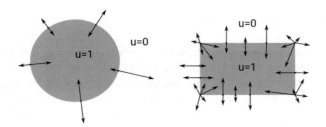

Fig. 1 An illustration of WF(u) when $u = \chi_D$ is the characteristic function of the unit disk D (left) and when $u = \chi_R$ is the characteristic function of a rectangle R (right)

as illustrated in Fig. 1 (left). Intuitively, these normal vectors point in the direction of greatest "non-smoothness."

If S is an arbitrary set with smooth boundary, then the wavefront set of χ_S consists of all normals to the boundary of S.

If the set S has a corner, then the wavefront set of χ_S will include all vectors at the corner. For example, the wavefront set of the characteristic function χ_R of a rectangle R will include all normal vectors along the edges of the rectangle and all vectors at the vertices of the rectangle, see Fig. 1 (right).

In general, if u is not smooth at a point x, then u has wavefront set above x; that is, for some $\xi \in \mathbb{R}^2 \setminus \{\mathbf{0}\}$, $(x, \xi) \in$ WF(u).

The following theorem will be important to analyze added artifacts.

Theorem 1 ([21, Theorem 8.2.10]) *Let Ω_1 be an open set in $\mathbb{R}^n$ and let $u \in \mathcal{E}'(\Omega_1)$. If the following non-cancellation condition holds*

$$\forall (z, \xi) \in \mathrm{WF}(u) : \quad (z, -\xi) \notin \mathrm{WF}(\chi_B), \tag{2}$$

then the product $\chi_B u$ can be defined as a distribution. In this case,

$$\mathrm{WF}(\chi_B u) \subset Q(B, \mathrm{WF}(u)),$$

where for $W \subset \Omega_1 \times \mathbb{R}^n \setminus \{\mathbf{0}\}$

$$Q(B, W) := \{(z, \xi + \eta) \mid z \in B, [(z, \xi) \in W \vee \xi = 0] \wedge [(z, \eta) \in \mathrm{WF}(\chi_B) \vee \eta = 0]\}.$$

2.2 Fourier Integral Operators

In this section, we define the fundamental classes of operators on which our analysis is based. Note that we do not give the general definitions but ones that are sufficient for our purposes. In particular, we consider two-dimensional imaging problems in this article, i.e. we set the dimension to $n = 2$ in the following. For other applications, one would use the definition for general spaces in [38, Chapter VI.2] or [20]. These operators are defined in terms of amplitudes and we start with this definition.

Definition 2 (Amplitude of Order k) Let Ω_1 and Ω_2 be open sets in $\mathbb{R}^2$ and let $m \in \{1, 2\}$. Now let $a(z, x, \tau)$ be a smooth function on $\Omega_1 \times \Omega_2 \times \mathbb{R}^m$. Then a is an *amplitude of order k* if it satisfies the following condition. For each compact subset K in $\Omega_1 \times \Omega_2$ and each $M \in \mathbb{N}$, there exists a positive constant $C_{K,M}$ such that

$$\left| D_z^{\overline{\alpha}} D_x^{\overline{\beta}} D_\tau^{\overline{\gamma}} a(z, x, \tau) \right| \leq C_{K,M} (1 + \|\tau\|)^{k - |\overline{\gamma}|} \tag{3}$$

for all $(z, x, \tau) \in K \times \mathbb{R}^m$ whenever $|\overline{\alpha}| + \left|\overline{\beta}\right| + |\overline{\gamma}| \leq M$.

We now define the general class of operators we consider in this article.

Definition 3 (Fourier Integral Operator (FIO)) Let $m \in \{1, 2\}$ and let Ω_1 and Ω_2 be open subsets of $\mathbb{R}^2$. The real-valued function $\Phi = \Phi(z, x, \tau) \in C^\infty (\Omega_1 \times \Omega_2 \times (\mathbb{R}^m \setminus \{0\}))$ is called a *phase function* if Φ is positive homogeneous of degree 1 in the phase variable τ. We define

$$\Sigma_\Phi = \left\{ (z, x, \tau) \in \Omega_1 \times \Omega_2 \times \mathbb{R}^m \setminus \{0\} \,\middle|\, D_\tau \Phi = 0 \right\} \tag{4}$$

and we call the phase function Φ *non-degenerate* if

$$D_z \Phi \text{ and } D_x \Phi \text{ are both nonzero for all } (z, x, \tau) \in \Sigma_\Phi. \tag{5}$$

Now let $a(z, x, \tau)$ be an amplitude (see Definition 2) of order k and let Φ be a non-degenerate phase function.

The operator $\mathcal{T}$ defined for $u \in \mathcal{E}'(\Omega_2)$ by

$$\mathcal{T}u(z) = \int e^{i\Phi(z,x,\tau)} a(z, x, \tau) u(x) \mathrm{d}x \, \mathrm{d}\tau \tag{6}$$

is a *Fourier Integral Operator (FIO)* of order $k + (m - 2)/2$.

The *canonical relation* for $\mathcal{T}$ is

$$C := \left\{ (z, D_z \Phi(z, x, \tau); x, -D_x \Phi(z, x, \tau)) \,\middle|\, (z, x, \tau) \in \Sigma_\Phi \right\}. \tag{7}$$

Since the phase function Φ satisfies (5), the sets Σ_Φ and C are smooth manifolds.

Example (Radon Transform)

The mathematical model of computerized tomography is given by the classical *Radon transform*

$$\mathcal{R}u(\varphi, s) = \int u(x)\delta(s - x^T\theta(\varphi))\,dx,$$

which integrates u along the straight lines

$$\left\{ x \in \mathbb{R}^2 \,\middle|\, x^T\theta(\varphi) = s \right\}$$

with $\theta(\varphi) = (\cos\varphi, \sin\varphi)^T$ and δ the delta-distribution. Note that $\Omega_1 = [0, 2\pi] \times \mathbb{R}$ with 0 and 2π identified in this case; therefore the data variable $z \in \Omega_1$ has been replaced by $(\varphi, s) \in [0, 2\pi] \times \mathbb{R}$. This operator is an FIO of order $-1/2$ with phase variable $\tau \in \mathbb{R} \setminus \{0\}$ and representation

$$\mathcal{R}u(\varphi, s) = \int e^{i\tau(s - x^T\theta(\varphi))}\frac{1}{2\pi}u(x)\,dx\,d\tau,$$

where the phase function is $\Phi(\varphi, s, x, \tau) = \tau(s - x^T\theta(\varphi))$ and the amplitude is $a(\varphi, s, x, \tau) = \frac{1}{2\pi}$, which is a symbol of order zero. Note that this Fourier representation of $\mathcal{R}$ is valid by the Fourier Slice Theorem (e.g., [26, Theorem 1.1]).

Example (Pseudodifferential Operators (PSIDOs))

We now define a special type of FIO. In this case, $m = 2$ and the phase variable will be denoted $\xi \in \mathbb{R}^2$. Let Ω be an open subset of $\mathbb{R}^2$.

Let the function $a(z, x, \xi)$ for $(z, x, \xi) \in \Omega \times \Omega \times \mathbb{R}^2$ be an amplitude satisfying Definition 2. Define

$$\Phi(z, x, \xi) = \xi \cdot (z - x),$$

then Φ is a phase function satisfying the non-degeneracy condition (5).

Under these conditions the *pseudodifferential operator (PSIDO)*

$$\mathcal{P}u(z) = \int e^{i\Phi(z,x,\xi)}a(z, x, \xi)u(x)\,dx\,d\xi$$

is an FIO satisfying Definition 3.

(continued)

Note that, if the amplitude a has order k, then $\mathcal{P}$ is an FIO of order k associated to the canonical relation

$$\Delta = \left\{ (x, \xi, x, \xi) \,\middle|\, (x, \xi) \in \Omega \times \left(\mathbb{R}^2 \setminus \{\mathbf{0}\} \right). \right\}$$

Every smooth differential operator is a PSIDO, and its order as a PSIDO is the same as its order as a differential operator.

2.3   FIO and Wavefront Sets

To state the theorems that describe how operators change wavefront sets, we need the following definitions. Let X and Y be sets and let $B \subset X \times Y$, $C \subset Y \times X$, and $D \subset X$. Then, we define

$$C^t := \left\{ (x, y) \,\middle|\, (y, x) \in C \right\}$$

$$C \circ D := \left\{ y \in Y \,\middle|\, \exists x \in D, (y, x) \in C \right\} \tag{8}$$

$$B \circ C := \left\{ (x', x) \in X \times X \,\middle|\, \exists y \in Y, (x', y) \in B, \ (y, x) \in C \right\},$$

and

$$\Pi_L : C \to Y, \quad \Pi_L(y, x) = y$$

$$\Pi_R : C \to X, \quad \Pi_R(y, x) = x$$

are the natural projections from C.

Next, we note that the formal dual of an FIO is an FIO.

Theorem 2 ([20, Theorem 4.2.1]) *Let $\mathcal{T}$ be an FIO of order k with canonical relation C. Then the formal dual operator, $\mathcal{T}^*$ to $\mathcal{T}$ is an FIO of order k with canonical relation C^t.*

The next definition is helpful to determine which singularities are visible, as we will discuss in the next section.

Definition 4 Let $\mathcal{T}$ be an FIO given by (6) with amplitude a of order k. Then $\mathcal{T}$ is *elliptic* if its amplitude a satisfies the following condition. For each compact set $K \subset \Omega_1 \times \Omega_2$ there are constants $C_K > 0$ and $S_K > 0$ such that for all $(z, x) \in K$ and for all $\tau \in \mathbb{R}^m$ such that $\|\tau\| > S_K$,

$$|a(z, x, \tau)| \geq C_K (1 + \|\tau\|)^k.$$

Our next definition is fundamental for our results.

Definition 5 Let $\mathcal{T}$ be an FIO with canonical relation C. Then, $\mathcal{T}$ satisfies the *semi-global Bolker Assumption* if the natural projection $\Pi_L : C \to \Omega_1 \times \mathbb{R}^2 \setminus \{0\}$ is an embedding–a smooth injective map with injective derivative.

Victor Guillemin [12, 14] called Definition 5 plus additional geometric conditions (including that $\mathcal{T}$ is a Radon transform defined by a double fibration for which the projection to X is proper, and Π_R is surjective) the Bolker Assumption. His extra conditions assure that one can compose $\mathcal{T}^*$ and $\mathcal{T}$ and that the composition is an elliptic pseudodifferential operator. This is not true in general without extra assumptions.

A straightforward calculation shows that PSIDOs satisfy the semi-global Bolker Assumption.

FIOs transform wavefront sets in precise ways, and our next theorem, a special case of the Hörmander-Sato Lemma, is a key to our analysis.

Theorem 3 ([20, Theorems 2.5.7 and 2.5.14], [38, Section 6.3, (6.22)]) *Let $\mathcal{T}$ be an FIO (Definition 3) with canonical relation C. Let $f \in \mathcal{E}'(\Omega_2)$. Then*

$$\mathrm{WF}(\mathcal{T}f) \subset C \circ \mathrm{WF}(f). \tag{9}$$

If $\mathcal{T}$ is elliptic and satisfies the semi-global Bolker Assumption, then equality holds in (9).

For PSIDOs, this theorem implies that $\mathrm{WF}(\mathcal{P}(f)) \subset \mathrm{WF}(f)$ and equality holds if $\mathcal{P}$ is elliptic since the canonical relation of PSIDOs is Δ.

We will need several continuity results for FIOs.

Theorem 4 ([21, Theorem 8.2.13]) *Let $\mathcal{T}$ be an FIO satisfying Definition 3. Then $\mathcal{T} : \mathcal{E}'(\Omega_2) \to \mathcal{D}'(\Omega_1)$ is weakly continuous.*

Therefore, if $\mathcal{P}$ is a PSIDO satisfying the conditions in the Example on Pseudodifferential Operators, then $\mathcal{P} : \mathcal{E}'(\Omega) \to \mathcal{D}'(\Omega)$ is weakly continuous.

Theorem 4 is valid because we assume (5) in the definition of FIO, and this condition holds for the phase function for PSIDO. In general, FIOs are continuous in Sobolev scale. Before stating our theorem, we provide some definitions.

Definition 6 Let Ω be an open subset of $\mathbb{R}^2$. The set $H_c^s(\Omega)$ is the set of all distributions u with compact support in Ω such that the Sobolev norm

$$\|u\|_s = \sqrt{\int_{\xi \in \mathbb{R}^2} \mathcal{F}u(\xi) \left(1 + \|\xi\|^2\right)^s \, d\xi}$$

is finite.

The set $H_{\mathrm{loc}}^s(\Omega)$ is the set of all distributions u supported in Ω such that for all cutoff functions $\varphi \in \mathcal{D}(\Omega)$, $\varphi u \in H_c^s(\Omega)$.

We say a linear operator $A : H_c^s(\Omega_2) \to H_{\text{loc}}^{s-k}(\Omega_1)$ is continuous if for each fixed compact set $K \subset \Omega_2$ and each $\varphi \in \mathcal{D}(\Omega_1)$, there is a constant $C_{K,\varphi} > 0$ such that for all $u \in \mathcal{E}'(\Omega_2)$ supported in K,

$$\|\varphi A u\|_{s-k} \leq C_{K,\varphi} \|u\|_s .$$

Theorem 5 ([20, Theorem 4.3.1]) *Let $\mathcal{T}$ be an FIO of order $k \in \mathbb{R}$ and assume the projection $\Pi_L : C \to \Omega_1 \times \mathbb{R}^2$ is an immersion (i.e. the derivative of Π_L is injective). Then*

$$\mathcal{T} : H_c^s(\Omega_2) \to H_{loc}^{s-k}(\Omega_1)$$

is continuous.

Therefore, if $\mathcal{A}$ is a PSIDO of order k then $\mathcal{A} : H_c^s(\Omega) \to H_{\text{loc}}^{s-k}(\Omega)$ is continuous.

Note that the condition in this theorem about Π_L will be true whenever $\mathcal{T}$ satisfies the semi-global Bolker Assumption.

2.4   Visible Singularities and Artifacts

In the rest of the article, the reconstruction operators we consider will be either regular PSIDOs or PSIDO-like operators that have discontinuous symbols, and we will use the theory of singularities and FIOs developed in this section to describe what these operators can do to singularities of the object in the reconstruction step. We now provide the basic terminology to describe this.

Definition 7 Let $\mathcal{L}$ be a reconstruction operator, $f \in \mathcal{E}'(\mathbb{R}^2)$, and $(x, \xi) \in \text{WF}(f)$.

Then, (x, ξ) will be a singularity of f that is visible in the reconstruction or *visible singularity* if $(x, \xi) \in \text{WF}(\mathcal{L}f)$.

On the other hand, (x, ξ) will be an *invisible singularity* of f if $(x, \xi) \notin \text{WF}(\mathcal{L}f)$.

Any singularity $(y, \eta) \in \text{WF}(\mathcal{L}f)$ that is not in $\text{WF}(f)$ will be called an *artefact*.

This terminology is illustrated using two examples from static 2D-CT. According to the inversion formula of the Radon transform, all singularities of f can be recovered via a filtered backprojection algorithm, if data are collected for $\varphi \in [0, \pi]$, [26]. However, if a only a smaller angular range can be covered, certain singularities will be invisible in the reconstruction and streak artifacts arise instead [31]. This is illustrated in Fig. 2 for the Shepp-Logan phantom with $\varphi \in [0, \frac{3}{4}\pi]$ and in Fig. 3 for a circular phantom with $\varphi \in [0, \frac{\pi}{2}]$ (left). For this circular phantom, the visible/invisible singularities and the added artifacts are highlighted in Fig. 3 (right).

A detailed analysis of visible and invisible singularities as well as added artifacts in dynamic image reconstruction is provided in Sect. 4.

Fig. 2 Shepp-Logan
Phantom reconstructed from
2D CT-data with limited
angular range $[0, \frac{3}{4}\pi]$

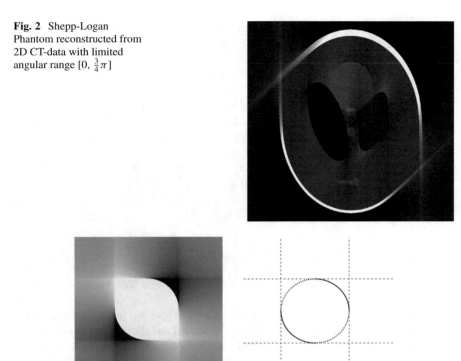

Fig. 3 Left: Circular phantom reconstructed from "D-CT data with limited angular range $[0, \frac{\pi}{2}]$. Right: Visible singularities (solid line), invisible singularities (dotted line) and added artifacts (dashed line) for a circular phantom

3   Encoding Object Singularities in Dynamic Imaging Data

In this section, we analyze how singularities of a moving object get encoded in dynamic imaging data. Therefore, we first recall the motion model developed in the chapter *Motion compensation strategies in tomography* [16] and the mathematical characterization for our moving object.

Let $[0, T]$, $T \in \mathbb{R}_{>0} = (0, \infty)$ denote the time interval required for the data acquisition process and let $\mathbb{R}_T$ be an open interval containing $[0, T]$.

A two-dimensional specimen that changes in time can be described by a time-dependent function $h : \mathbb{R}_T \times \mathbb{R}^2 \to \mathbb{R}$, where $h(t, \cdot)$ corresponds to the state of the searched-for quantity at a fixed time instance $t \in \mathbb{R}_T$. We define $f(x) := h(0, x)$ to be the *initial state* of the specimen. In tomographic applications, the object under investigation typically has compact support at all time instances. Thus, without loss of generality, we assume f (and all object states $h(t, \cdot)$, $t \in \mathbb{R}_T$) to be compactly supported in a fixed open set $\Omega_2 \subset \mathbb{R}^2$.

Let $\Gamma : \mathbb{R}_T \times \Omega_2 \to \Omega_2$ be the mapping which relates the state of the object at time t to its reference configuration f, more precisely

$$h(t, x) = f(\Gamma_t x).$$

Thus, Γ describes the motion of the object particles over time. More precisely, the vector $\Gamma(t, x)$ denotes which particle is at position x at time t.

Throughout the article, we make the following assumption on Γ.

Smooth Diffeomorphic Motion Model
We call a mapping $\Gamma : \mathbb{R}_T \times \Omega_2 \to \Omega_2$ a *smooth diffeomorphic motion model* with *motion functions* $\Gamma_t : \Omega_2 \to \Omega_2$, $\Gamma_t := \Gamma(t, \cdot)$, if the following conditions are satisfied:

- $\Gamma : \mathbb{R}_T \times \Omega_2 \to \Omega_2$ is smooth,
- $\Gamma_t : \Omega_2 \to \Omega_2$ is a diffeomorphism for all $t \in \mathbb{R}_T$.

Remark 1 The diffeomorphism condition guarantees that two particles cannot move into the same position, no particle gets lost (or added) and their relocation is smooth.

In practical applications, only discrete data sets are measured, i.e. the dynamic behavior is only ascertained for finitely many time instances, and the body does move continuously. This justifies the assumption that Γ is smooth with respect to time

3.1 Dynamic FIOs

Let $\mathcal{T}$ be a FIO according to Definition 3 for a static quantity $f \in \mathcal{E}'(\Omega_2)$, where we identify one of the data variables (without loss of generality z_1) as time instance t. Thus, we replace the data variable z by $(t, y) \in \mathbb{R}_T \times \Pi$ where Π is an open subset of $\mathbb{R}$. This results in the representation

$$\mathcal{T}f(t, y) = \int e^{i\Phi(t,y,x,\tau)} a(t, y, x, \tau) f(x) \mathrm{d}x \, \mathrm{d}\tau, \quad (t, y) \in \mathbb{R}_T \times \Pi.$$

In the dynamic setting, at time t, the state of the object $f(\Gamma_t \cdot)$ is encoded by $\mathcal{T}$, resulting (after a change of variables) in the associated dynamic forward operator

$$\mathcal{T}_\Gamma f(t, y) = \int e^{i\Phi(t,y,\Gamma_t^{-1}x,\tau)} a(t, y, \Gamma_t^{-1}x, \tau) \, |\det D_x \Gamma_t^{-1}x| \, f(x) \mathrm{d}x \, \mathrm{d}\tau.$$

We denote

$$a_\Gamma(t, y, x, \tau) := |\det D_x \Gamma_t^{-1} x| \, a(t, y, \Gamma_t^{-1} x, \tau)$$

and

$$\Phi_\Gamma(t, y, x, \tau) := \Phi(t, y, \Gamma_t^{-1} x, \tau)$$

for $(t, y) \in \mathbb{R}_T \times \Pi$, $x \in \Omega_2$, $\tau \in \mathbb{R}^m \setminus \{0\}$.

Theorem 6 *Let Γ be a smooth diffeomorphic motion model. Assume the static operator $\mathcal{T}$ is a FIO of order $k + (m - 2)/2$ (see Definition 3) with amplitude a of order k and non-degenerate phase function Φ. Further, assume*

$$\begin{aligned}
&if \ (t, y, \Gamma_t^{-1} x, \tau) \in \Sigma_\Phi \ and \ D_y \Phi = 0, \ then \\
&D_t \Phi(t, y, \Gamma_t^{-1} x, \tau) + D_x \Phi(t, y, \Gamma_t^{-1} x, \tau) \cdot D_t \Gamma_t^{-1} x \neq 0.
\end{aligned} \tag{10}$$

Then, $\mathcal{T}_\Gamma$ is a FIO of order $k + (m - 2)/2$ with amplitude a_Γ of order k and phase function Φ_Γ. If, in addition, $\mathcal{T}$ is elliptic, then $\mathcal{T}_\Gamma$ is elliptic.

Remark 2 Note that the condition (10) is satisfied if $D_y \Phi$ is never equal to zero on Σ_Φ. We will see in Sect. 3.2 that this is the case for generalized Radon transforms.

Proof Since a is an amplitude of order k, this property transfers to a_Γ due to the smoothness of the motion functions Γ_t and their inverse Γ_t^{-1}.

By the same argument, Φ_Γ inherits the smoothness property from Φ. Since Φ is positive homogeneous of degree 1 in τ, the same holds for Φ_Γ. Thus, Φ_Γ is a phase function.

Using the chain rule, we compute

$$D_{(t,y)} \Phi_\Gamma(t, y, x, \tau) = D_{(t,y)} \Phi(t, y, \Gamma_t^{-1} x, \tau) + D_x \Phi(t, y, \Gamma_t^{-1} x, \tau) \cdot D_{(t,y)} \Gamma_t^{-1} x,$$

$$D_x \Phi_\Gamma(t, y, x, \tau) = D_x \Phi(t, y, \Gamma_t^{-1} x, \tau) \cdot D_x \Gamma_t^{-1} x,$$

and

$$D_\tau \Phi_\Gamma(t, y, x, \tau) = D_\tau \Phi(t, y, \Gamma_t^{-1} x, \tau).$$

From the last property, it follows

$$\begin{aligned}
\Sigma_{\Phi_\Gamma} &= \left\{ (t, y, x, \tau) \in \mathbb{R}_T \times \Pi \times \Omega_2 \times \mathbb{R}^m \setminus \{0\} \, \middle| \, D_\tau \Phi(t, y, \Gamma_t^{-1} x, \tau) = 0 \right\} \\
&= \left\{ (t, y, x, \tau) \, \middle| \, (t, y, \Gamma_t^{-1} x, \tau) \in \Sigma_\Phi \right\}.
\end{aligned}$$

Since Φ is non-degenerate, $D_x\Phi$ is nonzero on Σ_Φ. Further, Γ_t^{-1} is a diffeomorphism for all t and therefore, $D_x\Gamma_t^{-1}x$ has nonzero determinant, i.e. the matrix is regular. Thus, on Σ_{Φ_Γ}, the derivative $D_x\Phi_\Gamma(t, y, x, \tau)$ is nonzero. Since $\Gamma_t^{-1}x$ is independent of y, it follows together with condition (10) that Φ_Γ is non-degenerate, and thus, according to Definition 3, $\mathcal{T}_\Gamma$ is a FIO.

The ellipticity of the static FIO $\mathcal{T}$ transfers to its dynamic counterpart $\mathcal{T}_\Gamma$ due to the smoothness of the motion functions Γ_t and inverse Γ_t^{-1} and that Γ_t is a diffeomorphism. $\square$

The next statement follows directly from Theorem 3,

Theorem 7 *Let $\mathcal{T}_\Gamma$ be a FIO (according to Definition 3) with canonical relation C_Γ. Then, for $f \in \mathcal{E}'(\Omega_2)$,*

$$\mathrm{WF}(\mathcal{T}_\Gamma f) \subset C_\Gamma \circ \mathrm{WF}(f).$$

Equality holds if $\mathcal{T}_\gamma$ is elliptic and satisfies the semi-global Bolker Assumption.

Thus, each singularity in the dynamic data stems from a singularity of the object.

Warning

Without additional assumptions on the motion model, the dynamic FIO $\mathcal{T}_\Gamma$ does not, in general, satisfy the semi-global Bolker condition, even if the static FIO $\mathcal{T}$ does. An example corresponds to $\mathcal{T}$ being the classical Radon transform and Γ describing a smooth rotational movement of the same speed than the radiation source, see [18].

In the next section, we state additional assumptions on Γ, under which the semi-global Bolker condition holds at least for dynamic generalized Radon transforms.

3.2 Dynamic Generalized Radon Transforms

The measurement process in many imaging modalities (such as CT, PAT, sonar, etc.) can be modelled by a generalized Radon transform, i.e. an operator that integrates over smooth curves in the plane. We assume the curves are defined as level sets in x of a smooth function $\Psi : \mathbb{R}_T \times \Omega_2 \to \mathbb{R}$. Specifically, we assume the following hypothesis.

Hypothesis 1 Let $\Psi : \mathbb{R}_T \times \Omega_2 \to \mathbb{R}$. If Ψ satisfies the following properties, then Ψ will be called a *defining function*.

1. Ψ is smooth and for all $(t, x) \in \mathbb{R}_T \times \Omega_2$, $D_x\Psi(t, x) \neq \mathbf{0}$.
2. There is an open interval Π such that for each $(t, y) \in \mathbb{R}_t \times \Pi$

$$S(t, y) = \{x \in \Omega_2 \mid y = \Psi(t, x)\}$$

defines a nontrivial smooth curve.

3. For each $t \in \mathbb{R}_T$, $\Omega_2 \subset \cup_{y \in \Pi} S_\Gamma(t, y)$ (so the curves $S(t, \cdot)$ cover Ω_2).
4. For each compact set $K \subset \Omega_2$, there is a compact subset L of Π such that $K \cap S(t, y) = \emptyset$ for all $(t, y) \in \mathbb{R}_T \times (\Pi \setminus L)$.

Each part of Hypothesis 1 puts more structure on the set of curves $S(t, y)$. Part 1 ensures that each curve is a smooth regular curve. Part 2 means that the curve $S(t, y)$ is defined for all $y \in \Pi$. Part 3 means that, for each t, the curves $S(t, \cdot)$ cover Ω_2, and part 4 will allow us to compose operators in Sect. 4.2 by assuming that $S(t, y)$ is "near" the boundary of Ω_2 if y is "near" the boundary of Π.

With this notation, the generalized Radon transform can be written

$$\mathcal{A}f(t, y) = \int a(t, y, x) f(x) \delta(y - \Psi(t, x)) \, dx, \quad (t, y) \in \mathbb{R}_T \times \Pi \qquad (11)$$

which integrates the quantity f weighted with the C^∞ function a on the curve in $\mathbb{R}^2$ defined by $y = \Psi(t, x)$. Because $D_x \Psi$ is never zero, $\mathcal{A}$ can be written

$$\mathcal{A}f(t, y) = \int e^{i\sigma(y - \Psi(t,x))} a(t, y, x) f(x) dx \, d\sigma \qquad (12)$$

with phase variable σ in $\mathbb{R}$ and phase $\Phi(t, y, x, \sigma) = \sigma(y - \Psi(t, x))$ and amplitude a. These statements follow from basic facts about the Fourier transform and arguments in [2, 13] and the calculation starting at (10) in [30]. By Theorem 8 below, $\mathcal{A}$ is an FIO.

Due to their practical relevance, we now study this type of operators in more detail.

With a smooth diffeomorphic motion model Γ, the associated dynamic forward operator becomes

$$\mathcal{A}_\Gamma f(t, y) = \int a(t, y, \Gamma_t^{-1}x) f(x) \delta(y - \Psi(t, \Gamma_t^{-1}x)) |\det D_x \Gamma_t^{-1} x| \, dx \qquad (13)$$

for all $(t, y) \in \mathbb{R}_T \times \Pi$. Then, the FIO version is

$$\mathcal{A}_\Gamma f(t, y) = \int e^{i\sigma(y - \Psi(t, \Gamma_t^{-1}x))} a(t, y, \Gamma_t^{-1}x) |\det D_x \Gamma_t^{-1} x| f(x) \, dx \, d\sigma. \qquad (14)$$

These are justified using a change of variable in (11) and in (12). To simplify the subsequent expressions, we set

$$\Psi_\Gamma(t, x) := \Psi(t, \Gamma_t^{-1}x), \quad (t, x) \in \mathbb{R}_T \times \Omega_2. \qquad (15)$$

The operator $\mathcal{A}_\Gamma$ integrates the weighted initial state f along the curves

$$S_\Gamma(t, y) := \left\{ x \in \Omega_2 \,\middle|\, y = \Psi_\Gamma(t, x) \right\}. \tag{16}$$

Our next theorem shows that $\mathcal{A}_\Gamma$ is an elliptic FIO under reasonable conditions.

Theorem 8 *Let Ψ be a defining function, and let $\mathcal{A}$ be the generalized Radon transform defined by (11) where a in (11) is smooth. Then $\mathcal{A}$ is an elliptic FIO of order $-1/2$.*

Let Γ be a smooth diffeomorphic motion model. Then, the dynamic operator $\mathcal{A}_\Gamma$ in (14) is an elliptic FIO of order $-1/2$ with amplitude

$$a_\Gamma(t, y, s, \sigma) = |\det D_x \Gamma_t^{-1} x| \, a(t, y, \Gamma_t^{-1} x),$$

phase function

$$\Phi_\Gamma(t, y, x, \sigma) = \sigma(y - \Psi(t, \Gamma_t^{-1} x)),$$

defining function Φ_Γ and canonical relation

$$C_\Gamma = \left\{ \left((t, \Psi_\Gamma(t, x)), \left(-\sigma D_t \Psi_\Gamma(t, x), \sigma \right); x, \sigma D_x \Psi_\Gamma(t, x) \right) \right.$$

$$\left. \middle|\, t \in \mathbb{R}_T, x \in \Omega_2, \sigma \in \mathbb{R} \setminus \{0\} \right\}.$$

Proof To show that $\mathcal{A}$ is an FIO, we first note that the phase is

$$\Phi(t, y, x, \sigma) = \sigma(y - \Psi(t, x)).$$

Since $D_x \Psi$ is never zero, $D_x \Phi$ is nowhere zero, and $D_y \Phi(t, y, x, \sigma) = \sigma$ is nonzero for all $\sigma \in \mathbb{R} \setminus \{0\}$, so Φ is a nondegenerate phase function. Therefore, $\mathcal{A}$ satisfies Definition 3 and $\mathcal{A}$ is an FIO. Since a is smooth, positive and independent of σ, a is an amplitude of order zero and so $\mathcal{A}$ is an elliptic FIO of order $-1/2$.

Now, we explain why $\mathcal{A}_\Gamma$ is an elliptic FIO. Since

$$D_y \Phi_\Gamma \Gamma = \sigma \quad \text{and} \quad D_\sigma \Phi_\Gamma(t, y, x, \sigma) = y - \Psi_\Gamma(t, x),$$

the set Σ_{Φ_Γ} is characterized by

$$\Sigma_{\Phi_\Gamma} = \left\{ (t, \Psi_\Gamma(t, x), x, \sigma) \,\middle|\, (t, x, \sigma) \in \mathbb{R}_T \times \Pi \times \Omega_2 \times \mathbb{R} \setminus \{0\} \right\}.$$

Further, we obtain the derivatives

$$D_x \Phi_\Gamma = -\sigma D_x \Psi_\Gamma,$$

$$D_{(t,y)} \Phi_\Gamma = (-\sigma D_t \Psi_\Gamma, \sigma).$$

In particular $D_y \Phi_\Gamma \Gamma$ is never equal to zero on Σ_{Φ_Γ}. Thus, $\mathcal{A}_\Gamma$ is an elliptic FIO of order $-1/2$ according to Theorem 6. The stated representation for the canonical relation C_Γ follows directly from the representation of the above derivatives.

The property that Ψ_Γ is a defining function follows from the respective property of Ψ. First, $D_x(\Psi_\Gamma) = D_x \phi D_x \Gamma_t^{-1}$ is nowhere zero by part 1 for Ψ. This proves part 1 for Ψ_Γ. The other parts of the proposition follow from the fact that, for all $t \in \mathbb{R}_T$, $\Gamma_t : \Omega_2 \to \Omega_2$ is a bijective diffeomorphism. $\qquad\square$

We now state conditions on the motion model and the phase function such that the dynamic operator $\mathcal{A}_\Gamma$ satisfies the semi-global Bolker assumption.

Theorem 9 *Let Ψ be a defining function, and let $\mathcal{A}$ be the generalized Radon transform defined by* (11) *where a in* (12) *is smooth.*

Let Γ be a smooth diffeomorphic motion model and let $\mathcal{A}_\Gamma$ be the dynamic FIO (13) *with Ψ_Γ given by (15).*

We further assume, that Ψ_Γ satisfies the following additional conditions:

- *The map*

$$x \mapsto \begin{pmatrix} \Psi_\Gamma(t, x) \\ D_t \Psi_\Gamma(t, x) \end{pmatrix} \tag{17}$$

  is one-to-one for each t.
- *For all $x \in \Omega_2$ and all $t \in \mathbb{R}_T$,*

$$\det \begin{pmatrix} D_x \Psi_\Gamma(t, x) \\ D_x D_t \Psi_\Gamma(t, x) \end{pmatrix} \neq 0. \tag{18}$$

Then, $\mathcal{A}_\Gamma$ satisfies the semi-global Bolker Assumption.

Condition (17) implies the injectivity of Π_L, and this ensures that the data, respectively the integration curves, can distinguish different points in the object. Condition (18) implies that $\Pi_L : C_\Gamma \to \mathbb{R}_T \times \Pi \times \mathbb{R}^2 \setminus \{\mathbf{0}\}$ is an immersion (i.e. its derivative is injective), and this guarantees that the integration curves vary sufficiently to detect the object singularities. For a more detailed interpretation, we refer to [18].

The proof has been stated in the literature, for instance in [33] for generic integration curves, in [18] for dynamic CT or in [4] for dynamic PAT. The argument applies by analogy to our case and is therefore omitted here.

Theorem 10 *Let Ψ be a defining function, and let $\mathcal{A}$ be the generalized Radon transform defined by* (11) *where a in* (11) *is smooth. Let Γ be a smooth diffeomorphic motion model. Then, for our dynamic imaging operator $\mathcal{A}_\Gamma$ in* (13),

$$\mathrm{WF}(\mathcal{A}_\Gamma f) \subset C_\Gamma \circ \mathrm{WF}(f).$$

If $\mathcal{A}$ is, in addition, elliptic (a is nowhere zero) and if the motion model satisfies the stronger conditions (17) *and* (18), *then*

$$\mathrm{WF}(\mathcal{A}_\Gamma f) = C_\Gamma \circ \mathrm{WF}(f).$$

This theorem is a direct consequence of Theorems 7 and 8.

Using the representation of the canonical relation C_Γ from Theorem 8, we obtain the following explicit correspondence between wavefront of $\mathcal{A}_\Gamma f$ and that of f. Let $(t, y) \in \mathbb{R}_T \times \Pi$ and $\sigma \neq 0, v \in \mathbb{R}$. If $\big((t, y), (-\sigma v, \sigma)\big) \in \mathrm{WF}(\mathcal{A}_\Gamma f)$, then there exists an element $x \in S_\Gamma(t, y)$ such that

$$(x, \sigma D_x \Psi_\Gamma(t, x)) \in \mathrm{WF}(f),$$

where $S_\Gamma(t, y)$ is the integration curve given by (16).

If $\mathcal{A}_\Gamma$ is in addition elliptic and satisfies the semi-global Bolker Assumption, then, for $t \in \mathbb{R}_T$, $\big((t, y), (-\sigma v, \sigma)\big) \in \mathrm{WF}(\mathcal{A}_\Gamma f)$ if and only if there exists an $x \in S_\Gamma(t, y)$ such that $D_t \Psi_\Gamma(t, x) = v$ and $(x, \sigma D_x \Psi_\Gamma(t, x)) \in \mathrm{WF}(f)$. In case such a point x exists, it is unique.

We conclude this section by stating smoothing properties of $\mathcal{A}_\Gamma$ between Sobolev spaces. Note that $\mathbb{R}_T \times \Pi$ is an open subset of $\mathbb{R}^2$ so one can use our definitions for Sobolev spaces on $\mathbb{R}_T \times \Pi$.

Theorem 11 *Let Ψ be a defining function, and let $\mathcal{A}$ be the generalized Radon transform defined by* (11) *where a in* (12) *is smooth. Assume Γ is a smooth diffeomorphic motion model, and assume the dynamic operator $\mathcal{A}_\Gamma$ in* (13) *satisfies the additional condition* (18). *Then,*

$$\mathcal{A}_\Gamma : H_c^s(\Omega_2) \to H_{loc}^{s+1/2}(\mathbb{R}_T \times \Pi)$$

is continuous.

Proof According to Theorem 8, $\mathcal{A}_\Gamma$ is a FIO of order $k = -1/2$. Additionally, condition (18) yields that the projection $\Pi_L : C_\Gamma \to T^*(\mathbb{R}_T \times \Pi) \setminus \{0\}$ is an immersion. Hence, we can apply Theorem 5 and obtain that $\mathcal{A}_\Gamma : H_c^s(\Omega_2) \to H_{loc}^{s+1/2}(\mathbb{R}_T \times \Pi)$ is continuous.   □

According to the above theorem, the data $\mathcal{A}_\Gamma f$ are smoother than f by 1/2 in Sobolev scale. In particular, for a smooth diffeomorphic motion model satisfying (17) and (18), $\mathcal{A}_\Gamma$ has the same smoothing property as the static operator $\mathcal{A}$.

After analyzing the overall information content of dynamic data, we now study which object singularities can be reliably reconstructed and which additional artifacts have to be expected.

4 Reconstruction Operators and Artefact Study

In this section, we apply the theory of microlocal analysis to define and analyze reconstruction operators to solve dynamic inverse problems $\mathcal{T}_\Gamma f = g$. From Sect. 2, we know that Fourier integral operators encode singularities of f in specific ways. The idea now is to construct reconstruction operators which recover the visible singularities from the measured data $g = \mathcal{T}_\Gamma f$.

4.1 An Ideal Scenario: Smoothly Periodic Motion

In practical applications, data can only be measured for t in a closed interval $[0, T] \subset \mathbb{R}_T$. From a theoretical point of view, this is troublesome since smooth function (and hence distributions) are defined on open sets in order for derivatives to be well defined.

For a specific type of functions, namely smoothly T-periodic functions, this does not impose a restriction. A function of t (and perhaps other variables) will be called *smoothly T−periodic* if it extends to $t \in \mathbb{R}$ as a smooth function that is T−periodic. This allows us to define $\mathcal{E}([0, T] \times \Pi)$ to be the set of functions on $[0, T] \times \Pi$ which extend to functions on $\mathbb{R} \times \Pi$ that are smooth and T−periodic in t. The set $\mathcal{D}([0, T] \times \Pi)$ then denotes the set of those functions with compact support.

We start our study of reconstruction operators within this idealized framework by assuming that the motion model Γ, the amplitude a and the phase function Φ (and Ψ in case of a generalized Radon transform) are all smoothly T−periodic. So, for example, Γ can be extended in t to a function on $\mathbb{R}$ and $\Gamma(t, \cdot) = \Gamma(t + T, \cdot)$ for all $t \in \mathbb{R}$.

The dual operator plays a crucial role, and it is defined by

$$\mathcal{T}_\Gamma^t g(x) = \int_{[0,T]} \int_\Pi \int_{\mathbb{R}^2} e^{i\,\Phi(t,y,\Gamma_t^{-1}x,\tau)} a(t, y, \Gamma_t^{-1}x, \tau) \,|\det D_x \Gamma_t^{-1}x| \, g(t, y) \, d\tau \, dy \, dt.$$

In particular, the operator $\mathcal{T}_\Gamma^t$ then corresponds to the formal dual of $\mathcal{T}_\Gamma$ for $g \in \mathcal{D}([0, T] \times \Pi)$.

For dynamic generalized Radon transforms as in (13), this corresponds to the *backprojection operator* $\mathcal{A}_\Gamma^t$ defined by

$$\mathcal{A}_\Gamma^t g(x) = \int_{t \in [0,T]} a_\Gamma(t, \Psi_\Gamma(t, x), x) g(t, \Psi_\Gamma(t, x)) dt, \tag{19}$$

for $g \in \mathcal{E}(\mathbb{R}_T \times \Pi)$. This is true by taking the expression (13) and calculating its dual and integrating the δ function with respect to y.

Our next theorem provides conditions under which $\mathcal{T}_\Gamma$ is an FIO in the T–periodic case, and this includes dynamic generalized Radon transforms satisfying Hypothesis 1.

Theorem 12 *Let Γ be a smooth diffeomorphic T–periodic motion model such that $\mathcal{T}_\Gamma$ is a dynamic FIO that satisfies the semi-global Bolker Assumption. In addition, assume that $\mathcal{T}_\Gamma$ and $\mathcal{T}_\Gamma{}^t$ are both strongly continuous as mappings*

$$\mathcal{T}_\Gamma : \mathcal{D}(\Omega_2) \to \mathcal{D}([0, T] \times \Pi), \quad \mathcal{T}_\Gamma^t : \mathcal{E}([0, T] \times \Pi) \to \mathcal{E}(\Omega_2). \quad (20)$$

Let $\mathcal{P}$ be a PSIDO. Then, the operator

$$\mathcal{L}_\Gamma := \mathcal{T}_\Gamma^t \mathcal{P} \mathcal{T}_\Gamma$$

is well defined for $f \in \mathcal{E}'(\Omega_2)$ and

$$\mathrm{WF}(\mathcal{L}_\Gamma f) \subset \mathrm{WF}(f). \quad (21)$$

Now, assume that $\mathcal{P}$ and $\mathcal{T}_\Gamma$ are elliptic with positive symbols and the natural projection $\Pi_R : C_\Gamma \to \Omega_2 \times \mathbb{R}^2 \setminus \{\mathbf{0}\}$ is surjective. Then

$$\mathrm{WF}(\mathcal{L}_\Gamma f) = \mathrm{WF}(f). \quad (22)$$

Let Ψ be smoothly T–periodic and satisfy Hypothesis 1 and let $\mathcal{A}$ be a generalized Radon transform with defining function Ψ. Then (20) holds for $\mathcal{A}_\Gamma$. Therefore, (21) and (22) hold under the appropriate hypotheses above.

Proof Let C_Γ denote the canonical relation of the FIO $\mathcal{T}_\Gamma$. Since, the operator $\mathcal{T}_\Gamma^t$ is the formal dual of $\mathcal{T}_\Gamma$, it is a FIO as well with canonical relation C_Γ^t. As $\mathcal{T}_\Gamma$ satisfies the semi-global Bolker Assumption,

$$C_\Gamma^t \circ C_\Gamma \subset \Delta := \big\{ (x, \xi; x, \xi) \,\big|\, (x, \xi) \in \Omega_2 \times \mathbb{R}^2 \setminus \{\mathbf{0}\} \big\}.$$

According to Theorem 4, the PSIDO $\mathcal{P} : \mathcal{E}'([0, T] \times \Pi) \to \mathcal{D}'([0, T] \times \Pi)$ is weakly continuous. By duality with their adjoints and the continuity assumptions (20), $\mathcal{T}_\Gamma : \mathcal{E}'(\mathbb{R}^2) \to \mathcal{E}'([0, T] \times \Pi)$ is weakly continuous as is $\mathcal{T}_\Gamma^t : \mathcal{D}'([0, T] \times \Pi) \to \mathcal{D}'(\Omega_2)$. Therefore, the composition $\mathcal{L}_\Gamma := \mathcal{T}_\Gamma^t \mathcal{P} \mathcal{T}_\Gamma$ is well-defined and weakly continuous for $f \in \mathcal{E}'(\Omega_2)$.

Using Theorem 3, we obtain

$$\mathrm{WF}(\mathcal{L}_\Gamma f) \subset C_\Gamma^t \circ C_\Gamma \circ \mathrm{WF}(f) \subset \mathrm{WF}(f).$$

Next we explain why equality holds in (21) if all operators are elliptic, $\mathcal{T}_\Gamma$ satisfies the semi-global Bolker Assumption and Π_R is surjective from C_Γ to $\Omega_2 \times \mathbb{R}^2 \setminus \{\mathbf{0}\}$. Since $\mathcal{P}$ is a PSIDO, its canonical relation is Δ. Since Π_R is surjective, $C_\Gamma^t \circ \Delta \circ C_\Gamma = \Delta$. By the semi-global Bolker Assumption, the operators can be composed as FIO, and since they are all elliptic the symbol of the composition, $\mathcal{L}_\Gamma$, is the product symbol on the product canonical relations pulled back to Δ (see the symbol calculation in [30]). In this case, $\mathcal{L}_\Gamma$ is an elliptic PSIDO and equality holds in (21).

Now let $\mathcal{A}$ be a $T-$periodic generalized Radon transform with defining function, Ψ. Since Ψ satisfies Hypothesis 1, the dynamic generalized Radon transform $\mathcal{A}_\Gamma$ is a $T-$periodic FIO with defining function Ψ_Γ.

We now outline the proof of (20) for $\mathcal{A}_\Gamma$. Let K be a compact subset of Ω_2 and let $L \subset \Pi$ be the compact set in Hypothesis 1 part 4 for K. Then, $\mathcal{A}_\Gamma$ maps functions supported in K to functions supported in $[0, T] \times L$. One proves that this map is continuous from $\mathcal{D}_K$ to $\mathcal{D}_{[0,T]\times L}$ (see [35, sections 6.1-6.6]) using the explicit expression (13) (or that $\mathcal{A}_\Gamma$ is a generalized Radon transform, e.g., [30]). To prove that $\mathcal{A}_\Gamma^t : \mathcal{E}([0, T] \times \Pi) \to \mathcal{E}(\Omega_2)$, one uses the expression (19) for $\mathcal{A}_\Gamma^t$ and the fact that $[0, T]$ is compact and all functions are smoothly $T-$periodic. This allows us to apply the statements for $\mathcal{T}_\Gamma$ for $\mathcal{A}_\Gamma$.   □

Theorem 12 provides a strategy to design suitable reconstruction operators. If we choose a PSIDO as above, then the operator

$$\mathcal{S}_\Gamma := \mathcal{T}_\Gamma^t \mathcal{P}$$

applied to the data $g = \mathcal{T}_\Gamma f$ provides an image $\mathcal{L}_\Gamma f$ whose singularities coincide with singularities of the searched-for quantity f. In particular, $\mathcal{L}_\Gamma f$ displays the object singularities that it reconstructs at their correct location with the correct direction, i.e. the motion is compensated for. If the operators are elliptic and the other assumptions of Theorem 12 hold, then $\mathcal{L}_\Gamma f$ reproduces all singularities of f. Therefore, $\mathcal{L}_\Gamma f = \mathcal{T}_\Gamma^t \mathcal{P} \mathcal{T}_\Gamma f = \mathcal{S}_\Gamma g$ can be interpreted as an approximate inversion formula for the purpose of motion compensation.

4.2 The Realistic Case: Non-periodic Motion

The T-periodicity assumption on Γ in the last section imposes a severe restriction regarding practical applications. This assumption implies that the data have to encode the same state of the object at beginning and end of the scanning–a condition which, in general, will not be met.

Therefore, we want to analyze this more realistic setting in the following. More generally, we consider the scenario that data $g(t, y) = \mathcal{T}_\Gamma f(t, y)$ are measured for $(t, y) \in [\alpha, \beta] \subset \mathbb{R}_T$ with $0 \leq \alpha < \beta \leq T$. This framework covers, for instance, also data acquisition protocols with limited angular range. Then, formally, the forward operator $\mathcal{T}_\Gamma$ needs to be restricted to the data set. This can be achieved by multiplying with the characteristic function $\chi_{[\alpha,\beta]\times\Pi}$ of $[\alpha, \beta] \times \Pi$.

In order to study the effect on the singularities in the data and under reconstruction, we can apply the paradigm given in [10] which characterizes a broad range of incomplete data problems. In particular, the study divides into the following steps:

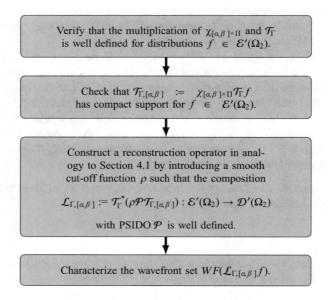

Note if $\mathcal{T}_\Gamma^* : \mathcal{D}' \to \mathcal{D}'$ then the cutoff ρ is not necessary.

With this general outline, we now perform the artefact study for operator $\mathcal{A}_\Gamma$ from Sect. 3.2 (13), i.e. for a dynamic generalized Radon transform with smooth diffeomorphic motion model Γ.

First, we verify that the multiplication with the characteristic function $\chi_{[\alpha,\beta] \times \Pi}$ is well-defined.

Proposition 1 *Assume Ψ satisfies Hypothesis 1 and Γ is a smooth motion model. Then, the operator $\mathcal{A}_{\Gamma,[\alpha,\beta]} f := \chi_{[\alpha,\beta] \times \Pi} \mathcal{A}_\Gamma f$ is well-defined for distributions $f \in \mathcal{E}'(\Omega_2)$.*

Proof Let $f \in \mathcal{E}'(\Omega_2)$. We apply Theorem 1 with the data set $B := [\alpha, \beta] \times \Pi$. Using the representation of the canonical relation C_Γ of $\mathcal{A}_\Gamma$ (see Theorem 8), it follows that

$$C_\Gamma \circ \mathrm{WF}(f) \subset \left\{ (t, y, \xi) \in \mathbb{R}_T \times \Pi \times \mathbb{R}^2 \setminus \{\mathbf{0}\} \,\middle|\, \xi_2 \neq 0 \right\}. \tag{23}$$

However, $\mathrm{WF}(\chi_{[\alpha,\beta]}) = \left\{ (t, y, \xi_1, 0) \,\middle|\, t \in \{\alpha, \beta\}, y \in \Pi, \xi_1 \neq 0 \right\}$. Therefore sums of such points are of the form (t, y, η_1, ξ_2) where $\xi_2 \neq 0$. Therefore the non-cancellation condition (2) holds, and Theorem 1 can be used to conclude that $\mathcal{A}_{\Gamma,[\alpha,\beta]} f$ is well-defined for $f \in \mathcal{E}'(\Omega_2)$. $\qquad\square$

Proposition 2 *Assume* Ψ *satisfies Hypothesis 1 and* Γ *is a smooth motion model. Then,* $\mathcal{A}_{\Gamma,[\alpha,\beta]} f$ *has compact support for all* $f \in \mathcal{E}'(\Omega_2)$, *i.e.* $\mathcal{A}_{\Gamma,[\alpha,\beta]} : \mathcal{E}'(\Omega_2) \to \mathcal{E}'(\mathbb{R}_T \times \Pi)$.

Proof Let $f \in \mathcal{E}'(\Omega_2)$. By Theorem 8, Ψ_Γ is a defining function. Thus, according to Hypothesis 1 part 4, there is a compact set $L \subset \Pi$ such that $\mathcal{A}_\Gamma f(t, y)$ is supported on $\mathbb{R}_T \times L$.

Because $\chi_{[\alpha,\beta]}$ is zero for $t \notin [\alpha, \beta]$ and all y and $\mathcal{A}_{\Gamma,[\alpha,\beta]} f$ is zero for all y outside a compact set, the product, $\mathcal{A}_{\Gamma,[\alpha,\beta]} f$ has compact support in $[\alpha, \beta] \times L$.
$\square$

The formal dual to $\mathcal{A}_\Gamma$ on $\mathbb{R}_T \times \Pi$ is given by

$$\mathcal{A}_\Gamma^* g(x) = \int_{\mathbb{R}_T} a_\Gamma(t, \Psi_\Gamma(t, x), x) \, g(t, \Psi_\Gamma(t, x)) \mathrm{d}t. \tag{24}$$

Since the domain of $\mathcal{A}_\Gamma^*$ is not, in general, $\mathcal{D}'(\mathbb{R}_T \times \Pi)$, we multiply by a smooth cutoff function. Therefore, let $\rho : \mathbb{R}_T \to \mathbb{R}$ be smooth and equal to one on $[\alpha, \beta]$ and be supported in $\mathbb{R}_T$. The corresponding restricted backprojection operator is then given by

$$\mathcal{A}_{\Gamma,\rho}^t g := \mathcal{A}_\Gamma^*(\rho g). \tag{25}$$

In analogy to Sect. 4.1, we would like to consider

$$\mathcal{L}_{\Gamma,[\alpha,\beta]} := \mathcal{A}_{\Gamma,\rho}^t \mathcal{P} \mathcal{A}_{\Gamma,[\alpha,\beta]}, \tag{26}$$

with a PSIDO $\mathcal{P}$ to build a reconstruction operator for the non-periodic case. Therefore, we have to prove that this composition is well-defined.

Proposition 3 *Let* $\mathcal{P}$ *be a pseudodifferential operator, then* $\mathcal{A}_{\Gamma,\rho}^t$, $\mathcal{P}$, *and* $\mathcal{A}_{\Gamma,[\alpha,\beta]}$ *can be composed and* $\mathcal{L}_{\Gamma,[\alpha,\beta]} : \mathcal{E}'(\Omega_2) \to \mathcal{D}'(\Omega_2)$ *is well-defined.*

Proof From Propositions 1 and 2, we know that $\mathcal{A}_{\Gamma,[\alpha,\beta]} f \in \mathcal{E}'(\mathbb{R}_T \times \Pi)$ for $f \in \mathcal{E}'(\Omega_2)$. Therefore, $\mathcal{P} \mathcal{A}_{\Gamma,[\alpha,\beta]} f$ is defined as a distribution in $\mathcal{D}'(\mathbb{R}_T \times \Pi)$.

Since $M_\rho : g \mapsto M_\rho := \rho g$ is a trivial pseudodifferential operator, which is continuous on $\mathcal{D}(\mathbb{R}_T \times \Pi)$, the operator $M_\rho \mathcal{A}_\Gamma = \rho \mathcal{A}_\Gamma$ is continuous from $\mathcal{D}(\Omega_2)$ to $\mathcal{D}(\mathbb{R}_T \times \Pi)$. Here we use Hypothesis 1 and the fact that ρ has compact support in t.

This implies, that the dual $(M_\rho \mathcal{A}_\Gamma)^* = (\rho \mathcal{A}_\Gamma)^* = \mathcal{A}_\Gamma^* \rho = \mathcal{A}_{\Gamma,\rho}^t$ is weakly continuous from $\mathcal{D}'(\mathbb{R}_T \times \Pi)$ to $\mathcal{D}'(\Omega_2)$. Therefore, $\mathcal{L}_{\Gamma,[\alpha,\beta]}$ is well-defined. $\square$

We now state the main result of this section, which provides a characterization of the visible singularities and the possible added artifacts from data above $[\alpha, \beta]$.

Theorem 13 *Let* $f \in \mathcal{E}'(\Omega_2)$ *and* Γ *be a smooth diffeomorphic motion which satisfies the additional conditions (17) and (18). Further, let* $\mathcal{P}$ *be a PSIDO and*

$\mathcal{L}_{\Gamma,[\alpha,\beta]}$ *be given by (26) where* Ψ *satisfies Hypothesis 1. Then,*

$$\mathrm{WF}(\mathcal{L}_{\Gamma,[\alpha,\beta]}f) \subset (\mathrm{WF}(f) \cap \mathcal{V}_{[\alpha,\beta]}) \cup \mathcal{Z}_{\{\alpha,\beta\}}(f),$$

where

$$\mathcal{V}_{[\alpha,\beta]} := \big\{(x, \sigma \partial_x \Psi_\Gamma(t, x)) \,\big|\, x \in \Omega_2, \ t \in [\alpha, \beta], \ \sigma \in \mathbb{R} \setminus \{0\}\big\}$$

is the set of all (potentially) visible singularities from data above $[\alpha, \beta]$, *and*

$$\mathcal{Z}_{\{\alpha,\beta\}}(f) := \big\{(x, \sigma D_x \Psi_\Gamma(t, x)) \,\big|\, t \in \{\alpha, \beta\}, \ y \in \Pi, \ x \in \mathcal{S}_\Gamma(t, y), \ \sigma \in \mathbb{R} \setminus \{0\},$$
$$\exists \tilde{x} \in \mathcal{S}_\Gamma(t, y) : \ (\tilde{x}, \sigma D_x \Psi_\Gamma(t, \tilde{x})) \in \mathrm{WF}(f)\big\}$$

denotes the set of (possible) added artifacts.

Proof Since ρ is a smooth cutoff, $\mathcal{A}^t_{\Gamma,\rho}$ is a FIO with the same canonical relation as $\mathcal{A}^*_\Gamma$. Thus, we have

$$\mathrm{WF}(\mathcal{L}_{\Gamma,[\alpha,\beta]}f) = \mathrm{WF}(\mathcal{A}^t_{\Gamma,\Phi}\mathcal{P}\mathcal{A}_{\Gamma,[\alpha,\beta]}f) \subset C^t_\Gamma \circ \mathrm{WF}(\mathcal{P}\mathcal{A}_{\Gamma,[\alpha,\beta]}f). \tag{27}$$

Further, $\mathcal{P}$ is a pseudodifferential operator, i.e. its canonical relation is Δ and therefore

$$\mathrm{WF}(\mathcal{P}\mathcal{A}_{\Gamma,[\alpha,\beta]}f) \subset \mathrm{WF}(\mathcal{A}_{\Gamma,[\alpha,\beta]}f).$$

Following Proposition 1 and Theorem 1, we obtain

$$\mathrm{WF}(\mathcal{A}_{\Gamma,[\alpha,\beta]}f) \subset Q([\alpha, \beta] \times \Pi, \mathrm{WF}(\mathcal{A}_\Gamma f)) \tag{28}$$

with Q defined in Theorem 1. From Theorem 10, we know $\mathrm{WF}(\mathcal{A}_\Gamma f) \subset C_\Gamma \circ \mathrm{WF}(f)$ and hence

$$\mathrm{WF}(\mathcal{L}_{\Gamma,[\alpha,\beta]}f) \subset C^t_\Gamma \circ Q([\alpha, \beta] \times \Pi, C_\Gamma \circ \mathrm{WF}(f)). \tag{29}$$

From the definition, we get

$$Q([\alpha, \beta] \times \Pi, C_\Gamma \circ \mathrm{WF}(f))$$
$$= \big\{(z, \xi + \eta) \,\big|\, z \in [\alpha, \beta] \times \Pi, [(z, \xi) \in C_\Gamma \circ \mathrm{WF}(f) \vee \xi = 0]$$
$$\wedge [(z, \eta) \in \mathrm{WF}(\chi_{[\alpha,\beta] \times \Pi}) \vee \eta = 0]\big\}.$$

This set can be written as a union of three sets

$$Q([\alpha, \beta] \times \Pi, C_\Gamma \circ \mathrm{WF}(f)) = (C_\Gamma \circ \mathrm{WF}(f)) \cap ([\alpha, \beta] \times \Pi \times \mathbb{R}^2 \setminus \{0\})$$

$$\cup \mathrm{WF}(\chi_{[\alpha,\beta]\times\Pi})$$
$$\cup \mathcal{W}_{\{\alpha,\beta\}}(f),$$

where $\mathcal{W}_{\{\alpha,\beta\}}(f)$ summarizes the case $\xi \neq 0 \wedge \eta \neq 0$, i.e. the set is defined by

$$\mathcal{W}_{\{\alpha,\beta\}}(f) := \big\{(z, \xi + \eta) \,\big|$$
$$z \in [\alpha,\beta] \times \Pi, \ (z,\xi) \in C_\Gamma \circ \mathrm{WF}(f), \ (z,\eta) \in \mathrm{WF}(\chi_{[\alpha,\beta]\times\Pi}))\big\}$$
$$= \big\{((t, y), (v, \sigma)) \,\big|\, t \in \{\alpha, \beta\}, \ v \in \mathbb{R}, \ y \in \Pi, \ \sigma \in \mathbb{R} \setminus \{0\},$$
$$\exists x \in \mathcal{S}_\Gamma(t, y) : \ (x, \sigma D_x \Psi_\Gamma(t, x)) \in \mathrm{WF}(f)\big\}.$$

Hence, we obtain

$$\mathrm{WF}(\mathcal{L}_{\Gamma,[\alpha,\beta]} f) \subset C_\Gamma^t \circ \big[(C_\Gamma \circ \mathrm{WF}(f)) \cap ([\alpha,\beta] \times \Pi \times \mathbb{R}^2 \setminus \{0\})\big] \qquad (30)$$
$$\cup\, C_\Gamma^t \circ \mathrm{WF}(\chi_{[\alpha,\beta]\times\Pi}) \qquad (31)$$
$$\cup\, C_\Gamma^t \circ \mathcal{W}_{\{\alpha,\beta\}}(f). \qquad (32)$$

Under the additional conditions (17) and (18) on the motion Γ, $\mathcal{A}_\Gamma$ satisfies the semi-global Bolker Assumption (see Theorem 9). Thus, $C_\Gamma^t \circ C_\Gamma \subset \Delta$ and $C_\Gamma^t \circ C_\Gamma \circ \mathrm{WF}(f) \subset \mathrm{WF}(f)$.

Therefore, the first component (30) yields

$$C_\Gamma^t \circ \big[(C_\Gamma \circ \mathrm{WF}(f)) \cap ([\alpha, \beta] \times \Pi \times \mathbb{R}^2 \setminus \{0\})\big]$$
$$\subset \mathrm{WF}(f) \cap (C_\Gamma^t \circ ([\alpha, \beta] \times \Pi \times \mathbb{R}^2 \setminus \{0\})).$$

Since

$$C_\Gamma^t = \big\{(x, \sigma D_x \Psi_\Gamma(t, x); (t, \Psi_\Gamma(t, x)), (-\sigma \partial_t \Psi_\Gamma(t, x), \sigma)) \,\big|$$
$$t \in [\alpha, \beta], \ x \in \Omega_2, \ \sigma \in \mathbb{R} \setminus \{0\}\big\}$$

we obtain

$$C_\Gamma^t \circ ([\alpha, \beta] \times \Pi \times \mathbb{R}^2 \setminus \{0\}) = \big\{(x, \sigma D_x \Psi_\Gamma(t, x)) \,\big|\, t \in [\alpha, \beta], \ x \in \Omega_2, \ \sigma \in \mathbb{R} \setminus \{0\}\big\}$$
$$= \mathcal{V}_{[\alpha,\beta]}.$$

For the second component (31), we have $C_\Gamma^t \circ \mathrm{WF}(\chi_{[\alpha,\beta]\times\Pi}) = \emptyset$, since for any $(x, \xi) \in \mathrm{WF}(\chi_{[\alpha,\beta]\times\Pi})$ it is $\xi_2 = 0$, but for all vectors $((z, \eta), (\tilde{x}, \tilde{\xi})) \in C_\Gamma^t$ we have $\tilde{\xi}_2 = \sigma \neq 0$.

Now, we consider the third set (32): $C_\Gamma^t \circ \mathcal{W}_{\{\alpha,\beta\}}(f)$.

Let $\rho = ((t, y), (v, \sigma)) \in \mathcal{W}_{\{\alpha,\beta\}}(f)$. Then, we obtain from the definition of the set $\mathcal{W}_{\{\alpha,\beta\}}(f)$, that $t \in \{\alpha, \beta\}$, $v \in \mathbb{R}$, $y \in \Pi$, $\sigma \in \mathbb{R} \setminus \{\mathbf{0}\}$ and that there exists an element $x \in \mathcal{S}_\Gamma(t, y)$ with $(x, \sigma D_x \Psi_\Gamma(t, x)) \in \mathrm{WF}(f)$. So any element of the set

$$C_\Gamma^t \circ \{\rho\} = \left\{ (\tilde{x}, \sigma D_x \Psi_\Gamma(t, \tilde{x})) \mid (\tilde{x}, \sigma D_x \Psi_\Gamma(t, \tilde{x}), \rho) \in C_\Gamma^t \right\} \tag{33}$$

has to fulfill $s = \Psi_\Gamma(t, \tilde{x})$ (by definition of C_Γ^t) and

$$v = -\sigma \mathrm{D}_t \Psi_\Gamma(t, \tilde{x}) \;\Leftrightarrow\; -\frac{v}{\sigma} = \mathrm{D}_t \Psi_\Gamma(t, \tilde{x}).$$

Since v is arbitrary, the set (33) is nonempty. Hence, for any $\tilde{x} \in \mathcal{S}_\Gamma(t, y)$, we have $(\tilde{x}, \sigma D_x \Psi_\Gamma(t, \tilde{x})) \in C_\Gamma^t \circ \mathcal{W}_{\{\alpha,\beta\}}(f)$ and

$$C_\Gamma^t \circ \mathcal{W}_{\{\alpha,\beta\}}(f) = \left\{ (\tilde{x}, \sigma D_x \Psi_\Gamma(t, \tilde{x})) \mid t \in \{\alpha, \beta\}, \, y \in \Pi, \, \tilde{x} \in \mathcal{S}_\Gamma(t, y), \, \sigma \in \mathbb{R} \setminus \{\mathbf{0}\}, \right.$$
$$\left. \exists x \in \mathcal{S}_\Gamma(t, y) : \, (x, \sigma D_x \Psi_\Gamma(t, x)) \in \mathrm{WF}(f) \right\}$$
$$= \mathcal{Z}_{\{\alpha,\beta\}}(f).$$

This concludes the proof.   □

The above theorem shows that only singularities, which are in the visibility range (i.e., in $\mathcal{V}_{[\alpha,\beta]}$) can be reconstructed from the dynamic data, whereas singularities outside of this range are smoothed. According to the computations within the proof, the singularities arising in a reconstruction $\mathcal{L}_\Gamma f$ can be divided into three categories:

- Visible singularities of f from data above $[\alpha, \beta]$
  (corresponding to the set $\mathrm{WF}_{[\alpha,\beta]}(f)$),
- Added artefacts that stem from the scanning geometry and that are independent of the object f
  (corresponding to the set $C_\Gamma^t \circ \mathrm{WF}(\chi_{[\alpha,\beta] \times \Pi})$),
- Added artefacts that stem from the object
  (corresponding to the set $\mathcal{Z}_{\Gamma,[\alpha,\beta]}$).

The proof further reveals that the artefact set $C_\Gamma^t \circ \mathrm{WF}(\chi_{[\alpha,\beta] \times \Pi})$ is empty in our case of generalized Radon transforms. However, for different cutoff functions than $\chi_{[\alpha,\beta] \times \Pi}$, this might no longer be the case, see [3].

The added artefacts stemming from the object can be descriptively characterized as follows. If $(x, \sigma \partial_x \Psi_\Gamma(t_0, x)) \in \mathrm{WF}(f)$, where $t_0 \in \{\alpha, \beta\}$, then this singularity of f can generate artifacts along the curve $\mathcal{S}_\Gamma(t_0, \Psi_\Gamma(t_0, x))$.

In the next section we illustrate this theoretical characterization by numerical examples.

5   Numerical Results

In this section, we want to illustrate our theoretical results at numerical examples from Photoacoustic tomography (PAT).

In this imaging modality, an organism is subjected to non-ionizing laser pulses. The biological tissue absorbs a part of the delivered energy and converts it into heat generating ultrasonic pressure waves (photoacoustic effect). The emitted waves propagate through the medium and are measured by transducers located outside the organism on an observation surface. The goal is to recover the initial pressure $f(x)$ from the measured response g since it encodes characteristic information about the biological tissue.

In practical PAT applications, the object is not entirely surrounded by transducers. We consider the case where the transducer rotates around the object, thus acquiring the data g sequentially in time [4, 17].

Under simplifying assumptions, the measured data g match in the two-dimensional setting with the circular Radon transform of the initial pressure $f(x)$, i.e.

$$g(t, y) = \frac{1}{2\pi y} \int f(x)\, \delta(y - \|\theta(t) - x\|)\, dx =: \mathcal{A}f(t, y),$$

for $f \in \mathcal{E}'(V_1(0))$, $V_1(0)$ being the open unit disk and where $(t, y) \in [0, 2\pi] \times (0, 2)$ and $\theta(t) = (\cos t, \sin t)^T$. In particular $\mathcal{A}$ represents a FIO with amplitude $a(y) = (2\pi y)^{-1}$ and phase function $\Phi(t, y, x, \sigma) = \sigma(y - \Psi(t, y, x))$, where $\Psi(t, y, x) = \|\theta(t) - x\|$ is a defining function according to our Hypothesis 1.

If the searched-for quantity changes during the sequential data acquisition according to a smooth diffeomorphic motion model Γ, the corresponding dynamic forward operator

$$\mathcal{A}_\Gamma f(t, y) = \frac{1}{2\pi y} \int f(x)\, |\det D\Gamma_t^{-1} x|\, \delta(y - \|\theta(t) - \Gamma_t^{-1} x\|)\, dx \tag{34}$$

is a FIO according to Theorem 8 which integrates the reference configuration $f(x)$ along the curves

$$S_\Gamma(t, y) := \{x \in V_1(0) \,|\, y = \Psi_\Gamma(t, x)\}$$

with

$$\Psi_\Gamma(t, x) = \|\theta(t) - \Gamma_t^{-1} x\|. \tag{35}$$

If the motion model fulfills the additional conditions (17) and (18), then $\mathcal{A}_\Gamma$ satisfies the semi-global Bolker Assumption.

In particular, we want to illustrate the reconstruction results obtained by applying an operator of type

$$\mathcal{S}_\Gamma := \mathcal{A}^t_{\Gamma,[0,2\pi]}\mathcal{P}$$

to the dynamic data $g(t, y) = \mathcal{A}_\Gamma f(t, y)$, $(t, y) \in [0, 2\pi] \times (0, 2)$ with

$$\mathcal{P}g(t, y) = \int \partial_r r \partial_r g(t, r) \log |r^2 - y^2| \, dr$$

stemming from the inversion formula for $\mathcal{A}$ in the static case, see [7, 17]. In the following examples, we simulate (dynamic) PAT data by discretizing $\mathcal{A}_\Gamma f$ with the trapezoidal rule with 1400 samples, where f is the respective phantom. The discrete dynamic data is then given by

$$g_{i,j} := (\mathcal{A}_{\Gamma_{t_i}} f)(t_i, y_j), \quad i = 1, \dots, N \text{ and } j = 1, \dots, M,$$

with $N = 300$ uniformly distributed angles t_i in $[0, 2\pi]$ and $M = 300$ uniformly distributed radii y_j in $(0, 2]$. The reconstructed images are computed on a 600×600 grid.

5.1 2π-Periodic Motion

For our first numerical example, we consider the ideal scenario (from Sect. 4.1), where we have a smooth and 2π-periodic motion model. More precisely, we consider the rotation matrix

$$B_t = \begin{pmatrix} \cos(2t) & \sin(2t) \\ -\sin(2t) & \cos(2t) \end{pmatrix},$$

which defines a 2π-periodic motion $\Gamma_t x = B_t x$ for $x \in \mathbb{R}^2$ and $t \in [0, 2\pi]$. The dynamic behavior of the object–namely a rotation in the same direction as the transducer but twice as fast–is illustrated in Fig. 4. In particular, with Γ describing such a rotational movement, the representation (34) of the dynamic FIO $\mathcal{A}_\Gamma$ simplifies to

$$\mathcal{A}_\Gamma f(t, y) = \frac{1}{2\pi y} \int f(x) \delta(y - \|B_t \theta(t) - x\|) \, dx,$$

i.e. the dynamic behavior of f can be equivalently expressed by adapting the rotation of the transducer, see [16] for more details.

From our theory (in particular Theorem 12), we expect to see no additional artefacts and, since all singularities are encoded in the data according to the theory developed in Sect. 3, we anticipate to see all singularities correctly reconstructed. These theoretical results are indeed confirmed by our numerical reconstruction result in Fig. 5.

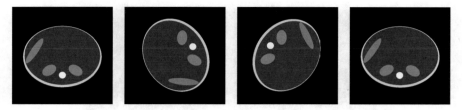

Fig. 4 Movement of phantom in $[0, \pi]$ for $t = 0$, $\frac{\pi}{2}$, $\frac{3\pi}{2}$, π. This movement is repeated in $[\pi, 2\pi]$

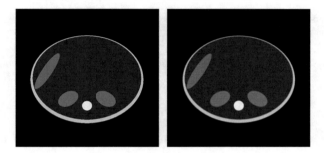

Fig. 5 Left: Ground-truth phantom. Right: Dynamic reconstruction under the 2π-periodic rotational motion model $\Gamma_t = B_t$

5.2 Non-periodic Motion

In our next examples, we consider the more realistic case of a non-periodic motion. In order to make clearly observable the visible (and invisible) singularities and the additional artifacts, which we expect from our theory (Sect. 4), we first consider a less complex phantom, namely a circle (see Fig. 6 (left)).

We start with an example that illustrates the following: Even in case of a full data, i.e. when all object singularities are encoded in the measured data, added artifacts can appear if the object is not in the same state at the beginning and the end of the scanning. As example, we consider the rotation matrix

$$G_t = \begin{pmatrix} \cos\left(-\frac{2}{3}t\right) & \sin\left(-\frac{2}{3}t\right) \\ -\sin\left(-\frac{2}{3}t\right) & \cos\left(-\frac{2}{3}t\right) \end{pmatrix}$$

and the associated motion model $\Gamma_t x = G_t x$, for $x \in \mathbb{R}^2$ and $t \in [0, 2\pi]$. In this example, the investigated object rotates in the opposite direction than the transducer. Thus, the transducer in relation to the object can perform at least one complete turn (even more) around the object, i.e. all object singularities are encoded in the dynamic data.

Fig. 6 Ground-truth phantom (left) and dynamic reconstruction (right)

Since $\Gamma_0 x = x$, the transducer position at $t = 0$ is $\theta(0) = (1, 0)^T$. Further, we have

$$\Psi_\Gamma(t, x) = \|\theta(t) - G_t^{-1} x\| = \|G_t \theta(t) - x\|,$$

since G_t is an isometry. Hence, the transducer position at $t = 2\pi$ with respect to the initial state of the object is given by $G_t \theta(t) = \theta\left(\frac{5t}{3}\right) = \theta\left(\frac{10}{3}\pi\right) = \theta\left(\frac{4}{3}\pi\right) = \left(-\frac{1}{2}, -\frac{\sqrt{3}}{2}\right)^T$. In the interval $[0, \frac{4}{3}\pi]$ the object is scanned twice and thus, all singularities of the object are visible in the reconstruction, see Fig. 6, where the contour of the circle is clearly visible.

However, we notice in the dynamic reconstruction Fig. 6 (right) the appearance of additional artifacts which occur because the motion is not 2π-periodic. We have

$$D_x \Psi_\Gamma(t, x) = \frac{x - \theta\left(\frac{5t}{3}\right)}{\left\|\theta\left(\frac{5t}{3}\right) - x\right\|},$$

so any singularity (x, ξ) of f with

$$\xi = \sigma D_x \Psi_\Gamma(t, x) = \sigma \frac{x - \theta\left(\frac{5t}{3}\right)}{\left\|\theta\left(\frac{5t}{3}\right) - x\right\|} = \tilde{\sigma}\left(x - \theta\left(\frac{5t}{3}\right)\right),$$

for $\tilde{\sigma} \in \mathbb{R} \setminus \{0\}$ and $t \in \{0, 2\pi\}$ can create artifacts along the curve

$$\mathcal{S}_\Gamma(t, \Psi_\Gamma(t, x)) = \left\{\tilde{x} \in \mathbb{R}^2 \; : \; \Psi_\Gamma(t, \tilde{x}) = \Psi_\Gamma(t, x)\right\}$$

$$= \left\{\tilde{x} \in \mathbb{R}^2 \; : \; \left\|\theta\left(\frac{5t}{3}\right) - \tilde{x}\right\| = \left\|\theta\left(\frac{5t}{3}\right) - x\right\|\right\}.$$

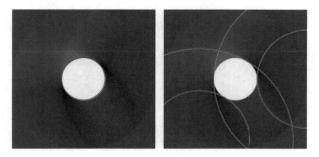

Fig. 7 Dynamic reconstruction with the visible singularities (left) and added artifacts (right) predicted by our theory highlighted in red

Hence, the artifacts appear along circle lines with centers $\theta(0) = (1,0)^T$ and $\theta\left(\frac{4}{3}\pi\right) = \left(-\frac{1}{2}, -\frac{\sqrt{3}}{2}\right)^T$, when a singularity of f is conormal to the circle. Figure 7 confirms that the additional artifacts predicted by our theory match the artifacts arising in our numerical reconstruction result.

Our next example shows that the dynamic behaviour of the investigated object can cause a limited-data problem. Here, we consider the rotation matrix

$$R_t = \begin{pmatrix} \cos\left(\frac{3}{4}t\right) & \sin\left(\frac{3}{4}t\right) \\ -\sin\left(\frac{3}{4}t\right) & \cos\left(\frac{3}{4}t\right) \end{pmatrix}$$

and the respective motion model $\Gamma_t x = R_t x$, for $x \in \mathbb{R}^2$ and $t \in [0, 2\pi]$. In this example, the object performs a rotational movement in the same direction as the rotation of the transducer.

The transducer position at $t = 0$ is $\theta(0) = (1,0)^T$ as in the example before and the source position at $t = 2\pi$ is now given by $R_t \theta(t) = \theta\left(\frac{t}{4}\right) = \theta\left(\frac{\pi}{2}\right) = (0,1)^T$. This scenario corresponds to the static limited angle case, where the object is only scanned for transducer locations associated to the interval $[0, \frac{3}{2}\pi] \subsetneq [0, 2\pi]$.

To validate our theory, we again compare the observed artifacts with their analytic characterization from Sect. 4.2. We have

$$D_x \Psi_\Gamma(t, x) = \frac{x - \theta\left(\frac{t}{4}\right)}{\left\| \theta\left(\frac{t}{4}\right) - x \right\|},$$

so any singularity (x, ξ) of f with

$$\xi = \sigma D_x \Psi_\Gamma(t, x) = \sigma \frac{x - \theta\left(\frac{t}{4}\right)}{\left\| \theta\left(\frac{t}{4}\right) - x \right\|} = \tilde{\sigma}\left(x - \theta\left(\frac{t}{4}\right)\right),$$

for $\tilde{\sigma} \in \mathbb{R} \setminus \{0\}$ and $t \in \{0, 2\pi\}$ can create artifacts along the curve

Fig. 8 Ground-truth phantom (left) and dynamic reconstruction (right)

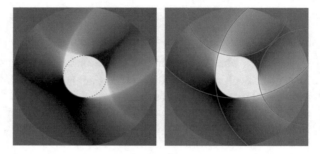

Fig. 9 Left: Visible (red solid line) and invisible (red dashed line) singularities. Right: Added artifacts (red)

$$\mathcal{S}_\Gamma(t, \Psi_\Gamma(t, x)) = \left\{ \tilde{x} \in \mathbb{R}^2 \ : \ \Psi_\Gamma(t, \tilde{x}) = \Psi_\Gamma(t, x) \right\}$$

$$= \left\{ \tilde{x} \in \mathbb{R}^2 \ : \ \|\theta\left(\tfrac{t}{4}\right) - \tilde{x}\| = \|\theta\left(\tfrac{t}{4}\right) - x\| \right\}.$$

I.e. here, the artifacts appear along circle lines with centers $\theta(0) = (1, 0)^T$ and $\theta\left(\tfrac{\pi}{2}\right) = (0, 1)^T$, when a singularity of f is conormal to the circle, see Fig. 9 (right).

Since the visible singularities are given by

$$\mathcal{V}_{[0,2\pi]} = \left\{ \left(x, \tilde{\sigma}\left(x - \theta\left(\tfrac{t}{4}\right)\right) dx\right) \ : \ t \in [0, 2\pi], \ x \in \mathbb{R}^2, \ \tilde{\sigma} \in \mathbb{R} \setminus \{0\} \right\}$$

$$= \left\{ \left(x, \tilde{\sigma}\left(x - \theta(t_v)\right) dx\right) \ : \ t_v \in [0, \tfrac{\pi}{2}] \cup [\tfrac{3\pi}{2}, 2\pi], \ x \in \mathbb{R}^2, \ \tilde{\sigma} \in \mathbb{R} \setminus \{0\} \right\},$$

singularities (x, ξ) of f with direction $\xi = \sigma(x - \theta(t_\xi))$, $t_\xi \in (\tfrac{\pi}{2}, \tfrac{3\pi}{2})$, cannot be reconstructed from the dynamic data, see Figs. 8 and 9 (left).

As we can see in Fig. 9 visible singularities and added artifacts appear as predicted from our theory.

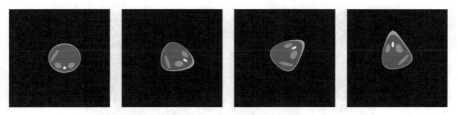

Fig. 10 Movement of phantom in $[0, 2\pi]$ for $t = 0, \frac{2\pi}{3}, \frac{4\pi}{3}, 2\pi$

This example shows, that the dynamic behavior can result in limited data problems (even though in the static case, the measured data would have been sufficient to recover all singularities).

After this detailed study, we finally want to provide one last example with the phantom from Sect. 5.1 and a more complex motion model, namely the non-affine (and non-periodic) deformation illustrated in Fig. 10.

Regarding the dynamic behavior, we consider

$$\Gamma_t x = Z_t A_t x,$$

with the rotation matrix

$$A_t = \begin{pmatrix} \cos\left(\frac{1}{2}t\right) & \sin\left(\frac{1}{2}t\right) \\ -\sin\left(\frac{1}{2}t\right) & \cos\left(\frac{1}{2}t\right) \end{pmatrix}$$

and the non-affine motion described by $Z_0 x = x$ and

$$(Z_t x)_i = \frac{(\sqrt[4]{5m_i(t)}x_i + 1)^5 - 1}{5\sqrt[4]{5m_i(t)}},$$

for $t \in (0, 2\pi]$ and $i = 1, 2$ with

$$m_1(t) = \sin\left(0.005\left(t\frac{N+1}{2\pi} - 1\right)\frac{3}{N}\right),$$

$$m_2(t) = \sin\left(0.007\left(t\frac{N+1}{2\pi} - 1\right)\frac{3}{N}\right).$$

Applying our reconstruction method to the corresponding dynamic data set provides an image showing the visible singularities of the initial object state as well as additional artifacts, see Fig. 11 (right), which are caused by the object singularities encoded at beginning and end of the scanning and which spread along the respective integration curves. In particular, we observe that the artifacts spread along deformed circle lines due to the non-affine motion model.

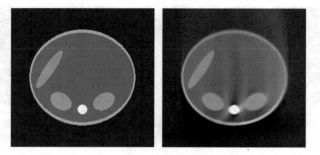

Fig. 11 Ground-truth phantom (left) and dynamic reconstruction (right)

6 Conclusion

In this chapter, we analysed the overall information content of dynamic tomography data using the framework of Fourier integral operators and microlocal analysis. In particular, we extended our previous results in [18] and [17] to a larger class of operators. Based on this analysis, we further provided a detailed characterization on what is visible in a respective reconstruction result assuming the motion is exactly known, which we illustrated with various numerical examples from dynamic photoacoustic tomography.

The developed theory can further be utilized to study the scenario where only incorrect motion information are available (accounting for instance for modelling or estimation errors). So far, this has been studied for instance in [15] for the specific example of dynamic computerized tomography. The gained insights could then serve as a guiding principle for the design of motion estimation protocols or additional artifact reduction strategies.

Acknowledgments The authors thank the referee for the thorough review and thoughtful comments. The work of the first and second authors was supported by the Deutsche Forschungsgemeinschaft under grant HA 8176/1-1. The first author thanks Tufts University for its hospitality and support during her visit which was essential to the work on this article. The work of the third author was partially supported by U.S. NSF grant DMS 1712207.

References

1. G. Ambartsoumian, R. Felea, V.P. Krishnan, C.J. Nolan, E.T. Quinto, Singular FIOs in SAR imaging, II: transmitter and receiver at different speeds. SIAM J. Math. Anal. **50**(1), 591–621 (2018)
2. G. Beylkin, The inversion problem and applications of the generalized Radon transform. Comm. Pure Appl. Math. **37**, 579–599 (1984)
3. L. Borg, J. Frikel, J.S. Jorgensen, E.T. Quinto, Analyzing reconstruction artifacts from arbitrary incomplete X-ray CT data. SIAM J. Imaging Sci. **11**, 2786–2814 (2018)

4. J. Chung, L. Nguyen, Motion estimation and correction in photoacoustic tomographic reconstruction. SIAM J. Imaging Sci. **10**(1), 216–242 (2017)
5. M. DeHoop, Microlocal analysis of seismic imaging, in *Inside - Out: Inverse Problems and Applications*, ed. by G. Uhlmann (MSRI Publications, Cambridge, 2003)
6. R. Felea, C. Nolan, Monostatic SAR with fold/cusp singularities. J. Fourier Anal. Appl. **21**(4), 799–821 (2015)
7. D. Finch, M. Haltmeier, Rakesh, Inversion of spherical means and the wave equation in even dimensions. SIAM J. Appl. Math. **68**(2), 392–412 (2007)
8. B. Frigyik, P. Stefanov, G. Uhlmann, The X-ray transform for a generic family of curves and weights. J. Geom. Anal. **18**, 81–97 (2008)
9. J. Frikel, E.T. Quinto, Characterization and reduction of artifacts in limited angle tomography. Inverse Probl. **29**(12), 125007 (2013)
10. J. Frikel, E.T. Quinto, Artifacts in incomplete data tomography with applications to photoacoustic tomography and sonar. SIAM J. Appl. Math. **75**, 703–725 (2015)
11. C. Grathwohl, P. Kunstmann, E.T. Quinto, A. Rieder, Microlocal analysis of imaging operators for effective common offset seismic reconstruction. Inverse Probl. **34**(11), 114001 (2018)
12. V. Guillemin, Some remarks on integral geometry. Technical Report, MIT (1975)
13. V. Guillemin, On some results of Gelfand in integral geometry. Proc. Symp. Pure Math. **43**, 149–155 (1985)
14. V. Guillemin, S. Sternberg, *Geometric Asymptotics* (American Mathematical Society, Providence, 1977)
15. B.N. Hahn, A motion artefact study and locally deforming objects in computerized tomography. Inverse Probl. **33**, 114001 (2017)
16. B.N. Hahn, Motion compensation strategies in tomography, in *Time-Dependent Problems in Imaging and Parameter Identification*, ed. by B. Kaltenbacher, T. Schuster, A. Wald (Springer, Cham, 2021), pp. 51–84
17. B.N. Hahn, M.L. Kienle-Garrido, An efficient reconstruction approach for a class of dynamic imaging operators. Inverse Probl. **35**, 094005 (2019)
18. B.N. Hahn, E.T. Quinto, Detectable singularities from dynamic Radon data. SIAM J. Imaging Sci. **9**, 1195–1225 (2016)
19. S. Holman, F. Monard, P. Stefanov, The attenuated geodesic X-ray transform. Inverse Probl. **34**(6), 064003 (2018)
20. L. Hörmander, Fourier integral operators, I. Acta Math. **127**, 79–183 (1971)
21. L. Hörmander, The analysis of linear partial differential operators. I: Distribution theory and Fourier analysis, in *Classics in Mathematics* (Springer, Berlin, 2003)
22. A. Katsevich, Motion compensated local tomography. Inverse Probl. **24**, 045012 (2008)
23. A. Katsevich, M. Silver, A. Zamyatin, Local tomography and the motion estimation problem. SIAM J. Imaging Sci. **4**, 200–219 (2011)
24. V.P. Krishnan, E.T. Quinto, Microlocal analysis in tomography, in *Handbook of Mathematical Methods in Imaging*, ed. by O. Scherzer, 2nd edn. (Springer, Berlin, 2015)
25. A.K. Louis, E.T. Quinto, Local tomographic methods in SONAR, in *Surveys on Solution Methods for Inverse Problems*, ed. by D. Colton, H. Engl, A. Louis, J. McLaughlin, W. Rundell (Springer, Vienna/New York, 2000), pp. 147–154
26. F. Natterer, *The Mathematics of Computerized Tomography* (B. G. Teubner, Stuttgart, 1986)
27. L.V. Nguyen, How strong are streak artifacts in limited angle computed tomography? Inverse Probl. **31**(5), 055003 (2015)
28. C.J. Nolan, M. Cheney, Synthetic Aperture inversion. Inverse Probl. **18**(1), 221–235 (2002)
29. T.J.P.M. Op 't Root, C.C. Stolk, M.V. de Hoop, Linearized inverse scattering based on seismic reverse time migration. J. Math. Pures Appl. **98**(2), 211–238 (2012)
30. E.T. Quinto, The dependence of the generalized Radon transform on defining measures. Trans. Amer. Math. Soc. **257**, 331–346 (1980)
31. E.T. Quinto, Singularities of the X-ray transform and limited data tomography in $\mathbb{R}^2$ and $\mathbb{R}^3$. SIAM J. Math. Anal. **24**, 1215–1225 (1993)

32. E.T. Quinto, O. Öktem, Local tomography in electron microscopy. SIAM J. Appl. Math. **68**, 1282–1303 (2008)
33. S. Rabieniaharatbar, Invertibility and stability for a generic class of Radon transforms with application to dynamic operators. J. Inverse Ill-Posed Probl. **27**, 469–486 (2018)
34. G. Rigaud, B.N. Hahn, 3D Compton scattering imaging and contour reconstruction for a class of Radon transforms. Inverse Probl. **34**(7), 075004 (2018)
35. W. Rudin, *Functional Analysis*. McGraw-Hill Series in Higher Mathematics (McGraw-Hill, New York, 1973)
36. P. Stefanov, G. Uhlmann, Is a curved flight path in SAR better than a straight one? SIAM J. Appl. Math. **73**(4), 1596–1612 (2013)
37. F. Trèves, *Introduction to Pseudodifferential and Fourier Integral Operators. Volume 1: Pseudodifferential Operators* (Plenum Press, New York, 1980)
38. F. Trèves, *Introduction to Pseudodifferential and Fourier Integral Operators. Volume 2: Fourier Integral Operators* (Plenum Press, New York, 1980)
39. J.W. Webber, S. Holman, Microlocal analysis of a spindle transform. Inverse Probl. Imaging **13**(2), 231–261 (2019)

The Tangential Cone Condition for Some Coefficient Identification Model Problems in Parabolic PDEs

Barbara Kaltenbacher, Tram Thi Ngoc Nguyen, and Otmar Scherzer

Abstract The tangential condition was introduced in Hanke et al. (Numer Math 72:21–37 1995) as a sufficient condition for convergence of the Landweber iteration for solving ill–posed problems.

In this paper we present a series of time dependent benchmark inverse problems for which we can verify this condition.

1 Introduction

We consider the problem of recovering a parameter θ in the evolution equation

$$\dot{u}(t) = f(t, \theta, u(t)) \quad t \in (0, T) \tag{1}$$

$$u(0) = u_0, \tag{2}$$

where for each $t \in (0, T)$ we consider $u(t)$ as a function on a bounded $C^{1,1}$ domain $\Omega \subset \mathbb{R}^d$. In (1), $\dot{u}$ denotes the first order time derivative of u and f is a nonlinear function. We here focus on the setting of θ not being time dependent. Problems with state equations of the form $\dot{u}(t) = f(t, \theta(t), u(t))$ could be treated analogously but this would lead to different requirements on the underlying function spaces.

B. Kaltenbacher (✉)
Department of Mathematics, Alpen-Adria-Universität Klagenfurt, Klagenfurt, Austria
e-mail: barbara.kaltenbacher@aau.at

T. T. N. Nguyen
Institute of Mathematics and Scientific Computing, University of Graz, Graz, Austria
e-mail: tram.nguyen@aau.at

O. Scherzer
Faculty of Mathematics, University of Vienna, Vienna, Austria

Johann Radon Institute for Computational and Applied Mathematics (RICAM), Linz, Austria
e-mail: otmar.scherzer@univie.ac.at

© Springer Nature Switzerland AG 2021
B. Kaltenbacher et al. (eds.), *Time-dependent Problems in Imaging and Parameter Identification*, https://doi.org/10.1007/978-3-030-57784-1_5

These model equations are equipped with additional data obtained from continuous observations over time

$$y(t) = C(t, u(t)),$$ (3)

with an observation operator C, which will be assumed to be linear; in particular, in most of what follows C is the continuous embedding $V \hookrightarrow Y$, with V and Y introduced below.

While formulating the requirements and results first of all in this general framework, we will also apply it to a number of examples as follows.

1.1 Identification of a Potential

We study the problem of identifying the space-dependent parameter c from observation of the state u in $\Omega \times (0, T)$ in

$$\dot{u} - \Delta u + cu = \varphi \qquad (t, x) \in (0, T) \times \Omega \tag{4}$$

$$u_{|\partial\Omega} = 0 \qquad t \in (0, T) \tag{5}$$

$$u(0) = u_0 \qquad x \in \Omega, \tag{6}$$

where $\varphi \in L^2(0, T; H^{-1}(\Omega))$ and $u_0 \in L^2(\Omega)$ are known. Here, $-\Delta$ could be replaced by any linear elliptic differential operator with smooth coefficients.

With this equation, known, among others, as diffusive Malthus equation [31], one can model the evolution of a population u with diffusion and with exponential growth as time progresses. The latter phenomenon is quantified by the growth rate c, which, in this particular case, depends only on the environment.

1.2 Identification of a Diffusion Coefficient

We further consider the problem of recovering the space-dependent parameter a from measurements of u in $\Omega \times (0, T)$, governed by the diffusion equation

$$\dot{u} - \nabla \cdot \left(a\nabla u\right) = \varphi \qquad (t, x) \in (0, T) \times \Omega \tag{7}$$

$$u_{|\partial\Omega} = 0 \qquad t \in (0, T) \tag{8}$$

$$u(0) = u_0 \qquad x \in \Omega, \tag{9}$$

where $\varphi \in L^2(0, T; H^{-1}(\Omega))$ and $u_0 \in L^2(\Omega)$ are known. This is, for instance, a simple model of groundwater flow, whose temporal evolution is driven by the divergence of the flux $-a\nabla u$ and the source term φ. The coefficient a represents the diffusivity of the sediment and u is the piezometric head [14].

Banks and Kunisch [3, Chapter I.2] discussed the more general model: $\dot{u} + \nabla \cdot \left(-a\nabla u + bu\right) + cu$, describing the sediment formation in lakes and deep seas, in particular, the mixture of organisms near the sediment-water interface.

1.3 An Inverse Source Problem with a Quadratic First Order Nonlinearity

Here we are interested in the problem of identifying the space-dependent source term θ from observation of the state u in $\Omega \times (0, T)$

$$\dot{u} - \Delta u - |\nabla u|^2 = \theta \qquad (t, x) \in (0, T) \times \Omega \qquad (10)$$

$$u_{|\partial\Omega} = 0 \qquad\qquad t \in (0, T) \qquad (11)$$

$$u(0) = u_0 \qquad\qquad x \in \Omega. \qquad (12)$$

This sort of PDE with a quadratic nonlinearity in ∇u arises, e.g., in stochastic optimal control theory [10, Chapter 3.8].

1.4 An Inverse Source Problem with a Cubic Zero Order Nonlinearity

The following nonlinear reaction-diffusion equation involves determining the space-dependent source term θ from observation of the state u in $\Omega \times (0, T)$, in a semiliear parabolic equation

$$\dot{u} - \Delta u + \Phi(u) = \varphi - \theta \qquad (t, x) \in (0, T) \times \Omega \qquad (13)$$

$$u_{|\partial\Omega} = 0 \qquad\qquad t \in (0, T) \qquad (14)$$

$$u(0) = u_0 \qquad\qquad x \in \Omega, \qquad (15)$$

where the possibly space and time dependent source term $\varphi \in L^2(0, T; H^{-1}(\Omega))$ and the initial data $u_0 \in H_0^1(\Omega)$ are known.

Here we selectively mention some applications for PDEs with cubic nonlinearity $\Phi(u)$:

- $\Phi(u) = u(1 - u^2)$: Ginzburg-Landau equations of superconductivity [4], Allen-Cahn equation for the phase separation process in a binary metallic alloy [1, 33], Newell-Whitehead equation for convection of fluid heated from below [11].

- $\Phi(u) = u^2(1 - u)$: Zel'dovich equation in combustion theory [11].
- $\Phi(u) = u(1 - u)(u - \alpha), 0 < \alpha < 1$: Fisher's model for population genetics [36], Nagumo equation for bistable transmission lines in electric circuit theory [32].

In part of the analysis we will also consider an additional gradient nonlinearity $\Psi(\nabla u)$ in the PDE, cf. (44) below.

Coming back to the general setting (1)–(3) we will make the following assumptions, where all the considered examples fit into. The operators defining the model and observation equations above are supposed to map between the function spaces

$$f : (0, T) \times X \times V \to W^* \tag{16}$$

$$C : (0, T) \times V \to Y, \tag{17}$$

where $X, Y, W, V \subseteq Y$ are Banach spaces. More precisely, X is the parameter space, Y the data space, W^* the space in which the equations is supposed to hold and V the state space. The latter three are first of all the spaces for the respective values at fixed time instances and will also be assigned a version for time-dependent functions, denoted by calligraphic letters. So X, Y, W^* will denote the parameter, data and equation spaces, respectively, and U or $\tilde{U}$ (to distinguish between the different versions in the reduced and all-at-once setting below) the state space. The initial condition $u_0 \in H$, where H is a Banach space as well, will in most of what follows be supposed to be independent of the coefficient θ here. Dependence of the initial data and also of the observation operator on θ can be relevant in some applications but leads to further technicalities, thus for clarity of exposition we shift consideration of these dependencies to future work.

For fixed θ, we assume that the Caratheodory mappings f and C as defined above induce Nemytskii operators [42, Section 4.3] (for which we will use the same notation f and C) on the function space

$$U = L^2(0, T; V) \cap H^1(0, T; W^*) \text{ or } \tilde{U} = L^\infty(0, T; V) \cap H^1(0, T; W^*),$$

cf. (34) and (48) respectively, in which the state u will be contained, and map into the image space W^* and observation space Y, respectively, where

$$W^* = L^2(0, T; W^*), \qquad Y = L^2(0, T; Y). \tag{18}$$

Moreover, $\tilde{U}$ or U, respectively, will be assumed to continuously embed into $C(0, T; H)$ in order to make sense out of (2).

We will consider formulation of the inverse problem on one hand in a classical way, as a nonlinear operator equation

$$F(\theta) = y \tag{19}$$

with a forward operator F mapping between Banach spaces $\mathcal{X}$ and $\mathcal{Y}$, and on the other hand also, alternatively, as a system of model and observation equation

$$\mathcal{A}(\theta, u) = 0; \tag{20}$$

$$C(u) = y. \tag{21}$$

Here,

$$\mathcal{A} : \mathcal{X} \times \mathcal{U} \to \mathcal{W}^* \times H, \quad (\theta, u) \mapsto \mathcal{A}(\theta, u) = (\dot{u} - f(\theta, u), u(0) - u_0)$$
$$C : \mathcal{U} \to \mathcal{Y}$$
$$\tag{22}$$

are the model and observation operators, so that with the parameter-to-state map $S : \mathcal{X} \to \mathcal{U}$ defined by

$$\mathcal{A}(\theta, S(\theta)) = 0 \tag{23}$$

and

$$F = C \circ S, \tag{24}$$

Equation (19) is equivalent to the all-at-once formulation (20) and (21). Defining

$$\mathbf{F} : \mathcal{X} \times \mathcal{U} \to \mathcal{W}^* \times H \times \mathcal{Y}$$

by

$$\mathbf{F}(\theta, u) = (\mathcal{A}(\theta, u), C(u)),$$

and setting $\mathbf{y} = (0, y)$, we can rewrite (20) and (21) analogously to (19), as

$$\mathbf{F}(\theta, u) = \mathbf{y}. \tag{25}$$

All-at-once approaches have been studied for PDE constrained optimization in, e.g., [24, 25, 29, 35, 40, 41, 43] and more recently, for ill-posed inverse problems in, e.g., [5, 6, 13, 18, 22, 43], particularly for time dependent models in [19, 34]. The fact that we are actually using different state spaces $\mathcal{U}$, $\tilde{\mathcal{U}}$ in these two settings is on one hand due to the requirements arising from the need for well-definedness and differentiability of the parameter-to-state map in the reduced setting. On the other hand, while these constraints are not present in the all-at-once setting and a quite general choice of the state space is possible there, whenever a Hilbert space setting is required—e.g., for reasons of easier implementation—this does not only apply to the parameter and data space but also to the state and equation spaces in the all-at-once setting, whereas in a reduced setting these spaces are "hidden" inside the forward operator.

Convergence proofs of iterative regularization methods for solving (19) (and likewise (25)) such as the Landweber iteration [15, 21] or the iteratively regularized Gauss-Newton method [2, 20, 21] require structural assumptions on the nonlinear forward operator F such as the tangential cone condition [39]

$$\|F(\theta) - F(\tilde{\theta}) - F'(\theta)(\theta - \tilde{\theta})\|_{\mathcal{Y}} \leq c_{tc} \|F(\theta) - F(\tilde{\theta})\|_{\mathcal{Y}} \quad \forall \theta, \tilde{\theta} \in B_{\rho}^{\mathcal{X}}(\theta^0), \quad (26)$$

for some sufficiently small constant c_{tc}. Here $F'(\theta)$ does not necessarily need to be the Fréchet or Gâteaux derivative of F, but it is just required to be some linear operator that is uniformly bounded in a neighborhood of the initial guess θ_0, i.e., $F'(\theta) \in L(\mathcal{X}, \mathcal{Y})$ such that

$$\|F'(\theta)\|_{L(\mathcal{X}, \mathcal{Y})} \leq C_F \quad \forall \theta, \tilde{\theta} \in B_{\rho}^{\mathcal{X}}(\theta^0), \quad (27)$$

for some $C_F > 0$.

The conditions (26) and (27) enforce certain local convexity conditions of the residual $\theta \mapsto \|F(\theta) - y\|^2$, cf.[23]. In this sense, the conditions are structurally similar to conditions used in the analysis of Tikhonov regularization, such as those in [7]. The tangential cone condition eventually guarantees convergence to the solution of (19) by a gradient descent method for the residual (and also for the Tikhonov functional). Therefore it ensures that the iterates are not trapped in local minima.

The key contribution of this chapter is therefore to establish (26) and (27) in the reduced setting (19) as well as its counterpart in the all-at-once setting (25) for the above examples (as well as somewhat more general classes of examples) of parameter identification in initial boundary value problems for parabolic PDEs represented by (1) and (2). In the reduced setting this also involves the proof of well-definedness and differentiability of the parameter-to-state map S, whereas in the all-at-once setting this is not needed, thus leaving more freedom in the choice of function spaces. Correspondingly, the examples classes considered in Sect. 2 will be more general than those in Sect. 3.

Some non-trivial static benchmark problems where the tangential condition has been verified can be found e.g., in [8, 17, 28].

We mention in passing that in view of existing convergence analysis for such iterative regularization methods for (19) or (25) in rather general Banach spaces we will formulate our results in general Lebesgue and Sobolev spaces. Still, we particularly strive for a full Hilbert space setting as preimage and image spaces X and Y, since derivation and implementation of adjoints is much easier then, and also the use of general Banach spaces often introduces additional nonlinearity or nonsmoothness. Moreover we point out that while in the reduced setting, we will focus on examples of parabolic problems in order to employ a common framework for establishing well-definedness of the parameter-to-state map, the all-at-once version of the tangential condition trivially carries over to the wave equation (or also fractional sub- or superdiffusion) context by just replacing the first time derivative by a second (or fractional) one.

The remainder of this paper is organized as follows. Section 2 provides results for the all-at-once setting, that are also made use of in the subsequent Sect. 3 for the reduced setting. The proofs of the propositions in Sect. 2 and the notation can be found in the Appendix.

2 All-at-Once Setting

The tangential cone condition and boundedness of the derivative in the all-at-once setting $\mathbf{F}(\theta, u) = \mathbf{y}$ (25) with

$$\mathbf{F} : \mathcal{X} \times \mathcal{U} \to \mathcal{W}^* \times H \times \mathcal{Y}, \quad \mathbf{F}(\theta, u) = \begin{pmatrix} \dot{u} - f(\theta, u) \\ u(0) - u_0 \\ C(u) \end{pmatrix} \tag{28}$$

and the norms

$$\|(\theta, u)\|_{\mathcal{X} \times \mathcal{U}} := \left(\|\theta\|_{\mathcal{X}}^2 + \|u\|_{\mathcal{X}}^2 \right)^{1/2},$$

$$\|(w, h, y)\|_{\mathcal{W}^* \times H \times \mathcal{Y}} := \left(\|w\|_{\mathcal{W}^*}^2 + \|h\|^2 + \|y\|_{\mathcal{Y}}^2 \right)^{1/2},$$

on the product spaces read as

$$\|f(\theta, u) - f(\tilde{\theta}, \tilde{u}) - f_\theta'(\theta, u)(\theta - \tilde{\theta}) - f_u'(\theta, u)(u - \tilde{u})\|_{\mathcal{W}^*}$$

$$\leq c_{tcc}^{AAO} \left(\|\dot{u} - \dot{\tilde{u}} - f(\theta, u) + f(\tilde{\theta}, \tilde{u})\|_{\mathcal{W}^*}^2 + \|u(0) - \tilde{u}(0)\|_H^2 + \|C(u - \tilde{u})\|_{\mathcal{Y}}^2 \right)^{1/2},$$

$$\forall (\theta, u), (\tilde{\theta}, \tilde{u}) \in B_\rho^{\mathcal{X} \times \mathcal{U}}(\theta^0, u^0), \tag{29}$$

and

$$\left(\|\dot{v} - f_\theta'(\theta, u)\chi - f_u'(\theta, u)v\|_{\mathcal{W}^*}^2 + \|v(0)\|_H^2 + \|Cv\|_Y^2 \right)^{1/2}$$

$$\leq C_{\mathbf{F}} \left(\|\chi\|_{\mathcal{X}}^2 + \|v\|_{\mathcal{U}}^2 \right)^{1/2}, \tag{30}$$

$$\forall (\theta, u) \in B_\rho^{\mathcal{X} \times \mathcal{U}}(\theta^0, u^0), \ \chi \in \mathcal{X}, \ v \in \mathcal{U}$$

where we have assumed linearity of C.

Since the right hand side terms $\|u(0) - \tilde{u}(0)\|_H$ and $\|f(\theta, u) - f(\tilde{\theta}, \tilde{u})\|_{\mathcal{W}^*}$ in (29) are usually too weak to help for verification of this condition, we will just skip it in the following and consider

$$\|f(\theta, u) - f(\tilde{\theta}, \tilde{u}) - f_\theta'(\theta, u)(\theta - \tilde{\theta}) - f_u'(\theta, u)(u - \tilde{u})\|_{\mathcal{W}^*}$$

$$\leq c_{tcc}^{AAO} \|C(u - \tilde{u})\|_{\mathcal{Y}}, \quad \forall (\theta, u), (\tilde{\theta}, \tilde{u}) \in B_\rho^{\mathcal{X} \times \mathcal{U}}(\theta^0, u^0) \tag{31}$$

which under these conditions is obviously sufficient for (29). Moreover, in order for the remaining right hand side term to be sufficiently strong in order to be able to dominate the left hand side, we will need to have full observations in the sense that

$$\overline{\mathcal{R}(C(t))} = Y. \tag{32}$$

In the next section, it will be shown that under certain stability conditions on the generalized ODE in (1), together with (32), the version (31) of the all-at-once tangential cone condition is sufficient for its reduced counterpart (26).

Likewise, we will further consider the following sufficient conditions for boundedness of the derivative,

$$\|f_\theta'(\theta, u)\|_{L(X, \mathcal{W}^*)} \le C_{\mathbf{F},1}, \qquad \|f_u'(\theta, u)\|_{L(\mathcal{U}, \mathcal{W}^*)} \le C_{\mathbf{F},2},$$

$$\|\partial_t\|_{L(\mathcal{U}, \mathcal{W}^*)} \le C_{\mathbf{F},0}, \qquad \|C\|_{L(\mathcal{U}, \mathcal{Y})} \le C_{\mathbf{F},3} \tag{33}$$

$$\forall(\theta, u) \in B_\rho^{X \times \mathcal{U}}(\theta^0, u^0).$$

The function space setting considered here will be

$$\mathcal{U} = \{u \in L^2(0, T; V) : \dot{u} \in L^2(0, T; W^*)\} \hookrightarrow C(0, T; H),$$
$$\mathcal{W} = L^2(0, T; W), \qquad \mathcal{Y} = L^2(0, T; Y), \tag{34}$$

so that the third bound in (33) is automatically satisfied with $C_{\mathbf{F},0} = 1$. We focus on Lebesgue and Sobolev spaces[1]

$$V = W^{s,m}(\Omega), \qquad W = W^{t,n}(\Omega), \qquad Y = L^q(\Omega), \tag{35}$$

with $s, t \in [0, \infty)$, $m, n \in [1, \infty]$, $q \in [1, \hat{q}]$, and $\hat{q}$ the maximal index such that V continuously embeds into $L^{\hat{q}}(\Omega)$, i.e. such that

$$s - \frac{d}{m} \succeq -\frac{d}{\hat{q}}, \tag{36}$$

so that with C defined by the embedding operator $\mathcal{U} \to \mathcal{Y}$, the last bound in (33) is automatically satisfied.[2] For the notation $\succeq$ we refer to the Appendix.

---

[1] In place of V, its intersection with $H_0^1(\Omega)$ might be considered in order to take into account homogeneous Dirichlet boundary conditions. For the estimates themselves, this does not change anything.

[2] One could possibly think of also extending to more general Lebesgue spaces instead of L^2 with respect to time. As long as the summability index is the same for $\mathcal{W}$ and $\mathcal{Y}$ this would not change anything in Sect. 2.1. As soon as the summability indices differ, one has to think of continuity of the embedding $\mathcal{U} = L^{r_1}(0, T; V) \cap W^{1,r_2}(0, T; W^*) \hookrightarrow \mathcal{Y} = L^{r_3}(0, T; Y)$ as a whole, possibly taking advantage of some interpolation between $L^{r_1}(0, T; V)$ and $W^{1,r_2}(0, T; W^*)$. This could become very technical but might pay off in specific applications.

The parameter space X may be very general at the beginning of Sect. 2.1 and in Sect. 2.4. We will only specify it in the particular examples of Sect. 2.1.

We will now verify the conditions (31) and (33) for some (classes of) examples.

2.1 Bilinear Problems

Many coefficient identification problems in linear PDEs, such as the identification of a potential or of a diffusion coefficient, as mentioned above, can be treated in a general bilinear context.

Consider an evolution driven by a bilinear operator, i.e.,

$$f(\theta, u)(t) = L(t)u(t) + ((B\theta)(t))u(t) - g(t),\tag{37}$$

where for almost all $t \in (0, T)$, and all $\theta \in X$, $v \in V$ we have $L(t), (B\theta)(t) \in \mathcal{L}(V, W^*)$, $\theta \mapsto (B\theta)(t)v \in \mathcal{L}(X, W^*)$, and $g(t) \in W^*$, with

$$\sup_{t \in [0,T]} \|L(t)\|_{\mathcal{L}(V,W^*)} \leq C_L, \qquad \sup_{t \in [0,T]} \|(B\theta)(t)\|_{\mathcal{L}(V,W^*)} \leq C_B \|\theta\|_X\tag{38}$$

so that the first and second bounds in (33) are satisfied, due to the estimates

$$\|f'_\theta(\theta, u)\chi\|_{W^*} = \left(\int_0^T \|((B\chi)(t))u(t)\|_{W^*}^2\right)^{1/2} \leq C_B \|\chi\|_X \left(\int_0^T \|u(t)\|_V^2\right)^{1/2}$$

$$\|f'_u(\theta, u)v\|_{W^*} = \left(\int_0^T \|L(t)v(t) + ((B\theta)(t))v(t)\|_{W^*}^2\right)^{1/2}$$

$$\leq (C_L + C_B \|\theta\|_X) \left(\int_0^T \|v(t)\|_V^2\right)^{1/2}.$$

For the left hand side in (31), we have

$$\left(f(\theta, u) - f(\tilde{\theta}, \tilde{u}) - f'_u(\theta, u)(u - \tilde{u}) - f'_\theta(\theta, u)(\theta - \tilde{\theta})\right)(t) = -((B(\theta - \tilde{\theta})(t))(u - \tilde{u})(t),$$

and (31) is satisfied if and only if

$$\|(B(\theta - \tilde{\theta}))(u - \tilde{u})\|_{W^*} \leq c_{tcc}^{AAO} \|C(u - \tilde{u})\|_Y, \quad \forall (\theta, u), (\tilde{\theta}, \tilde{u}) \in B_\rho^{X \times \mathcal{U}}(\theta^0, u^0),$$

hold. A sufficient condition for this to hold is

$$\|(B(\theta - \tilde{\theta}))(t)(v - \tilde{v})\|_{W^*} \leq c_{tcc}^{AAO} \|C(t)(v - \tilde{v})\|_Y,$$

$$\forall (\theta, v), (\tilde{\theta}, \tilde{v}) \in B_\rho^{X \times V}(\theta^0, u^0(t)), \quad t \in (0, T)\tag{39}$$

The proofs of the propositions for the following examples can be found in the Appendix. Likewise, the conditions on the summability and smoothness indices s, t, p, q, m, n of the used spaces, (A.108), (A.110), (A.113), (A.114), (A.116) and (A.120)–(A.124) as appearing in the formulation of the propositions, are derived there.

2.2 Identification of a Potential c

Problem (4)–(6) can be cast into the form (37) by setting $\theta = c$ and

$$L(t) = \Delta, \quad (Bc)(t)v = -cv, \tag{40}$$

(i.e., $(Bc)(t)$ is a multiplication operator with the multiplier c). We set

$$X = L^p(\Omega). \tag{41}$$

Proposition 1 *For $\mathcal{U}$, $\mathcal{W}$, $\mathcal{Y}$ according to (34) with (35) and (A.108), $-\Delta \in \mathcal{L}(V, W^*)$, the operator $\mathbf{F}$ defined by (28), (37) and (40), $C = id : \mathcal{U} \to \mathcal{Y}$ satisfies the tangential cone condition (31) with a uniformly bounded operator $\mathbf{F}'(c)$, i.e., the family of linear operators $(\mathbf{F}'(c))_{c \in M}$ is uniformly bounded in the operator norm, for c in a bounded subset M of X.*

Remark 1 A full Hilbert space setting can be achieved by setting $p = q = m = n = 2$ and choosing $s \geq 0, t > \frac{d}{2}$.

2.3 Identification of a Diffusion Coefficient a

The a problem (7)–(9) is defined by setting

$$L(t) \equiv 0, \quad (Ba)(t)v = \nabla \cdot (a\nabla v), \tag{42}$$

so that

$$\|(B(\hat{a})(t))\hat{v}\|_{W^*}$$

$$= \sup_{w \in W, \, \|w\|_W \leq 1} \int_\Omega \hat{a}\nabla\hat{v} \cdot \nabla w \, dx = \sup_{w \in W, \, \|w\|_W \leq 1} \int_\Omega \hat{v}\left(\nabla\hat{a} \cdot \nabla w + \hat{a}\Delta w\right) dx$$

$$\leq \|\hat{v}\|_{L^q}\left(\|\nabla\hat{a}\|_{L^p} \sup_{w \in W, \, \|w\|_W \leq 1} \|\nabla w\|_{L^{\frac{p^*q}{q-p^*}}} + \|\hat{a}\|_{L^r} \sup_{w \in W, \, \|w\|_W \leq 1} \|\Delta w\|_{L^{\frac{r^*q}{q-r^*}}}\right).$$

Note that since $Y = L^q(\Omega)$ we had to move all derivatives away from $\hat{v}$ by means of integration by parts, which forces us to use spaces of differentiability order at least two in W and at least one in X. Thus we here consider

$$X = W^{1,p}(\Omega). \tag{43}$$

Proposition 2 *For $\mathcal{U}$, $\mathcal{W}$, $\mathcal{Y}$ according to (34) with (35) and (A.110), the operator* **F** *defined by (28), (37) and (42), $C = id : \mathcal{U} \to \mathcal{Y}$ satisfies the tangential cone condition (31) with a uniformly bounded operator* $\mathbf{F}'(a)$.

Remark 2 A full Hilbert space setting $p = q = m = n = 2$ requires to choose

$$s \geq 0 \text{ and } t \begin{cases} \geq 2 \text{ if } d = 1 \\ > 2 \text{ if } d = 2 \\ > 1 + \frac{d}{2} \text{ if } d \geq 3 \end{cases}.$$

2.4 Nonlinear Inverse Source Problems

Consider nonlinear evolutions that are linear with respect to the parameter θ, i.e.

$$f(\theta, u)(t) = L(t)u(t) + \Phi(u(t)) + \Psi(\nabla u(t)) - B(t)\theta \tag{44}$$

where for almost all $t \in (0, T)$, $L(t) \in \mathcal{L}(V, W^*)$, $B(t) \in \mathcal{L}(X, W^*)$ and $\Phi, \Psi \in C^2(\mathbb{R})$ satisfy the Hölder continuity and growth conditions

$$|\Phi'(\lambda) - \Phi'(\tilde{\lambda})| \leq C_{\Phi''}(1 + |\lambda|^\gamma + |\tilde{\lambda}|^\gamma)|\tilde{\lambda} - \lambda|^\kappa \tag{45}$$

for all $\tilde{\lambda}, \lambda \in \mathbb{R}$

$$|\Psi'(\lambda) - \Psi'(\tilde{\lambda})| \leq C_{\Psi''}(1 + |\lambda|^{\hat{\gamma}} + |\tilde{\lambda}|^{\hat{\gamma}})|\tilde{\lambda} - \lambda|^{\hat{\kappa}} \tag{46}$$

for all $\tilde{\lambda}, \lambda \in \mathbb{R}^d$, where $\gamma, \hat{\gamma}, \kappa, \hat{\kappa} \geq 0$. We will show that the exponents $\gamma, \hat{\gamma}$ may actually be arbitrary as long as the smoothness s, t of V and W is chosen appropriately.

Proposition 3 *The operator* **F** *defined by (28), (37) and (42), $C = id : \mathcal{U} \to \mathcal{Y}$ in either of the four following cases*

(a) *Equation (45) and Ψ affinely linear and $\mathcal{U}$, $\mathcal{W}$, $\mathcal{Y}$ as in (34) with (35), (A.113) and (A.114);*

(b) *Equations (45) and (46) and $\mathcal{U}$, $\mathcal{W}$, $\mathcal{Y}$ as in (34) with (35), (A.113), (A.114), (A.116) and (A.120);*

(c) *Equation (45) and Ψ affinely linear, $\mathcal{W}$, $\mathcal{Y}$ as in (34), $\mathcal{U}$ as in (A.121) with (35), (36) and (A.123);*

(d) Equations (45) *and* (46), $\mathcal{W}$, $\mathcal{Y}$ *as in* (34), $\mathcal{U}$ *as in* (A.122) *with* (35), (36), (A.123) *and* (A.124);

satisfies the tangential cone condition (31) with a uniformly bounded operator $\mathbf{F}'(\theta)$.

Remark 3 A Hilbert space setting $p = q = m = n = 2$ is therefore possible for arbitrary γ, κ, $\hat{\gamma}$, provided t and s are chosen sufficiently large, cf. (A.113) and (A.114) in case $C_{\Psi''} = 0$, and additionally (A.116) and (A.120) otherwise.

3 Reduced Setting

In this section, we formulate the system (1)–(3) by one operator mapping from the parameter space to the observation space. To this end, we introduce the parameter-to-state map

$$S : \mathcal{D} \subseteq X \to \tilde{\mathcal{U}}, \qquad \text{where} \qquad u = S(\theta) \text{solves}$$

(1)-(2) then, with $\mathcal{D}(F) = \mathcal{D}$ the forward operator for the reduced setting can be expressed as

$$F : \mathcal{D}(F) \subseteq X \to \mathcal{Y}, \qquad \theta \mapsto C(S(\theta)) \tag{47}$$

and the inverse problem of recovering θ from y can be written as

$$F(\theta) = y.$$

Here, differently from the state space $\mathcal{U}$ in the all-at-once setting, cf., (34), we use a non Hilbert state space

$$\tilde{\mathcal{U}} = \{u \in L^\infty(0, T; V) : \dot{u} \in L^2(0, T; W^*)\} \tag{48}$$

as this appears to be more appropriate for applying parabolic theory.

We now establish a framework for verifying the tangential cone condition as well as boundedness of the derivative in this general setting.

For this purpose, we make the following assumptions.

Assumption 3.1

(R1) *Local Lipschitz continuity of* f

$$\forall M \geq 0, \exists L(M) \geq 0, \forall^{a.e.} t \in (0, T) :$$

$$\|f(t, \theta_1, v_1) - f(t, \theta_2, v_2)\|_{W^*} \leq L(M)(\|v_1 - v_2\|_V + \|\theta_1 - \theta_2\|_X),$$

$$\forall v_i \in V, \theta_i \in X : \|v_i\|_V, \|\theta_i\|_X \leq M, i = 1, 2.$$

(R2) *Well-definedness of the parameter-to-state map*

$$S : \mathcal{D}(F) \subseteq X \to \tilde{\mathcal{U}}$$

with $\tilde{\mathcal{U}}$ as in (48) as well as its boundedness in the sense that there exists $C_S > 0$ such that for all $\theta \in \mathcal{B}_\rho^X(\theta^0)$ the estimate

$$\|S(\theta)\|_{L^\infty(0,T;V)} \leq C_S$$

holds.

(R3) *Continuous dependence on data of the solution to the linearized problem with zero initial data, i.e., there exists a constant C_{lin} such that for all $\theta \in \mathcal{B}_\rho^X(\theta^0)$, $b \in \mathcal{W}^*$, and any z solving*

$$\dot{z}(t) = f_u'(\theta, S(\theta))(t)z(t) + b(t) \quad t \in (0, T) \tag{49}$$

$$z(0) = 0, \tag{50}$$

the estimate

$$\|z\|_{\mathcal{Y}} \leq C_{lin}\|b\|_{\mathcal{W}^*}. \tag{51}$$

holds.

(R4) *Tangential cone condition of the all-at-once setting* (31)

$$\exists \rho > 0, \forall (\theta, u), (\tilde{\theta}, \tilde{u}) \in \mathcal{B}_\rho^{X,\mathcal{U}}(\theta^0, u^0) :$$

$$\|f(\tilde{\theta}, \tilde{u}) - f(\theta, u) - f_u'(\theta, u)(\tilde{u} - u) - f_\theta'(\theta, u)(\tilde{\theta} - \theta)\|_{\mathcal{W}^*}$$

$$\leq c_{tcc}^{AAO}\|C\tilde{u} - Cu\|_{\mathcal{Y}}.$$

The main result of this section is as follows.

Theorem 3.2 *Suppose Assumption 3.1 holds and C is the embedding $V \hookrightarrow Y$. Then there exists a constant $\rho > 0$ such that for all $\theta, \tilde{\theta} \in \mathcal{B}_\rho^X(\theta_0) \subset \mathcal{D}(F)$,*

(i) $F'(\theta)$ is uniformly bounded:

$$\|F'(\theta)\|_{\mathcal{L}(X,\mathcal{Y})} \leq M$$

for some constant M, and

(ii) The tangential cone condition is satisfied:

$$\|F(\tilde{\theta}) - F(\theta) - F'(\theta)(\tilde{\theta} - \theta)\|_{\mathcal{Y}} \leq c_{tcc}^{Re}\|F(\tilde{\theta}) - F(\theta)\|_{\mathcal{Y}} \tag{52}$$

for some small constant c_{tcc}^{Re}.

This is a consequence of the following two propositions, in which we combine the all-at-once versions of the tangential cone and boundedness conditions, respectively, with the assumed stability of S and its linearization.

Proposition 4 *Given C is the embedding $V \hookrightarrow Y$ and u_0 is independent of θ, the tangential cone condition in the reduced setting (52) follows from the one in the all-at-once setting (R4) if the linearized forward operator is boundedly invertible as in (R3) and S is well defined according to (R2).*

Proof We begin by observing that the functions

$$v := S(\theta) - S(\tilde{\theta})$$

$$w := S'(\theta)h$$

$$z := S(\theta) - S(\tilde{\theta}) - S'(\theta)(\theta - \tilde{\theta})$$

solve the corresponding equations

$$\dot{v}(t) = f(\theta, S(\theta))(t) - f(\tilde{\theta}, S(\tilde{\theta}))(t) \qquad t \in (0, T), \qquad v(0) = 0 \qquad (53)$$

$$\dot{w}(t) = f'_u(\theta, S(\theta))w(t) + f'_\theta(\theta, S(\theta))h(t) \qquad t \in (0, T), \qquad w(0) = 0 \tag{54}$$

$$\dot{z}(t) = f'_u(\theta, S(\theta))z(t)$$

$$+ \left(- f'_u(\theta, S(\theta))v(t) - f'_\theta(\theta, S(\theta))(\theta - \tilde{\theta})(t) \right) \tag{55}$$

$$+ f(\theta, S(\theta))(t) - f(\tilde{\theta}, S(\tilde{\theta}))(t) \Big)$$

$$=: f'_u(\theta, S(\theta))z(t) + r(t) \qquad t \in (0, T), \qquad z(0) = 0. \tag{56}$$

Hence we end up with the following estimate, using the assumed bounded invertibility of the linearized problem (56) and the fact that C is the embedding $V \hookrightarrow Y$,

$$\|F(\theta) - F(\tilde{\theta}) - F'(\theta)(\theta - \tilde{\theta})\|_Y = \|S(\theta) - S(\tilde{\theta}) - S'(\theta)(\theta - \tilde{\theta})\|_Y$$

$$\leq C_{lin}\|r\|_{W^*} \tag{57}$$

$$\leq C_{lin}c_{tcc}^{AAO}\|F(\theta) - F(\tilde{\theta})\|_Y, \tag{58}$$

where $\|r\|_{W^*}$ and c_{tcc}^{AAO} are respectively the left hand side and the constant in the all-at-once tangential cone estimate, applied to $u = S(\theta)$ and $\tilde{u} = S(\tilde{\theta})$. $\square$

Remark 4 The inverse problem (19) with (22), (23) and (24) can be written as a composition of the linear observation operator C and the nonlinear parameter-to-state map S. Such problems have been considered and analyzed in [16], but as opposed to that the inversion of our observation operator is ill-posed so the theory of [16] does not apply here.

Note that in (58), c_{tcc}^{AAO} must be sufficiently small such that the tangential cone constant in the reduced setting $c_{tcc}^{Re} := C_{lin}c_{tcc}^{AAO}$ fulfills the smallness condition required in convergence proofs as well. Moreover we wish to emphasize that for the Proof of Proposition 4, the constant C_{lin} does not need to be uniform but could as well depend on θ. Also the uniform boundedness condition on S from (R2) is not yet needed here.

Under further assumptions on the defining functions f, we also get existence and uniform boundedness of the linear operator $F'(\theta)$ as follows.

Proposition 5 *Let S be well defined and bounded according to (R2), and let (R1), (R3) be satisfied.*
Then $F'(\theta)$ is Gâteaux differentiable and its derivative given by

$$F'(\theta) : \mathcal{X} \to \mathcal{Y}, \qquad where \qquad F'(\theta)h = w \; solves \qquad (59)$$

(54) is uniformly bounded in $\mathcal{B}_\rho^\mathcal{X}(\theta_0)$.

Proof For differentiability of F relying on conditions (R1)–(R3), we refer to [34, Proposition 4.2]. Moreover again using (R1)–(R3), for any $\theta \in \mathcal{B}_\rho^\mathcal{X}(\theta_0)$ we get

$$\|F'(\theta)h\|_\mathcal{Y} = \|S'(\theta)h\|_\mathcal{Y} \leq C_{lin}\|f_\theta'(\theta, S(\theta))h\|_{L^2(0,T;W^*)}$$

$$\leq C_{lin}\sqrt{T}\|f_\theta'(\theta, S(\theta))\|_{\mathcal{X}\to W^*}\|h\|_\mathcal{X}$$

$$\leq C_{lin}\sqrt{T}L(M)\|h\|_\mathcal{X}$$

for $M = C_S + \|\theta_0\|_\mathcal{X} + \rho$, where $L(M)$ is the Lipschitz constant in (R1) and C_{lin} is as in (R3). Above, we employ boundedness of S by C_S as assumed in (R2).

This proves uniform boundedness of $F'(\theta)$.

We now discuss Assumption 3.1 in more detail. □

Remark 5 For the case $V = W$.

We rely on the setting of a Gelfand triple $V \subseteq H \subseteq V^*$ for the general framework of nonlinear evolution equations. By this, (R2) can be fulfilled under the conditions suggested by Roubíček [38, Theorems 8.27, 8.31]:

For every $\theta \in \mathcal{D}(F)$

(S1) and for almost $t \in (0, T)$, the mapping $-f(t, \theta, \cdot)$ is pseudomonotone, i.e., $-f(t, \theta, \cdot)$ is bounded and

$$\left. \begin{array}{r} \liminf\limits_{k\to\infty} \langle f(t, \theta, u_k), u_k - u \rangle \geq 0 \\[2mm] u_k \rightharpoonup u \end{array} \right\} \Rightarrow \left\{ \begin{array}{l} \langle f(t, \theta, u), u - v \rangle \geq \limsup\limits_{k\to\infty} \langle f(t, \theta, u_k), u_k - v \rangle \\[2mm] \forall v \in V. \end{array} \right.$$

(S2) $-f(\cdot, \theta, \cdot)$ is semi-coercive, i.e.,

$$\forall v \in V, \forall^{a.e.} t \in (0, T) : \langle -f(t, \theta, v), v \rangle_{V^*,V} \geq C_0^\theta |v|_V^2 - C_1^\theta(t)|v|_V - C_2^\theta(t)\|v\|_H^2$$

for some $C_0^\theta > 0$, $C_1^\theta \in L^2(0, T)$, $C_2^\theta \in L^1(0, T)$ and some seminorm $|.|_V$ satisfying

$$\forall v \in V : \|v\|_V \leq c_{|.|}(|v|_V + \|v\|_H) \text{ for some } c_{|.|} > 0.$$

(S3) f satisfies the growth condition

$$\exists \gamma^\theta \in L^2(0, T), \hbar^\theta : \mathbb{R} \to \mathbb{R} \text{ increasing} : \|f(t, \theta, v)\|_{V^*} \leq \hbar^\theta(\|v\|_H)(\gamma^\theta(t) + \|v\|_V)$$

and a condition for uniqueness of the solution, e.g.,

$$\forall u, v \in V, \forall^{a.e.} t \in (0, T) : \langle f(t, \theta, u) - f(t, \theta, v), u - v \rangle_{V^*, V} \leq \rho^\theta(t)\|u - v\|_H^2$$

for some $\rho^\theta \in L^1(0, T)$

and further conditions for $S(\theta) \in L^\infty(0, T; V)$, e.g., [38, Theorem 8.16, 8.18].

In case of linear $f(t, \theta, \cdot)$, (S1)–(S3) boil down to boundedness and semi-coercivity (S2) of $-f(\cdot, \theta, \cdot)$ according to [38, Theorem 8.27, 8.31, 8.28]. Alternatively, one can observe that linear boundedness implies the growth condition in (S3) with $\gamma^\theta = 0$, $\hbar^\theta = \|f(\cdot, \theta)\|$, and (S2) implies the rest of (S3) with $\rho^\theta = C_2^\theta$ if $C_1^\theta \leq 0$ as $C_0^\theta < 0$. The pseudomonotonicity assumption (S1), which guarantees week convergence of $f(\cdot, \theta, u_k)$ to $f(\cdot, \theta, u)$ when the approximation solution sequence u_k converges weakly to u, can be replaced by weak continuity of $f(\cdot, \theta, \cdot)$ which holds in this linear bounded case.

Treating the linearized problem (49)–(50) as an independent problem, we can impose on $f_u'(\theta, S(\theta))$ the boundedness and semi-coercivity properties, then (R3) follows.

Remark 6 For general spaces V, W.

Some examples even in case $V \neq W$ allow to use the results quoted in Remark 5 with an appropriately chosen Gelfand triple, see, e.g., Sect. 3.1 below.

When dealing with linear and quasilinear parabolic problems, detailed discussions for unique existence of the solution are exposed in the books, e.g., of Evans [9], Ladyzhenskaya et al. [26], Pao [36]. If constructing the solution to the initial value problem through the semigroup approach, one can find several results, e.g., from Evans [9], Pazy [37] combined with the elliptic results from Ladyzhenskaya et al. [27].

Addressing (R3), a possible strategy is using the following dual argument.

Suppose W is reflexive and z is a solution to the problem (49)–(50), then by the Hahn-Banach Theorem

$$\|z\|_{L^2(0,T;V)} = \sup_{\|\phi\|_{L^2(0,T;V^*)} \leq 1} \int_0^T \langle z, \phi \rangle_{V,V^*} dt$$

$$= \sup_{\|\phi\|_{L^2(0,T;V^*)} \leq 1} \int_0^T \langle z, -\dot{p} - f_u'(\theta, S(\theta))^* p \rangle_{V,V^*} dt$$

$$= \sup_{\|\phi\|_{L^2(0,T;V^*)} \leq 1} \int_0^T \langle \dot{z} - f_u'(\theta, S(\theta))z, p \rangle_{W^*,W} dt$$

$$= \sup_{\|\phi\|_{L^2(0,T;V^*)} \leq 1} \int_0^T \langle b, p \rangle_{W^*,W} dt$$

$$\leq \sup_{\|\phi\|_{L^2(0,T;V^*)} \leq 1} \|b\|_{L^2(0,T;W^*)} \|p\|_{L^2(0,T;W)},$$

where

$$f_u'(\theta, S(\theta))(t) : V \to W^*, \qquad f_u'(\theta, S(\theta))(t)^* : W^{**} = W \to V^*,$$

and p solves the adjoint equation

$$-\dot{p}(t) = f_u'(\theta, S(\theta))^* p(t) + \phi(t) \quad t \in (0, T) \tag{60}$$

$$p(T) = 0. \tag{61}$$

If in the adjoint problem the estimate

$$\|p\|_{L^2(0,T;W)} \leq \tilde{C}_{lin} \|\phi\|_{L^2(0,T;V^*)} \tag{62}$$

holds for some uniform constant $\tilde{C}_{lin}$, then we obtain

$$\|z\|_{\mathcal{Y}} \leq \|C\|_{V \to Y} \|z\|_{L^2(0,T;V)} \leq \|C\|_{V \to Y} \tilde{C}_{lin} \|b\|_{\mathcal{W}^*}. \tag{63}$$

Thus (R3) is fulfilled.

So we can replace (R3) by

(R3-dual) *Continuous dependence on data of the solution to the adjoint linearized problem* associated with zero final condition, i.e., there exists a constant $\tilde{C}_{lin}$ such that for all $\theta \in \mathcal{B}_\rho^X(\theta^0)$, $\phi \in L^2(0, T, V^*)$, and any p solving (60) and (61), the estimate (62) holds.

In the following sections, we examine the specific examples introduced in the introduction, in the relevant function space setting

$$X = L^p(\Omega) \quad \text{or} \quad X = W^{1,p}(\Omega) \qquad\qquad p \in [1, \infty] \tag{64}$$

$$Y = L^q(\Omega) \qquad\qquad q \in [1, \bar{q}] \tag{65}$$

$$\tilde{\mathcal{U}} = \{u \in L^\infty(0, T; V) : \dot{u} \in L^2(0, T; W^*)\}, \tag{66}$$

where V, W will be chosen subject to the particular example, where $\hat{q}$ is the maximum power allowing $V \hookrightarrow L^{\hat{q}}(\Omega)$ and $\bar{q} \leq \hat{q}$ is the maximum power such that (51) in (R3) holds.

3.1 Identification of a Potential

We investigate this problem in the function spaces

$$\mathcal{D}(F) = X = L^p(\Omega), \quad Y = L^q(\Omega), \quad V = L^2(\Omega), \quad W = H^2(\Omega) \cap H_0^1(\Omega).$$

Now we verify the conditions proposed in Assumption 3.1.

(R1) *Local Lipschitz continuity of f:*
Applying Hölder's inequality, we have

$$\|f(\tilde{c}, \tilde{u}) - f(c, u)\|_{W*} = \|\tilde{c}\tilde{u} - cu\|_{W*} = \sup_{\|w\|_W \leq 1} \int_\Omega (\tilde{c}\tilde{u} - cu)w \, dx$$

$$\leq \sup_{\|w\|_W \leq 1} \|w\|_W C_{W \to L^{\bar{p}}} \left(\int_\Omega |\tilde{c}(\tilde{u} - u) + (\tilde{c} - c)u|^{\bar{p}^*} dx \right)^{\frac{1}{\bar{p}^*}}$$

$$\leq C_{W \to L^{\bar{p}}} (\|\tilde{c}\|_{L^p} \|\tilde{u} - u\|_{L^r} + \|\tilde{c} - c\|_{L^p} \|u\|_{L^r})$$

$$\leq L(M)(\|\tilde{u} - u\|_V + \|\tilde{c} - c\|_X)$$

with the dual index $\bar{p}^* = \frac{\bar{p}}{\bar{p}-1}$ and $r = \frac{\bar{p}p}{\bar{p}p-p-\bar{p}}$, $L(M) = C_{W \to L^{\bar{p}}} C_{V \to L^r}$ $(\|u\|_V + \|\tilde{u}\|_V + \|c\|_X + \|\tilde{c}\|_X) + 1$. Above, we invoke the continuous embeddings through the constants $C_{W \to L^{\bar{p}}}, C_{V \to L^r}$, where $\bar{p}$ denotes the maximum power allowing $W \subseteq L^{\bar{p}}$. Thus we are supposing

$$p \geq \max \left\{ \frac{2\bar{p}}{\bar{p} - 2}, \frac{\bar{p}}{\bar{p} - 1} \right\} = \frac{2\bar{p}}{\bar{p} - 2} \text{ and } 2 - \frac{d}{2} \geq -\frac{d}{\bar{p}} \tag{67}$$

in order to guarantee $V = L^2(\Omega) \hookrightarrow L^r(\Omega)$ and $W = H^2(\Omega) \cap H_0^1(\Omega) \hookrightarrow L^{\bar{p}}(\Omega)$

(R2) *Well-definedness and boundedness of the parameter-to-state map:*
Verifying boundedness and semi-coercivity conditions with the Gelfand triple $H_0^1(\Omega) \hookrightarrow L^2(\Omega) \hookrightarrow H^{-1}(\Omega)$ (while remaining with $V = L^2(\Omega)$ in the definition of the space $\tilde{\mathcal{U}}$) shows that, for $u_0 \in L^2(\Omega)$, $\varphi \in L^2(0, T; H^{-1}(\Omega))$ the initial value problem (4)–(6) admits a unique solution $u \in W(0, T) := \{u \in L^2(0, T; H_0^1(\Omega)) : \dot{u} \in L^2(0, T; H^{-1}(\Omega))\} \subset \{u \in L^\infty(0, T; L^2(\Omega)) : \dot{u} \in L^2(0, T; H^{-2}(\Omega))\} = \tilde{\mathcal{U}}$.
Indeed, coercivity is deduced as follows. For

$$p \geq 2, \quad d \leq 3, \tag{68}$$

we see

$$\int_\Omega cu^2 dx \le \|c\|_{L^2(\Omega)} \left(\int_\Omega u^4 dx\right)^{\frac{1}{2}} \le \|c\|_{L^2(\Omega)} \left(\int_\Omega u^2 dx\right)^{\frac{1}{4}} \left(\int_\Omega u^6 dx\right)^{\frac{1}{4}}$$

$$\le \|c\|_{L^2(\Omega)} \|u\|_{L^2(\Omega)}^{\frac{1}{2}} \|u\|_{L^6(\Omega)}^{\frac{3}{2}}$$

$$\le C_{H_0^1 \to L^6} \|c\|_{L^2(\Omega)} \|u\|_{L^2(\Omega)}^{\frac{1}{2}} \|u\|_{H_0^1(\Omega)}^{\frac{3}{2}} \tag{69}$$

$$\le C_{H_0^1 \to L^6} \|c\|_{L^2(\Omega)} \left(\frac{1}{4\epsilon} \|u\|_{L^2(\Omega)} \|u\|_{H_0^1(\Omega)} + \epsilon \|u\|_{H_0^1(\Omega)}^2\right)$$

$$\le C_{H_0^1 \to L^6} \|c\|_{L^2(\Omega)} \left(\frac{1}{16\epsilon\epsilon_1} \|u\|_{L^2(\Omega)}^2 + \frac{\epsilon_1}{4\epsilon} \|u\|_{H_0^1(\Omega)}^2 + \epsilon \|u\|_{H_0^1(\Omega)}^2\right),$$

which yields semi-coercivity

$$\langle -f(t,c,u), u\rangle_{H^{-1}, H_0^1} = \int_\Omega (-\Delta u + cu) u\, dx$$

$$\ge \left(1 - C_{H_0^1 \to L^6} \|c\|_{L^2(\Omega)} \left(\frac{\epsilon_1}{4\epsilon} + \epsilon\right)\right) \|u\|_{H_0^1(\Omega)}^2 - \frac{C_{H_0^1 \to L^6}}{16\epsilon\epsilon_1} \|c\|_{L^2(\Omega)} \|u\|_{L^2(\Omega)}^2,$$

$$=: C_0^c \|u\|_{H_0^1(\Omega)}^2 + C_2^c \|u\|_{L^2(\Omega)}^2,$$

where the constant C_0^c is positive if choosing $\epsilon_1 < \epsilon$ and ϵ, ϵ_1 sufficiently small.

Boundedness of f can be concluded from

$$\| -f(t,c,u)\|_{H^{-1}(\Omega)} = \sup_{\|v\|_{H_0^1} \le 1} \int_\Omega (-\Delta u + cu) v\, dx$$

$$\le \sup_{\|v\|_{H_0^1} \le 1} \left(\|u\|_{H^1(\Omega)} \|v\|_{H^1(\Omega)} + C_{H_0^1 \to L^6} C_{H_0^1 \to L^3} \|c\|_{L^2(\Omega)} \|u\|_{H^1(\Omega)} \|v\|_{H^1(\Omega)}\right)$$

$$\le C \|c\|_{L^2(\Omega)} \|u\|_{H^1(\Omega)}.$$

Moreover, by the triangle inequality: $\|c\|_{L^2(\Omega)} \le \|c^0\|_{L^2(\Omega)} + \|c - c^0\|_{L^2(\Omega)} \le \|c^0\|_{L^2(\Omega)} + \rho$, semi-coercivity of f is satisfied with the constants C_0, C_1 now depending only on the point c^0. This hence gives us uniform boundedness of S on the ball $\mathcal{B}_\rho^X(c^0)$.

(R3) *Continuous dependence on data of the solution to the linearized problem with zero initial data:*
We use the duality argument mentioned in Remark 6. To do so, we need to prove existence of the adjoint state $p \in L^2(0,T;W)$ and the associated estimate 6.

Initially, by the transformation $v = e^{-\lambda t} p$ and putting $\tau = T - t$, the adjoint problem (60)–(61) is equivalent to

$$\dot{v}(t) - \Delta v(t) + (\lambda + c)v(t) = e^{-\lambda t}\phi(t) \qquad t \in (0, T) \tag{70}$$

$$v(0) = 0. \tag{71}$$

We note that this problem with $c = \hat{c} \in L^\infty(\Omega)$, $\lambda + \hat{c} > -C_{PF}$, the constant in the Poincaré-Friedrichs inequality, $\phi \in L^2(0, T; L^2(\Omega))$, $\partial\Omega \in C^2$, admits a unique solution in $L^2(0, T; H^2(\Omega) \cap H_0^1(\Omega))$ [9, Section 7.1.3, Theorem 5][3] and the operator $\frac{d}{dt} - \Delta + (\lambda + \hat{c}) : L^2(0, T; H^2(\Omega) \cap H_0^1(\Omega)) \to L^2(0, T; L^2(\Omega)) \times H^1(\Omega)$, $p \mapsto (\phi, p_0)$ is boundedly invertible.

Suppose u solves (70)–(71), by the identity

$$\dot{u} - \Delta u + (\lambda + c)u = e^{-\lambda t}\phi \quad \Leftrightarrow \quad \dot{u} - \Delta u + (\lambda + \hat{c})u = e^{-\lambda t}\phi + (\hat{c} - c)u$$

$$u = \left(\frac{d}{dt} - \Delta + (\lambda + \hat{c})\right)^{-1} \left[e^{-\lambda t}\phi + (\hat{c} - c)u\right]$$

$$=: \mathbb{T}u,$$

we observe that $\mathbb{T} : L^2(0, T; H^2(\Omega) \cap H_0^1(\Omega)) \to L^2(0, T; H^2(\Omega) \cap H_0^1(\Omega))$ is a contraction

$$\|\mathbb{T}(u - v)\|_{L^2(0,T;H^2 \cap H_0^1)}$$

$$\leq \left\|\left(\frac{d}{dt} - \Delta + (\lambda + \hat{c})\right)^{-1}\right\|_{L^2(0,T;L^2(\Omega)) \to L^2(0,T;H^2 \cap H_0^1)} \|(\hat{c} - c)(u - v)\|_{L^2(0,T;L^2(\Omega))}$$

$$\leq C^{\hat{c}} \|\hat{c} - c\|_{L^p} \|u - v\|_{L^2(0,T;L^{\frac{2p}{p-2}}(\Omega))}$$

$$\leq C\epsilon \|u - v\|_{L^2(0,T;H^2 \cap H_0^1)}, \tag{72}$$

where $C\epsilon < 1$ if we assume $\hat{c} = c^0 \in L^\infty(\Omega)$ and ρ is sufficiently small. In some case, smallness of ρ can be omitted (discussed at the end of (R3)). Estimate (72) holds provided

$$W = H^2(\Omega) \cap H_0^1(\Omega) \hookrightarrow L^{\frac{2p}{p-2}}(\Omega) \text{ i.e., } \quad p \geq \frac{2\bar{p}}{\bar{p} - 2}. \tag{73}$$

---

[3] Where smoothness of the domain can be slightly relaxed to $C^{1,1}$ as assumed here, see, e.g., [12].

Thus, for $\phi \in L^2(0, T; L^2(\Omega))$ there exists a unique solution $v \in L^2(0, T; H^2(\Omega) \cap H_0^1(\Omega))$ to the problem (70)–(71), which implies $p = e^{\lambda t} v \in L^2(0, T; H^2(\Omega) \cap H_0^1(\Omega))$ is the solution to the adjoint problem (60)–(61).

Observing that p solves

$$\dot{p}(t) - \Delta p(t) + \hat{c} p(t) = (\hat{c} - c)p(t) + \phi(t) \qquad t \in (0, T)$$

$$p(0) = 0,$$

employing again [9, Section 7.1.3 , Theorem 5] and smallness of ρ yields

$$\|p\|_{L^2(0,T;W)} \leq C(\|(\hat{c} - c)p\|_{L^2(0,T;L^2(\Omega))} + \|\phi\|_{L^2(0,T;L^2(\Omega))})$$

$$\leq C(2\rho \|p\|_{L^2(0,T;H^2 \cap H_0^1)} + \|\phi\|_{L^2(0,T;L^2(\Omega))}) \qquad (74)$$

$$\leq C \|\phi\|_{L^2(0,T;V^*)}$$

with some constant C independent of $\theta \in \mathcal{B}_\rho^X(c^0)$. This yields (R3-dual) with $\bar{q} = 2$.

If $d = 1, p = 2$ or $d = 2, p > 2$ or $d = 3, p \geq \frac{12}{5}$, the smallness condition on ρ can be omitted. Indeed, for $d = 3, p \geq \frac{12}{5}$ testing the adjoint equation by $-\Delta p$ yields

$$\int_\Omega -\dot{p}\Delta p + (\Delta p)^2 dx = \int_\Omega (cp - \phi)\Delta p \, dx$$

$$\frac{1}{2}\frac{d}{dt}\|\nabla p\|_{L^2(\Omega)}^2 + \|\Delta p\|_{L^2(\Omega)}^2 \leq \frac{1}{2}\|\Delta p\|_{L^2(\Omega)}^2 + \|\phi\|_{L^2(\Omega)}^2 + \|cp\|_{L^2(\Omega)}^2 \qquad (75)$$

$$\frac{1}{2}\frac{d}{dt}\|\nabla p\|_{L^2(\Omega)}^2 + \frac{1}{2}\|\Delta p\|_{L^2(\Omega)}^2 \leq \|\phi\|_{L^2(\Omega)}^2 + \|c\|_{L^p(\Omega)}^2 \left(\int_\Omega p^{\frac{p}{p-2} + \frac{p}{p-2}} dx\right)^{\frac{p-2}{p}}$$

$$\leq \|\phi\|_{L^2(\Omega)}^2 + \|c\|_{L^p(\Omega)}^2 \|p\|_{L^6(\Omega)} \|p\|_{L^\infty(\Omega)} |\Omega|^{\frac{5p-12}{6p}}$$

$$\leq \|\phi\|_{L^2(\Omega)}^2 + (\|c^0\|_X^2 + \rho^2)|\Omega|^{\frac{5p-12}{6p}} \left(\frac{C_{H_0^1 \to L^6}^2}{4\epsilon}\|\nabla p\|_{L^2(\Omega)}^2 \right.$$

$$\left. + \epsilon C_{H^2 \cap H_0^1 \to L^\infty}^2 \left(\|\Delta p\|_{L^2(\Omega)}^2 + \|\nabla p\|_{L^2(\Omega)}^2\right)\right),$$

where in the last estimate we apply Young's inequality. Choosing ϵ sufficiently small allows us to subtract the term involving $\|\Delta p\|_{L^2(\Omega)}^2$ on the right hand side from the one on the left hand side and get a positive coefficient in front. Here, the choice of ϵ depends only on the constants $c^0, \rho, \Omega, C_{H^2 \cap H_0^1 \to L^\infty}$.

It is also obvious that, if $d < 3$, in the second line of the above calculation, we can directly estimate as follow

$$d = 1, p = 2 : \|cp\|^2_{L^2(\Omega)} \leq \|c\|^2_{L^2(\Omega)} \|p\|^2_{L^\infty(\Omega)} \leq C^2_{H_0^1 \to L^\infty} \|c\|^2_{L^2(\Omega)} \|\nabla p\|^2_{L^2(\Omega)}$$

$$d = 2, p > 2 : \|cp\|^2_{L^2(\Omega)} \leq \|c\|^2_{L^p(\Omega)} \|p\|^2_{L^{\frac{2p}{p-2}}(\Omega)} \leq C^2_{H_0^1 \to L^{\frac{2p}{p-2}}} \|c\|^2_{L^p(\Omega)} \|\nabla p\|^2_{L^2(\Omega)}.$$

$$(76)$$

Employing firstly Gronwall-Bellman inequality with initial data $\nabla p(0) = 0$, then taking the integral on $[0, T]$, we obtain

$$\|p\|_{L^\infty(0,T;H^1(\Omega))} + \|\Delta p\|_{L^2(0,T;L^2(\Omega))} \leq C \|\phi\|_{L^2(0,T;L^2(\Omega))} \tag{77}$$

with the constant C depending only on c^0, ρ. This estimate is valid for all $c \in \mathcal{B}_\rho^X(c^0)$. Since the adjoint problem has the same form as the original problem, applying (77) in (72) we can relax $\hat{c}$, by means of without fixing $\hat{c} = c^0$ but chossing it sufficiently close to c since $\overline{L^\infty(\Omega)} = L^p(\Omega), |\Omega| < \infty$ to have $C^{\hat{c}}\epsilon \leq C\epsilon$ arbitrarily small with constant C as in (77). Therefore the constraint on smallness of ρ can be omitted in these cases.

(R4) *All-at-once tangential cone condition:*
According to (36) and (A.108) with $s = 0, t = 2, m = n = 2$, this follows if

$$\frac{p}{p-1} \leq q \leq \hat{q} \geq 2 \text{ and } 2 - \frac{d}{2} \geq -\frac{d(p-1)}{p} + \frac{d}{q}.$$

Corollary 1 *Assume* $u_0 \in L^2(\Omega), \varphi \in L^2(0, T; H^{-1}(\Omega))$, *and*

$$\mathcal{D}(F) = X = L^p(\Omega), \quad Y = L^q(\Omega), \quad V = L^2(\Omega), \quad W = H^2(\Omega) \cap H_0^1(\Omega)$$

$$p \geq 2, \quad q \in \left[\underline{q}, 2\right], \quad d \leq 3 \tag{78}$$

with $\underline{q} = \max \left\{ \frac{p}{p-1}, \min_{q \in [1,\infty]} \left\{ 2 - \frac{d}{2} \geq -\frac{d(pq-p-q)}{pq} \right\} \right\}$.

Then F *defined by* $F(c) = u$ *solving* (4)–(6) *satisfies the tangential cone condition* (52) *with a uniformly bounded operator* $F'(c)$ *defined by* (59), *see also* [15] *for the static case.*

Remark 7 This allows a full Hilbert space setting of X and Y by choosing $p = q = 2$ as long as $d \leq 3$.

3.2 Identification of a Diffusion Coefficient

We pose this problem in the function spaces

$$X = W^{1,p}(\Omega), \quad Y = L^q(\Omega), \quad V = L^2(\Omega), \quad W = H^2(\Omega) \cap H_0^1(\Omega) \qquad p > d \tag{79}$$

so that $X \hookrightarrow L^\infty(\Omega)$ and define the domain of F by

$$\mathcal{D}(F) = \{a \in X : a \geq \underline{a} > 0 \text{ a.e. on } \Omega\}. \tag{80}$$

Now we examine the conditions (R1)–(R3).

(R1) *Local Lipschitz continuity of* f:

$$\| - \nabla \cdot (\tilde{a} \nabla \tilde{u}) + \nabla \cdot (a \nabla u) \|_{W^*}$$

$$= \sup_{\|w\|_W \leq 1} \int_\Omega (\tilde{a} \nabla \tilde{u} - a \nabla u) \nabla w \, dx$$

$$= \sup_{\|w\|_W \leq 1} \int_\Omega (a \nabla (\tilde{u} - u) + (\tilde{a} - a) \nabla \tilde{u}) \nabla w \, dx$$

$$= \sup_{\|w\|_W \leq 1} \int_\Omega (\tilde{u} - u)(\nabla a \nabla w + a \Delta w) + \tilde{u}(\nabla (\tilde{a} - a) \nabla w + (\tilde{a} - a) \Delta w) \, dx$$

$$\leq \sup_{\|w\|_W \leq 1} \int_\Omega (\|\tilde{u} - u\|_{L^2} \|\nabla a\|_{L^p} + \|\tilde{u}\|_{L^2} \|\nabla (a - a)\|_{L^p}) \|\nabla w\|_{L^{\frac{2p}{p-2}}}$$

$$+ (\|\tilde{u} - u\|_{L^2} \|a\|_{L^\infty} + \|\tilde{u}\|_{L^2} \|a - a\|_{L^\infty}) \|\Delta w\|_{L^2} \, dx$$

$$\leq L(M)(\|\tilde{u} - u\|_V + \|\tilde{a} - a\|_X)$$

with $M = \left(C_{W \to W^{1, \frac{2p}{p-2}}} + C_{X \to L^\infty} \right) (\|u\|_V + \|\tilde{u}\|_V + \|c\|_X + \|\tilde{c}\|_X)$, subject to the constraint

$$W = H^2(\Omega) \cap H_0^1(\Omega) \hookrightarrow L^{\frac{2p}{p-2}}(\Omega) \text{ i.e.,} \quad p \geq \frac{2\bar{p}}{\bar{p} - 2}. \tag{81}$$

(R2) *Well-definedness and boundedness of the parameter-to-state map:*
A straightforward verification of boundedness and coercivity gives unique existence of the solution $u \in W(0, T) \subset \tilde{\mathcal{U}}$ for $a \in \mathcal{D}(F) \subset X \hookrightarrow L^\infty(\Omega), \varphi \in L^2(0, T; H^{-1}(\Omega)), u_0 \in L^2(\Omega)$.
Similarly to the c-problem, the fact that the coercivity property of f holds

$$\langle -f(t, a, u), u \rangle_{H^{-1}, H_0^1} = \int_\Omega -\nabla \cdot (a\nabla u)u dx \geq \underline{a} \|u\|_{H_0^1(\Omega)}$$

with the coefficient $\underline{a}$ being independent of a shows uniform boundedness of S.

(R3) *Continuous dependence on data of the solution to the linearized problem with zero initial data:*

We employ the result in [9, Section 7.1.3, Theorem 5] with noting that the actual smoothness condition needed for the coefficient is that, a is differentiable a.e on Ω and $a \in W^{1,\infty}(\Omega)$ rather than $a \in C^1(\Omega)$. From the observation $a \in \mathcal{D}(F) = W^{1,p}(\Omega)$, $p > d$ is differentiable a.e and the fact that $W^{1,\infty}(\Omega)$ is dense in $W^{1,p}(\Omega)$, it enables us to imitate the contraction scenario and the dual argument as in the c-problem.

Taking u, v solving (7)–(9), we see

$$\mathbb{T} : L^2(0, T; H^2(\Omega) \cap H_0^1(\Omega)) \to L^2(0, T; H^2(\Omega) \cap H_0^1(\Omega))$$

$$\mathbb{T} = \left(\frac{d}{dt} - \nabla \cdot (\hat{a}\nabla) \right)^{-1} \nabla \cdot \left((a - \hat{a})\nabla \right)$$

is a contraction

$$\|\mathbb{T}(u - v)\|_{L^2(0,T;H^2 \cap H_0^1)}$$

$$\leq \left\| \left(\frac{d}{dt} - \nabla \cdot (\hat{a}\nabla) \right)^{-1} \right\|_{L^2(0,T;L^2(\Omega)) \to L^2(0,T;H^2 \cap H_0^1)} \|\nabla \cdot \left((a - \hat{a})\nabla(u - v) \right)\|_{L^2(0,T;L^2(\Omega))}$$

$$\leq C^{\hat{a}} \|\hat{a} - a\|_X \|u - v\|_{L^2(0,T;H^2 \cap H_0^1)}$$

$$\leq C\epsilon \|u - v\|_{L^2(0,T;H^2 \cap H_0^1)}, \tag{82}$$

where $C\epsilon < 1$ if we assume $\hat{a} = a^0 \in W^{1,\infty}(\Omega)$ and ρ is sufficiently small. If the index p is large enough, smallness of ρ can be omitted (discussed at the end of (R3)). Therefore, given $\phi \in L^2(0, T; L^2(\Omega))$, the adjoint state $p \in L^2(0, T; H^2 \cap H_0^1)$ uniquely exists.

We also have the estimate

$$\|p\|_{L^2(0,T;W)} \leq C \|\nabla \cdot \left((a - \hat{a})\nabla p \right)\|_{L^2(0,T;L^2(\Omega))} + \|\phi\|_{L^2(0,T;L^2(\Omega))})$$

$$\leq C(2\rho \|p\|_{L^2(0,T;H^2 \cap H_0^1)} + \|\phi\|_{L^2(0,T;L^2(\Omega))})$$

$$\leq C\|\phi\|_{L^2(0,T;V^*)},$$

which proves continuous dependence of p on $\phi \in L^2(0, T; V^*)$, consequently, continuous dependence of the solution $z \in L^2(0, T; V)$ on the data $b \in L^2(0, T; W^*)$ in (49) and (50). Here smallness of ρ is assumed.

If $p \geq 4$, smallness of ρ is not required. To verify this, we test the adjoint equation by $-\Delta p$

$$\int_\Omega -\dot{p}\Delta p + a(\Delta p)^2 dx = \int_\Omega (-\nabla a \nabla p - \phi)\Delta p dx$$

$$\frac{1}{2}\frac{d}{dt}\|\nabla p\|^2_{L^2(\Omega)} + \underline{a}\|\Delta p\|^2_{L^2(\Omega)} \leq \frac{a}{2}\|\Delta p\|^2_{L^2(\Omega)} + \frac{1}{\underline{a}}\|\phi\|^2_{L^2(\Omega)} + \frac{1}{\underline{a}}\|\nabla a \nabla p\|^2_{L^2(\Omega)}$$

$$\frac{1}{2}\frac{d}{dt}\|\nabla p\|^2_{L^2(\Omega)} + \frac{a}{2}\|\Delta p\|^2_{L^2(\Omega)} \leq \frac{1}{\underline{a}}\|\phi\|^2_{L^2(\Omega)} + \frac{1}{\underline{a}}\|\nabla a \nabla p\|^2_{L^2(\Omega)}, \tag{83}$$

where the last term on the right hand side can be estimated as in (69) of the c-problem with $(\nabla a)^2$ in place of c, ∇p in place of u and the assumption $X \hookrightarrow W^{1,4}(\Omega)$

$$\frac{1}{\underline{a}}\|\nabla a \nabla p\|^2_{L^2(\Omega)}$$

$$\leq \frac{C_{H^1_0 \to L^6}}{\underline{a}}\|\nabla a\|^2_{L^4(\Omega)}\left(\frac{1}{16\epsilon_1}\|\nabla p\|^2_{L^2(\Omega)} + \left(\frac{\epsilon_1}{4\epsilon} + \epsilon\right)\|\nabla p\|^2_{H^1_0(\Omega)}\right)$$

$$\leq \frac{2C_{H^1_0 \to L^6}}{\underline{a}}(\|a^0\|^2_X + \rho^2)\left(\frac{1}{16\epsilon_1}\|\nabla p\|^2_{L^2(\Omega)} + \left(\frac{\epsilon_1}{4\epsilon} + \epsilon\right)\|\Delta p\|^2_{L^2(\Omega)}\right). \tag{84}$$

Choosing $\epsilon_1 < \epsilon$, and ϵ_1, ϵ sufficiently small such that we can move the term involving $\|\Delta p\|^2_{L^2(\Omega)}$ from the right hand side to the left hand side of (83). Note that, this choice of ϵ_1, ϵ is just subject to a^0 and ρ.

Proceeding similarly to the c-problem, meaning applying Gronwall-Bellman inequality then taking the integral on $[0, T]$, we obtain

$$\|p\|_{L^\infty(0,T;H^1(\Omega))} + \|\Delta p\|_{L^2(0,T;L^2(\Omega))} \leq C\|\phi\|^2_{L^2(0,T;L^2(\Omega))} \tag{85}$$

with a constant C depending only on a^0, ρ.

Observing the similarity in the form of the adjoint problem and the original problem, invoking the uniform bound (85) w.r.t parameter a and the fact $W^{1,\infty}(\Omega) = W^{1,p}(\Omega)$ one can eliminate the need of smallness of ρ.

(R4) *All-at-once tangential cone condition:*
According to (36) and (A.110) with $s = 0$, $t = 2$, $m = n = 2$, we require

$$\frac{p}{p-1} \leq q \leq \hat{q} \geq 2 \text{ and } 1 - \frac{d}{2} \geq -\frac{d(p-1)}{p} + \frac{d}{q} \text{ and } -\frac{d}{2} \geq -d + \frac{d}{p} - 1.$$

Corollary 2 *Assume $u_0 \in L^2(\Omega)$, $\varphi \in L^2(0, T; H^{-1}(\Omega))$, and*

$$X = W^{1,p}(\Omega), \qquad Y = L^q(\Omega), \qquad V = L^2(\Omega), \qquad W = H^2(\Omega) \cap H_0^1(\Omega)$$

$$p \geq 2, \quad q \in \left[\underline{q}, 2\right], \quad d < p,$$

$$(86)$$

where $\underline{q} = \max \left\{ \frac{p}{p-1}, \min_{q \in [1,\infty]} \left\{ 1 - \frac{d}{2} \geq -\frac{d(p-1)}{p} + \frac{d}{q} \wedge -\frac{d}{2} \geq -d + \frac{d}{p} - 1 \right\} \right\}.$

Then F defined by $F(a) = u$ solving (7)–(9) satisfies the tangential cone condition (52) with a uniformly bounded operator $F'(a)$ defined by (59).

Remark 8 This yields the possibility of a full Hilbert space setting $p = q = 2$ of X and Y in case $d = 1$, see also [14] and, for the static case, [15].

3.3 An Inverse Source Problem with a Quadratic First Order Nonlinearity

By the transformation $U := e^u$, the initial-value problem (10)–(12) can be converted into an inverse potential problem as considered in Sect. 3.1

$$\dot{U} - \Delta U + \theta U = 0 \qquad (t, x) \in (0, T) \times \Omega \tag{87}$$

$$U_{|\partial\Omega} = 1 \qquad t \in (0, T) \tag{88}$$

$$U(0) = U_0 \qquad x \in \Omega \tag{89}$$

with $U_0 = e^{u_0}$. Thus, in principle it is covered by the analysis from the previous section, as long as additionally positivity of U can be established. So the purpose of this section is to investigate whether we can allow for different function spaces X, Y by directly considering (10)–(12) instead of (87)–(89).

We show that f verifies the hypothesis proposed for the tangential cone condition in the reduced setting on the function spaces

$$X = L^p(\Omega), \qquad Y = L^q(\Omega), \qquad V = W = H^2(\Omega) \cap H_0^1(\Omega). \tag{90}$$

(R1) *Local Lipschitz continuity of f*:

$$\| - |\nabla \tilde{u}|^2 + |\nabla u|^2 - \tilde{\theta} + \theta \|_{W^*} = \sup_{\|w\|_W \leq 1} \int_\Omega \left(\nabla(u - \tilde{u}) \cdot \nabla(u + \tilde{u}) - \tilde{\theta} + \theta \right) w \, dx$$

$$\leq C_{W \to L^{\tilde{p}}} \left(\|(\nabla(u - \tilde{u}) \cdot \nabla(u + \tilde{u})\|_{L^{\frac{\tilde{p}}{\tilde{p}-1}}} + \|\theta - \tilde{\theta}\|_{L^{\frac{\tilde{p}}{\tilde{p}-1}}} \right)$$

$$\leq C_{W\to L^{\bar{p}}}\left(\|\nabla(u-\tilde{u})\|_{L^{\frac{2\bar{p}}{\bar{p}-1}}}\|\nabla(u+\tilde{u})\|_{L^{\frac{2\bar{p}}{\bar{p}-1}}}+\|\theta-\tilde{\theta}\|_{L^{\frac{\bar{p}}{\bar{p}-1}}}\right)$$

$$\leq C_{W\to L^{\bar{p}}}\left(C^2_{V\to W^{1,\frac{2\bar{p}}{\bar{p}-1}}}\|u-\tilde{u}\|_V\|u+\tilde{u}\|_V+C_{X\to L^{\frac{\bar{p}}{\bar{p}-1}}}\|\theta-\tilde{\theta}\|_X\right).$$

We can chose $L(M)=C_{W\to L^{\bar{p}}}\left(C^2_{V\to W^{1,\frac{2\bar{p}}{\bar{p}-1}}}(\|u\|_V+\|\tilde{u}\|_V)+C_{X\to L^{\frac{\bar{p}}{\bar{p}-1}}}\right)+$ 1, under the conditions

$$V = H^2(\Omega)\cap H^1_0(\Omega) \hookrightarrow W^{1,\frac{2\bar{p}}{\bar{p}-1}}(\Omega) \text{ i.e., } 1-\frac{d}{2}\geq -\frac{d(\bar{p}-1)}{2\bar{p}}$$

$$X = L^p(\Omega)\hookrightarrow L^{\frac{\bar{p}}{\bar{p}-1}}(\Omega) \text{ i.e., } p\geq \frac{\bar{p}}{\bar{p}-1}.$$

(91)

(R2) *Well-definedness and boundedness of parameter-to-state map:*
We argue unique existence of the solution to (10)–(12) via the transformed problem (87)–(89) for $U = e^u$.

To begin, by a similar argument to (72) with the elliptic operator $A = -\Delta + \theta, \theta \in L^p(\Omega)$ in place of the parabolic operator, we show that the corresponding elliptic problem admits a unique solution in $H^2(\Omega)\cap H^1_0(\Omega)$ if the index p satisfies (73). Employing next the semigroup theory in [9, Section 7.4.3, Theorem 5] or [37, Chapter 7, Corollary 2.6] with assuming that $U_0 \in D(A) = H^2(\Omega)\cap H^1_0(\Omega)$ implies unique existence of a solution $U \in C^1(0, T; H^2(\Omega))$ to (87)–(89).

Let $U, \hat{U}$ respectively solve (87)–(89) associated with the coefficients $\theta \in X, \hat{\theta} \in L^\infty(\Omega)$ with the same boundary and initial data, then $v = U - \hat{U}$ solves

$$\dot{v}(t) - \Delta v(t) + \hat{\theta}v(t) = (\hat{\theta} - \theta)U(t) \qquad t \in (0, T)$$

$$v(0) = 0.$$

Owing to the regularity from [9, Section 7.1.3, Theorem 5] and estimating similarly to (72), we obtain

$$\|U - \hat{U}\|_{L^\infty(0,T;H^2(\Omega))} \leq C^{\hat{\theta}}\|(\hat{\theta} - \theta)U\|_{H^1(0,T;L^2(\Omega))}$$

$$\leq C\|\hat{\theta} - \theta\|_X\|U\|_{H^1(0,T;H^2(\Omega))}$$

(92)

with positive $\hat{U}$ since $\hat{\theta} \in L^\infty(\Omega)$ and the constant C depending only on θ^0, ρ. Here we assume $\hat{\theta} = \theta^0 \in L^\infty(\Omega)$ and ρ is sufficiently small such that the right hand side is sufficiently small. Then $U \in L^\infty(0, T; H^2(\Omega)) \subseteq L^\infty((0, T)\times\Omega)$ is close to $\hat{U}$ and therefore positive as well. This assertion is valid if $0 < U_0 = e^{u_0} \in H^2(\Omega)\cap H^1_0(\Omega), 0 < U|_{\delta\Omega}$, which is chosen as

$U|_{\delta\Omega} = 1$ in this case (such that $\log(U|_{\delta\Omega}) = 0$) and

$$H^2(\Omega) \hookrightarrow L^{\frac{2p}{p-2}}(\Omega) \text{ i.e.,} \quad p \geq \frac{2\bar{p}}{\bar{p} - 2}$$

$$V = H^2(\Omega) \hookrightarrow L^\infty(\Omega) \text{ i.e.,} \quad d \leq 3.$$

(93)

This leads to unique existence of the solution $u := \log(U)$ to the problem (10)–(12), moreover $0 < \underline{c} \leq U \in C^1(0, T; H^2(\Omega))$ allows $u = \log(U) \in C^1(0, T; H^2(\Omega) \cap H_0^1(\Omega))$.

If $d = 1$, $p \geq 2$, no assumption on smallness of ρ is required since

$$\|U - \hat{U}\|_{L^\infty(0,T;H^1(\Omega))} \leq C^\theta \|(\hat{\theta} - \theta)\hat{U}\|_{L^2(0,T;L^2(\Omega))}$$

$$\leq C\|\hat{\theta} - \theta\|_X \|\hat{U}\|_{L^2(0,T;H^2(\Omega))}$$

(94)

due to the estimates (75)–(77) in Sect. 3.1. Here the constant C depends only on θ^0, ρ as claimed in (77). This and the fact $\overline{L^\infty(\Omega)} = L^p(\Omega)$ allow us to chose $\hat{\theta} \in L^\infty(\Omega)$ being sufficiently close to $\theta \in L^p(\Omega)$ to make the right hand side of (94) arbitrarily small without the need of smallness of ρ.

We have observed that, with the same positive boundary and initial data, the solution $U = U(\theta)$ to (87)–(89) is bounded away from zero for all $\theta \in \mathcal{B}_\rho^X(\theta^0)$. Besides, $S : \theta \mapsto U$ is a bounded operator as proven in (R2) of Sect. 3.1. Consequently, $u = \log(U)$ with $\Delta u = -\frac{|\nabla U|^2}{U^2} + \frac{\Delta U}{U}$ is uniformly bounded in $L^2(0, T; H^2(\Omega) \cap H_0^1(\Omega))$ for all $\theta \in \mathcal{B}_\rho^X(\theta^0)$, thus $S : \theta \mapsto u$ is a bounded operator on $\mathcal{B}_\rho^X(\theta^0)$.

Moreover, we can derive a uniform bound for U in $H^1(0, T; H^2(\Omega))$ with respect to θ. From

$$(\dot{U} - \dot{\hat{U}}) - \Delta(U - \hat{U}) + (\theta - \hat{\theta})(U - \hat{U}) = -\hat{\theta}(U - \hat{U}) - (\theta - \hat{\theta})\hat{U},$$

by taking the time derivative of both sides then test them with $-\Delta(\dot{U} - \dot{\hat{U}})$ we have

$$\frac{1}{2}\frac{d}{dt}\|\nabla(\dot{U} - \dot{\hat{U}})\|^2_{L^2(\Omega)} + \|\Delta(\dot{U} - \dot{\hat{U}})\|^2_{L^2(\Omega)}$$

$$\leq C_{H^2 \hookrightarrow L^\infty}\|\theta - \hat{\theta}\|_{L^2(\Omega)}\|\Delta(\dot{U} - \dot{\hat{U}})\|^2_{L^2(\Omega)}$$

$$+ \|\hat{\theta}\|_{L^\infty(\Omega)}\|\dot{U} - \dot{\hat{U}}\|_{L^2(\Omega)}\|\Delta(\dot{U} - \dot{\hat{U}})\|_{L^2(\Omega)}$$

$$+ C_{H^2 \hookrightarrow L^\infty}\|\theta - \hat{\theta}\|_{L^2(\Omega)}\|\Delta\dot{\hat{U}}\|_{L^2(\Omega)}\|\Delta(\dot{U} - \dot{\hat{U}})\|_{L^2(\Omega)}$$

$$\frac{1}{2}\frac{d}{dt}\|\nabla(\dot{U}-\dot{\hat{U}})\|^2_{L^2(\Omega)} + (1-\rho C_{H^2\hookrightarrow L^\infty}-\epsilon)\|\Delta(\dot{U}-\dot{\hat{U}})\|^2_{L^2(\Omega)}$$

$$\leq \frac{1}{2\epsilon}\left(\|\hat{\theta}\|^2_{L^\infty(\Omega)}\|\dot{U}-\dot{\hat{U}}\|^2_{L^2(\Omega)} + C^2_{H^2\hookrightarrow L^\infty}\rho^2\|\Delta\dot{\hat{U}}\|^2_{L^2(\Omega)}\right),$$

where $\|\Delta\dot{\hat{U}}\|_{L^2(\Omega)}$ is attained by estimating with the same technique for (87)–(89) with the coefficient $\hat{\theta} \in L^\infty(\Omega)$. Since ϵ is arbitrarily small, if ρ is sufficiently small and the following condition holds

$$X = L^p(\Omega) \hookrightarrow L^2(\Omega) \text{ i.e., } p \geq 2, \tag{95}$$

applying Gronwall's inequality then integrating on $[0, T]$ yields

$$\|U - \hat{U}\|_{H^1(0,T;H^2(\Omega))} \leq C\|\hat{\theta} - \theta\|_X\|\hat{U}\|_{H^1(0,T;H^2(\Omega))} \tag{96}$$

for fixed $\hat{U} = S(\hat{\theta}) = S(\theta^0)$. So, $S(\mathcal{B}^X_\rho(\theta^0))$ is bounded in $H^1(0, T; H^2(\Omega))$ and its diameter can be controlled by ρ. In case $d = 1$, smallness of ρ can be omitted if one uses the estimate (94).

(R3) *Continuity of the inverse of the linearized model:*
Now we consider the linearized problem

$$\dot{z}(t) - \Delta z(t) + 2\nabla u(t) \cdot \nabla z(t) = r(t) \qquad t \in (0, T) \tag{97}$$

$$z(0) = 0, \tag{98}$$

whose adjoint problem after transforming $t = T - \tau$ is

$$\dot{p}(t) - \Delta p(t) - 2\nabla \cdot (\nabla u(t)p(t)) = \phi(t) \qquad t \in (0, T) \tag{99}$$

$$p(0) = 0. \tag{100}$$

Since $u \in C^1(0, T; H^2(\Omega) \cap H^1_0(\Omega))$ as proven in (R2), this equation with the coefficients $m := -2\nabla u \in C^1(0, T; H^1(\Omega)), n := -2\Delta u \in C^1(0, T; L^2(\Omega))$ is feasible to attain the estimate (R3) by the contraction argument.

Indeed, let us take p solving (99)–(100), then

$$\dot{p} - \Delta p + \hat{m} \cdot \nabla p + \hat{n}p = \phi + (\hat{m} - m) \cdot \nabla p + (\hat{n} - n)p$$

$$p = \left(\frac{d}{dt} - \Delta + \hat{m} \cdot \nabla + \hat{n}\right)^{-1}[\phi + (\hat{m} - m) \cdot \nabla p + (\hat{n} - n)p]$$

$$=: \mathbb{T}p$$

with some $\hat{m} \in L^{\infty}((0, T) \times \Omega)$ and some $\hat{n} \in L^{\infty}((0, T) \times \Omega)$ approximating m and n. Then for $d \leq 3$, $\mathbb{T} : L^2(0, T; H^2(\Omega) \cap H_0^1(\Omega)) \rightarrow L^2(0, T; H^2(\Omega) \cap H_0^1(\Omega))$ is a contraction

$$\|\mathbb{T}(p - q)\|_{L^2(0,T;H^2 \cap H_0^1)}$$

$$\leq \left\| \left(\frac{d}{dt} - \Delta + \hat{m} \cdot \nabla + \hat{n} \right)^{-1} \right\|_{L^2(0,T;L^2(\Omega)) \rightarrow L^2(0,T;H^2 \cap H_0^1)}$$

$$\cdot \left(\|(\hat{m} - m) \cdot \nabla(p - q)\|_{L^2(0,T;L^2(\Omega))} + \|(\hat{n} - n)(p - q)\|_{L^2(0,T;L^2(\Omega))} \right)$$

$$\leq C^{\hat{\theta}} \left(\|\hat{m} - m\|_{L^{\infty}(0,T;H^1(\Omega))} \|\nabla(p - q)\|_{L^2(0,T;H^1(\Omega))} \right.$$

$$\left. + \|\hat{n} - n\|_{L^{\infty}(0,T;L^2(\Omega))} \|p - q\|_{L^2(0,T;L^{\infty}(\Omega))} \right)$$

$$\leq C\epsilon \|p - q\|_{L^2(0,T;H^2 \cap H_0^1)}, \tag{101}$$

where $H_0^1(\Omega) \hookrightarrow L^6(\Omega)$, $H^2(\Omega) \cap H_0^1(\Omega) \hookrightarrow L^{\infty}(\Omega)$ for $d \leq 3$. Above, we apply from [9, Section 7.1.3 , Theorem 5] the continuity of $\left(\frac{d}{dt} - \Delta + \hat{m} \cdot \nabla + \hat{n} \right)^{-1}$ with noting that, although the theorem is stated for time-independent coefficients, the proof reveals it is still applicable for $\hat{m} = \hat{m}(t, x), \hat{n} = \hat{n}(t, x)$ being bounded in time and space.

The above constant $C^{\hat{\theta}}$, which depends on $\hat{m} \in \nabla \cdot S(\mathcal{B}_\rho^X(\theta^0)) \cap L^{\infty}(0, T; L^{\infty}(\Omega))$, $\hat{n} \in \Delta S(\mathcal{B}_\rho^X(\theta^0)) \cap L^{\infty}(0, T; L^{\infty}(\Omega))$ can be bounded by some constant C depending only on $S(\theta^0)$ and the diameter of $S(\mathcal{B}_\rho^X(\theta^0))$ similarly to Sects. 3.1 and 3.2 if choosing $\hat{\theta} = \theta^0$. In order to make $C\epsilon$ less than one, we require $\|\hat{m} - m\|_{L^{\infty}(0,T;H^1(\Omega))}$ and $\|\hat{n} - n\|_{L^{\infty}(0,T;L^2(\Omega))}$ to be sufficiently small. Those conditions turn out to be uniform boundedness of $\|\hat{U} - U\|_{L^{\infty}(0,T;H^2(\Omega))}$ (or the diameter of $S(\mathcal{B}_\rho^X(\theta^0))$), which can be seen as smallness of ρ as in (96) since $H^1(0, T) \hookrightarrow L^{\infty}(0, T)$. From that, existence of the dual state $p \in L^2(0, T; H^2(\Omega) \cap H_0^1(\Omega))$ for given $\phi \in L^2(0, T; L^2(\Omega))$ is shown.

Then (R3-dual) follows without adding further constraints on p

$$\|p\|_{L^2(0,T;H^2(\Omega))}$$

$$\leq C(\|(\hat{m} - m) \cdot \nabla p\|_{L^2(0,T;L^2(\Omega))} + \|(\hat{n} - n)p\|_{L^2(0,T;L^2(\Omega))} + \|\phi\|_{L^2(0,T;L^2(\Omega))})$$

$$\leq C\|\phi\|_{L^2(0,T;L^2(\Omega))}$$

with constant C depending only on some fixed $\hat{m}, \hat{n}$ and the assumption on smallness of ρ. Here with the L^2-norm on the right hand side, the maximum q is limited by $\bar{q} = 2$.

Observing that the problem (99)–(100) has the form of the a-problem written in (83), with $\underline{a} = 1$, $\nabla a = -2\nabla u(t) \in L^6(\Omega)$ and the additional term in the last line of the right hand side, namely,

$$\frac{1}{\underline{a}}\|np\|^2_{L^2(\Omega)} = \|\Delta up\|^2_{L^2(\Omega)} \leq \|\Delta u\|^2_{L^2(\Omega)}\|p\|^2_{L^\infty(\Omega)}$$

$$\leq C^2_{H^1_0 \to L^\infty}\|\Delta u\|^2_{L^2(\Omega)}\|\nabla p\|^2_{L^2(\Omega)} \tag{102}$$

if the dimension $d = 1$.

The solution $u = S(\theta)$ also lies in some ball in $C^1(0, T; H^2(\Omega) \cap H^1_0(\Omega))$ for all $\theta \in \mathcal{B}^X_\rho(\theta^0)$, as in (R2) we have shown boundedness of the operator S.

It allows us to evaluate analogously to (83)–(84) with taking into account the additional term (102) to eventually get

$$\|\Delta p\|_{L^2(0,T;L^2(\Omega))} \leq C\|\phi\|^2_{L^2(0,T;L^2(\Omega))}$$

with the constant C depending only on θ^0, ρ. Hence, if $d = 1$, ρ is not required to be small.

(R4) *All-at-once tangential cone condition:*
According to (36) and (A.124) with $s = t = 2$, $m = n = 2$, $\hat{\gamma} = 0$, $\rho = 2$ this follows if

$$2 - \frac{d}{2} \geq 1 - \frac{d}{q^*} + \frac{d}{R} \text{ and}$$

$$1 \leq \frac{R}{q^*} \text{ and } q \leq \hat{q} \text{ and } 2 - \frac{d}{2} \geq \max\left\{-\frac{d}{\hat{q}}, 1 - \frac{d}{R}\right\},$$

where the latter conditions come from the requirements $V = H^2(\Omega) \cap H^1_0(\Omega) \hookrightarrow W^{1,R}(\Omega)$.

Corollary 3 *Assume $u_0 \in V$ and*

$$\mathcal{D}(F) = X = L^p(\Omega), \qquad Y = L^q(\Omega), \qquad V = W = H^2(\Omega) \cap H^1_0(\Omega)$$

$$p \geq 2, \quad q \in \left[\underline{q}, 2\right], \quad d \leq 3 \tag{103}$$

with $\underline{q} = \min\limits_q \left\{2 - \frac{d}{2} \geq 1 - d + \frac{d}{q} + \frac{d}{p} \wedge q \geq 1 + \frac{1}{p-1}\right\}$.

Then F defined by $F(\theta) = u$ solving (10)–(12) satisfies the tangential cone condition (52) with a uniformly bounded operator $F'(\theta)$ defined by (59).

Remark 9 To achieve a Hilbert space setting for X and Y, one can choose $p = q = 2$ if $d \leq 3$, see also [34].

3.4 An Inverse Source Problem with a Cubic Zero Order Nonlinearity

We investigate this problem in the function spaces

$$X = L^p(\Omega), \qquad\qquad Y = L^q(\Omega), \qquad\qquad V = W = H_0^1(\Omega).$$

In the following we examine the conditions required for deriving the tangential cone condition and boundedness of the derivative of the forward operator.

(R1) *Local Lipschitz continuity of* f:

$$\|\tilde{u}^3 - u^3 + \tilde{\theta} - \theta\|_{W^*} = \sup_{\|w\|_W \leq 1} \int_\Omega (\tilde{u} - u)(\tilde{u}^2 + \tilde{u}u + u^2)w + (\tilde{\theta} - \theta)w\,dx$$

$$\leq C_{W \to L^{\bar{p}}} \left(\|(\tilde{u} - u)(\tilde{u}^2 + \tilde{u}u + u^2)\|_{L^{\frac{\bar{p}}{\bar{p}-1}}} + \|\tilde{\theta} - \theta\|_{L^{\frac{\bar{p}}{\bar{p}-1}}} \right)$$

$$\leq C_{W \to L^{\bar{p}}} \left(2\|\tilde{u} - u\|_{L^{\bar{p}}} (\|\tilde{u}\|^2_{L^{\frac{2\bar{p}}{\bar{p}-2}}} + \|u\|^2_{L^{\frac{2\bar{p}}{\bar{p}-2}}}) + \|\tilde{\theta} - \theta\|_{L^{\frac{\bar{p}}{\bar{p}-1}}} \right)$$

$$\leq C_{W \to L^{\bar{p}}} \left(2C_{V \to L^{\bar{p}}} C^2_{V \to L^{\frac{2\bar{p}}{\bar{p}-2}}} \|\tilde{u} - u\|_V (\|\tilde{u}\|^2_V + \|u\|^2_V) + \|\tilde{\theta} - \theta\|_X C_{X \to L^{\frac{\bar{p}}{\bar{p}-1}}} \right).$$

We chose $L(M) = C_{W \to L^{\bar{p}}} \left(2C_{V \to L^{\bar{p}}} C^2_{V \to L^{\frac{2\bar{p}}{\bar{p}-2}}} (\|\tilde{u}\|^2_V + \|u\|^2_V) + C_{X \to L^{\frac{\bar{p}}{\bar{p}-1}}} \right)$
$+1$, subject to the conditions

$$V = W = H_0^1(\Omega) \hookrightarrow L^{\bar{p}}(\Omega) \text{ i.e., } 1 - \frac{d}{2} \geq -\frac{d}{\bar{p}}$$

$$V = H_0^1(\Omega) \hookrightarrow L^{\frac{2\bar{p}}{\bar{p}-2}}(\Omega) \text{ i.e., } d \leq 4$$

$$X = L^p(\Omega) \hookrightarrow L^{\frac{\bar{p}}{\bar{p}-1}}(\Omega) \text{ i.e., } p \geq \frac{\bar{p}}{\bar{p}-1}.$$

$$(104)$$

(R2) *Well-definedness and boundedness of the parameter-to-state map:*

Verifying the conditions (S1)–(S3) with the Gelfand triple $H_0^1(\Omega) \hookrightarrow L^2(\Omega) \hookrightarrow H^{-1}(\Omega)$ shows that the problem (13)–(15) admits a unique solution in the space $W(0, T)$. Subsequently, [38, Theorem 8.16] strengthens the solution to belong to $L^\infty(0, T; V)$. To validate this regularity result, the following additional assumptions are made

$$X = L^p(\Omega) \hookrightarrow L^2(\Omega) \text{ i.e., } p \geq 2, \tag{105}$$

the initial data $u_0 \in V$ and the known source term $\varphi \in L^2(0, T; L^2(\Omega))$.

From [34, Proposition 4.2, Section 6.1], we have

$$\|S(\theta)\|_{L^\infty(0,T;V)} \le N\left(\|\theta + \varphi\|_{L^2(0,T;L^2(\Omega))} + \sqrt{\int_\Omega \frac{1}{2}|\nabla u_0|^2 + \frac{1}{4}u_0^4 dx}\right)$$

$$\le N\left(\sqrt{T}(\|\theta_0\|_{L^2(\Omega)} + \rho) + \|\varphi\|_{L^2(0,T;L^2(\Omega))} + \sqrt{\int_\Omega \frac{1}{2}|\nabla u_0|^2 + \frac{1}{4}u_0^4 dx}\right)$$

for some N depending only on $c_0^\theta = c_0 = \frac{1}{2}$. This thus implies uniform boundedness of S on $\mathcal{B}_\rho^X(\theta_0)$.

(R3) *Continuous dependence on data of the solution to the linearized problem with zero initial data:*

For this purpose, semi-coercivity of the linearized forward operator is obvious

$$\langle -f_u'(t,\theta,v), v\rangle_{V^*,V} = \int_\Omega (-\Delta v + 3u^2 v)v dx$$

$$\ge \|\nabla v\|_{L^2(\Omega)}^2 = \|v\|_V^2.$$

(R4) *All-at-once tangential cone condition:*

According to (36) and (A.123), with $s = t = 1$, $m = n = 2$, $\gamma = \kappa = 1$, $r = \hat{q} = \bar{p}$ this follows if

$$2 \le \frac{\bar{p}}{q^*} \text{ and } 1 - \frac{d}{2} \ge -\frac{d}{q^*} + \frac{2d}{\bar{p}} \text{ and } q \le \bar{p} \text{ and } 1 - \frac{d}{2} \ge -\frac{d}{\bar{p}},$$

where the latter condition comes from the requirement $V = H_0^1(\Omega) \hookrightarrow L^{\bar{p}}(\Omega)$.

Corollary 4 *Assume $u_0 \in H_0^1(\Omega)$, $\varphi \in L^2(0, T; L^2(\Omega))$, and*

$$\mathcal{D}(F) = X = L^p(\Omega), \qquad Y = L^q(\Omega), \qquad V = W = H_0^1(\Omega)$$

$$p \ge 2, \quad q \in \left[\underline{q}, \bar{q}\right], \quad d \le 4,$$

(106)

where $\underline{q} = \min_q \left\{1 - \frac{d}{2} \ge -d + \frac{d}{q} + \frac{2d}{\bar{p}} \wedge q \ge 1 + \frac{2}{\bar{p}-2}\right\}$ with

$$d = 1 \text{ and } \bar{q} = \infty, \qquad d = 2 \text{ and } \bar{q} < \infty, \qquad d \ge 3 \text{ and } \bar{q} = \frac{2d}{d-2}.$$

(107)

Then F defined by $F(\theta) = u$ solving (13)–(15) satisfies the tangential cone condition (52) with a uniformly bounded operator $F'(\theta)$ defined by (59).

Remark 10 Here X and Y can be chosen as Hilbert spaces with $p = q = 2$ and $d \le 3$.

Appendix

Notation

- For $a, b \in \mathbb{R}$, the notation $a \succeq b$ means: $a \geq b$ with strict inequality if $b = 0$.
- For normed spaces A, B, the notation $A \hookrightarrow B$ means: A is continuously embedded in B.
- For a normed space A, an element $a \in A$ and $\rho > 0$, we denote by $\mathcal{B}_\rho^A(a)$ the closed ball of radius ρ around a in A.
- For vectors $\mathbf{a}, \mathbf{b} \in \mathbb{R}^n$, $\mathbf{a} \cdot \mathbf{b}$ denotes the Euclidean inner product. Likewise, $\nabla \cdot \mathbf{v}$ denotes the divergence of the vector field $\mathbf{v}$.
- C denotes a generic constant that may take different values whenever it appears.
- For $p \in [1, \infty]$, we denote by $p^* = \frac{p}{p-1}$ the dual index.
- The norm of some embedding $H^s(\Omega) \to L^p(\Omega)$ will be denoted by $C_{H^s \to L^p}^\Omega$.[4]

Proof of Proposition 1

On (41) for some $p \in [1, \infty]$, we can estimate by applying Hölder's inequality, once with exponent p and once with exponent $\frac{q}{p^*}$ (where $p^* = \frac{p}{p-1}$ is the dual index)

$$\|(B\hat{c})(t)\hat{v}\|_{W^*} = \sup_{w \in W, \, \|w\|_W \leq 1} \int_\Omega \hat{c}\,\hat{v}\,w\,dx \leq \|\hat{c}\|_{L^p}\|\hat{v}\|_{L^q} \sup_{w \in W, \, \|w\|_W \leq 1} \|w\|_{L^{\frac{p^*q}{q-p^*}}},$$

where we need to impose $q \geq p^*$ and in case of equality formally set $\frac{p^*q}{q-p^*} = \infty$.
In order to guarantee continuity of the embedding $W \hookrightarrow L^{\frac{p^*q}{q-p^*}}(\Omega)$ as needed here, we therefore, together with (36), require the conditions

$$s - \frac{d}{m} \succeq -\frac{d}{\hat{q}} \quad \text{and} \quad \hat{q} \geq q \geq p^* \quad \text{and} \quad t - \frac{d}{n} \succeq -\frac{d(q - p^*)}{p^*q}. \tag{A.108}$$

Proof of Proposition 2

With X as in (43), in order to guarantee the required boundedness of the embeddings

$$X \hookrightarrow L^r(\Omega), \quad W \hookrightarrow W^{1, \frac{p^*q}{q-p^*}}(\Omega), \quad W \hookrightarrow W^{2, \frac{r^*q}{q-r^*}}(\Omega),$$

for some $r \in [1, \infty]$ such that $r^* \leq q$

---

[4]Note that the assumed $C^{1,1}$ smoothness of Ω suffices for all embeddings used here, see, e.g., [30].

we impose, additionally to (36), the conditions

$$(a)\ \hat{q} \geq q \geq \max\{p^*, r^*\} \text{ and} \qquad (b)\ t - 1 - \frac{d}{n} \geq -\frac{d(q - p^*)}{p^* q} \text{ and}$$

$$(c)\ t - 2 - \frac{d}{n} \geq -\frac{d(q - r^*)}{r^* q} \text{ and} \qquad (d)\ 1 - \frac{d}{p} \geq -\frac{d}{r}$$

for some $r \in [1, \infty]$. To eliminate r, observe that the requirement (c), i.e., $t - 2 - \frac{d}{n} \geq -\frac{d}{r^*} + \frac{d}{q}$ gets weakest when r^* is chosen minimal, which, subject to requirement (d) is

$$r \begin{cases} = \infty \text{ if } p > d \\ < \infty \text{ if } p = d \\ = \frac{dp}{d-p} \text{ if } p < d \end{cases} \text{, i.e., } \quad r^* \begin{cases} = 1 \text{ if } p > d \\ > 1 \text{ if } p = d \\ = \frac{dp}{dp-d+p} \text{ if } p < d \end{cases} . \qquad \text{(A.109)}$$

Inserting this into (c) and taking into account (36), we end up with the following requirements on s, t, p, q, m, n (using the fact that $q \geq p^*$ implies $q \geq \frac{dp}{dp-d+p}$):

$$s - \frac{d}{m} \geq -\frac{d}{\hat{q}} \text{ and } \hat{q} \geq q \geq p^* \text{ and}$$

$$t - 1 - \frac{d}{n} \geq -\frac{d(q - p^*)}{p^* q} \text{ and } t - 2 - \frac{d}{n} \begin{cases} \geq -d + \frac{d}{q} \text{ if } p > d \\ > -d + \frac{d}{q} \text{ and } q > 1 \text{ if } p = d \\ \geq -\frac{dp-d+p}{p} + \frac{d}{q} \text{ if } p < d . \end{cases}$$
$$\text{(A.110)}$$

Proof of Proposition 3

Here we have

$$\Big(f(\theta, u) - f(\tilde{\theta}, \tilde{u}) - f_u'(\theta, u)(u - \tilde{u}) - f_\theta'(\theta, u)(\theta - \tilde{\theta})\Big)(t)$$

$$= \int_0^1 \Big(\Phi'(u(t)) + \sigma(\tilde{u}(t) - u(t)) - \Phi'(u(t))\Big) d\sigma\,(\tilde{u}(t) - u(t))$$

$$+ \int_0^1 \Big(\Psi'(\nabla u(t) + \sigma(\nabla \tilde{u}(t) - \nabla u(t)) - \Psi'(\nabla u(t))\Big) d\sigma\,\nabla(\tilde{u}(t) - u(t)).$$

This shows that the only condition which has to be taken into account when choosing the space X is that $B(t) \in \mathcal{L}(X, W^*)$. Again we assume $C(t)$ to be the embedding operator $V \hookrightarrow Y$.

As opposed to Sect. 2.1, where we could do the estimates pointwise in time, we will now also have to use Hölder estimates with respect to time. To this end, we dispose over the following continuous embeddings

$$\mathcal{U} \hookrightarrow L^2(0, T; W^{s,m}(\Omega))$$

$$\mathcal{U} \hookrightarrow L^\infty(0, T; H^{\tilde{s}}(\Omega)) \text{ provided } W^{s-\tilde{s},m}(\Omega) \hookrightarrow W^{t+\tilde{s},n}(\Omega),$$

where the first holds just by definition of $\mathcal{U}$ and the second follows from [38, Lemma 7.3][5] with $\widetilde{W} = W^{t+\tilde{s},n}(\Omega)$, using the fact that

$$u \in L^2(0, T; W^{s,m}(\Omega)) \cap H^1(0, T; (W^{t,n}(\Omega))^*)$$

$$\Leftrightarrow D^{\tilde{s}}u \in L^2(0, T; W^{s-\tilde{s},m}(\Omega)) \cap H^1(0, T; (W^{t+\tilde{s},n}(\Omega))^*),$$

where $D^{\tilde{s}}v = \sum_{|\alpha| \le \tilde{s}} D^\alpha v$.

We first consider the case of an affinely linear (or just vanishing) function Ψ, which still comprises, e.g., models with linear drift and diffusion, so that $C_{\Psi''}$ can be set to zero. We can then estimate

$$\|f(\theta, u) - f(\tilde{\theta}, \tilde{u}) - f'_u(\theta, u)(u - \tilde{u}) - f'_\theta(\theta, u)(\theta - \tilde{\theta})\|_{L^2(0,T;W^*)}$$

$$\le C_{\Phi''} \left(\int_0^T \left(\sup_{w \in W, \|w\|_W \le 1} \int_\Omega (1 + |u(t)|^\gamma + |\tilde{u}(t)|^\gamma) |\tilde{u}(t) - u(t)|^{1+\kappa} w \, dx \right)^2 dt \right)^{1/2},$$

where, using Hölder's inequality three times ($P = q$, $P = \frac{r}{q^*(\gamma+\kappa)}$, $P = \frac{\gamma+\kappa}{\gamma}$) and continuity of the embedding $H^{\tilde{s}}(\Omega) \hookrightarrow L^r(\Omega)$ provided $\tilde{s} - \frac{d}{2} \ge -\frac{d}{r}$

$$\left(\int_0^T \left(\sup_{w \in W, \|w\|_W \le 1} \int_\Omega |u(t)|^\gamma |\tilde{u}(t) - u(t)|^{1+\kappa} w \, dx \right)^2 dt \right)^{1/2}$$

$$\le \|\tilde{u} - u\|_{L^2(0,T;L^q(\Omega))} \sup_{w \in W, \|w\|_W \le 1} \left\| |u|^\gamma |\tilde{u} - u|^\kappa w \right\|_{L^\infty(0,T;L^{q^*}(\Omega))}$$

$$\le \|\tilde{u} - u\|_y \left\| \left(|u|^\gamma |\tilde{u} - u|^\kappa \right)^{\frac{1}{\gamma+\kappa}} \right\|^{\gamma+\kappa}_{L^\infty(0,T;L^r(\Omega))} \sup_{w \in W, \|w\|_W \le 1} \|w\|_{L^{\frac{rq^*}{r-q^*(\gamma+\kappa)}}(\Omega)}$$

$$\le \|\tilde{u} - u\|_y \|u\|^\gamma_{L^\infty(0,T;L^r(\Omega))} \|\tilde{u} - u\|^\kappa_{L^\infty(0,T;L^r(\Omega))} \sup_{w \in W, \|w\|_W \le 1} \|w\|_{L^{\frac{rq^*}{r-q^*(\gamma+\kappa)}}(\Omega)}$$

$$\le (C^\Omega_{H^{\tilde{s}} \to L^r})^{\gamma+\kappa} \|u\|^\gamma_{L^\infty(0,T;H^{\tilde{s}}(\Omega))} \|\tilde{u} - u\|^\kappa_{L^\infty(0,T;H^{\tilde{s}}(\Omega))}$$

$$\|\tilde{u} - u\|_y \sup_{w \in W, \|w\|_W \le 1} \|w\|_{L^{\frac{rq^*}{r-q^*(\gamma+\kappa)}}(\Omega)}$$

$$\text{(A.111)}$$

---

[5] $L^2(0, T; \widetilde{W}) \cap H^1(0, T; \widetilde{W}^*) \hookrightarrow L^\infty(0, T; L^2(\Omega))$.

(and likewise for the term containing $|\tilde{u}(t)|^\gamma$) for some $r \in [1, \infty]$ with $\frac{r}{q^*} \geq \gamma + \kappa$. In order to get finiteness of the $L^\infty(0, T; H^{\tilde{s}}(\Omega))$ norms appearing here by means of [38, Lemma 7.3], we assume the embedding $W^{s-\tilde{s},m}(\Omega) \hookrightarrow W^{t+\tilde{s},n}(\Omega)$ to be continuous, which leads to the condition

$$s - \tilde{s} - \frac{d}{m} \geq t + \tilde{s} - \frac{d}{n} \text{ and } s - \tilde{s} \geq t + \tilde{s}.$$

Moreover, in order to guarantee continuity of the embedding $W \hookrightarrow L^{\frac{rq^*}{r-q^*(\gamma+\kappa)}}(\Omega)$ and for the above Hölder estimate to make sense we impose

$$\gamma + \kappa \leq \frac{r}{q^*} \text{ and } t - \frac{d}{n} \geq -\frac{d(r - q^*(\gamma + \kappa))}{rq^*}$$

for some $r \in [1, \infty]$. Summarizing, we have the following conditions

$$\tilde{s} - \frac{d}{2} \geq -\frac{d}{r} \text{ and } s - \tilde{s} - \frac{d}{m} \geq t + \tilde{s} - \frac{d}{n} \text{ and } s - \tilde{s} \geq t + \tilde{s} \text{ and}$$

$$\gamma + \kappa \leq \frac{r}{q^*} \text{ and } t - \frac{d}{n} \geq -\frac{d(r - q^*(\gamma + \kappa))}{rq^*} = -\frac{d}{q^*} + \frac{d(\gamma + \kappa)}{r}, \tag{A.112}$$

which imply

$$s \geq \frac{d}{m} + d - \frac{d}{q^*} + d\frac{\gamma + \kappa - 2}{r}.$$

This lower bound on s gets weakest for maximal r, if $\gamma + \kappa > 2$ and for minimal r if $\gamma + \kappa < 2$. We therefore make the following case distinction.

If $\gamma + \kappa > 2$ or $\gamma + \kappa = 2$ and $q = 1$ we set $r = \infty$, which leads to $\tilde{s} > \frac{d}{2}$, hence, according to (A.112), we can choose

case $\gamma + \kappa > 2$ or ($\gamma + \kappa = 2$ and $q = 1$):

$$t > \frac{d}{n} - \frac{d}{q^*}, \quad q \leq \hat{q},$$

$$s > \max\left\{t + d + \max\left\{0, \frac{d}{m} - \frac{d}{n}\right\}, \frac{d}{m} - \frac{d}{\hat{q}}\right\}. \tag{A.113}$$

If $\gamma + \kappa < 2$ or $\gamma + \kappa = 2$ and $q > 1$ we set $r = \max\{1, q^*(\gamma + \kappa)\} < \infty$, $\tilde{s} := \max\{0, \frac{d}{2} - \frac{d}{r}\}$ and, according to (A.112), can therefore choose case $\gamma + \kappa < 2$ or ($\gamma + \kappa = 2$ and $q > 1$):

$$t > \frac{d}{n} + \min\left\{0, -\frac{d}{q^*} + d(\gamma + \kappa)\right\}, \quad q \le \hat{q}, \tag{A.114}$$

$$s > \max\left\{t + \max\left\{0, d - \frac{2d}{\max\{1, q^*(\gamma + \kappa)\}}\right\}, \frac{d}{m} - \frac{d}{\hat{q}}\right\}.$$

Now we consider the situation of nonvanishing gradient nonlinearities $C_{\Psi''} > 0$ where we additionally need to estimate terms of the form

$$\left(\int_0^T \left(\sup_{w \in W,\, \|w\|_W \le 1} \int_\Omega |\nabla u(t)|^{\hat{\gamma}} |\nabla \tilde{u}(t) - \nabla u(t)|^{1+\hat{\kappa}} w\, dx\right)^2 dt\right)^{1/2},$$

which, in order to end up with an estimate in terms of $\|\tilde{u} - u\|_{L^2(0,T;L^q(\Omega))}$ requires us to move the gradient by means of integration by parts. Assuming for simplicity that $\hat{\kappa} = 1$ we get

$$\left(\int_0^T \left(\sup_{w \in W,\, \|w\|_W \le 1} \int_\Omega |\nabla u(t)|^{\hat{\gamma}} |\nabla \tilde{u}(t) - \nabla u(t)|^2 w\, dx\right)^2 dt\right)^{1/2}$$

$$= \left(\int_0^T \left(\sup_{w \in W,\, \|w\|_W \le 1} \int_\Omega (\tilde{u}(t) - u(t))\, g^w(t)\, dx\right)^2 dt\right)^{1/2}$$

$$\le \|\tilde{u} - u\|_{L^2(0,T;L^q(\Omega))} \sup_{w \in W,\, \|w\|_W \le 1} \|g^w\|_{L^\infty(0,T;L^{q^*}(\Omega))},$$

where

$$g^w(t) = \nabla \cdot \left(|\nabla u(t)|^{\hat{\gamma}} \nabla(\tilde{u}(t) - u(t))\, w\right)$$

$$= \hat{\gamma} |\nabla u(t)|^{\hat{\gamma}-2} (\nabla^2 u(t) \nabla u(t)) \cdot \nabla(\tilde{u}(t) - u(t))\, w$$

$$+ |\nabla u(t)|^{\hat{\gamma}} \Delta(\tilde{u}(t) - u(t))\, w + |\nabla u(t)|^{\hat{\gamma}} \nabla(\tilde{u}(t) - u(t)) \cdot \nabla w$$

$$=: g_1(t) + g_2(t) + g_3(t),$$

where ∇^2 denotes the Hessian. For the last term we proceed analogously to above (basically replacing u by ∇u and w by ∇w) to obtain

$$\|g_3\|_{L^\infty(0,T;L^{q^*}(\Omega))} = \||\nabla u(t)|^{\hat{\gamma}} \nabla(\tilde{u}(t) - u(t)) \cdot \nabla w\|_{L^\infty(0,T;L^{q^*}(\Omega))}$$

$$\le \|\nabla u\|_{L^\infty(0,T;L^R(\Omega))}^{\hat{\gamma}} \|\nabla(\tilde{u} - u)\|_{L^\infty(0,T;L^R(\Omega))} \sup_{w \in W,\, \|w\|_W \le 1} \|\nabla w\|_{L^{\frac{Rq^*}{R-q^*(\hat{\gamma}+1)}}(\Omega)} \tag{A.115}$$

and use [38, Lemma 7.3] with $\nabla u \in L^2(0, T; W^{s-1,m}(\Omega)) \cap H^1(0, T; (W^{t+1,n}(\Omega))^*)$, which under the conditions

$$t - \frac{d}{n} \geq 1 - \frac{d(R - q^*(\hat{\gamma} + 1))}{Rq^*},$$

$$s - 1 - \tilde{s} - \frac{d}{m} \geq t + 1 + \tilde{s} - \frac{d}{n}, \quad s - 1 - \tilde{s} \geq t + 1 + \tilde{s}, \quad \tilde{s} - \frac{d}{2} \geq -\frac{d}{R}$$

(A.116)

yields $\nabla u \in L^\infty(0, T; H^{\tilde{s}}(\Omega)) \subseteq L^\infty(0, T; L^R(\Omega))$ and $W \hookrightarrow W^{1, \frac{Rq^*}{R-q^*(\hat{\gamma}+1)}}(\Omega)$.
The other two terms can be bounded by

$$|g_1(t) + g_2(t)| \leq \left(\hat{\gamma}|\nabla^2 u(t)| |\nabla u(t)|^{\hat{\gamma}-1} |\nabla(\tilde{u}(t) - u(t))|\right.$$
$$\left. + |\nabla^2(\tilde{u}(t) - u(t))| |\nabla u(t)|^{\hat{\gamma}}\right)|w|$$

(note that here $|\cdot|$ denotes the Frobenius norm of a matrix) so that it suffices to find an estimate on expressions of the form

$$\||\nabla^2 z| |\nabla v|^{\hat{\gamma}-1} |\nabla y| |w|\|_{L^\infty(0,T;L^2(\Omega))}$$

for $z, v, y \in \mathcal{U}$, $w \in W$. To this end, we will again employ [38, Lemma 7.3], making use of the fact that for any $\varrho, R \in [1, \infty)$, due to Hölder's inequality with $P = \frac{\varrho}{2}$ and with $P = \frac{R(\varrho-2)}{2\varrho\hat{\gamma}}$, the estimate

$$\||\nabla^2 z| |\nabla v|^{\hat{\gamma}-1} |\nabla y| |w|\|_{L^2(\Omega)}$$

$$\leq \||\nabla^2 z|\|_{L^\varrho(\Omega)} \|\left(|\nabla v|^{\hat{\gamma}-1}|\nabla y|\right)^{\frac{1}{\hat{\gamma}}}\|_{L^R(\Omega)}^{\hat{\gamma}} \|w\|_{L^{\frac{2R\varrho}{R(\varrho-2)-2\varrho\hat{\gamma}}}(\Omega)}$$

$$\leq C^\Omega_{H^{\tilde{s}} \to L^\varrho} (C^\Omega_{H^{\tilde{s}} \to L^\varrho})^{\hat{\gamma}} \||\nabla^2 z|\|_{H^{\tilde{s}}(\Omega)} \|\left(|\nabla v|^{\hat{\gamma}-1}|\nabla y|\right)^{\frac{1}{\hat{\gamma}}}\|_{H^{\tilde{s}}(\Omega)}^{\hat{\gamma}} \|w\|_{L^{\frac{2R\varrho}{R(\varrho-2)-2\varrho\hat{\gamma}}}(\Omega)}$$

(A.117)

holds. To make sense of these Hölder estimates and to guarantee continuity of the embedding $W \hookrightarrow L^{\frac{2R\varrho}{R(\varrho-2)-2\varrho\hat{\gamma}}}(\Omega)$ we impose

$$\varrho \geq 2 \text{ and } R \geq \frac{2\varrho\hat{\gamma}}{\varrho - 2} \text{ and } t - \frac{d}{n} \geq -\frac{d(R(\varrho - 2) - 2\varrho\hat{\gamma})}{2R\varrho} = -\frac{d}{2} + \frac{d}{\varrho} + \frac{d\hat{\gamma}}{R}$$

(A.118)

Taking into account the fact that here $\nabla^2 z$ contains second and $\left(|\nabla v|^{\hat{\gamma}-1}|\nabla y|\right)^{\frac{1}{\hat{\gamma}}}$ first derivatives of elements of $\mathcal{U}$, we therefore aim at continuity of the embeddings

$$L^2(0, T; W^{s-2,m}(\Omega)) \cap H^1(0, T; W^{t+2,n}(\Omega)) \hookrightarrow L^\infty(0, T; H^{\hat{s}}(\Omega)) \hookrightarrow L^\infty(0, T; L^\varrho(\Omega))$$

$$L^2(0, T; W^{s-1,m}(\Omega)) \cap H^1(0, T; W^{t+1,n}(\Omega)) \hookrightarrow L^\infty(0, T; H^{\check{s}}(\Omega)) \hookrightarrow L^\infty(0, T; L^R(\Omega)),$$

which can be achieved by means of [38, Lemma 7.3] under the conditions

$$s - 2 - \hat{s} - \frac{d}{m} \succeq t + 2 + \hat{s} - \frac{d}{n} \text{ and } s - 2 - \hat{s} \geq t + 2 + \hat{s} \text{ and } \hat{s} - \frac{d}{2} \succeq -\frac{d}{\varrho}$$

$$s - 1 - \check{s} - \frac{d}{m} \succeq t + 1 + \check{s} - \frac{d}{n} \text{ and } s - 1 - \check{s} \geq t + 1 + \check{s} \text{ and } \check{s} - \frac{d}{2} \succeq -\frac{d}{R}.$$

$$\text{(A.119)}$$

For instance, we may set $\varrho = 2$, $R = \infty$ to obtain, inserting into (A.116), (A.118) and (A.119), that $\hat{s} \geq 0$, $\check{s} > \frac{d}{2}$ hence

$$t > \frac{d}{n}, \ t - \frac{d}{n} \geq 1 - \frac{d}{q^*}, \ s \geq t + 2 + \max\{2, d\}$$

$$+ \frac{d}{m} - \frac{d}{n}, \ s \geq t + 2 + \max\{2, d\}, \qquad\qquad \text{(A.120)}$$

$$s - \frac{d}{m} \succeq -\frac{d}{\hat{q}}, \ q \leq \hat{q}.$$

In order to avoid the use of too high values of s and t, we can alternatively skip the use of [38, Lemma 7.3] and instead set

$$\mathcal{U} = \{u \in L^\infty(0, T; L^r(\Omega)) \cap L^2(0, T; V) : \dot{u} \in L^2(0, T; W^*)\} \qquad \text{(A.121)}$$

in case $C_{\Psi''} = 0$, or

$$\mathcal{U} = \{u \in L^\infty(0, T; L^r(\Omega) \cap W^{1,R}(\Omega) \cap W^{2,\varrho}(\Omega)) \cap L^2(0, T; V) : \dot{u} \in L^2(0, T; W^*)\}$$
$$\text{(A.122)}$$

otherwise. This can also be embedded in a Hilbert space setting by replacing $L^\infty(0, T)$ with $H^\sigma(0, T)$ for some $\sigma > \frac{1}{2}$. Going back to estimate (A.111) in case $C_{\Phi''} = 0$ we end up with the conditions

$$\gamma + \kappa \leq \frac{r}{q^*} \text{ and } t - \frac{d}{n} \succeq -\frac{d}{q^*} + \frac{d(\gamma + \kappa)}{r}, \qquad\qquad \text{(A.123)}$$

cf. (A.112), and in case $C_{\Psi''} > 0$, considering estimates (A.115) and (A.117) otherwise, we require

$$t - \frac{d}{n} \succeq \max\left\{ 1 - \frac{d}{q^*} + \frac{d(\hat{\gamma} + 1)}{R}, -\frac{d}{2} + \frac{d}{\varrho} + \frac{d\hat{\gamma}}{R} \right\} \text{ and }$$

$$\hat{\gamma} + 1 \leq \frac{R}{q^*} \text{ and } \varrho \geq 2 \text{ and } \hat{\gamma} \leq \frac{R(\varrho - 2)}{2\varrho},$$

(A.124)

cf. (A.116) and (A.118), and in both cases we additionally need to impose (36).

Acknowledgments BK is supported by the Austrian Science Fund (FWF) with project P30054 (Solving Inverse Problems without Forward Operators). OS is supported by the Austrian Science Fund (FWF) with project F6807-N36 (Tomography with Uncertainties) and with project I3661-N27 (Novel Error Measures and Source Conditions of Regularization Methods for Inverse Problems).

BK and OS acknowledge the support of BIRS for a stay at the Banff center, Canada, where the paper has been finished.

This article was written during Tram Nguyen's employment at Alpen-Adria-Universität Klagenfurt.

References

1. S.M. Allen, J.W. Cahn, Ground state structures in ordered binary alloys with second neighbor interactions. Acta Met. **20**, 423 (1972)
2. A.B. Bakushinsky, M.Y. Kokurin, *Iterative Methods for Approximate Solution of Inverse Problems*. Mathematics and Its Applications (Springer, Dordrecht, 2004)
3. H.T. Banks, K. Kunisch, *Estimation Techniques for Distributed Parameter Systems* (Birkhäuser, Boston, 1989)
4. L. Bronsard, B. Stoth, The Ginzburg-Landau equations of superconductivity and the one-phase Stefan problem. Ann. Inst. Henri Poincaré **15**(3), 371–397 (1998)
5. M. Burger, W. Mühlhuber, Iterative regularization of parameter identification problems by sequential quadratic programming methods. Inverse Probl. **18**, 943–969 (2002)
6. M. Burger, W. Mühlhuber, Numerical approximation of an SQP-type method for parameter identification. SIAM J. Numer. Anal. **40**, 1775–1797 (2002)
7. G. Chavent, K. Kunisch, On weakly nonlinear inverse problems. SIAM J. Appl. Math. **56**, 542–572 (1996)
8. F. Dunker, T. Hohage. On parameter identification in stochastic differential equations by penalized maximum likelihood. Inverse Probl. **30**, 095001 (2014)
9. L.C. Evans, *Partial Differential Equations*. Graduate Studies in Mathematics (AMS, Providence, 1998)
10. W.H. Fleming, H.M. Soner, *Controlled Markov Processes and Viscosity Solutions* (Springer, Berlin, 2006)
11. B.H. Gilding, R. Kersner, *Travelling Waves in Nonlinear Diffusion-Convection Reaction* (Springer Basel AG, Switzerland, 2004)
12. P. Grisvard, *Elliptic Problems in Nonsmooth Domains* (Pitman Advanced Publication Program Boston, Boston, 1985)
13. E. Haber, U.M. Ascher, Preconditioned all-at-once methods for large, sparse parameter estimation problems. Inverse Probl. **17**, 1847 (2001)

14. M. Hanke, A regularizing Levenberg-Marquardt scheme, with applications to inverse groundwater filtration problems. Inverse Probl. **13**, 79–95 (1997)
15. M. Hanke, A. Neubauer, O. Scherzer, A convergence analysis of the Landweber iteration for nonlinear ill-posed problems. Numer. Math. **72**, 21–37 (1995)
16. B. Hofmann, On the degree of ill-posedness for nonlinear problems. J. Inverse Ill-Posed Prob. **2**, 61–76 (1994)
17. S. Hubmer, E. Sherina, A. Neubauer, O. Scherzer, Lamé parameter estimation from static displacement field measurements in the framework of nonlinear inverse problems. SIAM J. Imaging Sci. **11**, 1268–1293 (2018)
18. B. Kaltenbacher, Regularization based on all-at-once formulations for inverse problems. SIAM J. Numer. Analy. **54**, 2594–2618 (2016)
19. B. Kaltenbacher, All-at-once versus reduced iterative methods for time dependent inverse problems. Inverse Probl. **33**, 064002 (2017)
20. B. Kaltenbacher, M.L. Previatti de Souza, Convergence and adaptive discretization of the IRGNM Tikhonov and the IRGNM Ivanov method under a tangential cone condition in Banach space. Numer. Math. **140**, 449–478 (2018)
21. B. Kaltenbacher, A. Neubauer, O. Scherzer, *Iterative Regularization Methods for Nonlinear Problems*. Radon Series on Computational and Applied Mathematics (de Gruyter, Berlin, 2008)
22. B. Kaltenbacher, A. Kirchner, B. Vexler, Goal oriented adaptivity in the IRGNM for parameter identification in PDEs II: all-at once formulations. Inverse Probl. **30**, 045002 (2014)
23. S. Kindermann, Convergence of the gradient method for ill-posed problems. Inverse Probl. Imaging **11**, 703–720 (2017)
24. K. Kunisch, E.W. Sachs, Reduced SQP methods for parameter identification problems. SIAM J. Numer. Analy. **29**, 1793–1820 (1992)
25. F. Kupfer, E. Sachs, Numerical solution of a nonlinear parabolic control problem by a reduced SQP method. Comput. Optim. Appl. **1**, 113–135 (1992)
26. O.A. Ladyzhenskaya, V. Solonnikov, N.N. Ural'tseva, *Linear and Quasilinear Equations of Parabolic Type* (Izd. Nauka, Moscow, 1967). (Engl. Transl.: AMS, Providence, 1968)
27. O.A. Ladyzhenskaya, N.N. Ural'tseva, *Linear and Quasilinear Equations of Elliptic Type* (Izd. Nauka, Moscow, 1964). (Engl. Transl.: Academic, New York, 1968)
28. A. Lechleiter, A. Rieder, Newton regularizations for impedance tomography: convergence by local injectivity. Inverse Probl. **24**, 065009 (2008)
29. F. Leibfritz, E.W. Sachs, Inexact SQP interior point methods and large scale optimal control problems. SIAM J. Control Optim. **38**, 272–293 (1999)
30. G. Leoni, *A First Course in Sobolev Spaces*. Graduate Studies in Mathematics (American Mathematical Society, Providence, 2009)
31. T. Malthus, *An Essay on the Principles of Population* (J. Johnson, London, 1798)
32. J. Nagumo, S. Yoshizawa, S. Arinomoto, Bistable transmission lines. IEEE Trans. Circuit Theory **CT-12**(3), 400–412 (1965)
33. A.A. Nepomnyashchy, Coarsening versus pattern formation. C. R. Phys. **16**, 1–14 (2016)
34. T.T.N. Nguyen, Landweber-Kaczmarz for parameter identification in time-dependent inverse problems: all-at-once versus reduced version. Inverse Probl. **35**, 035009 (2019)
35. C.E. Orozco, O.N. Ghattas, A reduced SAND method for optimal design of non-linear structures. Int. J. Numer. Methods Eng. **40**, 2759–2774 (1997)
36. C.V. Pao, *Nonlinear Parabolic and Elliptic Equations* (Plenum Press, New York, 1992)
37. A. Pazy, *Semigroups of Linear Operators and Applications to Partial Differential Equations* (Springer, New York, 1983)
38. T. Roubíček, *Nonlinear Partial Differential Equations with Applications*. International Series of Numerical Mathematics (Springer, Berlin, 2013)
39. O. Scherzer, Convergence criteria of iterative methods based on Landweber iteration for nonlinear problems. J. Math. Anal. Appl. **194**, 911–933 (1995)
40. A.R. Shenoy, M. Heinkenschloss, E.M. Cliff, Airfoil design by an all-at-once method. Int. J. Comput. Fluid Mech. **11**, 3–25 (1998)

41. S. Ta'asan, "One shot" methods for optimal control of distributed parameter systems I: finite dimensional control, Technical Report, Institute for Computer Applications in Science and Engineering: NASA Langley Research Center, 1991
42. F. Tröltzsch, *Optimal Control of Partial Differential Equations Theory, Methods and Applications*. Graduate Studies in Mathematics (American Mathematical Society, Providence, 2010)
43. T. van Leeuwen, F.J. Herrmann, A penalty method for PDE-constrained optimization in inverse problems. Inverse Probl. **32**, 015007 (2016)

Sequential Subspace Optimization for Recovering Stored Energy Functions in Hyperelastic Materials from Time-Dependent Data

Rebecca Klein, Thomas Schuster, and Anne Wald

Abstract Monitoring structures of elastic materials for defect detection by means of ultrasound waves (Structural Health Monitoring, SHM) demands for an efficient computation of parameters which characterize their mechanical behavior. Hyperelasticity describes a nonlinear elastic behavior where the second Piola-Kirchhoff stress tensor is given as a derivative of a scalar function representing the stored (strain) energy. Since the stored energy encodes all mechanical properties of the underlying material, the inverse problem of computing this energy from measurements of the displacement field is very important regarding SHM. The mathematical model is represented by a high-dimensional parameter identification problem for a nonlinear, hyperbolic system with given initial and boundary values. Iterative methods for solving this problem, such as the Landweber iteration, are very time-consuming. The reason is the fact that such methods demand for several numerical solutions of the hyperbolic system in each iteration step. In this contribution we present an iterative method based on sequential subspace optimization (SESOP) which in general uses more than only one search direction per iteration and explicitly determines the step size. This leads to a significant acceleration compared to the Landweber method, even with only one search direction and an optimized step size. This is demonstrated by means of several numerical tests.

1   Introduction

Monitoring structures consisting of materials like fiber-reinforced plastics or metal laminates is of utmost importance regarding the early detection of defects such as cracks and delaminations or to estimate the structure's lifetime. Such materials play an important role in the construction of wind power stations, aircrafts and automobiles. A Structural Health Monitoring (SHM) system consists of a number of

R. Klein · T. Schuster (✉) · A. Wald
Department of Mathematics, Saarland University, Saarbrücken, Saarland, Germany
e-mail: klein@num.uni-sb.de; thomas.schuster@num.uni-sb.de; anne.wald@num.uni-sb.de

© Springer Nature Switzerland AG 2021
B. Kaltenbacher et al. (eds.), *Time-dependent Problems in Imaging and Parameter Identification*, https://doi.org/10.1007/978-3-030-57784-1_6

actuators and sensors that are applied to the structure. We refer to the seminal book of Giurgiutiu [9] for a comprehensive outline of piezoelectric sensor based SHM systems and their mechanics. A comprehensive monograph on Lamb wave based SHM in polymer composites is given by Gabbert et al. [7]. The mechanical waves that are generated by the actuators propagate through the structure, interact with a possible damage and are measured at the sensors. The inverse problem then consists in recovering the damage from the given sensor measurements. The mathematical model of wave propagation in solids is represented by Cauchy's equation of motion

$$\rho \ddot{u} - \nabla \cdot P = f,$$

where ρ denotes the mass density, P the first Piola-Kirchhoff stress tensor, f an external volume force vector, u is the displacement field of the wave and $\ddot{u}$ the acceleration vector. Materials such as fiber-reinforced plastics or metal laminates are elastic and, depending on the respective response function for P, we obtain a corresponding system of hyperbolic partial differential equations for the displacement field u. The response function for P in turn encodes macroscopic mechanical properties of the material, such as, e.g., the Poisson number or Young's modulus, yielding pointers to hidden damages. There is a vast amount of literature concerning inverse problems connected to Cauchy's equation of motion in elasticity and we refer here only to recent works that have a close relation to the topic of this contribution. Inverse problems in linear elasticity are, e.g., considered in [2, 4, 13, 16]. A nice overview on inverse problems in elasticity is [3]. An important material class in elasticity is given by *hyperelastic* materials, which are characterized by the fact that the stress tensor is given as a derivative of a scalar function with respect to the strain tensor. This scalar function is the stored (strain) energy function and its integral equals the total strain energy which is necessary to deform the body. Since all relevant material properties can be deduced from the stored energy function, its computation should reveal valuable pointers to damages in the structure. The corresponding Cauchy equation then is nonlinear. In [21] the authors investigate higher harmonics of Lamb waves in hyperelastic isotropic materials. Inverse problems in nonlinear elasticity are, e.g., considered in [6, 25, 27–29]. Nonlinear elastic inversion especially in seismics is considered in [8, 30, 31]. In the present contribution we consider the nonlinear inverse problem of reconstructing the stored energy function from the knowledge of the full displacement field u. The stable solution of nonlinear, dynamic inverse problems is currently counted among the most demanding mathematical challenges.

Nonlinear inverse problems are usually solved by iterative regularization techniques. Standard methods such as the Landweber iteration scheme prove to be tremendously slow when applied to such a high-dimensional nonlinear inverse problem. To increase numerical efficiency Sequential Subspace Optimization (SESOP) techniques have been developed and analyzed for various settings, see [19, 23, 24, 33, 35]. The general idea is to reduce the number of iterations until the stopping criterion is fulfilled. To this end, the classical Landweber method is extended by two features. First, a finite number of search directions is used in each iteration.

Second, the length of each search direction is explicitly calculated. This is done in such a way that the method admits a very intuitive interpretation: The iterate is sequentially projected onto subsets that contain the solution set of the inverse problem. These subsets are intersections of stripes that correspond to the respective search directions. The calculation of the projection yields a regulation of the step widths.

This technique has been successfully applied, for example in parameter identification [26, 34, 35], demonstrating a significant increase in efficiency. In addition, they have been used and analyzed in combination with, e.g., sparsity constraints, total variation or Nesterov methods [10, 17, 32].

This contribution delivers a proof-of-concept by demonstrating that RESESOP applied to a high-dimensional nonlinear and dynamic inverse problem leads to a significantly faster convergence as well as less computation time with at the same time higher accuracy compared to Landweber's method.

Outline In Sect. 2 we briefly summarize essential concepts of continuum mechanics for elastic solids and deduce the exact mathematical setting for identifying the stored energy function of a hyperelastic material from measurements of the displacement field. In order to guarantee that the reconstructed energy is physically meaningful we use a dictionary of finitely many elements. The inverse problem subsequently reduces to the computation of the corresponding coefficients with respect to the given dictionary. Section 3 outlines the introduction and analysis of the Landweber method and RESESOP. In Sect. 4 we finally present several numerical experiments using three different damage scenarios for a structure consisting of a Neo-Hookean material showing the superiority of RESESOP compared to the Landweber method.

2 Hyperelastic Materials

In this chapter we briefly discuss some basic facts from continuum mechanics and especially on Cauchy's equation of motion and hyperelastic constitutive equations. For deeper insights we refer to the standard literature [5, 12, 18].

The considered elastic structure is described by a bounded, open, connected subset $\Omega \subset \mathbb{R}^3$ with a sufficiently smooth boundary. We start with the mathematical definition of a deformation of Ω.

Definition 1 A *deformation* of a body Ω is an invertible, continuously differentiable mapping $\varphi : [0, T] \times \Omega \to \mathbb{R}^3$, which is orientation-preserving such that

$$det(\nabla \varphi(t, x)) > 0 \qquad \forall (t, x) \in [0, T] \times \Omega,$$

where

$$\nabla\varphi(t,x) = \left(\frac{\partial\varphi_i}{\partial x_j}(t,x)\right)_{i,j=1,2,3} = \left(\partial_{x_j}\varphi_i(t,x)\right)_{i,j=1,2,3} \in \mathbb{R}^{3\times3}.$$

Definition 1 implies that the body will not be torn apart or penetrate itself during the deformation. Since $\nabla\varphi$ is invertible, any two points in Ω can be separated at any time $t \in [0,T]$. If Ω undergoes a deformation, then a fixed point x is shifted to a point $\varphi(t,x)$. Their difference defines the displacement field.

Definition 2 Let $\varphi : [0,T] \times \Omega \to \mathbb{R}^3$ be a deformation. Then the *displacement field* $u : [0,T] \times \Omega \to \mathbb{R}^3$ is given by

$$u(t,x) = \varphi(t,x) - x.$$

The set Ω is also called the *reference configuration* whereas $\Omega(t) := \varphi(t,\Omega) \subset \mathbb{R}^3$ is called the *deformed configuration* and represents the body after deformation at time t. A guided wave that is generated by actuators and propagates through the structure Ω will cause a displacement field u which subsequently can be measured by applied sensors. This is the key idea of an SHM system (c.f. [9]).

Definition 3 Let $\varphi : [0,T] \times \Omega \to \mathbb{R}^3$ be a deformation and u be the corresponding displacement field. The *displacement gradient* is given by

$$\nabla u(t,x) = \nabla\varphi(t,x) - I$$

with the identity matrix $I \in \mathbb{R}^{3\times3}$. The gradient ∇ refers to the spatial coordinates.

The propagation of ultrasound waves in Ω is mathematically described by *Cauchy's equation of motion*, which follows from the stress principle of Euler and Cauchy and the axioms of force and moment balance. For all $(t,x) \in [0,T] \times \Omega$ we have

$$\rho(x)\ddot{u}(t,x) - \nabla \cdot P(t,x) = f(t,x). \tag{1}$$

Here $\rho : \Omega \to \mathbb{R}^+$ denotes the mass density, $f : [0,T] \times \Omega \to \mathbb{R}^3$ the external body force and $P : [0,T] \times \Omega \to \mathbb{R}^{3\times3}$ the first Piola-Kirchhoff stress tensor. This is a differential equation for the unknowns u and P and obviously not uniquely solvable in its present form. But by now we did not include the phenomenon of elasticity to Ω and Eq. (1). Elasticity means that there is a stress-strain relation which is implied by the existence of a so called *response function* for the Cauchy stress tensor. To be short: a deformation of the body Ω causes strain which again causes stress. Postulating the existence of a response function will furthermore reduce the degrees of freedom in (1).

Before we formulate the principle of elasticity we introduce by $\sigma : [0,T] \times \Omega(t) \to \mathbb{R}^{3\times3}$ the *Cauchy stress tensor*. This is a continuously differentiable, symmetric tensor field whose existence follows from Cauchy's theorem. In some sense this is the counterpart to the first Piola-Kirchhoff stress tensor P: The Cauchy stress tensor σ is defined on the deformed configuration $\Omega(t)$ whereas P is defined on the reference configuration Ω. Of course, each of these can be transformed into the respective other one, and the specific relation between σ and P is given by Eq. (3).

Definition 4 A material is called *elastic*, if a mapping

$$\tilde{\sigma} : \overline{\Omega} \times GL_+(3) \to Sym(3), \qquad (x, Y) \mapsto \tilde{\sigma}(x, Y)$$

exists, such that the Cauchy stress tensor satisfies

$$\sigma(t, \varphi(t, x)) = \tilde{\sigma}(x, \nabla\varphi(t, x)) \tag{2}$$

for every deformation φ, where

$$GL_+(3) := \{Y \in \mathbb{R}^{3 \times 3} | \det(Y) > 0\}$$

denotes the set of 3×3 matrices with a positive determinant and $Sym(3)$ is the set of symmetric 3×3 matrices. The function $\tilde{\sigma}$ is called the *response function* for σ. Equation (2) is called a *constitutive equation of the material*.

The first Piola-Kirchhoff stress tensor can be computed from σ by applying the *Piola transform*

$$P(t, x) = \det(\nabla\varphi(t, x))\sigma(t, x)\nabla\varphi(t, x)^{-\top}. \tag{3}$$

So, if there exists a response function $\tilde{\sigma}$ for σ, then we easily obtain a response function $\tilde{P}$ for P from (3) via

$$\tilde{P}(x, Y) := \det Y \tilde{\sigma}(x, Y)Y^{-\top}, \qquad x \in \Omega, \ Y \in GL_+(3).$$

Remark 1 The *Cauchy-Green strain tensor B* is defined as $B = \nabla\varphi^\top \nabla\varphi$ and we have $B = I$ if and only if the deformation is rigid. Thus, B measures the 'deviation' between a deformation φ and a rigid motion. It is quite obvious from (2), that the existence of a response function $\tilde{\sigma}$ implies the existence of a function $\hat{\sigma}$ with

$$\sigma(t, x) = \tilde{\sigma}(x, \varphi(t, x)) = \hat{\sigma}(x, B(t, x)).$$

In this way (2) can be interpreted as a relation between stress and strain which is the reason why (2) is also called *stress-strain relation*. Hence, elasticity in fact means that a material replies to strain with stress.

A large class of physically very important elastic materials is represented by the *hyperelastic materials*. For this class the response functions have a very specific form.

Definition 5 An elastic body is called *hyperelastic* if the response function of the first Piola-Kirchhoff stress tensor is given by

$$\tilde{P}(x, Y) = \nabla_Y \hat{C}(x, Y), \qquad x \in \Omega, \ Y \in M,$$

and a scalar function $\hat{C} : \Omega \times \mathrm{GL}_+(3) \to \mathbb{R}$. This function $\hat{C}$ is called *stored (strain) energy function.*

The derivative ∇_Y used in Definition 5 is to be understood as

$$\nabla_Y g(x, Y) = \left[\partial_{Y_{i,j}} g(x, Y) \right]_{1 \leq i, j \leq 3} \in \mathbb{R}^{3 \times 3}, \qquad x \in \Omega, \ Y \in \mathrm{GL}_+(3),$$

for a differentiable function $g : \Omega \times M \to \mathbb{R}$ and $M \subset \mathbb{R}^{3 \times 3}$.

Remark 2

(a) If φ is a deformation and the body Ω consists of a hyperelastic material, then the integral

$$E(t) = \int_\Omega \hat{C}\big(x, \nabla\varphi(t, x)\big) \, \mathrm{d}x$$

denotes the *strain energy* $E(t)$ which is necessary to perform the deformation at time t. This explains the term stored (strain) energy function for $\hat{C}$.

(b) The fourth order *elasticity tensor* $\mathbb{C}$ can directly be computed from $\hat{C}$ by

$$\mathbb{C}(x) = \nabla_Y \nabla_Y \hat{C}(x, I), \qquad x \in \Omega.$$

It plays a crucial role in linear elasticity and its entries are important functions describing material properties such as Young's modulus and the Poisson number. In this sense $\hat{C}$ encodes all important material properties and yields pointers for defects in hyperelastic structures.

(c) We consider hyperelastic materials since on the one hand linear elastic models do often not accurately enough describe the stress-strain behavior of the considered structure and on the other hand the derived methodology is appropriate for a broader range of applications (composites, rubber, biological tissues).

Let Ω be hyperelastic. Then Cauchy's equation of motion reads

$$\rho(x)\ddot{u}(t, x) - \nabla \cdot \nabla_Y \hat{C}(x, \nabla u(t, x)) = f(t, x), \qquad (t, x) \in [0, T] \times \Omega. \tag{4}$$

Note that in (4) we silently used the identity $\nabla u = \nabla\varphi - I$ to write, in slight misuse of notation, $\hat{C}(x, \nabla u(t, x))$. This means that, by assuming Ω to be hyperelastic and $\hat{C}$ to be known explicitly, Cauchy's equation of motion is no longer underdetermined since we have three equations and three unknowns, i.e., the three components of the displacement vector u. To ensure uniqueness one furthermore has to postulate initial and boundary values for u (c.f. [36]).

The inverse problem which is numerically solved in this contribution consists in computing the stored energy function $\hat{C}$ from measurements of the displacement field u. To specify this we follow the idea of computing $\hat{C}$ as a conical combination with respect to a given dictionary consisting of physically reasonable stored energy functions C_K, $K = 1, \ldots, N$., c.f. [14, 25]. Let $\{C_K : \Omega \times \mathbb{R}^{3\times3} \to \mathbb{R} : K = 1, \ldots, N\}$ be such a dictionary. Then we write

$$\hat{C}(x, Y) = \sum_{K=1}^{N} \alpha_K C_K(x, Y), \qquad x \in \Omega, \ Y \in \mathbb{R}^{3\times3},$$

for certain coefficients $\alpha_K \geq 0$. Equipped with appropriate initial and boundary values we obtain Cauchy's equation of motion in its final form: The balance equation reads

$$\rho \ddot{u}(t, x) - \sum_{K=1}^{N} \alpha_K \nabla \cdot \nabla_Y C_K(x, \nabla u(t, x)) = f(t, x), \qquad (t, x) \in [0, T] \times \Omega. \tag{5}$$

We furthermore assume initial values

$$u(0, \cdot) = u_0 \in H^2(\Omega, \mathbb{R}^3), \tag{6}$$

$$\dot{u}(0, \cdot) = u_1 \in H^1(\Omega, \mathbb{R}^3) \tag{7}$$

as well as homogeneous boundary values

$$u(t, \xi) = 0, \quad \xi \in \partial\Omega. \tag{8}$$

The respective inverse problem is formulated as follows:

(IP) Given (f, u_0, u_1) and the displacement field $u(t, x)$ for $t \in [0, T]$ and $x \in \Omega$, determine the coefficients $\alpha = (\alpha_1, \ldots, \alpha_N) \in \mathbb{R}_+^N$, such that u satisfies the initial boundary value problem (5)–(8).

If we define by $F : \mathcal{D}(F) \subset \mathbb{R}_+^N \to X$ the forward operator which maps, for fixed given (f, u_0, u_1), a vector $\alpha \in \mathbb{R}_+^N$ to the unique solution $u \in X$, then the inverse problem demands for solving the nonlinear operator equation

$$F(\alpha) = u.$$

Here $\mathcal{D}(F)$ denotes the domain of F consisting of those $\alpha \in \mathbb{R}_+^N$ admitting a unique solution and $X = L^2\big(0, T; H_0^1(\Omega, \mathbb{R}^3)\big) \cap H^1\big(0, T; L^2(\Omega, \mathbb{R}^3)\big)$ denotes the image space of F containing all admissible solutions. For more details regarding existence and uniqueness of solutions for the IBVP (5)–(8) we refer the reader to [27, 36].

In Sect. 4 we will see that a convenient approach to define the dictionary elements C_K is to use tensor products

$$C_K(x, Y) = v_K(x)\hat{C}(Y), \qquad K = 1, \dots, N,$$

with B-splines v_K that are also used for the Finite Element solution of (5) and physically reasonable stored energy functions $\hat{C}$ depending only on Y. This idea is taken from [28].

3 Sequential Subspace Optimization

In this contribution we present numerical results that are obtained with both the attenuated Landweber as well as the RESESOP method as a solver for the inverse problem (IP). In [28] some results using the attenuated Landweber method, implemented in C++ together with the finite element library deal.II [1], have already been presented. For the reader's convenience, we will introduce some notation and briefly summarize the attenuated Landweber method.

Consider a (nonlinear) problem

$$F(x) = y, \quad F : \mathcal{D}(F) \subset X \to Y,$$

with Hilbert spaces X and Y. Then the respective attenuated Landweber iteration reads

$$x_{k+1}^\delta = x_k^\delta + \omega F'(x_k^\delta)^*(y^\delta - F(x_k^\delta)), \qquad k = 0, 1, \dots \tag{9}$$

where the parameter $\omega > 0$ is called a *relaxation* or *damping parameter*. Since ω is fixed, there is no strategy to adapt the step width in each individual iteration. It is assumed that we only have disturbed data y^δ with $\|y^\delta - y\| < \delta$ and noise level $\delta > 0$ at our disposal. The convergence of the Landweber method is guaranteed by selecting

$$\omega \in \left(0, \frac{1}{C_\rho^2}\right)$$

with the constant

$$C_\rho := \sup\{\|F'(x)\| : x \in B_\rho(x_0)\}.$$

In case of noisy data, the iteration is stopped by the discrepancy principle, which turns it into a regularization method [11, 15, 22].

However, the Landweber method is known to be very slowly converging, and it often takes a lot of iterations to obtain a suitable regularized solution. Particularly in view of an application in parameter identification, where the calculation of each gradient involves the numerical evaluation of the forward operator as well as the adjoint of its linearization, a reconstruction via the Landweber method is too time-consuming and hardly practicable, see, e.g., [28, 34].

In contrast to the attenuated Landweber method, the SESOP method not only involves a regulation of the step width, it also potentially uses multiple search directions per iteration. This of course requires additional (but numerically cheap) calculations in each iteration step, such that a SESOP step will take slightly longer. However, we anticipate that the SESOP and RESESOP methods will need far less iterations and thus lead to a faster convergence of the iteration.

In this section we will give a short introduction to sequential subspace optimization (SESOP) and regularizing sequential subspace optimization (RESESOP). From the RESESOP method we derive the algorithm which we will use for our later experiments, where we solve (IP) numerically from simulated noisy data.

The idea behind the SESOP method and its regularizing version RESESOP is to reduce the number of iteration steps by sequentially projecting the current iterate onto suitable subsets of the source space X that are hyperplanes or stripes in X and contain the solution set of the respective inverse problem $F(x) = y$. This approach is inspired by the fact that in the case of linear problems, the solution set itself is an affine subspace. More detailed information about the SESOP method for linear problems can be found in [19, 23, 24]. Results concerning the SESOP method as a solution technique for nonlinear problems are presented in [10, 32, 33, 35].

3.1 Basics

We will first state some basics for the RESESOP method, in particular the definitions of hyperplanes, half-spaces and stripes, as well as the metric projection.

Definition 6 (Hyperplanes, Half-Spaces and Stripes) Let $u \in X \setminus \{0\}$ and $\alpha, \xi \in \mathbb{R}, \xi \geq 0$. For these parameters, we define the *hyperplane*

$$H(u, \alpha) := \{x \in X \ : \ \langle u, x \rangle = \alpha\},$$

the *half-space*

$$H_\leq(u, \alpha) := \{x \in X \ : \ \langle u, x \rangle \leq \alpha\},$$

and the *stripe*

$$H(u, \alpha, \xi) := \{x \in X \ : \ |\langle u, x \rangle - \alpha| \leq \xi\}.$$

The half-spaces $H_>(u, \alpha)$, $H_<(u, \alpha)$ and $H_>(u, \alpha)$ are defined analogously. We see that the half space $H_<(u, \alpha)$ is simply the space beneath the hyperplane $H(u, \alpha)$. The stripe $H(u, \alpha, \xi)$ emerges from the hyperplane $H(u, \alpha)$ by admitting a width that is determined by ξ. Hyperplanes, half-spaces as well as stripes are convex, non-empty sets according to their definition. In addition, the sets $H(u, \alpha)$, $H_{\leq}(u, \alpha)$, $H_{\geq}(u, \alpha)$ and $H(u, \alpha, \xi)$ are closed.

The solution set $M_{Fx=y}$ of a linear operator equation $Fx = y$ can be described by

$$M_{Fx=y} := \{x \in X \, : \, Fx = y\} = x_0 + \mathcal{N}(F)$$

for some $x_0 \in \mathcal{N}(F)^{\perp}$.

Another tool that plays an important role is the metric projection.

Definition 7 The *metric projection* of $x \in X$ onto a non-empty closed convex set $C \subset X$ is the unique element $P_C(x) \in C$, such that

$$\|x - P_C(x)\|^2 = \min_{z \in C} \|x - z\|^2.$$

The metric projection P_C onto a convex set fulfills the descent property of the form

$$\|z - P_C(x)\|^2 \leq \|z - x\|^2 - \|P_C(x) - x\|^2 \tag{10}$$

for all $z \in C$.

Since hyperplanes and stripes are, by definition, closed and convex non-empty sets, the metric projection of $x \in X$ onto these specific subsets is well-defined. For example, if $C := H(u, \alpha)$ is a hyperplane of X, then the metric projection of $x \in X$ onto C corresponds to the orthogonal projection, i.e., we have

$$P_{H(u,\alpha)}(x) = x - \frac{\langle u, x \rangle - \alpha}{\|u\|^2} u \tag{11}$$

and (10) turns into an equation, see, e.g., [24, 26].

By the following theorem we want to provide some tools that will later be essential to define the sequential subspace optimization techniques we use to obtain faster reconstructions of the stored energy function. Essentially, these techniques consist of sequential metric projections onto (intersections of) hyperplanes or stripes. By Definition 7 we already know that a metric projection onto a non-empty, closed convex set can be formulated as a minimization problem. The special case of metric projections onto intersections of hyperplanes is summarized in the following theorem. A proof can be found in [26] for the more general setting of Bregman projections in (convex and uniformly smooth) Banach spaces X and Y.

Theorem 1

(a) *Let $H(u_i, \alpha_i)$ be hyperplanes for $i = 1, \ldots, N$ with non-empty intersection*

$$H := \bigcap_{i=1}^{N} H(u_i, \alpha_i).$$

The projection of x onto H is given by

$$P_H(x) = x - \sum_{i=1}^{N} \tilde{t}_i u_i,$$

where $\tilde{t} := (\tilde{t}_1, \ldots, \tilde{t}_N) \in \mathbb{R}^N$ minimizes the convex function

$$h(t) = \frac{1}{2} \left\| x - \sum_{i=1}^{N} t_i u_i \right\|^2 + \sum_{i=1}^{N} t_i \alpha_i, \quad t = (t_1, \ldots, t_N) \in \mathbb{R}^N.$$

The partial derivatives of the function $h(t)$ are given by

$$\frac{\partial}{\partial t_j} h(t) = -\left\langle u_j, x - \sum_{i=1}^{N} t_i u_i \right\rangle + \alpha_j. \tag{12}$$

If the vectors u_i, $i = 1, \ldots, N$, are linearly independent, h is strictly convex and $\tilde{t}$ is unique.

(b) Let $H_i := H_{\leq}(u_i, \alpha_i)$, $i = 1, 2$, be two half-spaces with linear independent vectors u_1 and u_2. Then $\tilde{x}$ is the projection of x onto $H_1 \cap H_2$ if $\tilde{x}$ satisfies the Karush-Kuhn-Tucker conditions for

$$\min_{z \in H_1 \cap H_2} \|z - x\|^2.$$

The Karush-Kuhn-Tucker conditions are given by

$$\tilde{x} = x - t_1 u_1 - t_2 u_2 \qquad \text{for any } t_1, t_2 \geq 0,$$
$$\alpha_i \geq \langle u_i, \tilde{x} \rangle, \qquad i = 1, 2,$$
$$0 \geq t_i (\alpha_i - \langle u_i, \tilde{x} \rangle), \qquad i = 1, 2.$$

(c) For $x \in H_{>}(u, \alpha)$ the projection of x onto $H_{\leq}(u, \alpha)$ is given by

$$P_{H_{\leq}(u,\alpha)}(x) = P_{H(u,\alpha)}(x) = x - t_+ u$$

with

$$t_+ = \frac{\langle u, x \rangle - \alpha}{\|u\|^2} > 0.$$

(d) The projection of $x \in X$ onto the stripe $H(u, \alpha, \xi)$ is given by

$$
P_{H(u,\alpha,\xi)}(x) = \begin{cases} P_{H_{\leq}(u,\alpha+\xi)}(x) & \text{if } x \in H_{>}(u, \alpha + \xi), \\ x, & \text{if } x \in H(u, \alpha, \xi), \\ P_{H_{\geq}(u,\alpha-\xi)}(x) & \text{if } x \in H_{<}(u, \alpha - \xi). \end{cases}
$$

Part (a) of Theorem 1 allows us to use tools from optimization (see also, e.g., [20]) to determine the parameters $t = (t_1, \ldots, t_N)$. The fact that the minimization of the function $h(t)$ corresponds to the projection onto the intersection of the hyperplanes $H(u_i, \alpha_i)$ for $i = 1, \ldots, N$ can be seen by taking a look at the partial derivatives (12) of $h(t)$. Let us assume that the parameters $\tilde{t} = (\tilde{t}_1, \ldots, \tilde{t}_N)$ represent the local minimum of the function $h(t)$. Then,

$$
\frac{\partial}{\partial t_j} h(\tilde{t}) = -\left\langle u_j, x - \sum_{i=1}^{N} \tilde{t}_i u_i \right\rangle + \alpha_j = 0.
$$

Since by definition we have

$$
P_H(x) = x - \sum_{i=1}^{N} \tilde{t}_i u_i,
$$

we obtain

$$
\langle u_j, P_H(x) \rangle = \alpha_j
$$

for all $j = 1, \ldots N$, which shows that $P_H(x) = x - \sum_{i=1}^{N} \tilde{t}_i u_i$ is an element of each hyperplane $H(u_i, \alpha_i)$, $i = 1, \ldots, N$ and, as a direct consequence, we have

$$
P_H(x) \in H.
$$

Remark 3 If F is a linear operator and the given data y^δ are noisy with noise level $0 \leq \|y^\delta - y\| \leq \delta$, then the solution set $\mathcal{M}_{Fx=y}$ of the linear operator equation $Fx = y$ is contained in the stripes $H(u, \alpha, \xi)$, where

$$
u := F^* w
$$

$$
\alpha := \langle w, y^\delta \rangle
$$

$$
\xi := \delta \|w\|
$$

with arbitrary $w \in Y$, since for each $x \in \mathcal{M}_{Fx=y}$ we have

$$|\langle u, x \rangle - \alpha| = \left|\langle F^*w, x \rangle - \langle w, y^\delta \rangle\right|$$
$$= \left|\langle w, Fx - y^\delta \rangle\right| = \left|\langle w, y - y^\delta \rangle\right|$$
$$\leq \delta \|w\| = \xi.$$

This observation is the basis to derive an iteration of the form

$$x^\delta_{n+1} = P_{H^\delta_n}(x^\delta_n), \quad n \in \mathbb{N},$$

where $H^\delta_n := \bigcap_{i \in I_n} H(u^\delta_n, \alpha^\delta_n, \xi^\delta_n)$ is the intersection of stripes containing the solutions of $Fx = y$. For each solution x, a reasonable choice of the parameters that define the stripes yields the descent property

$$\left\| x - x^\delta_{n+1} \right\|^2 \leq \left\| x - x^\delta_n \right\|^2 - C \left\| Fx^\delta_n - y^\delta \right\|^2.$$

This property is used to show convergence and regularization properties of the method, see [26].

3.2 RESESOP for Nonlinear Problems

We turn to the regularizing sequential subspace optimization (RESESOP) technique for nonlinear inverse problems

$$F(x) = y, \quad F : \mathcal{D}(F) \subset X \to Y. \tag{13}$$

in Hilbert spaces X, Y and noisy data y^δ with known noise level $\delta > 0$. The respective SESOP method that is applicable to unperturbed data can easily be derived by setting $\delta = 0$, see also [33].

In order to adapt the methods for linear operators to the nonlinear case, we must ensure that we project sequentially onto subsets of X that contain the solution set

$$\mathcal{M}_{F(x)=y} := \{ x \in \mathcal{D}(F) : F(x) = y \}$$

of the operator equation (13). In contrast to linear problems, we have to take into account the local character of nonlinear operators, i.e., we have to incorporate information on the local nonlinear behaviour of the forward operator into the definition of the stripes onto which we project in each iteration. To do this appropriately, we need the following assumptions on the operator F.

Let $F : \mathcal{D}(F) \subset X \to Y$ be continuous and Fréchet differentiable in an open ball

$$B_\rho(x_0) := \{ x \in X : \|x - x_0\| < \rho \} \subset \mathcal{D}(F)$$

around the starting value $x_0 \in \mathcal{D}(F)$ with radius $\rho > 0$ and let the mapping

$$B_\rho(x_0) \ni x \mapsto F'(x)$$

from $B_\rho(x_0)$ into the space $L(X, Y)$ of linear and continuous mappings be continuous.

We assume there exists a solution $x^+ \in X$ of (13) that satisfies $x^+ \in B_\rho(x_0)$. This ensures that we start the iteration close to a solution, which is a mandatory requirement for nonlinear problems.

Furthermore, we assume that the forward operator F satisfies the tangential cone condition

$$\left\| F(x) - F(\tilde{x}) - F'(x)(x - \tilde{x}) \right\| \le c_{\text{tc}} \left\| F(x) - F(\tilde{x}) \right\| \tag{14}$$

with a positive constant

$$0 < c_{\text{tc}} < 1$$

and the estimate (continuity of the Fréchet derivative)

$$\left\| F'(x) \right\| < c_F$$

with $c_F > 0$ for all $x, \tilde{x} \in B_\rho(x_0)$.

We also assume that the operator F is weakly sequentially closed. That is, for a weakly convergent sequence $\{x_n\}_{n \in \mathbb{N}}$ with $x_n \rightharpoonup x$ and $F(x_n) \to y$ holds

$$x \in \mathcal{D}(F) \quad \text{and} \quad F(x) = y.$$

If all these properties are fulfilled, we can formulate the RESESOP method as proposed in [33] and obtain a regularization technique.

Remark 4 The goal of general SESOP methods is to use multiple search directions $u_{n,i}^\delta$, $i \in I_n$, $|I_N| < \infty$, in each step $n \in \mathbb{N}$ of the iteration in combination with a regulation of the step width. We have $M_{F(x)=y} \subset H(u_{n,i}^\delta, \alpha_{n,i}^\delta, \xi_{n,i}^\delta)$ if we set

$$u_{n,i}^\delta := F'(x_i^\delta)^* w_{n,i}^\delta$$

$$\alpha_{n,i}^\delta := \left\langle w_{n,i}^\delta, F(x_i^\delta) - y^\delta \right\rangle - \left\langle F'(x_i^\delta)^* w_{n,i}^\delta, x_i^\delta \right\rangle$$

$$\xi_{n,i}^\delta := \| w_{n,i}^\delta \| \left(c_{\text{tc}} \left(\| R_i^\delta \| + \delta \right) + \delta \right),$$

see also [33].

These definitions show that each hyperplane is related to the properties of F close to the respective iterate. In particular, the noise level δ and the constant c_{tc}

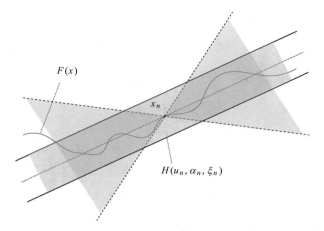

$F(x)$

x_n

$H(u_n, \alpha_n, \xi_n)$

Fig. 1 Illustration of a nonlinear function F with stripe $H(u_n, \alpha_n, \xi_n)$

from (14) determine the width of the stripe: the higher the noise level and the larger the opening angle of the cone, the larger we have to choose the width of the stripe.

Figure 1 illustrates the tangential cone condition (14) and its relevance for the choice of the stripes in the case $I_n := \{n\}$ for a function F in two dimensions and exact data y. The graph of F is plotted in red and for the point x_n the linearization $F'(x_n)$ of F in x_n is represented by the red dotted line. The graph is contained in the cone, determined by the tangential cone condition, highlighted in gray. The size of c_{tc} directly corresponds to the opening angle of the cone: The better F is approximated by its linearization, the smaller is c_{tc} and thus also the opening angle of the grey cone. Figure 1 also shows that the cone condition can be used to define a stripe $H(u_n, \alpha_n, \xi_n)$ (marked in blue), such that the graph of F is locally contained in $H(u_n, \alpha_n, \xi_n)$, i.e., in a neighborhood of x_n.

In the following we formulate the regularizing SESOP iteration for the special case of a single search direction per iteration, i.e., we set $I_n := \{n\}$ for all $n \in \mathbb{N}$. Furthermore, we define the n-th search direction as

$$u_n^\delta := F'(x_n^\delta)^* \left(F(x_n^\delta) - y^\delta \right),$$

such that we essentially obtain a Landweber-type method with an adaptation of the step size. In comparison to the attenuated Landweber method, we thus have a dynamic relaxation parameter that adapts to the projection in each iteration step. Together with the discrepancy principle, we obtain a regularization method for which several convergence results could be shown (see [33]).

Algorithm (RESESOP with One Search Direction) We choose a starting value $x_0^\delta = x_0 \in \mathcal{D}(F)$. For all $n \geq 0$ we select the search direction u_n^δ such that

$$u_n^\delta := F'(x_n^\delta)^* w_n^\delta,$$

$$w_n^\delta := R_n^\delta := F(x_n^\delta) - y^\delta.$$

We define the stripe H_n^δ by

$$H_n^\delta := H(u_n^\delta, \alpha_n^\delta, \xi_n^\delta)$$

with

$$\alpha_n^\delta := \langle u_n^\delta, x_n^\delta \rangle - \|R_n^\delta\|^2,$$

$$\xi_n^\delta := \|R_n^\delta\| \left(\delta + c_{tc} \left(\|R_n^\delta\| + \delta \right) \right).$$

As tolerance parameter for the discrepancy principle we choose

$$\tau > \frac{1 + c_{tc}}{1 - c_{tc}} > 1. \tag{15}$$

As long as $\|R_n^\delta\| > \tau\delta$ is valid, we have

$$x_n^\delta \in H_>(u_n^\delta, \alpha_n^\delta + \xi_n^\delta) \tag{16}$$

and we calculate the new iterate x_{n+1}^δ by

$$x_{n+1}^\delta := P_{H(u_n^\delta, \alpha_n^\delta, \xi_n^\delta)}(x_n^\delta) = P_{H(u_n^\delta, \alpha_n^\delta + \xi_n^\delta)}(x_n^\delta) \tag{17}$$

$$= x_n^\delta - \frac{\langle u_n^\delta, x_n^\delta \rangle - (\alpha_n^\delta + \xi_n^\delta)}{\|u_n^\delta\|^2} u_n^\delta. \tag{18}$$

Remark 5 Note that due to (16), the iterate x_n^δ lies *above* the stripe H_n^δ and, according to Theorem 1 (d), we obtain the identity (17). This projection is explicitly formulated in (18).

The choice of τ in (15) depends strongly on the constant c_{tc} of the cone condition. The smaller c_{tc}, the better the approximation of F by its linearization. However, if c_{tc} is large, this also means that τ is large and the algorithm is usually stopped for larger residuals $\|R_n^\delta\|$.

Remark 6 We want to state some observations for the RESESOP method in comparison to the Landweber iteration from previous research [23, 24, 34, 35].

(a) For RESESOP, good estimates of the noise level δ and the constant c_{tc} are required. This is because both of these constants directly influence the calculation of the optimization parameters $t_{n,i}$, whereas in the Landweber iteration they only influence the parameter τ in the discrepancy principle.

Hence, Landweber is more robust to errors in these constants than RESESOP. This is, however, a small price to pay in order to obtain a faster regularization method.

(b) Previous research has shown that the use of multiple search directions per iteration yields faster convergence. However, if the noise level δ is high or if the forward operator is not well approximated by its linearization, i.e., c_{tc} is not close to 0, the stripes have to be chosen with a large width $\xi(\delta, c_{tc})$. In this case, the RESESOP algorithm often automatically skips using multiple search directions since the projection of the current iterate onto the first stripe is often already contained in the other stripes due to their large width.

This is precisely what we observed for the parameter identification problem that is addressed in this work, which is why we put the emphasis on RESESOP with a single search direction.

For an analysis and a detailed discussion of general SESOP methods with multiple search directions in Hilbert and Banach space settings, we refer to the literature [23, 24, 26, 33, 35].

4   Numerical Results

In this section we present some numerical results to solve the inverse problem (IP) from Sect. 2. In all tests we use data that are simulated by solving the initial boundary value problem (5)–(8) using the θ-method with respect to time and the Finite Element method in space. The resulting system of nonlinear equations is then solved by Newton's method. A detailed outline of the numerical forward solver for (5) is contained in [28].

The experimental setup for the numerical tests consists of a plate with measures $1\,\text{m} \times 1\,\text{m}$ and a thickness of 6.7 mm. These measures can be numerically transferred to values of $\Omega = [-0.1, 0.1] \times [-15, 15]^2$. The plate is discretized using $5 \times 31 \times 31$ knots with respect to x and trilinear Finite Elements that are given by tensor products of linear B-splines. The time interval is given by $[0\,\mu s, 133\,\mu s]$, which we numerically represent as $[0, T] = [0, 4]$, and is discretized by $t_j = j\Delta t$, $j = 0, \ldots, 15$, and step size $\Delta t = 0.25$. We assume that the plate is at rest at $t = 0$ yielding $u_0 = u_1 = 0$. The excitation signal $f(t, x)$ is compactly supported with respect to t and hence not band limited. It is emitted at the center of the plate acting in x_3-direction. The reason for using a broad band signal is the fact that different defects are sensitive to different frequencies. In this way we avoid a frequency dependent selectivity of the defect-wave interaction. The chosen sampling in t corresponds to a sampling frequency of 120 kHz. We refer to [2, 28] for more details. Actuators and sensors of SHM systems in real world applications generate signals that have an essential frequency range in 100–600 kHz, see [7, Ch. 17].

As already mentioned in Sect. 2 the dictionary of stored energy functions $\{C_K : K = 1, \ldots, N\}$ is defined as tensor products

$$C_K(x, Y) = v_K(x)\hat{C}(Y).$$

For our simulations we use the stored energy of a Neo-Hookean material model

$$\hat{C}(Y) = c(I_1 - 3) + \frac{c}{\beta}(D^{-2\beta} - 1),$$

where $I_1 = \|\nabla\varphi\|_F^2$, $D = \det(\nabla\varphi)$ and the constants are given by $\beta = \frac{3v-2\mu}{6\mu} > 0$ and $c = \frac{\mu}{2} > 0$ with specific values $v = 68.6$ GPa and $\mu = 26.32$ GPa taken from [21]. The functions v_K are exactly the linear tensor product B-splines that are used for the Finite Element discretization of the forward solver. Since linear tensor product B-splines have small compact support and represent a partition of unity, i.e.

$$\sum_{K=1}^{N} v_K(x) = 1, \qquad x \in \Omega, \tag{19}$$

any defects can be appropriately modeled by coefficients $\alpha_K \neq 1$ whereas for the undamaged plate we set $\alpha_K = 1$, $K = 1, \ldots, N$.

If we denote by b_i, b_j the linear B-splines corresponding to the given discretizations in the (x_2, x_3)-plane, then we can simulate a delamination at the upper surface of the plate by defining the stored energy as

$$C(x, Y) := \sum_{i=0}^{30} \sum_{j=0}^{30} \alpha_{ij} b_i(x_2) b_j(x_3)\hat{C}(Y) \qquad \text{at } x_1 = 0.05 \tag{20}$$

and setting $\alpha_{ij} \neq 1$ for locations of the delamination. Due to (19), $\alpha_{ij} = 1$ corresponds to regions of the (x_2, x_3)-plane that are unaffected by the damage. Setting $\alpha_{ij} = 1$ for all i, j yields $C(x, Y) = \hat{C}(Y)$ for all $x \in \Omega$ and thus models a homogeneous material. Note, that in (20) we use double indices in α_{ij} according to the tensor product structure of the Finite Elements $b_i \otimes b_j$, i.e., we have $\alpha_K = \alpha_{ij}$ with $K = 31 \cdot i + j$.

An implementation of the RESESOP method needs the adjoint of the Fréchet derivative $F'(\alpha)^* : X^* \to \mathbb{R}^N$. For completeness we state the representation, a deduction is found in [27]. For $\alpha \in \mathbb{R}_+^N$ we have

$$\left[F'(\alpha)^* w\right]_K = -\int_0^T \int_\Omega \nabla_Y C_K(x, \nabla u(t, x)) : \nabla p(t, x)\, dx\, dt, \qquad K = 1, \ldots, N,$$

where $p \in L^2(0, T; L^2(\Omega, \mathbb{R}^3))$ is the weak solution of the hyperbolic backward IBVP

$$\rho \ddot{p}(t, x) - \nabla \cdot \left[\nabla_Y \nabla_Y C_\alpha(x, \nabla u(t, x)) : \nabla p(t, x)\right] = w(t, x)$$

$$p(T, x) = \dot{p}(T, x) = 0, \qquad x \in \Omega$$

$$p(t, \xi) = 0, \qquad (t, \xi) \in [0, T] \times \partial\Omega$$

with $A : B = \sum_{ij} A_{ij} B_{ij}$ and

$$C_\alpha(x, Y) = \sum_{K=1}^{N} \alpha_K C_K(x, Y).$$

Remark 7 It should be mentioned that we use simulated noise-free data in our experiments. However, we assume a small noise level, which is necessary for applying the discrepancy principle, for both methods in order to guarantee the robustness of the algorithms and thus convergence. In addition, the noise level in the RESESOP method determines the width of the stripes. Using a noise level for simulated exact data is justified by disturbances in the data due to discretization and potential inaccuracies in the model. The appropriate value was determined by trial and error.

The first series of experiments examines a plate with a delamination whose center is located at $(x_2, x_3) = (-1.5, -1.5)$, see Fig. 2. The corresponding coefficients α_{ij}, $i, j \in \{0, \ldots, 30\}$ in (20) are given by

$$\alpha_{13,13} = 2, \quad \alpha_{13,14} = 3, \quad \alpha_{14,13} = 4, \quad \alpha_{14,14} = 2,$$

and $\alpha_{i,j} = 1$ elsewhere (Experiment 1). This setting for α_{ij} in fact corresponds to the damage in Fig. 2 (left picture), which is emphasized in the right picture of Fig. 2 where the coefficient matrix $\alpha = (\alpha_{i,j})_{i,j=0,\ldots 30}$ is plotted. There as well as in all reconstruction plots we apply linear interpolation to α to obtain a picture of

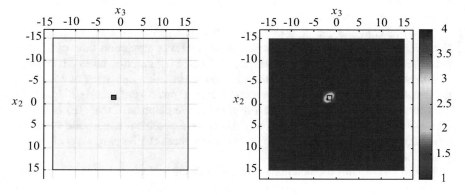

Fig. 2 Left picture: plate with damage at $(-1.5, -1.5)$ (Experiment 1). Right picture: exact coefficient matrix α for experiment 1

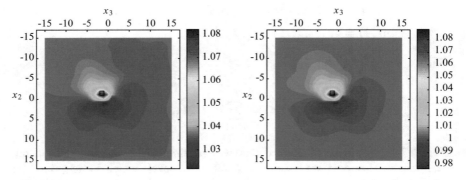

Fig. 3 Result of experiment 1 after 200 iterations with the Landweber method (left) and after 9 iterations with the RESESOP method (right)

higher resolution. The inverse problem consists of computing the coefficient matrix $\alpha \in \mathbb{R}^{31 \times 31}$ from full field data $u(t_j, x_m)$ where the discrete points x_m correspond to the knots of the Finite Element solver.

In the tests we compare different solution methods regarding the residual, the number of necessary iteration steps and computation time. We implemented the Landweber iteration (9) as well as RESESOP (18) with the Landweber descent as single search direction and optimized step size in each iteration. Figure 3 illustrates the results that are obtained after 9 iterations of RESESOP and 200 iterations of Landweber's method. The RESESOP iteration was stopped by the discrepancy principle, whereas the Landweber iteration was stopped before the discrepancy principle was fulfilled. In both cases the defect is detected at the correct location, but the coefficients α_{ij} are underestimated. We conclude that the same reconstruction quality is achieved with both methods but that RESESOP needs a significantly smaller number of iterations compared to the Landweber scheme. That means that RESESOP with only one search direction and optimized step size converges much faster than Landweber's method.

Next we compare the computing time that is needed for each iteration. One Landweber iteration needs 2.8 h, resulting in a total computation time of 23 days until the discrepancy principle is fulfilled. A RESESOP iteration takes 3 h and thus a bit more than a Landweber step. But, since only 9 iterations are necessary to satisfy the discrepancy principle, the entire reconstruction process only needs 27 h in total. This means an acceleration by a factor of $\sim$51. We emphasize that (IP) is a high-dimensional parameter identification problem for a nonlinear hyperbolic system in time and space and thus belongs to the currently most challenging class of inverse problems at all.

Figure 4 compares the residuals of the RESESOP technique and the Landweber method for Experiment 1. The red curve shows a typical behavior of the Landweber method. We see a strong decrease in the residual $\| R_n^\delta \|$ until iteration 15, followed by a very slow decrease afterwards. This phenomenon is the reason why Landweber's method needs so much time until the discrepancy criterion is fulfilled. We note also that the residual $\| R_n^\delta \|$ is not monotonically decreasing for RESESOP, in contrast to

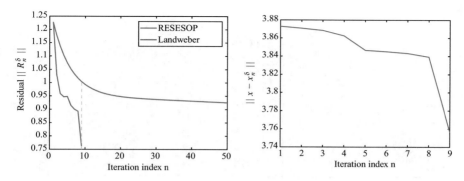

Fig. 4 Behavior of the respective residuals $\|R_n^\delta\|$ (left) and errors $\|x - x_n^\delta\|$ of the RESESOP method (right)

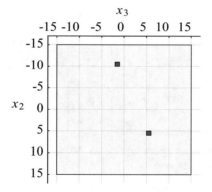

Fig. 5 Plate with damages A $(-1.5, -10.5)$ and B $(5.5, 5.5)$ (Experiment 2)

the Landweber iteration. The reason is that RESESOP is constructed such that the sequence $\|x - x_n^\delta\|$ is monotonically decreasing, but not the sequence of residuals $\|R_n^\delta\|$, where x denotes the exact solution of the underlying inverse problem and x_n^δ the n-th iterate for noisy data. This is also shown in the right-hand plot of Fig. 4 and is in accordance with the analysis of the method outlined in [33].

In the second experiment we consider a setting consisting of two damages that are not located at the plate's center. Note that the center is also the region of wave excitation by $f(t, x)$. We assume that the damage which is closer to the center is the first to interact with the wave and thus is more pronounced in the reconstruction. The experimental setup is illustrated in Fig. 5 (Experiment 2).

Figure 6 depicts the reconstruction from 50 iterations of the Landweber procedure (left picture). Then we terminated the iteration process because of its outrageous computation time. The values of the coefficient matrix α are contained in the very small interval $[1.0072, 1.0088]$. The situation is different for the RESESOP technique. RESESOP stopped after iteration 17 according to the discrepancy principle. The result is visualized in Fig. 6 (right picture). The entries $\alpha_{i,j}$ of the coefficient matrix are contained in $[0.94, 1.12]$ making it easier to distinguish

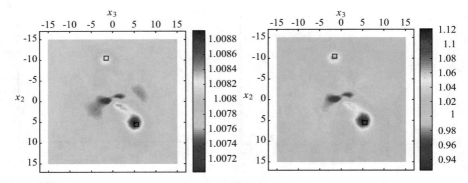

Fig. 6 Result of Experiment 2 after 50 iterations using the Landweber method (left) and after 17 iterations with the RESESOP method (right)

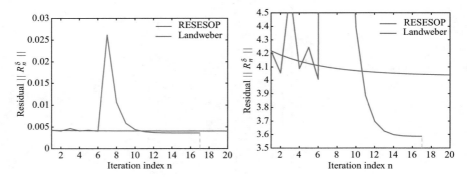

Fig. 7 Residuals $\|R_n^\delta\|$ from the RESESOP and Landweber method from Experiment 2 (left picture) and a re-scaling of the y-axis (right picture)

defects from undamaged parts of the structure. As expected the damage which is located closer to the center is more pronounced due to the excitation in the middle of the plate in both reconstructions.

In Fig. 7 we compare the residuals $\|R_n^\delta\|$ of the two methods applied to Experiment 2. The RESESOP iteration stops after iteration 17 according to the discrepancy principle, whereas the residual for the Landweber method seems to be almost constant. The right-hand plot in Fig. 7 shows a re-scaling to emphasize the oscillations of $\|R_n^\delta\|$ for RESESOP in the first few iterations as well as the monotonic decrease of $\|R_n^\delta\|$ for Landweber's method. Furthermore both figures demonstrate again a faster convergence of RESESOP compared to the Landweber procedure.

We consider a further numerical experiment where damage A is moved closer to the center of the plate compared to Experiment 2 and damage B remains fixed (Experiment 3). This scenario is illustrated in Fig. 8. The corresponding coefficient matrix α remains unchanged, only the locations of the entries $\alpha_{i,j}$ are adjusted to the damages A and B.

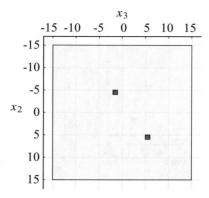

Fig. 8 Plate with damages A $(-1.5, -4.5)$ and B $(5.5, 5.5)$ (Experiment 3)

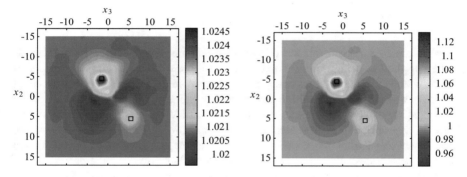

Fig. 9 Result of experiment 3 using the Landweber method after 50 iterations (left) and the RESESOP method after 14 iterations (right)

Figure 9 shows the reconstructed coefficient matrix α using Landweber and RESESOP. In both cases the locations of the defects are accurately detected, while again the damage that is located closer to the center is highlighted stronger. The Landweber iteration has been stopped after 50 iterations (yielding 140 h computation time) without having fulfilled the discrepancy principle. The RESESOP method, however, satisfied the discrepancy principle after 14 iterations (42 h computation time) only, showing that it is significantly more efficient in spite of the additional computation time due to the step size optimization in each iteration.

The RESESOP technique outperforms the Landweber method in other respects as well. Considering the reconstructed values $\alpha_{i,j}$, we observe that the contrast in the Landweber reconstructions is very low, whereas an application of RESESOP results in larger differences of the absolute values.

Figure 10 shows the residuals of the two methods when applied to Experiment 3. Similarly to Experiment 2, the figures clearly demonstrate the superiority of RESESOP compared to Landweber.

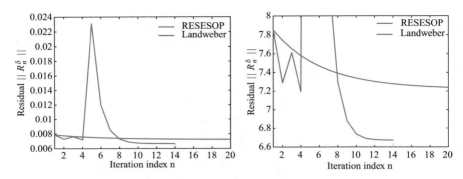

Fig. 10 Residuals in experiment 3 of the Landweber and the RESESOP method (left) and adapted scaling of the y-axis (right)

5   Conclusion

We presented the performance of two different iterative regularization methods when applied to a high-dimensional inverse problem from the class of parameter identification problems that is based on a system of nonlinear, hyperbolic differential equations equipped with initial and boundary values. The system describes the propagation of elastic waves in a three-dimensional structure whose constitutive law is appropriately represented by a hyperelastic material model, i.e. where the first Piola-Kirchhoff stress tensor is given as the derivative of the stored energy with respect to strain. The nonlinearity allows also for large deformations. The considered inverse problem is the computation of the stored energy from measurements of the full displacement field depending on space and time. Since the stored strain energy encodes virtually all essential mechanical properties of the structure on a macro-scale, it might yield useful pointers for possible damages and thus might be important for simulations in the area of Structural Health Monitoring (SHM). Moreover, since hyperelasticity describes the elastic behavior accurately for a large class of materials such as composites, rubber or biological tissues, the concepts that have been developed in this chapter are appropriate for a broad range of applications and not only for SHM systems.

To solve this inverse problem we implemented the well-known Landweber method and Regularized Sequential Subspace Optimization (RESESOP) technique. The latter consists of iterative metric projections on hyperplanes that are determined by the used search directions, the nonlinearity of the forward mapping (via the constant in the tangential cone condition) and the noise level. RESESOP uses in each iteration step a finite number of search directions where the Landweber direction, i.e. the negative gradient of the current residual, is included. Using only one search direction, RESESOP coincides with Landweber where the step size is optimized to minimize the norm-distance of the current iterate to a (locally unique) exact solution. Both numerical methods have been evaluated by means of three different damage scenarios for a Neo-Hookean material model and the usage of simulated measurement data. In all three cases RESESOP outperforms Landweber with respect to a faster convergence, a significant decrease of computation time and higher contrasts.

Future research could include model reduction techniques or the application of methods from Machine Learning. Both concepts could help to achieve a further significant improvement with respect to computation time that is necessary for an implementation of the method in real-world SHM scenarios.

References

1. D. Arndt, W. Bangerth, T.C. Clevenger, D. Davydov, M. Fehling, D. Garcia-Sanchez, G. Harper, T. Heister, L. Heltai, M. Kronbichler, R.M. Kynch, M. Maier, J.-P. Pelteret, B. Turcksin, D. Wells, The `deal.II` Library, Version 9.1. J. Numer. Math. **27**(4), 203–213 (2019)
2. F, Binder, F. Schöpfer, T. Schuster, Defect localization in fibre-reinforced composites by computing external volume forces from surface sensor measurements. Inverse Probl. **31**, 025006 (2015)
3. M. Bonnet, A. Constantinescu, Inverse problems in elasticity. Inverse Probl. **21**, R1–R50 (2005)
4. L. Bourgeois, F. Le Louer, E. Lunéville, On the use of Lamb modes in the linear sampling method for elastic waveguides. Inverse Probl. **27**, 055001 (2011)
5. P.G. Ciarlet, *Mathematical Elasticity, Volume I: Three-Dimensional Elasticity* (Elsevier Science Publishers B. V., Amsterdam, 2004)
6. M. de Hoop, G. Uhlmann, Y. Wang, Nonlinear interaction of waves in elastodynamics and an inverse problem. Math. Ann. **376**, 765–795 (2020)
7. U. Gabbert, R. Lammering, T. Schuster, M. Sinapius, P. Wierach, Lamb-wave based structural health monitoring in polymer composites, in *Research Topics in Aerospace* (Springer, Heidelberg, 2018)
8. O. Gauthier, J. Virieux, A. Tarantola, Two-dimensional nonlinear inversion of seismic waveforms: numerical results. Geophysics **51**(7), 1387–1403 (1986)
9. V. Giurgiutiu, *Structural Health Monitoring with Piezoelectric Wafer Active Sensors* (Academic, New York, 2008)
10. R. Gu, B. Han, Y. Chen, Fast subspace optimization method for nonlinear inverse problems in Banach spaces with uniformly convex penalty terms. Inverse Probl. **35**(12), 125011 (2019)
11. M. Hanke, A. Neubauer, O. Scherzer, A convergence analysis of the Landweber iteration for nonlinear ill-posed problems. Numer. Math. **72**, 21–37 (1995)
12. G.A. Holzapfel, *Nonlinear Solid Mechanics II* (Wiley, New York, 2000)
13. S. Hubmer, E. Sherina, A. Neubauer, O. Scherzer, Lamé parameter estimation from static displacement field measurements in the framework of nonlinear inverse problems. SIAM J. Imaging **11**(2), 1268–1293 (2018)
14. B. Kaltenbacher, A. Lorenzi, A uniqueness result for a nonlinear hyperbolic equation. Appl. Anal. **86**(11), 1397–1427 (2007)
15. B. Kaltenbacher, A. Neubauer, O. Scherzer, *Iterative Regularization Methods for Nonlinear Ill-Posed Problems* (De Gruyter, Berlin, 2008)
16. A. Lechleiter, J.W. Schlasche, Identifying Lamé parameters from time-dependent elastic wave measurements. Inverse Probl. Sci. Eng. **25**, 2–26 (2017)
17. P. Maaß, R. Strehlow, An iterative regularization method for nonlinear problems based on Bregman projections. Inverse Probl. **11**, 115013 (2018)

18. J.E. Marsden, T.J.R. Hughes, *Mathematical Foundations of Elasticity* (Courier Corporation, North Chelmsford, 1994)
19. G. Narkiss, M. Zibulevsky, Sequential subspace optimization method for large-scale unconstrained optimization. Technical report, Technion – The Israel Institute of Technology, Department of Electrical Engineering (2005)
20. J. Nocedal, S. Wright, *Numerical Optimization* (Springer Science & Business Media, New York, 2006)
21. N. Rauter, R. Lammering, Investigation of the higher harmonic Lamb wave generation in hyperelastic isotropic material. Phys. Procedia **70**, 309–313 (2015)
22. O. Scherzer, An iterative multi level algorithm for solving nonlinear ill-posed problems. Numer. Math. **80**, 579–600 (1998)
23. F. Schöpfer, T. Schuster, Fast regularizing sequential subspace optimization in Banach spaces. Inverse Probl. **24**, 015013 (2008)
24. F. Schöpfer, T. Schuster, A.K. Louis, Metric and Bregman projections onto affine subspaces and their computation via sequential subspace optimization methods. J. Inverse Ill-Posed Probl. **16**, 479–506 (2008)
25. T. Schuster, A. Wöstehoff, On the identifiability of the stored energy function of hyperelastic materials from sensor data at the boundary. Inverse Probl. **30**, 105002 (2014)
26. T. Schuster, B. Kaltenbacher, B. Hofmann, K.S. Kazimierski, *Regularization Methods in Banach Spaces* (De Gruyter, Berlin, 2012)
27. J. Seydel, T. Schuster, On the linearization of identifying the stored energy function of a hyperelastic material from full knowledge of the displacement field. Math. Meth. Appl. Sci. **40**, 183–204 (2016)
28. J. Seydel, T. Schuster, Identifying the stored energy of a hyperelastic structure by using an attenuated Landweber method. Inverse Probl. **33**, 124004 (2017)
29. S.L. Sridhar, Y. Mei, S. Goenezen, Improving the sensitivity to map nonlinear parameters for hyperelastic problems. Comput. Methods Appl. Mech. **331**, 474–491 (2018)
30. A. Tarantola, A strategy for nonlinear elastic inversion of seismic reflection data. Geophysics **51**(10), 1893–1903 (1986)
31. A. Tarantola, Theoretical background for the inversion of seismic waveforms including elasticity and attenuation. Pure Appl. Geophys. **128**, 365–399 (1988)
32. S. Tong, B. Han, H. Long, R. Gu, An accelerated sequential subspace optimization method based on homotopy perturbation iteration for nonlinear ill-posed problems. Inverse Probl. **35**(12), 125005 (2019)
33. A. Wald, T. Schuster, Sequential subspace optimization for nonlinear inverse problems. J Inverse Ill-Posed Probl. **25**(1), 99–117 (2017)
34. A. Wald, T. Schuster, Tomographic terahertz imaging using sequential subspace optimization, in *New Trends in Parameter Identification for Mathematical Models*, ed. by B. Hofmann, A. Leitao, J. Zubelli (Birkhäuser, Basel, 2018)
35. A. Wald, A fast subspace optimization method for nonlinear inverse problems in Banach spaces with an application in parameter identification. Inverse Probl. **34**, 085008 (2018)
36. A. Wöstehoff, T. Schuster, Uniqueness and stability result for Cauchy's equation of motion for a certain class of hyperelastic materials. Appl. Anal. **94**(8), 1561–1593 (2015)

Joint Motion Estimation and Source Identification Using Convective Regularisation with an Application to the Analysis of Laser Nanoablations

Lukas F. Lang, Nilankur Dutta, Elena Scarpa, Bénédicte Sanson,
Carola-Bibiane Schönlieb, and Jocelyn Étienne

Abstract We propose a variational method for joint motion estimation and source identification in one-dimensional image sequences. The problem is motivated by fluorescence microscopy data of laser nanoablations of cell membranes in live Drosophila embryos, which can be conveniently—and without loss of significant information—represented in space-time plots, so called kymographs. Based on mechanical models of tissue formation, we propose a variational formulation that is based on the nonhomogenous continuity equation and investigate the solution of this ill-posed inverse problem using convective regularisation. We show existence of a minimiser of the minimisation problem, derive the associated Euler–Lagrange equations, and numerically solve them using a finite element discretisation together with Newton's method. Based on synthetic data, we demonstrate that source estimation can be crucial whenever signal variations can not be explained by advection alone. Furthermore, we perform an extensive evaluation and comparison of various models, including standard optical flow, based on manually annotated kymographs that measure velocities of visible features. Finally, we present results for data generated by a mechanical model of tissue formation and demonstrate that our approach reliably estimates both a velocity and a source.

L. F. Lang (✉) · C.-B. Schönlieb
Department of Applied Mathematics and Theoretical Physics, University of Cambridge,
Cambridge, UK
e-mail: ll542@cam.ac.uk; cbs31@cam.ac.uk

N. Dutta · J. Étienne
Laboratoire Interdisciplinaire de Physique, Université Grenoble Alpes, Grenoble, France
e-mail: nilankur.dutta@univ-grenoble-alpes.fr; jocelyn.etienne@univ-grenoble-alpes.fr

E. Scarpa · B. Sanson
Department of Physiology, Development and Neuroscience, University of Cambridge,
Cambridge, UK
e-mail: es697@cam.ac.uk; bs251@cam.ac.uk

© Springer Nature Switzerland AG 2021 191
B. Kaltenbacher et al. (eds.), *Time-dependent Problems in Imaging and Parameter Identification*, https://doi.org/10.1007/978-3-030-57784-1_7

1 Introduction

1.1 *Motivation*

Motion estimation is a ubiquitous and fundamental problem in image analysis, see e.g. [5]. It is concerned with the efficient and accurate estimation of displacement fields in spatio-temporal data and has a wide range of applications, not necessarily limited to natural images. Optical flow [30] is one popular example of motion estimation, which designates the apparent motion of brightness patterns in a sequence of images and is based on the assumption of constant brightness. Recently, optical flow methods have been used for the quantitative analysis of biological image sequences on cellular and subcellular level. See, for instance, [2, 9, 10, 21, 31, 34, 39, 44, 50]. While the concept is in practice well-suited for natural scenes, the use of the less restrictive continuity equation, which arises from mass conservation, can be more favourable in certain scenarios. For instance, in [15, 16] it is used for fluid flow estimation.

In developmental biology, the study of the morphogenesis of model organisms is specifically calling for image analysis methods that are able to extract time-dependent deformations and flow velocities from microscopy image sequences. Morphogenesis is the process that leads to an organism developing its shape as a result of the implementation of a genetic programme [28] and includes, among other mechanisms, tissue deformations. These deformations can be observed through video microscopy by fluorescently labelling molecules that are associated with compounds of mechanical relevance within an embryo [32].

In many cases, these molecules are organised spatially in discrete structures, such as cell membranes. To compute deformations on the level of these molecular structures, segmentation or detection, and subsequent tracking are the methods of choice [22]. As a result, detailed knowledge of the mechanics of morphogenetic processes can be gained [7]. In case the recorded tissue lacks structure, particle image velocimetry (PIV) is generally used to compute (sparse) displacement fields in image sequences. See, for example, [37, 47].

One difficulty in abovementioned approaches is that the observed structures often have a short life time and are being degraded during the observation, while new structures of the same type are being created [47]. Indeed, these molecular structures, and thus their fluorescent signal response, can be described by an advection–reaction equation rather than pure advection [42]. The signal variations due to reaction are a source of error in the estimation of motion that we propose to address in this paper.

Beyond the need of measurements of dense velocities of moving fluorescently-labelled molecular structures that prove robust with respect to the reaction term, it is of general interest to quantify the reaction term itself [42], and of general interest to extract information about all the physical quantities and processes, such as diffusion, that govern the observed tissue flow. For this, an accurate estimation of the velocity

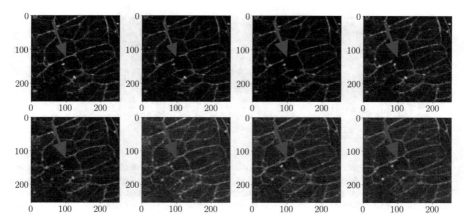

Fig. 1 Depicted are frames no. 4, 6, 10, 20, 40, 60, 80, and 90 (left to right, top to bottom) of a 2D image sequence of fluorescently labelled cell membranes of Drosophila during a laser nanoablation. The entire sequence contains 100 frames recorded over roughly 6.5 s, and the imaged section spans approximately $42.2 \times 42.2\,\mu m^2$. The laser ablation is applied at frame number five, i.e. between the first and the second image shown. Its location is indicated with a magenta arrow. Observe the instantaneous recoiling and tissue loss of the cut membrane, and the subsequent growth of tissue in the cut region. Moreover, note the changes in contrast over time and the line artefacts

field corresponding to the time evolution of the distribution of fluorescent molecules is crucial [42, 47].

In this article, we argue that utilising variational motion estimation can help to identify physical quantities by estimating velocities in real data. For simplicity, we choose to demonstrate this method in a case where the biophysical data can be reduced to one space dimension. This allows to create convenient space-time representations, so-called kymographs.

One-dimensional data are indeed relevant in tissue dynamics when the cell-cell junctions are found to be aligned along a straight line called a supracellular actomyosin cable [8, 40]. A common experiment to investigate the function of these cables is to cut them locally using intense laser illumination [25] and to observe the dynamics that follow. Figure 1 illustrates a prototypical two-dimensional (time-lapse) fluorescence microscopy image sequence where a cable is being severed by such a laser ablation.

Most of the relevant dynamics occur along the cable itself, thus projecting the recorded signal in a narrow stripe of less than two micrometers along its average direction preserves most of the information. See Fig. 2 for an example of a kymograph obtained from the image sequence shown in Fig. 1. Variations in fluorescence intensity are clearly visible and displacements of features can easily be measured, e.g. with existing (tracking) tools [12, 13, 38, 41]. See also Fig. 8 for an example of manually created tracks.

Analysing such data is challenging for many reasons. First, simultaneous estimation of velocity and decay or increase of the signal renders the problem ill-posed, as

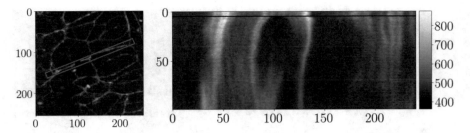

Fig. 2 The left image shows one frame of the image sequence in Fig. 1. A supracellular cable is indicated with a magenta rectangle. To reduce the problem to one dimension, the data is summed along the transverse direction within the rectangular zone. The right image depicts the kymograph obtained from this dimension reduction. Time runs from top to bottom and the horizontal black line at frame five indicates the time of the laser ablation, during which the signal acquisition was paused

we will illustrate below, and suitable (qualitative) assumptions on favoured solutions are required. Second, the obtained kymographs are very noisy, contain artefacts due to the acquisition technique, and sometimes suffer from off-plane motion of the cables. Third, the velocity field potentially contains discontinuities at the time of the laser ablation. Fourth, data is missing during the application of the laser cut and only a limited field of view is available due to the nature of the kymograph. Moreover, bleaching of tissue leads to a decrease in contrast when being exposed over long periods of time. See Figs. 1 and 2 for illustration of these issues. In this work we address mainly issues one and two.

Motivated by the laser nanoablation problem we restrict ourselves to the one-dimensional case and denote by $\Omega \subset \mathbb{R}$ the spatial domain. For $T > 0$, we model the actomyosin concentration as a function $f : (0, T) \times \Omega \to \mathbb{R}$ that is proportional to the observed fluorescence response. In the following, we assume that it solves the Cauchy problem for the non-homogeneous continuity equation, i.e. the right-hand side is a source (or sink), in one dimension:

$$\partial_t f + \partial_x (f v) = k \quad \text{in } (0, T) \times \Omega,$$
$$f(0, \cdot) = f_0 \quad \text{in } \Omega. \tag{1}$$

Here, $v : (0, T) \times \Omega \to \mathbb{R}$ is a given velocity field, $k : (0, T) \times \Omega \to \mathbb{R}$ a given source, and $f_0 : \Omega \to \mathbb{R}$ a given initial condition.

Solution theory for problem (1) is closely related to solutions of the initial value problem

$$\partial_t \phi(t, x_0) = v(t, \phi(t, x_0)), \quad \text{in } (0, T),$$
$$\phi(0, x_0) = x_0, \quad \text{in } \Omega, \tag{2}$$

via the method of characteristics, see e.g. [24, Chap. 3.2]. Here, the map $\phi: [0, T) \times \Omega \to \Omega$ denotes a so-called flow of v. Existence and uniqueness of problem (1) can be established by application of the Picard–Lindelöf theorem to (2), provided that both v and k are sufficiently regular. In particular, it can be shown that (2) has a unique (global) solution and, moreover, that this solution is a diffeomorphism. For an introduction and for details we refer, for instance, to [18, 60].

In this work we are concerned with the solution of the inverse problem associated with (1), which is to estimate a pair $(v, k)^\top$ from (potentially noisy) observations f. Its ill-posedness is immediate because the problem is underdetermined and, given a solution $(v_1, k_1)^\top$, the pair $(v_2, k_1 - \partial_x(f v_1) + \partial_x(f v_2))^\top$ for differentiable v_2 denotes a solution as well. In addition, it is easy to see that—without further assumptions on v and on k—the pair $(0, \partial_t f)^\top$ is always a solution, albeit not a desired one.

In view of this ill-posedness and this ambiguity we consider the variational form

$$\min_{(v,k)^\top} \|\partial_t f + \partial_x(f v) - k\|_{L^2}^2 + \alpha \mathcal{R}(v, k), \tag{3}$$

where the data term is the squared L^2 norm of the first equation in (1), $\mathcal{R}(v, k)$ a suitable regularisation functional, and $\alpha > 0$ is a regularisation parameter. In this article we consider different choices of $\mathcal{R}$. While source identification has been treated before in the literature, e.g. in [4], the main goal of this article is to recover a source (or sink) k which is constant along characteristics of the flow (2). In other words, we are interested in utilising the *convective derivative*

$$\frac{d}{dt} k(t, \phi(t, x_0)), \tag{4}$$

for regularisation. Here, the flow ϕ is a solution to (2).

Its use is inspired by the work in [33], where the convective derivative of the velocity along itself was used. From a physical perspective this choice seems natural as, in comparison to using, for instance, the L^2 norm or the H^1 seminorm for regularisation of k, it is consistent with the movement of the tracked cell tissue, which can be assumed to be the main origin of changes in the observed fluorescence intensity. However, from a numerical point of view this choice comes at the expense of having to solve nonlinear optimality conditions.

1.2 Contributions

The main contributions of this article are as follows. First, we propose a variational model based on the non-homogeneous continuity equation for joint motion estimation and source identification in kymographs. Second, we study the variational properties of utilising the convective derivative (4) for regularisation. Following

[33], we establish existence of minimisers of the nonconvex functional. Third, for the numerical solution of the corresponding nonlinear Euler–Lagrange equations we propose to use Newton's method and a finite element discretisation. Fourth, we present numerical results based on kymographs of laser nanoablation experiments conducted in live Drosophila embryos. Moreover, we provide an extensive experimental evaluation of different data fidelity and regularisation functionals based on manually created tracks, and evaluate our approach using synthetic data generated by solving a mechanical model of tissue formation.

1.3 Related Work

In [30], Horn and Schunck were the first to pursue a variational approach for dense motion estimation between a pair of images. They considered a quadratic Tikhonov-type functional that relies on conservation of brightness and used (squared) H^1 Sobolev seminorm regularisation. This isotropic regularisation incorporates a preference for spatially regular vector fields. Well-posedness of the functional was proved in [51] and the problem was solved numerically with a finite element method. For a general introduction to variational optical flow see, for instance, [5].

In [56], the problem was treated on the space-time domain and extended to incorporate both spatial as well as temporal isotropic regularisation. In [55], a unifying framework for a family of convex functionals was established, and both isotropic and anisotropic variants were considered. We refer to [54] for more details on nonlinear diffusion filtering, and to [57] for an overview of numerous optical flow models and a taxonomy of isotropic and anisotropic regularisation functionals.

The convective derivative has already been used in several works. For instance, in [14] for simultaneous image inpainting and motion estimation. In [43], an optical flow term was incorporated in a Mumford–Shah-type functional for joint image denoising and edge detection in image sequences. Moreover, in [11] it was used for joint motion estimation and image reconstruction in a more general inverse problems setting. In [33] the convective acceleration was used for regularisation together with a contrast invariant Horn–Schunck-type functional. The corresponding nonlinear Euler–Lagrange equations were solved using a finite element method and alternating minimisation.

According to [15], the article [52] is credited for being the first to propose the use of the less restrictive continuity equation for motion estimation. Later it was used, for instance, to find 3D deformations in medical images [20, 53], to analyse meteorological satellite images [6, 15, 61], and to estimate fluid [16, 59] and blood flow [3] in image sequences. For a general survey on variational methods for fluid flow estimation see [29].

Whenever mass conservation is not satisfied exactly, e.g. due to illumination changes, it can be beneficial to account for these violations. For instance, in [27] they incorporated physical models. In [4] they simultaneously estimated image intensity, flux, and a potential source. In contrast to our work, only L^2 integrability of the source was assumed and the constraint was enforced exactly, leading to an optimal control formulation, which was solved with a finite element method.

2 Problem Formulation

2.1 Preliminaries

2.1.1 Notation

For $T > 0$ and $\Omega \subset \mathbb{R}$ a bounded, connected, and open set we denote by $E = (0, T) \times \Omega$ the spatio-temporal domain and by ∂E its boundary. For a smooth function $f\colon E \to \mathbb{R}$ we denote by $\partial_t f$, respectively, $\partial_x f$ the partial derivatives with respect to time and with respect to space, and by $\nabla f = (\partial_t f, \partial_x f)^\top$ its spatio-temporal gradient. The space-time Laplacian of f is denoted by Δf. Analogously, for a smooth vector field $w\colon E \to \mathbb{R}^2$ with $w = (w^1, w^2)^\top$, its gradient is denoted by $\nabla w = (\nabla w^1, \nabla w^2)^\top$ and its spatio-temporal divergence is given by $\nabla \cdot w = \partial_t w^1 + \partial_x w^2$. Moreover, we will write $A \lesssim B$ whenever there exists a constant $c > 0$ such that $A \le cB$ holds. Finally, by the Cauchy–Schwarz inequality and application of Young's inequality, we have

$$\|a + b\|_{L^2}^2 \lesssim \|a\|_{L^2}^2 + \|b\|_{L^2}^2. \tag{5}$$

Here, and in the following, we use $\|\cdot\|_{L^2}$ instead of $\|\cdot\|_{L^2(E,\mathbb{R})}$ and $\|\cdot\|_{L^2(E,\mathbb{R}^2)}$ for simplicity.

2.1.2 Convective Derivative

Let $\phi\colon E \to \Omega$ be a flow through the domain Ω, i.e. for every $t \in (0, T)$ the map $\phi(t, \cdot)\colon \Omega \to \Omega$ is a diffeomorphism and, for a fixed starting point $x_0 \in \Omega$, the trajectory $\phi(\cdot, x_0)$ is smooth.

With every trajectory $\phi(\cdot, x_0)$ that originates at $x_0 \in \Omega$ we can associate a velocity at every time $t \in (0, T)$ via (2). Thus, a flow ϕ gives rise to a scalar (velocity) field $v\colon E \to \mathbb{R}$ by means of

$$v(t, x) = \partial_t \phi(t, x_0)\Big|_{x_0 = \phi^{-1}(t,x)}, \tag{6}$$

where $\phi^{-1}(t, x)$ denotes the inverse of $\phi(t, \cdot)$ at $x \in \Omega$, which is the starting point of the curve that passes through x at time t.

For a scalar quantity $k\colon E \to \mathbb{R}$, we define the *convective derivative* of k along a flow ϕ, denoted by $D_v k$, as

$$D_v k(t, x) := \frac{d}{dt} k(t, \phi(t, x_0)) \bigg|_{x_0 = \phi^{-1}(t,x)}$$

$$= \partial_t k(t, x) + \partial_x k(t, x) v(t, x).$$

For convenience we adopt the notation $\bar{v} = (1, v)^\top$. Clearly, $D_v k = \nabla k \cdot \bar{v}$ vanishes for pairs $(v, k)^\top$ such that $\bar{v} \perp \nabla k$ in $\mathbb{R}^2$. Moreover, from a straightforward calculation we obtain

$$|D_v k|^2 = \nabla k^\top \bar{v} \bar{v}^\top \nabla k, \tag{7}$$

where $\bar{v} \bar{v}^\top$ is the matrix

$$\bar{v} \bar{v}^\top = \begin{pmatrix} 1 & v \\ v & vv \end{pmatrix}.$$

As noted in [33], the action of $D_v k$ becomes more clear by writing (7) as

$$|D_v k|^2 = |\bar{v}|^2 \left(\nabla k \cdot \frac{\bar{v}}{|\bar{v}|} \right)^2 = |\nabla k|^2 \left(\bar{v} \cdot \frac{\nabla k}{|\nabla k|} \right)^2.$$

The first identity states that, for fixed $\bar{v}$, minimisation of the functional

$$\mathcal{F}: (v, k)^\top \mapsto \|D_v k\|_{L^2}^2 \tag{8}$$

promotes functions k that vary little in the direction of $\bar{v}$. In addition, the weighting factor gives importance to regions of large $\bar{v}$. On the other hand, the second identity states that, for fixed k, minimisation of (8) favours functions $\bar{v}$ that are tangent to the level lines of k. This time, importance is given to regions where ∇k is large. See Fig. 3 for illustration.

Fig. 3 Illustration of a pair $(v, k)^\top$ minimising (8) where the velocity v is constant

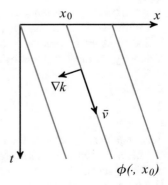

The connection to anisotropic diffusion [54] is immediate when considering the Euler–Lagrange equations associated with (8). They read

$$\partial_x k D_v k = 0,$$
$$\nabla \cdot (\bar{v} \bar{v}^\top \nabla k) = 0. \tag{9}$$

Note that the system (9) is nonlinear.

2.2 Variational Model and Existence of a Minimiser

In this section we formulate the joint motion estimation and source identification problem and study the use of (8) as regularisation functional. For simplicity, we will denote a velocity-source pair by $w = (v, k)^\top$. In the following, we intend to minimise a variational formulation of the form

$$\mathcal{E}(w) := \|\partial_t f + \partial_x (fv) - k\|_{L^2}^2 + \alpha \mathcal{R}(w), \tag{10}$$

where $\mathcal{R}$ is yet to be defined and $\alpha > 0$ is a regularisation parameter.

In order to fix a concrete choice of $\mathcal{R}$ let us discuss two issues. First, it should be selected such that the functional is well-defined for an appropriate function space. In particular, we require the weak derivative of v in the data term in (10) to exist and to be bounded with respect to an appropriate norm. Second, $\partial_x (fv)$ and k need to differ qualitatively in order to obtain a meaningful decomposition of the signal $\partial_t f$.

Before we state the model, let us follow the ideas in [33, Chap. 5.2.2] and discuss some issues arising with the choice

$$\alpha \mathcal{R}(w) := \alpha \|D_v k\|_{L^2}^2.$$

To ensure well-definedness of the functional (10) we derive, by application of (5) and Hölder's inequality, the estimates

$$\|\partial_t f + \partial_x (fv) - k\|_{L^2}^2 \lesssim \|\partial_t f\|_{L^2}^2 + \|f\|_{L^\infty}^2 \|\partial_x v\|_{L^2}^2 + \|\partial_x f\|_{L^\infty}^2 \|v\|_{L^2}^2 + \|k\|_{L^2}^2,$$

and

$$\|D_v k\|_{L^2}^2 \lesssim \|\partial_t k\|_{L^2}^2 + \|v\|_{L^\infty}^2 \|\partial_x k\|_{L^2}^2.$$

As a consequence, minimisation over the space

$$X = \{(v, k)^\top : v \in L^\infty(E), \partial_x v \in L^2(E), k \in H^1(E)\}, \tag{11}$$

seems appropriate for data $f \in W^{1,\infty}(E)$ in the space of functions with essentially bounded weak derivatives up to first order.

However, in order to show existence of minimisers of (10) by application of the direct method [19], one requires a coercivity estimate of the form

$$\|w\|_X^2 - b \lesssim \mathcal{E}(w), \tag{12}$$

where $b \geq 0$ is some constant and the norm of the space X is defined via

$$\|w\|_X^2 = \|v\|_{L^\infty}^2 + \|\partial_x v\|_{L^2}^2 + \|k\|_{H^1}^2.$$

Without further restriction of the data f, the above functional is not coercive with respect to this space as the following example shows.

Example

Let $f = ax$ for some $0 < a < +\infty$ and, consequently, we have that $\partial_x f = a$. Let $\{w_n\}$ be the sequence with $w_n = (n, an)^\top$, for $n \in \mathbb{N}$, and observe that $\|w_n\|_X \to \infty$ as $n \to +\infty$, while the value of the functional $\mathcal{E}$ stays bounded. In fact, $\mathcal{E}(w_n) = 0$, since all non-zero terms in the data fidelity functional cancel and all terms in the regularisation functional vanish.

Moreover, it is clear that, for pairs $w = (0, k)^\top$, the inequality (12) cannot hold in general. As a remedy, we consider minimising over all pairs $w = (v, k)^\top$ arising from the Sobolev space $H^1(E, \mathbb{R}^2)$. Its norm is defined via

$$\|w\|_{H^1}^2 = \|w\|_{L^2}^2 + \|\nabla w\|_{L^2}^2.$$

In further consequence, our goal is to find a minimiser $w \in H^1(E, \mathbb{R}^2)$ of the functional $\mathcal{E}$ in (10) with

$$\alpha \mathcal{R}_{\text{CR}}(w) := \alpha \|\nabla w\|_{L^2}^2 + \beta \|D_v k\|_{L^2}^2,$$

and regularisation parameters $\alpha, \beta > 0$. Our final model thus reads

$$\mathcal{E}(w) := \|\partial_t f + \partial_x(fv) - k\|_{L^2}^2 + \alpha \|\nabla w\|_{L^2}^2 + \beta \|D_v k\|_{L^2}^2. \tag{13}$$

Let us establish the existence of a minimiser to the problem $\min_{w \in H^1(E, \mathbb{R}^2)} \mathcal{E}(w)$. The following lemma will be used to show coercivity of $\mathcal{E}$ and is along the lines of [51, p. 29].

Lemma 1 *Let $w \in L^2(E, \mathbb{R}^2)$ be constant. Then, for $\nabla_{-1} f := (\partial_x f, -1)^\top$ with $\partial_x f \not\equiv$ const. the inequality*

$$\|w\|_{L^2}^2 \lesssim \|\nabla_{-1} f \cdot w\|_{L^2}^2 \tag{14}$$

holds.

Proof First, observe that, by the Cauchy–Schwarz inequality, the assumption $\partial_x f \not\equiv \text{const.}$ is equivalent to

$$|\langle \partial_x f, 1\rangle_{L^2}| < |E| \|\partial_x f\|_{L^2}, \tag{15}$$

where $|E|$ denotes the measure of E. Here, the assumption $\partial_x f \not\equiv \text{const.}$ is required to obtain a strict inequality.

Next, suppose to the contrary that there is no $C > 0$ such that (14) holds. Then, for all $n \in \mathbb{N}$ there exists $w_n \in L^2(E, \mathbb{R}^2)$ such that

$$\|\nabla_{-1} f \cdot w_n\|_{L^2}^2 < \frac{1}{n} \|w_n\|_{L^2}^2.$$

Let $\tilde{w}_n := w_n / \|w_n\|_{L^2}$ with $\tilde{w}_n = (\tilde{v}_n, \tilde{k}_n)^\top$ and obtain $\|\nabla_{-1} f \cdot \tilde{w}_n\|_{L^2}^2 < 1/n$. But then,

$$
\begin{aligned}
\frac{1}{n} &> \|\nabla_{-1} f \cdot \tilde{w}_n\|_{L^2}^2 \\
&= \int_E (\partial_x f \tilde{v}_n - \tilde{k}_n)^2 \, dE \\
&= \int_E (\partial_x f^2 \tilde{v}_n^2 + \tilde{k}_n^2 - 2\partial_x f \tilde{v}_n \tilde{k}_n) \, dE \\
&\geq \int_E (\partial_x f^2 \tilde{v}_n^2 + \tilde{k}_n^2) \, dE - 2|\langle \partial_x f \tilde{v}_n, \tilde{k}_n\rangle_{L^2}| \\
&= \int_E (\partial_x f^2 \tilde{v}_n^2 + \tilde{k}_n^2) \, dE - 2\|\partial_x f \tilde{v}_n\|_{L^2} \|\tilde{k}_n\|_{L^2} \frac{|\langle \partial_x f \tilde{v}_n, \tilde{k}_n\rangle_{L^2}|}{\|\partial_x f \tilde{v}_n\|_{L^2} \|\tilde{k}_n\|_{L^2}} \\
&\geq \left(\|\partial_x f \tilde{v}_n\|_{L^2}^2 + \|\tilde{k}_n\|_{L^2}^2 \right) \left(1 - \frac{|\langle \partial_x f \tilde{v}_n, \tilde{k}_n\rangle_{L^2}|}{\|\partial_x f \tilde{v}_n\|_{L^2} \|\tilde{k}_n\|_{L^2}} \right),
\end{aligned}
$$

implies that $\tilde{v}_n, \tilde{k}_n \to 0$ as $n \to +\infty$, since (15) allows us to conclude that

$$1 - \frac{|\langle \partial_x f \tilde{v}_n, \tilde{k}_n\rangle_{L^2}|}{\|\partial_x f \tilde{v}_n\|_{L^2} \|\tilde{k}_n\|_{L^2}} = 1 - \frac{|\langle \partial_x f, 1\rangle_{L^2}|}{|E| \|\partial_x f\|_{L^2}} > 0.$$

This contradicts the assumption and, therefore, (14) holds. $\square$

In particular, Lemma 1 holds for the (componentwise) average w_E of w, defined as

$$w_E = \frac{1}{|E|} \int_E w \, dE.$$

The following proposition is a straightforward adaptation of [33, Prop. 1] and utilises the direct method in the calculus of variations [19].

Proposition 1 *For* $f \in W^{1,\infty}(E)$ *satisfying* (15), *the functional* $\mathcal{E}$ *admits a minimiser in* $H^1(E, \mathbb{R}^2)$.

Proof The functional is proper and bounded from below since all terms are nonnegative and, for w identically zero,

$$\mathcal{E}(w) = \|\partial_t f\|_{L^2}^2 \lesssim \|\partial_t f\|_{L^\infty}^2 < +\infty,$$

since $f \in W^{1,\infty}(E)$.

Next, we show coercivity of (13) with respect to $H^1(E, \mathbb{R}^2)$. Observe that

$$
\begin{aligned}
\|w\|_{L^2}^2 &\lesssim \|w_E\|_{L^2}^2 + \|w - w_E\|_{L^2}^2 \\
&\lesssim \|\nabla_{-1} f \cdot w_E\|_{L^2}^2 + \|\nabla w\|_{L^2}^2 \\
&\lesssim \|\nabla_{-1} f \cdot w\|_{L^2}^2 + \|\nabla_{-1} f \cdot (w - w_E)\|_{L^2}^2 + \|\nabla w\|_{L^2}^2 \\
&\lesssim \|v \partial_x f - k\|_{L^2}^2 + \|\nabla w\|_{L^2}^2 \\
&\lesssim \|\partial_x (f v) - k\|_{L^2}^2 + \|\nabla w\|_{L^2}^2 \\
&\lesssim \mathcal{E}(w) + \|\partial_t f\|_{L^2}^2.
\end{aligned}
$$

The chain of inequalities follows from (5) and (14), the Poincaré–Wirtinger inequality [24, Chap. 5.8],

$$\|w - w_E\|_{L^2} \lesssim \|\nabla w\|_{L^2},$$

and the assumption $f \in W^{1,\infty}(E)$. Coercivity of $\mathcal{E}$ with respect to $H^1(E, \mathbb{R}^2)$ then follows since $\|\nabla w\|_{L^2}^2 \lesssim \mathcal{E}(w)$.

Next, we discuss sequential weak lower-semicontinuity of $\mathcal{E}$. Let $\{w_n\} \subset H^1(E, \mathbb{R}^2)$ such that $w_n \rightharpoonup \hat{w}$ in $H^1(E, \mathbb{R}^2)$. In particular, we have that $\nabla w_n \rightharpoonup \nabla \hat{w}$ in $L^2(E, \mathbb{R}^4)$. For $1 < p < 2$, the compact embedding $H^1(E, \mathbb{R}^2) \subset W^{1,p}(E, \mathbb{R}^2) \subset\subset L^2(E, \mathbb{R}^2)$ holds, see [24, Chap. 5.7]. As a consequence, there exists a subsequence, also denoted by w_n, such that $w_n \to \hat{w}$ in $L^2(E, \mathbb{R}^2)$. Then, weak lower-semicontinuity of $\mathcal{E}$ follows by application of [19, Thm. 3.23] since, for fixed w, we have that all terms are convex in ∇w. In particular, $|D_v k|^2$ is a quadratic form and therefore convex in ∇w, since $\bar{v} \bar{v}^\top$ in (7) is symmetric positive semidefinite.

Finally, by application of [19, Thm. 3.30], the functional $\mathcal{E}$ admits a minimiser.

$\square$

Let us add, however, that the convective regularisation functional $\mathcal{F}$ is nonconvex, as the following example shows. We pick $w_1 = (0, x)^\top$ and $w_2 = (1, 0)^\top$, and obtain

$$0 = \mathcal{F}(w_1) = \mathcal{F}(w_2) < \mathcal{F}((w_1 + w_2)/2) = \frac{|E|}{16}.$$

Therefore, $\mathcal{E}$ is nonconvex in general and several minima might exist. However, for $\beta = 0$ a unique minimiser exists.

3   Numerical Solution

In this section we derive necessary conditions for minimisers of (13) and discuss the numerical solution of a weak formulation by means of Newton's method.

3.1   Euler–Lagrange Equations

For convenience let us abbreviate $F := \partial_t f + \partial_x (f v) - k$. The Euler–Lagrange equations [17, Chap. IV] associated with minimisation of the functional $\mathcal{E}$ in (13) then read

$$
\begin{aligned}
f\partial_x F + \alpha \Delta v - \beta \partial_x k D_v k &= 0, \\
F + \nabla \cdot ((\alpha \mathrm{Id} + \beta \bar{v}\bar{v}^\top)\nabla k) &= 0,
\end{aligned}
\tag{16}
$$

where Id denotes the identity matrix of size two. Recall from Sect. 2.1 that Δ and ∇ are spatio-temporal operators. Moreover, the natural boundary conditions at ∂E are given by

$$
\begin{aligned}
n \cdot \left(\begin{pmatrix} 0 \\ fF \end{pmatrix} + \alpha \nabla v \right) &= 0, \\
n \cdot ((\alpha \mathrm{Id} + \beta \bar{v}\bar{v}^\top)\nabla k) &= 0,
\end{aligned}
\tag{17}
$$

where $n \in \mathbb{R}^2$ is the outward unit normal to the space-time domain E.

Let us highlight two aspects of (16). First, note that the system is nonlinear in the unknown $w = (v, k)^\top$ due to the convective regularisation. Second, as already mentioned in Sect. 2.1, there is a connection of the second set of equations in (16) with anisotropic diffusion with the diffusion tensor given by $\alpha \mathrm{Id} + \beta \bar{v}\bar{v}^\top$. The investigation of existence and regularity of solutions of (16) is left for future research.

3.2 Weak Formulation and Newton's Method

We minimise $\mathcal{E}$ by applying Newton's method to the weak formulation associated
with (16) together with boundary conditions (17). It is derived as follows.

Multiplying with a test function $\varphi = (\varphi^1, \varphi^2)^\top \in H^1(E, \mathbb{R}^2)$, integrating by
parts under consideration of (17), and adding both equations leads to the following
variational problem: Find $w \in H^1(E, \mathbb{R}^2)$ such that

$$\mathcal{G}(w; \varphi) := (\mathcal{G}_1 + \mathcal{G}_2)(w; \varphi) = 0, \quad \forall \varphi \in H^1(E, \mathbb{R}^2), \tag{18}$$

with

$$\mathcal{G}_1(w; \varphi) = -\int_E F \partial_x (f \varphi^1) \, dx - \alpha \int_E \nabla v \cdot \nabla \varphi^1 \, dx - \beta \int_E \partial_x k D_v k \varphi^1 \, dx,$$

$$\mathcal{G}_2(w; \varphi) = \int_E F \varphi^2 \, dx - \alpha \int_E \nabla k \cdot \nabla \varphi^2 \, dx - \beta \int_E (\bar{v} \bar{v}^\top \nabla k) \cdot \nabla \varphi^2 \, dx.$$

The Gâteaux derivative $D\mathcal{G}(w; \delta w, \varphi)$ of $\mathcal{G}$ at $w \in H^1(E, \mathbb{R}^2)$ in the direction
of $\delta w = (\delta v, \delta k)^\top \in H^1(E, \mathbb{R}^2)$ is given by $(D\mathcal{G}_1 + D\mathcal{G}_2)(w; \delta w, \varphi)$ with

$$DG_1(w; \delta w, \varphi) = -\int_E (\partial_x (f \delta v) - \delta k) \, \partial_x (f \varphi^1) \, dx - \alpha \int_E \nabla \delta v \cdot \nabla \varphi^1 \, dx$$

$$- \beta \int_E \left(\partial_x k D_v \delta k + (\partial_x k)^2 \delta v + \partial_x \delta k D_v k \right) \varphi^1 \, dx,$$

$$DG_2(w; \delta w, \varphi) = \int_E (\partial_x (f \delta v) - \delta k) \, \varphi^2 \, dx - \alpha \int_E \nabla \delta k \cdot \nabla \varphi^2 \, dx$$

$$- \beta \int_E (V \nabla k + \bar{v} \bar{v}^\top \nabla \delta k) \cdot \nabla \varphi^2 \, dx.$$

$$\tag{19}$$

Here, V is the Gâteaux derivative of $\bar{v} \bar{v}^\top$ at v in the direction δv and reads

$$V = \begin{pmatrix} 0 & \delta v \\ \delta v & 2v \delta v \end{pmatrix}.$$

We solve the nonlinear problem (18) with Newton's method, which proceeds as
follows. Starting from an initial solution $w^{(0)} := (0, 0)^\top$ we update the solution
according to the rule

$$w^{(n+1)} = w^{(n)} + \delta w, \tag{20}$$

where δw is the update and $n \in \mathbb{N}_0$. Computing δw in each step requires to solve a
linear variational problem: Find $\delta w \in H^1(E, \mathbb{R}^2)$ such that

$$a(\delta w, \varphi) = \ell(\varphi), \quad \forall \varphi \in H^1(E, \mathbb{R}^2), \tag{21}$$

with

$$a(\delta w, \varphi) = D\mathcal{G}(w^{(n)}; \delta w, \varphi),$$

$$\ell(\varphi) = -\mathcal{G}(w^{(n)}; \varphi).$$

Note that the dependence on the current iterate $w^{(n)}$ is through $\mathcal{G}$ and $D\mathcal{G}$, defined in (18) and (19), respectively. In our experiments we observed that (20) typically converges within only a few iterations.

3.3 Discretisation

We have implemented the weak formulation in FEniCS [1], which can also handle Newton's method automatically. The formulation (18) was discretised using multi-linear finite elements and the linear variational problem (21) was solved accordingly using FEniCS. Since kymographs serve as input data we have discretised the rectangular space-time domain E with a triangular mesh based on the regular grid so that every vertex of the mesh corresponds to one pixel of the image f and to one pair of values of the unknown w.

Moreover, in the implementation we penalise weak derivatives differently in space and time. This results in four regularisation parameters α_j^i, with $i \in \{v, k\}$ and $j \in \{t, x\}$, for the H^1 seminorm in (13) and one additional parameter β for the convective regularisation. Due to the equivalence of norms the existence result in Prop. 1 still holds true.

Integrals are computed exactly with an appropriate Gauss quadrature, which is automatically selected by FEniCS. Similarly, since the image f is represented by a piecewise multilinear function, products of partial derivatives of f that appear on the right-hand side of (21) are automatically projected onto the correct space.

As termination criterion for Newton's method we used the default criteria of FEniCS with both the absolute and the relative residual of (18) set to 10^{-10}. The maximum number of iterations was set to 15. Convergence was typically achieved within only a few iterations, which usually amounted to just a few seconds of computing time on a standard consumer laptop. It needs to be mentioned that, in our experiments we found that, the method fails to converge when the parameters α_t^k and α_x^k are chosen too small in comparison to β. This is, however, in line with the theoretical results in Sect. 2.2, which require H^1 regularity of k.

Both the source code[1] of our Python implementation and the microscopy data[2] used in the experiments are available online.

4   Experimental Results

In this section we present numerical results. In the first part, we demonstrate the importance of estimating a source based on synthetic data. Then, in the second part we show results for nonsynthetic microscopy data. After briefly discussing the data, we investigate the effects of varying the regularisation parameter β, which controls the amount of convective regularisation. We then qualitatively compare the (standard) 1D variational optical flow model with our continuity equation-based formulation for several choices of the regularisation functional $\alpha\mathcal{R}$ in (10). In addition, we present a quantitative evaluation of the considered models based on recoil velocities obtained from manual tracking of features in kymographs. Finally, in the last part, we evaluate our approach based on data coming from the solution of a mechanical model of tissue formation.

In all results, the computed velocity fields are presented visually with the help of streamlines, see e.g. [58]. These are integral curves computed by numerically solving the ordinary differential equation (2) for the estimated velocity v and a selected number of initial points. The resulting curves are then colour coded according to their velocities and shown superimposed with the kymograph data. This representation is more comprehensible and allows to visually check whether the estimated velocities are approximately correct.

4.1   Analytical Example

In order to demonstrate the necessity of estimating a source when the changes in the signal cannot be explained using mass conservation we conducted experiments for synthetic data. To this end, we generated a signal f, given as

$$f(t, x) = e^{-\frac{t}{\tau}} \cos\left(\frac{x - v_0 t}{\lambda}\right),$$

on a periodic domain $\Omega = (0, 1)$ and for the time interval $[0, 1]$. The parameters were set to $\tau = 1$ and to $\lambda = 1/(4\pi)$. This signal shifts to the right with constant velocity $v_0 = 0.1$ and decays exponentially over time in its magnitude. It can easily be verified that the source is given by $k(t, x) = -f(t, x)/\tau$.

---

[1]https://doi.org/10.5281/zenodo.3740696.
[2]https://doi.org/10.5281/zenodo.3257654.

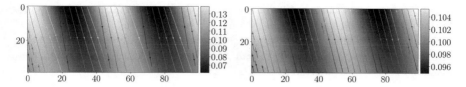

Fig. 4 In this example, we demonstrate the necessity of source estimation for decaying data. Here, we have generated a signal that shifts to the right with constant velocity $v_0 = 0.1$ and simultaneously decays exponentially (in magnitude). Shown are the signal f and streamlines computed from the estimated velocities using the homogenous (left) and the non-homogenous (right) continuity equation

We then solved the variational problem using in one case the homogenous, i.e. $k = 0$, and in one case the non-homogenous continuity equation. Periodic boundary conditions in space were enforced, and regularisation parameters were set to $\alpha_j^v = 10^{-3}$, $\alpha_j^k = 10^{-4}$, for $j \in \{t, x\}$, and to $\beta = 10^{-3}$. Since k is not constant along characteristics in this example we haven chosen β quite small.

Figure 4 illustrates the results of this experiment. It can clearly be seen that not accounting for the decay of the signal leads to a velocity that is significantly different to v_0 in this example, whereas using the non-homogenous continuity equation estimates both the velocity and the source (not shown) very well.

4.2 Microscopy Data and Acquisition of Kymographs

The data at hand are 2D image sequences of living Drosophila embryos recorded with two-photon laser-scanning microscopy. We refer to [48] for the used microscopy technique and for the preparation of flies, as well as for the details of the laser ablation method. For this study we recorded 15 image sequences, all of which feature cell membranes that have been fluorescently labelled with Myosin II-GFP, see [48]. The image sequences feature a square region of approximately $42.2 \times 42.2 \, \mu m^2$ at a spatial resolution of 250×250 pixels. A typical sequence contains between 60 and 100 frames that were recorded at a temporal interval of 727.67 ms, and the recorded image intensities f^{2D+T} are in the range $\{0, \ldots, 255\}$.

Each of the sequence shows a single plasma-induced laser nanoablation, which led to the controlled destruction of tissue in a linear region of $2 \, \mu m$ length that is approximately orthogonal to the cable. This ablation is expected to have a width of the order of the size of one pixel. Recall that in Fig. 1 we show such a typical dataset.

In order to obtain a kymograph from each microscopy sequence, we first labelled the location of the intersection between the ablation line and the actomyosin cable with a point $c \in \mathbb{R}^2$. Then, we visually determined an approximate orientation, given by a unit vector $e \in \mathbb{R}^2$, of the selected cable by defining a straight line of length $2L + 1$ pixels which passes through c. Typically, $L = 100$ pixels is sufficient for the considered datasets.

To create a one-dimensional image sequence we used standard nearest neighbour sampling along the abovementioned line. A nearest neighbour in terms of the pixel locations P of f^{2D+T} is given by

$$N(x) \in \{x_p \in P : |x - x_p| \le |x - x_q|, \forall p \ne q\}.$$

Then, the sampling points that are separated by distance one and lie along the abovementioned straight line are given by $p_i = c + ie$, for $i = -L, \ldots, L$. A nearest neighbour interpolation of f^{2D+T} along this line is given by $f^{2D+T}(t, N(p_i))$.

However, due to noise, the spatial extent, and minor displacements of the cable in orthogonal direction, we also considered sampling points that lie on the parallel line of distance $j = -h, \ldots, h$. These points are given by

$$p_{i,j} = c + ie + je^{\perp}.$$

Then, at $i = 1, \ldots, 2L + 1$, we define the intensity of a kymograph $f^{\delta}(t, i)$ as

$$f^{\delta}(t, i) = \sum_{j=-h,\ldots,h} f^{2D+T}(t, N(c + (i - L - 1)e + je^{\perp})).$$

In other words, the intensity at i is given as the sum of nearest neighbour interpolations at points that lie on an orthogonal straight line. The superscript δ indicates noise in the created kymograph. We found that $h = 5$ pixels generates satisfactory kymographs.

For this process, we used the reslice tool in Fiji [49] with a slice count of $2h$ and no interpolation selected, after manually placing a straight segment along a selected actomyosin cable, see Fig. 2 for illustration. Subsequently, a projection with the SUM option selected creates the final kymograph.

Since the image acquisition is paused during the laser ablation, see Fig. 2, we simply replaced the missing frame with the previous one. Moreover, we applied a Gaussian filter to the kymograph f^{δ} to guarantee the requirements specified in Sect. 2.2 and scaled the image intensities to the interval $[0, 1]$. The kernel size of the Gaussian filter was chosen as 10×10 pixels and the standard deviation set to $\sigma = 1$. The filtered and normalised kymograph is denoted in the following by f.

4.3   Qualitative Comparison

In the first experiment, we investigated the effect of the convective regularisation for one chosen kymograph. To this end, we solved the necessary conditions (16) as outlined in Sect. 3 for varying regularisation parameter β. Figure 5 shows streamlines for the estimated velocities together with the computed sources. Since

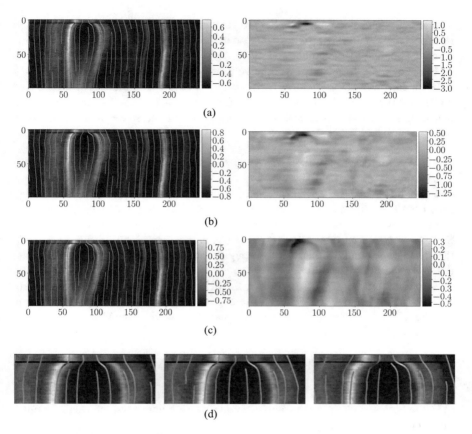

Fig. 5 Visualisation of the effect of the convective regularisation. Shown in Fig. 5a–c are kymographs f^δ (left) with streamlines superimposed and the estimated source k (right) for increasing regularisation parameter β (from top to bottom). The other parameters were fixed and set to $\alpha_j^v = 5 \cdot 10^{-3}$ and $\alpha_j^k = 10^{-4}$, for $j \in \{t, x\}$. (**a**) $\beta = 10^{-4}$. (**b**) $\beta = 10^{-3}$. (**c**) $\beta = 10^{-2}$. (**d**) Shown is a magnified view of the cut region of the above results (ordered left to right)

we did not find any significant difference between the results for $\beta = 10^{-4}$ and for $\beta = 0$, we simply omit the latter.

The two main findings of this experiment are as follows. First, for the chosen dataset, the convective regularisation can help to estimate more accurate velocities shortly after the ablation was applied. As can be seen in Fig. 5c (left) and 5d (right), this leads to a more accurate estimation of the recoil velocities at the cut ends. In Fig. 5d we display a magnified view of the results in the cut region.

Second, as expected, with increasing β the estimated source gets more regular in the direction of the flow, see Fig. 5a–c (right). In particular, the oscillations in space and time in the estimated source, which can be most likely attributed to noise and to artefacts created during the acquisition, decrease significantly. This certainly can lead to better interpretability and help to get a better understanding of the estimated reaction term k. Observe also the decrease in the magnitude of k as β increases. This is made apparent by the colour coding.

In the next experiment, we qualitatively compare the standard variational optical flow model with H^1 seminorm regularisation and the continuity equation-based model in (10) paired with different regularisation functionals. The first model we consider is based on the optical flow equation [30] in one space dimension, which reads

$$\partial_t f + \partial_x f v = 0, \tag{22}$$

and assumes that f is constant along characteristics of the flow (2). Even though no source k is estimated in this model, the main motivation for its inclusion in the evaluation is that it can serve as a baseline method. The quantitative evaluation in Sect. 4.4 is based on manually created tracks that follow highly visible features in the kymographs and, in many cases, roughly preserve their intensity as the tissue deforms. See Fig. 8 for illustration.

We highlight that, given $\partial_x f \neq 0$, Eq. (22) admits a unique solution, namely $v = -\partial_t f / \partial_x f$. However, due to noise degradation and aforementioned artefacts it is beneficial to solve (22) in a variational framework. Therefore, we consider minimising the functional

$$\|\partial_t f + \partial_x f v\|_{L^2}^2 + \alpha \|\nabla v\|_{L^2}^2, \tag{23}$$

analogously to the solution method outlined in Sect. 3. In contrast to the functional (13), the corresponding Euler–Lagrange equations are linear and the weak formulation can be solved directly.

All other functionals we investigate are based on (10) and read

$$\|\partial_t f + \partial_x (f v) - k\|_{L^2}^2 + \alpha \mathcal{R}_i(w).$$

Here, $\alpha \mathcal{R}_i$ represent different choices in the regularisation and, consequentially, also in the function space we minimise over. Since the above functional doesn't require any Sobolev regularity of the source k, we investigate the setting where $v \in H^1(E)$ and $k \in L^2(E)$, that is

$$\alpha \mathcal{R}_{H^1\text{-}L^2}(w) := \alpha \|\nabla v\|_{L^2}^2 + \gamma \|k\|_{L^2}^2, \tag{24}$$

with $\gamma > 0$. Moreover, we also consider the choice $v \in H^1(E)$ and $k \in H^1(E)$. The regularisation functional then reads

$$\alpha \mathcal{R}_{H^1}(w) := \alpha \|\nabla w\|_{L^2}^2. \tag{25}$$

Finally, we will present results for the setting $v \in H^1(E)$ and $k \in H^1(E)$ together with the convective regularisation. This is the main model investigated in Sect. 2.2 and is given by

$$\alpha \mathcal{R}_{CR}(w) := \alpha \|\nabla w\|_{L^2}^2 + \beta \|D_v k\|_{L^2}^2. \tag{26}$$

In Fig. 6 we show minimising functions for models (23)–(26) and in Fig. 7 we display a magnified view of the cut region. For all models, the regularisation param-

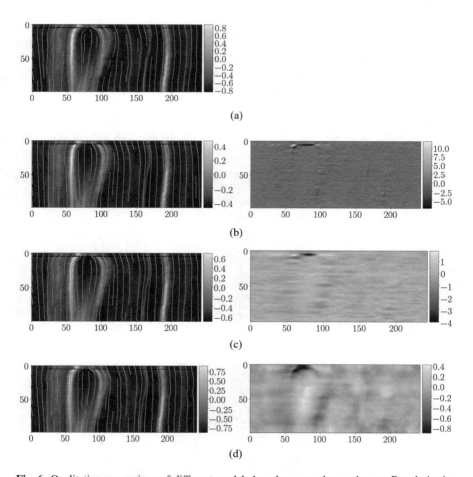

Fig. 6 Qualitative comparison of different models based on one chosen dataset. Regularisation parameters were chosen manually so that the recovered streamlines best matched the cut ends. (**a**) Streamline representation of the velocity obtained by solving the variational optical flow problem (23) with spatio-temporal H^1 seminorm regularisation. Parameters were set to $\alpha_j^v = 5 \cdot 10^{-3}$, for $j \in \{t, x\}$. (**b**) Result obtained using $\alpha \mathcal{R}_{H^1 \text{-} L^2}$ with parameters set to $\alpha_j^v = 5 \cdot 10^{-3}$ and $\gamma = 10^{-1}$. (**c**) Result obtained using $\alpha \mathcal{R}_{H^1}$ with parameters set to $\alpha_j^v = 5 \cdot 10^{-3}$ and $\alpha_j^k = 10^{-4}$. (**d**) Result obtained using $\alpha \mathcal{R}_{CR}$ with parameters set to $\alpha_j^v = 5 \cdot 10^{-3}, \alpha_j^k = 10^{-4}$, and $\beta = 2.5 \cdot 10^{-3}$

eters were chosen manually so that the streamlines obtained from the computed velocities best matched the cut ends which resulted from the laser ablation. For comparison, we also refer the reader to Fig. 8, which shows manually tracked cut ends. The four main observations from this experiment were as follows.

First, in Fig. 6a, which shows a minimising function of the optical flow model (23), it is clearly visible that many characteristics follow paths of constant fluorescence intensity. See also the two outermost markers in Fig. 7 (top, left). However, this model underestimate the recoil velocity shortly after the cut and may lead to wrong characteristics, see the middle marker in Fig. 7 (top, left).

Second, in Fig. 6b we illustrate a minimising pair $(v, k)^\top$ of the continuity equation-based model using $\alpha\mathcal{R}_{H^1-L^2}$. It is apparent that k captures both noise and artefacts, and leads to an undesirable underestimation of the velocities in the cut region. See also Fig. 7 (top, right). While this model unsurprisingly results in the smallest residual error (in norm) of the non-homogenous continuity equation, cf. also Table 1, it is insufficient for a meaningful quantification of tissue loss and growth.

Third, in Fig. 6c we depict a minimising pair for the model using $\alpha\mathcal{R}_{H^1}$ as regularisation functional. Less noise is picked up by the source k and the recovered

Fig. 7 Shown is a magnified view of the cut region of the results shown in Fig. 6 (ordered left to right, top to bottom). In the first image, the magenta arrows point at streamlines following paths of constant intensities. In the last image, the top-most markers indicate more accurately estimated velocities right after the laser cut

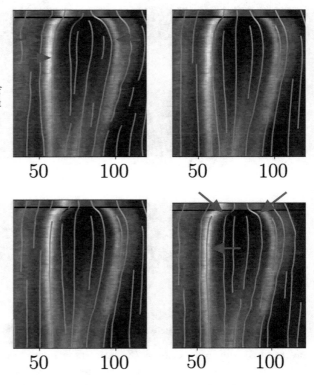

Table 1 Residual error for every dataset and every model investigated. For each dataset boldface values show the smallest residual in L^1 norm that is obtained during the parameter search

Dataset	OF	$\alpha\mathcal{R}_{H^1-L^2}$	$\alpha\mathcal{R}_{H^1}$	$\alpha\mathcal{R}_{CR}$
1	0.45	**0.16**	0.34	0.34
2	0.65	**0.27**	0.52	0.53
3	0.44	**0.17**	0.36	0.36
4	0.36	**0.13**	0.26	0.26
5	0.62	**0.26**	0.49	0.49
6	0.49	**0.17**	0.36	0.37
7	0.56	**0.20**	0.42	0.43
8	0.50	**0.16**	0.38	0.40
9	0.29	**0.12**	0.24	0.25
10	0.43	**0.17**	0.33	0.34
11	0.42	**0.16**	0.30	0.31
12	0.32	**0.12**	0.23	0.23
13	0.52	**0.20**	0.37	0.38
14	0.64	**0.26**	0.51	0.52
15	0.33	**0.11**	0.22	0.23
Average	0.47	**0.18**	0.36	0.36

velocities v are closer to what one would expect in the cut region, see also Fig. 7 (bottom, left). However, undesired oscillatory patterns are present in the source k.

Finally, in Fig. 6d we show a minimiser of the model using convective regularisation, that is $\alpha\mathcal{R}_{CR}$. This particular choice leads to significant visual improvement both in the recovered velocity and the estimated source. Specifically, as the streamlines in the cut region right after the laser ablation in Fig. 7 (bottom, right) show (see top two markers), the movement of the tissue is captured very well. In addition, bright features such as the left cut end, cf. the bottom marker in Fig. 7 (bottom, right), are followed accurately. Moreover, the changes in the fluorescence intensities are indicated nicely in the visualisation of the source k, see Fig. 6d (right). In particular, the significant increase between the cut ends towards the end of the sequence, possibly due to wound healing, is indicated adequately.

In summary, it can be said that our variational model based on the non-homogenous continuity equation together with convective regularisation can lead to improved results compared to existing models when parameters are selected by hand and comparison is performed visually.

4.4   Quantitative Comparison Based on Measured Recoil Velocities

In this section, we compare the models (23)–(26) presented in the previous section from a quantitative point of view. Our evaluation is based on manually measured

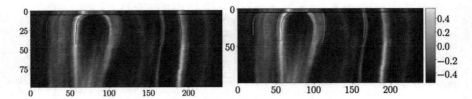

Fig. 8 Visualisation of manually created tracks of features (left) used for the evaluation superimposed with the corresponding kymograph and their corresponding third-order spline representation (right). Colour in the spline representation indicates velocity

recoil velocities of clearly visible features in the kymographs, and include the cut ends resulting from the laser nanoablation.

For this comparison, we have annotated 15 kymographs in Fiji [49] to obtain discrete trajectories to compare to. See Fig. 8 (left) for an example. These tracks can then be used to compare to either the estimated velocities or to the computed characteristics that solve (2) numerically. Observe in Fig. 8 that all trajectories start only after the ablation, which is the main region of interest from a tissue mechanics point of view.

Before presenting the comparison, let us briefly discuss the methodology and the used evaluation criteria. Since some created tracks do not feature a coordinate for each time instant we interpolated each track i with a third-order spline ϕ_i. As a result, velocities $\partial_t \phi_i$ can be computed conveniently and used for comparison to the velocities estimated by the variational approach. See Fig. 8 for an example of tracks (left) and their corresponding spline interpolations (right), which are colour coded according to their velocities.

In our experiments we found that each kymograph requires the regularisation parameters to be adjusted individually. Therefore, in the experimental comparison, we performed a search over all parameter combinations

$$\alpha_j^i, \gamma, \beta \in \{10^{-3}, 5 \cdot 10^{-3}, 10^{-2}, 5 \cdot 10^{-2}, 10^{-1}\}, \text{ with } i \in \{v, k\} \text{ and } j \in \{t, x\}, \tag{27}$$

for whichever parameters are applicable to the respective model. In the case of the most complex model, i.e. the one that includes the convective regularisation (26), this amounted to probing 5^5 parameter combinations per dataset. For each of the criteria listed below, we recorded the best result that was obtained with each model and for each kymograph during the parameter search.

4.4.1 Error in Residual

In the first comparison, the goal was to see which model best fits the recorded data. In Table 1 we report the L^1 norm of the smallest observed residual of the underlying model equation, that is

Fig. 9 Plotted is the mean error (28) (vertical axis) for the model $\alpha\mathcal{R}_{H^1}$ for each of the 15 datasets (in different colours) and all probed parameter combinations (horizontal axis). It can be seen that no single tested parameter settings led to a small error for all datasets

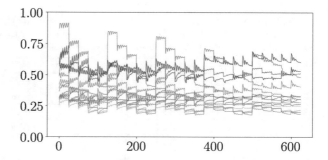

$$\|\partial_t f + \partial_x f v\|_{L^1} \quad \text{and} \quad \|\partial_t f + \partial_x (fv) - k\|_{L^1}.$$

As expected, the model $\alpha\mathcal{R}_{H^1\text{-}L^2}$ results on average and for every single dataset in the smallest residual. However, as already mentioned in Sect. 4.3, due to the high noise level it is only of limited use for quantifying the reaction term as k captures a significant amount of noise. The main finding of this experiment is that the optical flow model is by far not capturing the entire essence of the dataset, which is indicated by the high residual in comparison to the continuity equation-based models. Moreover, let us also highlight that the residual is on average not significantly increased in comparison to the $\alpha\mathcal{R}_{H^1}$ model when the convective regularisation is used in addition.

4.4.2 Error in Velocity

In the second comparison, we looked at the absolute error between the velocity of each manually created track and the velocities estimated with our models. For a particular track ϕ_i of a dataset, we define this error at time $t \in [0, T_i]$ as

$$|\partial_t \phi_i(t) - v(t, \phi_i(t))|.$$

Here, $T_i > 0$ is the length of the track and we have assumed for simplicity that all trajectories start at $t = 0$. The velocity v needs to be interpolated, since ϕ_i is a spline representation.

In further consequence we computed, for each dataset and for each parameter configuration, the mean squared L^2 norm of the error in velocity along all its N tracks. It is given by

$$E(v) := \frac{1}{N} \sum_{i=1}^{N} \frac{1}{|T_i|} \|\partial_t \phi_i - v(\cdot, \phi_i)\|^2_{L^2([0, T_i])}. \tag{28}$$

In our experiments, in none of the tested models we could find a single parameter combination that worked well for most datasets in terms of the mean error $E(v)$. This can be seen, for example, in Fig. 9, where we have plotted exemplary for each dataset and for each parameter setting in (27) the error $E(v)$ for the model $\alpha\mathcal{R}_{H^1}$.

Table 2 Tables depict the average error $E(v)$ in the L^2 norm (left), as in (28), and the maximum error (right) for every dataset and every model investigated. For each dataset boldface values show the smallest error that is obtained during the parameter search

Dataset	OF	$\alpha\mathcal{R}_{H^1-L^2}$	$\alpha\mathcal{R}_{H^1}$	$\alpha\mathcal{R}_{CR}$	Dataset	OF	$\alpha\mathcal{R}_{H^1-L^2}$	$\alpha\mathcal{R}_{H^1}$	$\alpha\mathcal{R}_{CR}$
1	**0.33**	0.45	0.40	0.38	1	**0.20**	0.25	0.23	0.23
2	**0.16**	0.17	0.17	0.17	2	**0.16**	**0.16**	0.21	0.21
3	0.30	0.35	**0.29**	**0.29**	3	**0.22**	0.31	**0.22**	**0.22**
4	**0.39**	0.44	0.44	0.43	4	**0.16**	0.18	0.17	0.17
5	0.26	0.31	0.25	**0.25**	5	0.20	0.23	**0.15**	**0.15**
6	**0.41**	0.50	0.45	**0.41**	6	**0.24**	0.46	0.41	0.39
7	**0.28**	0.34	0.31	0.31	7	**0.19**	0.29	0.23	0.22
8	0.29	0.33	0.24	**0.22**	8	0.20	0.19	0.15	**0.14**
9	0.21	0.22	**0.18**	**0.18**	9	0.10	0.10	**0.08**	0.09
10	**0.18**	0.24	**0.18**	**0.18**	10	0.18	0.16	**0.09**	**0.09**
11	**0.17**	0.21	0.19	0.19	11	0.17	0.29	**0.14**	**0.14**
12	0.25	0.31	0.24	**0.23**	12	**0.13**	0.18	0.14	0.14
13	**0.20**	0.31	0.25	0.25	13	**0.19**	0.27	**0.19**	**0.19**
14	0.55	0.53	**0.49**	**0.49**	14	0.39	0.45	**0.38**	**0.38**
15	**0.24**	0.30	0.28	0.28	15	**0.28**	0.44	0.30	0.30
Average	**0.28**	0.33	0.29	**0.28**	Maximum	**0.39**	0.46	0.41	**0.39**

As a consequence, we report in Table 2 (left) the best mean error $E(v)$ that we obtained for each dataset by the grid search. The main findings are as follows. First, and most importantly, the continuity equation-based model with convective regularisation performed on average as well as the optical flow-based model when using $E(v)$ as evaluation criterion, with the advantage of simultaneously yielding an estimate of the source.

A possible explanation for the comparably good performance of the optical flow-based model is that the manually created tracks approximately constitute trajectories of constant intensities rather than the characteristics associated with (1). However, in combination with the findings presented in Table 1, which show that the average residual is much smaller when using a continuity equation-based model with H^1 seminorm or convective regularisation, we are confident to state that these models are capable of estimating a meaningful source that can explain significantly more details of the observed signal.

In addition, we also evaluated (28) with $\|\cdot\|_{L^2}^2$ replaced by $\|\cdot\|_{L^\infty}$, see Table 2 (right). Qualitatively, this leads to slightly different results for some datasets but still supports our main findings.

Let us illustrate the advantage of the convective regularisation on the basis of one particular dataset. In Fig. 10 we show the best result obtained for dataset number eight for three models. For this particular dataset, the model $\alpha\mathcal{R}_{CR}$ outperforms all other models according to Table 2.

Figure 10a (left) shows the best result for the optical flow model and Fig. 10a (right) the manually created tracks for this kymograph used to evaluate the computed

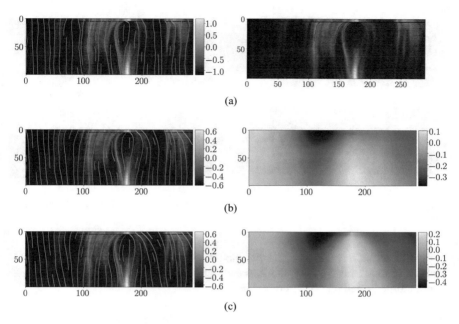

Fig. 10 Best results obtained for dataset number eight in terms of the mean error (28) in the estimated velocity, see Table 2 (left). (**a**) Streamline representation of the velocity obtained by solving the variational optical flow problem (23) with spatio-temporal H^1 seminorm regularisation with parameters $\alpha_t^v = 10^{-1}$ and $\alpha_x^v = 5 \cdot 10^{-2}$. (**b**) Result obtained using $\alpha \mathcal{R}_{H^1}$ with parameters $\alpha_t^v = 10^{-1}$, $\alpha_x^v = 5 \cdot 10^{-2}$, $\alpha_t^k = 10^{-1}$, and $\alpha_x^k = \cdot 10^{-1}$. (**c**) Result obtained using $\alpha \mathcal{R}_{CR}$ with parameters $\alpha_t^v = 10^{-1}$, $\alpha_x^v = 5 \cdot 10^{-2}$, $\alpha_t^k = 10^{-1}$, $\alpha_x^k = \cdot 10^{-1}$, and $\beta = 10^{-1}$

velocities. Notice the inaccurate velocity between the cut ends shortly after the laser ablation. Moreover, towards the end of the sequence, where the both cut ends meet again, the characteristics seem inappropriate.

In Fig. 10b we display the result for the model $\alpha \mathcal{R}_{H^1}$. As can be seen in Fig. 10b (left), the estimated velocity is improved significantly in the cut region and in the problematic region towards the end of the sequence.

In Fig. 10c (left) we show the result for the model including the convective regularisation. i.e. for $\alpha \mathcal{R}_{CR}$. The estimated velocity appears visually as good as in the previous model with the additional advantage that it allows a larger magnitude shortly after the laser ablation. Observe that in both cases the estimated source is both spatially and temporally very regular, and apart from *beta* the same parameters α_j^i were selected. However, in Fig. 10c (right) the effect of the anisotropic regularisation is clearly visible in the cut region.

Moreover, in Fig. 11 we show the best obtained results for two other datasets. For comparison, the top row shows the same dataset as in Figs. 5 and 6.

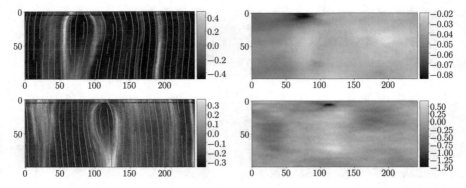

Fig. 11 Best results obtained for dataset number three and five in terms of the mean error (28) in the estimated velocity, see Table 2 (left)

Finally, let us mention that, for the model $\alpha\mathcal{R}_{CR}$, which we solve via Newton's method, in total only 53 computations did not converge out of the 3125 parameter combinations tested on 15 different datasets.

4.5   Comparison Based on a Mechanical Model of Tissue Formation

In Sect. 1 we have motivated the use of the non-homogenous continuity equation (1) mainly through mechanical models that are known to describe tissue formation, for example, in Drosophila. In order to see whether our variational formulation can reliably estimate velocity and source that both stem from such a process, we have implemented the partial differential equation-based model proposed in [26] to create synthetic data. In this model, the triplet (m, v, σ) solves the system

$$\partial_t m + \partial_x (mv) = k_{\mathrm{on}} - k_{\mathrm{off}} m, \tag{29}$$

$$\partial_x \sigma = \xi v, \tag{30}$$

$$\sigma = \eta \partial_x v + \chi m, \tag{31}$$

for $(t, x) \in (0, T) \times (0, 1)$, subject to the initial and boundary conditions

$$m(0, x) = m_0, \quad \text{in } (0, 1),$$

$$v(0, x) = 0, \quad \text{in } (0, 1), \tag{32}$$

$$v(t, x) = 0, \quad \text{in } (0, 1) \times \{0, 1\}.$$

We refer to [45] for a derivation of this model.

Briefly, (29) is an advection–reaction equation modelling mass conservation of myosin molecules m, subject to rates of adsorption $k_{on} > 0$ and desorbtion $k_{off} > 0$, which we assume to be constants. The advection is determined by a mechanical problem, where mechanical balance (30) involves the stress $\sigma(t, x)$ in the actin at the cell-cell junction and a friction force ξv exerted by the surrounding material, which we assume to be proportional to the velocity. Here, $\xi > 0$ is the coefficient of viscous drag. Finally, a constitutive model (31) of the junctional actin is proposed following e.g. [23, 26, 35] and involves viscous stresses and a pre-stress χm generated by myosin molecules. Again, $\eta, \chi > 0$ are constants. In addition, (32) enforces zero flux at the spatial boundaries, and m_0 is an initial concentration. In our experiments we set it to

$$m_0(x) := 20 - \frac{\sin(40x + \cos(40x))}{5}.$$

While such a model captures the essential features of actomyosin behaviour [23, 35, 47], its mean field approach means that away from the perturbation caused by the laser cut, the concentration m will equilibrate to the trivial solution $(m, v, \sigma) = (k_{on}/k_{off}, 0, \chi k_{on}/k_{off})$, which means that these parameters are uniform in space, and there will be no feature to track for an image analysis technique. This is also in contradiction with the experimental observations, where some material points along the cell-cell junctions exhibit accumulations of myosin that persist over time. One possible biophysical explanation for these accumulations is a locally larger density of actin binding sites.

This can be incorporated in the above model by introducing an additional variable $\rho(t, x)$ and a constant $k_{off}^0 > 0$ that modulates the off rate of myosin

$$k_{off}(t, x) = k_{off}^0 \rho(t, x).$$

In addition, the density ρ obeys the conservation equation

$$\partial_t \rho + \partial_x (\rho v) = 0, \tag{33}$$

and satisfies an initial condition. In our experiments we set the initial ρ at $t = 0$ as

$$\rho_0(x) := 1 + \frac{1 + \sin(40x + \cos(40x))}{10}. \tag{34}$$

We solve the system (29)–(31) numerically, under additional consideration of (29), with a standard upwind finite volume discretisation paired with the forward Euler method. See, for instance, [46, Appx. B] for a brief description. For completeness, we briefly outline our implementation here.

We discretise the space-time domain $[0, T] \times [0, 1]$ using N_t and N_x equally spaced discretisation points in time and in space, respectively. For the discretisation of the unknowns we make use of a centred and a staggered grid, denoted by $\mathcal{G}_c$ and

$\mathcal{G}_s$, respectively. They are defined as

$$\mathcal{G}_c = \{(i\Delta t, (j - 1/2)\Delta x) : 0 \le i \le N_t,\ 1 \le j \le N_x\},$$

and as

$$\mathcal{G}_s = \{(i\Delta t, (j - 1)\Delta x) : 0 \le i \le N_t,\ 1 \le j \le N_x + 1\},$$

where $\Delta t = T/N_t$ and $\Delta x = 1/N_x$. The concentrations and the stress are then discretised on the centred grid and the velocity on the staggered grid, leading to

$$(m, \rho, v, \sigma) \in \mathbb{R}^{|\mathcal{G}_c|} \times \mathbb{R}^{|\mathcal{G}_c|} \times \mathbb{R}^{|\mathcal{G}_s|} \times \mathbb{R}^{|\mathcal{G}_c|}.$$

In further consequence, the finite volume discretisation, see e.g. [36, Chap. 4], of (29) reads

$$\frac{d}{dt}m_j + \frac{1}{\Delta x}\left(F(m)|_{j+\frac{1}{2}} - F(m)|_{j-\frac{1}{2}}\right) = k_{\text{on}} - k_{\text{off}}^0 \rho_j m_j,$$

where $F(m)|_{j\pm\frac{1}{2}}$ is the flux at the cell boundaries, that is, at the nodes of the staggered grid. Assuming that m is constant on each cell and using the upstream value of m, the flux $F(m)(t)|_{j\pm\frac{1}{2}} := F^{(i)}_{j\pm\frac{1}{2}}$ at time $t = i\Delta t$ at the boundaries can be written as

$$F^{(i)}_{j+\frac{1}{2}} = \frac{1}{2}v^{(i)}_{j+\frac{1}{2}}\left(m^{(i)}_{j+1} + m^{(i)}_j\right) - \frac{1}{2}|v^{(i)}_{j+\frac{1}{2}}|\left(m^{(i)}_{j+1} - m^{(i)}_j\right).$$

Similarly, we obtain $F^{(i)}_{j-\frac{1}{2}}$ and, moreover, (17) results in zero flux at the boundaries. Then, approximating d/dt with forward finite differences yields

$$m^{(i+1)}_j = m^{(i)}_j - \frac{\Delta t}{\Delta x}\left(F^{(i)}_{j+\frac{1}{2}} - F^{(i)}_{j-\frac{1}{2}}\right) + \Delta t\left(k_{\text{on}} - k_{\text{off}}\rho^{(i)}_j m^{(i)}_j\right). \tag{35}$$

Analogously, an update equation for ρ is obtained.

The velocity at the current time step (i) is determined as follows. Using (30) to obtain $v = \partial_x\sigma/\xi$ and substituting it into (31) yields the second-order elliptic partial differential equation

$$\sigma - \frac{\eta}{\xi}\partial_{xx}\sigma = \chi m, \quad \text{in } (0, 1),$$

$$\partial_x\sigma = 0, \qquad \text{in } \{0, 1\}, \tag{36}$$

for the stress σ. Here, the zero Neumann boundary conditions follow from (32).

In each time step we solve (36), given the concentration $m^{(i)}$ from the previous time instant, with a standard finite-difference scheme using centred differences on $\mathcal{G}_c$. Then, with the help of (30) the velocity at nodes $j + 1/2$ can be approximated with

$$v_{j+1/2}^{(i)} \approx \frac{1}{\xi \Delta x} \left(\sigma_{j+1}^{(i)} - \sigma_j^{(i)} \right),$$

where we also use the fact that v is zero outside the spatial domain.

Finally, the concentration $m_i^{(t+1)}$ is updated according to (35) and ρ to its corresponding equation, and the procedure is repeated for the next time instant. The time interval is adjusted in each step so that the Courant–Friedrichs–Lewy condition is satisfied. This typically leads to intermediate results that are not recorded.

In our experiments we set $N_t = 300$ and $N_x = 300$, and the parameters controlling the time stepping were set to $T = 0.1$ and to $\Delta t = 2.5 \cdot 10^{-6}$. The mechanical parameters in (30) and (31) were chosen as $\eta = 1$, $\xi = 0.1$, and as $\chi = 1$. The parameters in (31) related to the source were set to $k_{\text{on}} = 200$ and to $k_{\text{off}}^0 = 10$.

In order to evaluate our approach described in Sect. 2.2, we conducted two experiments based on this mechanical model. In the first experiment, we used the solution method outlined above to generate a concentration m, which was then used as input to our variational formulation defined in (13). In the second experiment, we generated a concentration by solving (29) for a set velocity v, effectively removing the mechanical part of the model.

4.5.1 Unknown Velocities

In this experiment, we solved the mechanical model (29)–(31) together with (33) numerically as outlined above. In order to simulate a laser ablation, the concentration m_0 is set to zero at nodes within the interval $[0.495, 0.505]$. In this way, a disruption (or loss) of concentration is simulated. Figure 12 (top) shows the solution (m, v) and the resulting source $k = k_{\text{on}} - k_{\text{off}}^0 \rho m$.

We then solved (13) with $\alpha \mathcal{R}_{\text{CR}}$ numerically based on the generated concentration. This is achieved by setting $f := m$. However, in order to match the boundary conditions in (32) we also used zero Dirichlet boundary conditions for v in (18) at $x \in \{0, 1\}$ and at $t = 0$. In Fig. 12 (bottom), we depict an approximate minimiser. The parameters were set to $a_j^v = 10^{-4}$, $a_j^k = 5 \cdot 10^{-5}$, and to $\beta = 10^{-6}$.

Observe that both the velocity and the source are estimated approximately and are within the correct order of magnitude. However, let us add that the estimated source appears quite regular in comparison to the simulation, even though the regularisation parameters α_j^k were set comparably small.

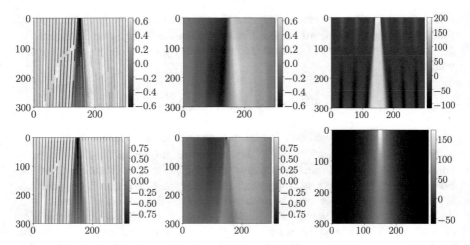

Fig. 12 Shown is in the top row the solution (m, v, k) of the extended mechanical model (29)–(31). For simplicity, ρ and σ are omitted. The bottom images depict an approximate minimiser $w = (v, k)^\top$ obtained using (10) with $\alpha \mathcal{R}_{CR}$. The left column shows the myosin concentration m together with streamlines obtained from the velocities, the middle column depicts the velocity v, and the right column illustrates the source k

4.5.2 Predefined Velocities

In the next experiment, we removed the mechanical part and solved just (29) and (33) with a predefined velocity field v that could potentially resemble a laser nanoablation in cell membranes as pictured, for example, in Fig. 2.

We generated a velocity profile as follows. First, we define a characteristic $\phi : [0, T] \to \mathbb{R}$ that is supposed to follow a cut end via the ordinary differential equation

$$\partial_t \phi(t) = v_0 e^{-\frac{t}{\tau}},$$

where the constant $v_0 > 0$ is the initial velocity at time zero. Integrating with respect to time yields the integral curve

$$\phi(t) = c_0 + v_0 \tau (1 - e^{-\frac{t}{\tau}})$$

where $c_0 \geq 0$ is a constant defining the starting point of the curve. We then define a velocity field

$$\bar{v}(t, x) := \partial_t \phi(t) \begin{cases} \frac{x}{\phi(t)}, & \text{if } x \leq \phi(t), \\ e^{-\frac{|x - \phi(t)|}{\ell}}, & \text{if } x > \phi(t), \end{cases}$$

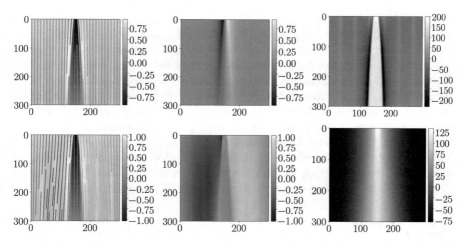

Fig. 13 The top row shows functions (m, k) obtained by solving (29) and (33) given a set velocity v. The bottom images depict an approximate minimiser $w = (v, k)^\top$ obtained using (10) with $\alpha\mathcal{R}_{\mathrm{CR}}$. The left column shows the myosin concentration m together with streamlines obtained from the velocities, the middle column depicts the velocity v, and the right column illustrates the source k

so that $\bar{v}(t, \cdot)$ is linear in the interval $[0, \phi(t)]$ and decays exponentially in $(\phi(t), +\infty)$. Here, $\tau, \ell > 0$ are constants and control the decay. Finally, we shift the origin to $1/2$, reflect $\bar{v}$, and obtain

$$
v(t, x) := \begin{cases} \bar{v}(t, x - \tfrac{1}{2}) & \text{if } x - \tfrac{1}{2} \geq 0, \\ -\bar{v}(t, -(x - \tfrac{1}{2})) & \text{if } x - \tfrac{1}{2} < 0. \end{cases} \tag{37}
$$

In Fig. 13 (top) we illustrate the solution (m, v, k) for this scenario, where v is set as in (37). The parameter c_0 was set to $c_0 = 0.05$ to match the width of the simulated laser ablation. The other parameters were set to $v_0 = 1$, $\tau = 0.075$, and to $\ell = 0.05$. All other settings and parameters were kept as in the previous experiment.

Then, we minimised (13) together with convective regularisation numerically based on the resulting concentration m. Figure 13 (bottom) shows the estimated velocity and source pair for the synthetic data. A similar behaviour as in the previous experiment can be observed.

5 Conclusions

In this article, we have investigated a variational model for joint velocity estimation and source identification in challenging fluorescence microscopy data of live Drosophila embryos that show the controlled destruction of tissue. We exploited the fact that a large proportion of tissue deformation occurs along one space dimension and allows to create kymographs. Our formulation is grounded on one-dimensional mechanical models of tissue formation and is based on the non-homogenous continuity equation. We have discussed the ill-posedness of this problem and devised a well-posed variational formulation using convective regularisation of the source. Moreover, we have shown the connection of convective regularisation of the source to anisotropic diffusion. In a thorough experimental evaluation, we have demonstrated that motion estimation can benefit from simultaneously estimating a source and that convective regularisation may help to estimate velocities more accurately. Our numerical results show that this method could potentially help to quantify the reaction term in biological models of tissue formation. The extension of our models to more than one space dimension is left for future research.

Acknowledgments LFL and CBS acknowledge support from the Leverhulme Trust project "Breaking the non-convexity barrier", the EPSRC grant EP/M00483X/1, the EPSRC Centre Nr. EP/N014588/1, the RISE projects ChiPS and NoMADS, the Cantab Capital Institute for the Mathematics of Information, and the Alan Turing Institute. ND and JE were supported by ANR-11-LABX-0030 "Tec21", by a CNRS Momentum grant, and by IRS "AnisoTiss" of Idex Univ. Grenoble Alpes. ND and JE are members of GDR 3570 MecaBio and GDR 3070 CellTiss of CNRS. Some of the computations were performed using the Cactus cluster of the CIMENT infrastructure, supported by the Rhône-Alpes region (GRANT CPER07_13 CIRA) and the authors thank Philippe Beys, who manages the cluster. Overall laboratory work was supported by Wellcome Trust Investigator Awards to BS (099234/Z/12/Z and 207553/Z/17/Z). ES was also supported by a University of Cambridge Herchel Smith Fund Postdoctoral Fellowship. The authors also wish to thank Pierre Recho for fruitful discussions and the re-use of his numerical simulation code.

References

1. M.S. Alnæs, J. Blechta, J. Hake, A. Johansson, B. Kehlet, A. Logg, C. Richardson, J. Ring, M.E. Rognes, G.N. Wells, The FEniCS project version 1.5. Arch. Numer. Softw. **3**(100), 9–23 (2015)
2. F. Amat, E.W. Myers, P.J. Keller, Fast and robust optical flow for time-lapse microscopy using super-voxels. Bioinformatics **29**(3), 373–380 (2013)
3. A.A. Amini, A scalar function formulation for optical flow, in *Proceedings of the 3rd European Conference on Computer Vision*, ed. by J.-O. Eklundh, pp. 123–131 (Springer, Berlin, 1994)
4. R. Andreev, O. Scherzer, W. Zulehner, Simultaneous optical flow and source estimation: space–time discretization and preconditioning. Appl. Numer. Math. **96**, 72–81 (2015)
5. G. Aubert, P. Kornprobst, *Mathematical Problems in Image Processing*. Applied Mathematical Sciences, 2nd edn. (Springer, New York, 2006). Partial differential equations and the calculus of variations
6. D. Béréziat, I. Herlin, L. Younes, A generalized optical flow constraint and its physical interpretation, in *Proceedings of the IEEE Conference on Computer Vision and Pattern Recognition* (2000), pp. 487–492
7. G.B. Blanchard, A.J. Kabla, N.L. Schultz, L.C. Butler, B. Sanson, N. Gorfinkiel, L. Mahadevan, R.J. Adams, Tissue tectonics: morphogenetic strain rates, cell shape change and intercalation. Nat. Meth. **6**(6), 458–464 (2009)

8. J.T. Blankenship, S.T. Backovic, J.S. P. Sanny, O. Weitz, J.A. Zallen, Multicellular rosette formation links planar cell polarity to tissue morphogenesis. Dev. Cell **11**(4), 459–470 (2006)
9. A. Boquet-Pujadas, T. Lecomte, M. Manich, R. Thibeaux, E. Labruyère, N. Guillén, J.-C. Olivo-Marin, A.C. Dufour, BioFlow: a non-invasive, image-based method to measure speed, pressure and forces inside living cells. Sci. Rep. **7**(1), 9178 (2017)
10. K. Boric, P. Orio, T. Viéville, K. Whitlock, Quantitative analysis of cell migration using optical flow. PLOS ONE **8**, 1–11 (2013)
11. M. Burger, H. Dirks, C.-B. Schönlieb, A variational model for joint motion estimation and image reconstruction. SIAM J. Imag. Sci. **11**(1), 94–128 (2018)
12. A.R. Chaphalkar, K. Jain, M.S. Gangan, C.A. Athale, Automated multi-peak tracking kymography (amtrak): a tool to quantify sub-cellular dynamics with sub-pixel accuracy. PLOS ONE **11**(12), 12 (2016)
13. N. Chenouard, J. Buisson, I. Bloch, P. Bastin, J.C. Olivo-Marin, Curvelet analysis of kymograph for tracking bi-directional particles in fluorescence microscopy images, in *IEEE International Conference on Image Processing* (2010), pp. 3657–3660
14. J.P. Cocquerez, L. Chanas, J. Blanc-Talon, Simultaneous inpainting and motion estimation of highly degraded video-sequences, in *Image Analysis*, ed. by J. Bigun, T. Gustavsson. Lecture Notes in Computer Science (Springer, Berlin, 2003), pp. 685–692
15. T. Corpetti, É. Mémin, P. Pérez, Dense estimation of fluid flows. IEEE Trans. Pattern Anal. Mach. Intell. **24**(3), 365–380 (2002)
16. T. Corpetti, D. Heitz, G. Arroyo, É. Mémin, A. Santa-Cruz, Fluid experimental flow estimation based on an optical-flow scheme. Exp. Fluids **40**(1), 80–97 (2006)
17. R. Courant, D. Hilbert, *Methods of Mathematical Physics. Vol. I* (Interscience Publishers, New York, 1953)
18. G. Crippa, The flow associated to weakly differentiable vector fields. Ph.D. Thesis, Classe di Scienze Matematiche, Fisiche e Naturali, Scuola Normale Superiore di Pisa/Institut für Mathematik, Universität Zürich (2007)
19. B. Dacorogna, *Direct Methods in the Calculus of Variations*. Applied Mathematical Sciences, 2nd edn. (Springer, New York, 2008)
20. M. Dawood, C. Brune, O. Schober, M. Schäfers, K.P. Schäfers, A continuity equation based optical flow method for cardiac motion correction in 3D PET data, in *Medical Imaging and Augmented Reality*, ed. by H. Liao, P.J. Edwards, X. Pan, Y. Fan, G.-Z. Yang. Lecture Notes in Computer Science (Springer, Berlin, 2010), pp. 88–97
21. M. Drechsler, L.F. Lang, H. Dirks, M. Burger, C.-B. Schönlieb, I.M. Palacios, Optical flow analysis reveals that kinesin-mediated advection impacts on the orientation of microtubules. bioRxiv (2019)
22. S.J. England, G.B. Blanchard, L. Mahadevan, R.J. Adams, A dynamic fate map of the forebrain shows how vertebrate eyes form and explains two causes of cyclopia. Development **133**(23), 4613–4617 (2006)
23. J. Étienne, J. Fouchard, D. Mitrossilis, N. Bufi, P. Durand-Smet, A. Asnacios, Cells as liquid motors: mechanosensitivity emerges from collective dynamics of actomyosin cortex. Proc. Nat. Acad. Sci. U.S.A. **112**(9), 2740–2745 (2015)
24. L.C. Evans, *Partial Differential Equations*. Graduate Studies in Mathematics, 2nd edn. (American Mathematical Society, Providence, 2010)
25. R. Fernandez-Gonzalez, S. de Matos Simoes, J.-C. Röper, S. Eaton, J.A. Zallen, Myosin II dynamics are regulated by tension in intercalating cells. Dev. Cell **17**(5), 736–743 (2009)
26. E. Hannezo, B. Dong, P. Recho, J.-F. Joanny, S. Hayashi, Cortical instability drives periodic supracellular actin pattern formation in epithelial tubes. Proc. Nat. Acad. Sci. U.S.A. **112**(28), 8620–8625 (2015)
27. H.W. Haussecker, D.J. Fleet, Computing optical flow with physical models of brightness variation. IEEE Trans. Pattern Anal. Mach. Intell. **23**(6), 661–673 (2001)
28. C.-P. Heisenberg, Y. Bellaïche, Forces in tissue morphogenesis and patterning. Cell **153**(5), 948–962 (2013)

29. D. Heitz, E. Mémin, Ch. Schnörr, Variational fluid flow measurements from image sequences: synopsis and perspectives. Exp. Fluids **48**(3), 369–393 (2010)

30. B.K.P. Horn, B.G. Schunck, Determining optical flow. Artif. Intell. **17**, 185–203 (1981)

31. Y. Huang, L. Hao, H. Li, Z. Liu, P. Wang, Quantitative analysis of intracellular motility based on optical flow model. J. Healthc. Eng. **2017**, 1–10 (2017)

32. J. Huisken, D.Y.R. Stainier, Selective plane illumination microscopy techniques in developmental biology. Development **136**(12), 1963–1975 (2009)

33. J.A. Iglesias, C. Kirisits, Convective regularization for optical flow, in *Variational Methods in Imaging and Geometric Control*, Radon Series on Computational and Applied Mathematics (Walter de Gruyter GmbH & Co. KG, 2016), pp. 184–201

34. C. Kirisits, L.F. Lang, O. Scherzer, Optical flow on evolving surfaces with space and time regularisation. J. Math. Imag. Vision **52**(1), 55–70 (2015)

35. K. Kruse, J.F. Joanny, F. Jülicher, J. Prost, K. Sekimoto, Generic theory of active polar gels: a paradigm for cytoskeletal dynamics. Eur. Phys. J. E **16**(1), 5–16 (2005)

36. R.J. LeVeque, *Finite Volume Methods for Hyperbolic Problems*. Cambridge Texts in Applied Mathematics (Cambridge University Press, Cambridge, 2002)

37. C.M. Lye, G.B. Blanchard, H.W. Naylor, L. Muresan, J. Huisken, R.J. Adams, B. Sanson, Mechanical coupling between endoderm invagination and axis extension in drosophila. PLOS Biol. **13**(11), e1002292 (2015)

38. P. Mangeol, B. Prevo, E.J.G. Peterman, E. Holzbaur, KymographClear and KymographDirect: two tools for the automated quantitative analysis of molecular and cellular dynamics using kymographs. Mol. Biol. Cell **27**(12), 1948–1957 (2016)

39. C. Melani, M. Campana, B. Lombardot, B. Rizzi, F. Veronesi, C. Zanella, P. Bourgine, K. Mikula, N. Peyriéras, A. Sarti, Cells tracking in a live zebrafish embryo, in *Proceedings of the 29th Annual International Conference of the IEEE Engineering in Medicine and Biology Society (EMBS 2007)*, (2007), pp. 1631–1634

40. B. Monier, A. Pélissier-Monier, A.H. Brand, B. Sanson, An actomyosin-based barrier inhibits cell mixing at compartmental boundaries in drosophila embryos. Nat. Cell Biol. **12**(1), 60–65 (2010)

41. S. Neumann, R. Chassefeyre, G.E. Campbell, S.E. Encalada, KymoAnalyzer: a software tool for the quantitative analysis of intracellular transport in neurons. Traffic **18**(1), 71–88 (2016)

42. M. Nishikawa, S.R. Naganathan, F. Jülicher, S.W. Grill, Controlling contractile instabilities in the actomyosin cortex. eLife **6**, e30537 (2017)

43. T. Preusser, M. Droske, C.S. Garbe, A. Telea, M. Rumpf, A phase field method for joint denoising, edge detection, and motion estimation in image sequence processing. SIAM J. Appl. Math. **68**(3), 599–618 (2008)

44. P. Quelhas, A.M. Mendonça, A. Campilho, Optical flow based arabidopsis thaliana root meristem cell division detection, in *Image Analysis and Recognition*, ed. by A. Campilho, M. Kamel. Lecture Notes in Computer Science (Springer, Berlin, 2010), pp. 217–226

45. P. Recho, T. Putelat, L. Truskinovsky, Mechanics of motility initiation and motility arrest in crawling cells. J. Mech. Phys. Solids **84**, 469–505 (2015)

46. P. Recho, J. Ranft, P. Marcq, One-dimensional collective migration of a proliferating cell monolayer. Soft. Matter **12**(8), 2381–2391 (2016)

47. A. Saha, M. Nishikawa, M. Behrndt, C.-P. Heisenberg, F. Jülicher, S.W. Grill, Determining physical properties of the cell cortex. Biophys. J. **110**(6), 1421–1429 (2016)

48. E. Scarpa, C. Finet, G.B. Blanchard, B. Sanson, Actomyosin-driven tension at compartmental boundaries orients cell division independently of cell geometry In Vivo. Dev. Cell **47**(6), 727–740.e6 (2018)

49. J. Schindelin, I. Arganda-Carreras, E. Frise, V. Kaynig, M. Longair, T. Pietzsch, S. Preibisch, C. Rueden, S. Saalfeld, B. Schmid, J.-Y. Tinevez, D.J. White, V. Hartenstein, K. Eliceiri, P. Tomancak, A. Cardona, Fiji: an open-source platform for biological-image analysis. Nature **9**(7), 676–682 (2012)

50. B. Schmid, G. Shah, N. Scherf, M. Weber, K. Thierbach, C. Campos Pérez, I. Roeder, P. Aanstad, J. Huisken, High-speed panoramic light-sheet microscopy reveals global endodermal cell dynamics. Nat. Commun. **4**, 2207 (2013)
51. Ch. Schnörr, Determining optical flow for irregular domains by minimizing quadratic functionals of a certain class. Int. J. Comput. Vision **6**, 25–38 (1991)
52. B.G. Schunck, The motion constraint equation for optical flow, in *Proceedings of the 7th International Conference on Pattern Recognition* (1984), pp. 29–22
53. S.M. Song, R.M. Leahy, Computation of 3-D velocity fields from 3-D cine CT images of a human heart. IEEE Trans. Med. Imag. **10**(3), 295–306 (1991)
54. J. Weickert, *Anisotropic Diffusion in Image Processing* (Teubner, Stuttgart, 1998). European Consortium for Mathematics in Industry
55. J. Weickert, Ch. Schnörr, A theoretical framework for convex regularizers in PDE-based computation of image motion. Int. J. Comput. Vis. **45**(3), 245–264 (2001)
56. J. Weickert, Ch. Schnörr, Variational optic flow computation with a spatio-temporal smoothness constraint. J. Math. Imag. Vis. **14**, 245–255 (2001)
57. J. Weickert, A. Bruhn, T. Brox, N. Papenberg, A survey on variational optic flow methods for small displacements, in *Mathematical Models for Registration and Applications to Medical Imaging*, ed. by O. Scherzer. Mathematics in Industry (Springer, Berlin 2006), pp. 103–136
58. D. Weiskopf, G. Erlebacher, Overview of flow visualization, in *The Visualization Handbook*, ed. by C.D. Hansen, C.R. Johnson (Elsevier, Amsterdam, 2005), pp. 261–278
59. R.P. Wildes, A.M. Amabile, M.J. Lanzillotto, T.S. Leu, Recovering estimates of fluid flow from image sequence data. Comput. Vis. Image Underst. **80**(2), 246–266 (2000)
60. L. Younes, *Shapes and Diffeomorphisms*. Applied Mathematical Sciences (Springer, Berlin, 2010)
61. L. Zhou, C. Kambhamettu, D.B. Goldgof, Fluid structure and motion analysis from multi-spectrum 2D cloud image sequences, in *Proceedings of the IEEE Conference on Computer Vision and Pattern Recognition* (2000), pp. 744–751

Quantitative OCT Reconstructions for Dispersive Media

Peter Elbau, Leonidas Mindrinos, and Leopold Veselka

Abstract We consider the problem of reconstructing the position and the time-dependent optical properties of a linear dispersive medium from OCT measurements. The medium is multi-layered described by a piecewise inhomogeneous refractive index. The measurement data are from a frequency-domain OCT system and we address also the phase retrieval problem. The parameter identification problem can be formulated as an one-dimensional inverse problem. Initially, we deal with a non-dispersive medium and we derive an iterative scheme that is the core of the algorithm for the frequency-dependent parameter. The case of absorbing medium is also addressed.

1 Introduction

Optical Coherence Tomography (OCT) is nowadays considered as a well-established imaging modality producing high-resolution images of biological tissues. Since it first appeared in the beginning of the 1990s [11, 17, 30], OCT has gained increasing acceptance because of its non-invasive nature and the use of non-harmful radiation. Main applications remain tissue diagnostics and ophthalmology. It operates at the visible and near-infrared spectrum and the measurements consist mainly of the backscattered light from the sample. OCT is analogous to Ultrasound Tomography where acoustic waves are used and differs from Computed Tomography (where electromagnetic waves are also used) because of its limited penetration depth (few millimeters) due to the lower energy radiation. As OCT data we consider the measured intensity of the backscattered light at some detector area usually far from the medium.

P. Elbau · L. Veselka
Faculty of Mathematics, University of Vienna, Vienna, Austria
e-mail: peter.elbau@univie.ac.at; leopold.veselka@univie.ac.at

L. Mindrinos (✉)
Johann Radon Institute for Computational and Applied Mathematics (RICAM), Linz, Austria
e-mail: leonidas.mindrinos@ricam.oeaw.ac.at

© Springer Nature Switzerland AG 2021
B. Kaltenbacher et al. (eds.), *Time-dependent Problems in Imaging and Parameter Identification*, https://doi.org/10.1007/978-3-030-57784-1_8

However, the intensity of light, undergoing few scattering events, is not measured directly, but the OCT setup is based on low coherence interferometry. The incoming broadband and continuous wave light passes through a beam-splitter and it is split into two identical beams. One part travels in a reference path and is totally back-reflected by a mirror and the second part is incident on the sample. The backscattered from the sample and the back-reflected light are recombined and their superposition is then measured at a detector. The maximum observed intensity refers to constructive interference, and this happens when the two beams travel equal lengths. For a detailed explanation of the experimental setup we refer to [10, 33] and to the book [4].

The way the measurements are performed characterizes and differentiates an OCT system. We summarize here the different setups considered in this work:

Time-domain OCT: The reference mirror is moving and for each position a measurement is performed. By scanning the reference arm, different depth information from the sample is obtained.

Frequency-domain OCT: The mirror is placed at a fixed position and the detector is replaced by a spectrometer, which captures the whole spectrum of the interference pattern.

State-of-the-art OCT: The incoming light is focused, through objective lenses, to a specific region at a certain depth in the sample. The backscattered light is measured at a point detector.

Standard OCT: The vector nature of light is ignored and the electromagnetic wave is treated as a scalar quantity. Then, only the total intensity is measured.

Time- and Frequency-domain OCT provide equivalent measurements that are connected through a Fourier transform. The advantage of the later is that no mechanical movement of the mirror is required, improving the acquisition time. The last two cases simplify the following mathematical analysis since we can consider scalar quantities and depth-dependent optical parameters. For an overview of the different mathematical models that can be used in OCT we refer to the book chapter [5].

We consider Maxwell's equations to model the light propagation in the sample, which is assumed to be a linear isotropic dielectric medium. We deal with dispersive and non-dispersive media. Firstly, using a general representation for the initial illumination, we present the direct problem of computing the OCT data, given the optical properties of the sample. Then, we derive reconstruction methods for solving the inverse problem of recovering the refractive index, real or complex valued. Motivated by the layer stripping algorithms [28, 31], we present a layer-by-layer reconstruction method that alternates between time and frequency domain and holds for dispersive media.

Without loss of generality, the OCT system can be simplified by placing the beam-splitter and the detector at the same position. The medium is contained in a bounded domain $\Omega \subset \mathbb{R}^3$, such that $\operatorname{supp}\chi (t, \cdot) \subset \Omega$, for all $t \in \mathbb{R}$, where χ is the electric susceptibility, a scalar quantity describing the optical properties of a linear dielectric medium. We set χ to zero for negative times. Also, the medium for $t \leq 0$

is assumed to be in a stationary state with zero stationary fields. Then, the electric field $E \in C^\infty(\mathbb{R} \times \mathbb{R}^3; \mathbb{R}^3)$ and the magnetic field $H \in C^\infty(\mathbb{R} \times \mathbb{R}^3; \mathbb{R}^3)$, in the absence of charges and currents, satisfy the Maxwell's equations

$$\nabla \times E(t, x) + \frac{1}{c}\frac{\partial H}{\partial t}(t, x) = 0, \quad \nabla \times H(t, x) - \frac{1}{c}\frac{\partial D}{\partial t}(t, x) = 0, \tag{1}$$

where c is the speed of light and D is the electric displacement, given by

$$D(t, x) = E(t, x) + 4\pi \int_{\mathbb{R}} \chi(\tau, x) E(t - \tau, x) d\tau. \tag{2}$$

This relation models a linear dielectric, dispersive medium with inhomogeneous, isotropic and non-stationary parameter.

The two identical laser pulses, one incident on the sample and the other on the mirror, are described initially, before the time $t = 0$, as vacuum solutions of the Maxwell's equations, meaning (1) for $D \equiv E$, defined by E_0, $H_0 \in C^\infty(\mathbb{R} \times \mathbb{R}^3; \mathbb{R}^3)$. In practice, the medium is illuminated by a Gaussian light, however at the scale of the sample the laser pulse can be approximated by a linearly polarized plane wave [9]. We assume that the incident wave does not interact with the medium until $t = 0$, resulting in the condition

$$E(t, x) = E_0(t, x), \quad H(t, x) = H_0(t, x), \quad t < 0, x \in \mathbb{R}^3. \tag{3}$$

The mirror is modeled as a medium with (infinitely) large constant electric susceptibility, with surface given by the hyperplane placed at distance $r \in \mathbb{R}$ from the source. Given the form of the incident wave, the reference field (back-reflected field), denoted by E_r, can be explicitly calculated.

The sample wave (backscattered wave) is given as a solution of the system (1)–(3). Then, the two backward traveling waves are recombined at the beam splitter, assumed to be at the detector position. In time-domain OCT, the sum of these two fields, integrated over all times, is measured at each point of the two-dimensional detector array $\mathcal{D} \subset \mathbb{R}^2$. Thus, as observed quantity we consider

$$\int_{\mathbb{R}} |(E - E_0)(t, x) + (E_r - E_0)(t, x)|^2 dt, \quad r \in \mathbb{R}, x \in \mathcal{D}. \tag{4}$$

Under some assumptions on the incident field [5], we may recover from the above measurements, the quantity

$$(\hat{E} - \hat{E}_0)(\omega, x), \quad \omega \in \mathbb{R}, x \in \mathcal{D}, \tag{5}$$

where $\hat{f} = \mathcal{F}(f)$ denotes the Fourier transform of f with respect to time

$$\mathcal{F}(f)(\omega) = \int_{\mathbb{R}} f(t)e^{i\omega t}\,dt.$$

In frequency-domain OCT, the detecting scheme is different. The mirror is not moving (r is fixed) and the detector is replaced by a spectrometer. Then, the intensity of the sum of the Fourier transformed fields at every available frequency (corresponding to different pixels at the CCD camera) is measured

$$\hat{m}(\omega, \mathrm{x}) = |(\hat{E} - \hat{E}_0)(\omega, \mathrm{x}) + (\hat{E}_r - \hat{E}_0)(\omega, \mathrm{x})|^2, \quad \omega \in \mathbb{R}, \; \mathrm{x} \in \mathcal{D}. \tag{6}$$

In practice, we obtain data only for few frequencies restricted by the limited bandwidth of the spectrometer. The OCT system allows also for measurements of the intensities of the two fields independently, by blocking one arm at a time. Thus, we assume that the quantity

$$\hat{m}_s(\omega, \mathrm{x}) = |(\hat{E} - \hat{E}_0)(\omega, \mathrm{x})|^2, \quad \omega \in \mathbb{R}, \; \mathrm{x} \in \mathcal{D}, \tag{7}$$

is also available. The main difference between the two setups is that (5) provides us with the full information of the backscattered field, amplitude and phase, which is not the case in (7), where we get phase-less data. We address later the problem of phase retrieval, meaning how to obtain (5) from (7).

Up to now, what we have modeled is known as full-field OCT where the whole sample is illuminated by an extended field. The main problem is that we want to reconstruct a $(1+3)$-dimensional function χ from OCT data, either (4) or (6), which are $(1+2)$-dimensional. Thus, we have to impose additional assumptions in order to compensate for the lack of dimension. To solve this problem, we consider a medium which admits a multi-layer structure. This assumption is not far from reality since OCT is mainly used in ophthalmology (imaging the retina) and human skin imaging. In both cases the imaging object consists of multiple layers with varying properties and thicknesses [14, 20].

If the medium is non-dispersive, meaning that the optical parameter is stationary, the function χ can be modeled as a $\delta-$distribution in time, so that its Fourier transform (temporal) does not depend on frequency. Then, even if we have enough information (theoretically), in OCT, as in any tomographic imaging technique, we deal with the problem of inverting partial and limited-angle data. This is the result of measuring only the back-scattered light for a limited frequency spectrum. In OCT, a narrow beam is used, resulting to an almost monochromatic illumination centered around a frequency.

In the following, we focus on data provided from a state-of-the-art and standard OCT system, where point-like illumination is used. In this case, only a small region inside the object is illuminated so that the function χ can be assumed depth-dependent and constant in the other two directions. Again we assume that locally the illumination is still properly described by a plane wave.

Let $\mathrm{x} = (x, y, z)$, where the z-direction denotes the depth direction. We model the light as a transverse electric polarized electromagnetic wave of the form

$$E(t, x) = \begin{pmatrix} 0 \\ u(t, z) \\ 0 \end{pmatrix}, \quad H(t, x) = \begin{pmatrix} v(t, z) \\ 0 \\ 0 \end{pmatrix}.$$

Then, the Maxwell's equations (1) together with (2) are simplified to

$$\Delta u(t, z) - \frac{1}{c^2} \frac{\partial^2}{\partial t^2} \int_{\mathbb{R}} \epsilon(\tau, z) u(t - \tau, z) d\tau = 0, \tag{8}$$

for the scalar valued function u, where $\Delta = \partial^2/\partial z^2$. Here, we define the time-dependent electric permittivity $\epsilon(t, z) = \delta(t) + 4\pi \chi(t, z)$, which varies also with respect to depth. The condition (3) is replaced by

$$u(t, z) = u_0(t, z), \quad t < 0, \ z \in \mathbb{R}. \tag{9}$$

The medium admits a multi-layered structure with N layers orthogonal to the z−direction, having spatial-independent but frequency-dependent refractive index $\hat{n} = \sqrt{\hat{\epsilon}}$, and varying lengths. We define $L = \cup_{j=1}^{N} L_j$ and we set

$$\hat{n}(\omega, z) = \begin{cases} n_0, & z \in \mathbb{R} \setminus \overline{L}, \\ \hat{n}_j(\omega), & z \in L_j. \end{cases} \tag{10}$$

This setup is commonly used for modeling the problem of parameter identification from OCT data. The volumetric OCT data consist of multiple A-scans, which are one-dimensional cross-sections of the medium across the z-direction. Under the assumption of a layered medium, the multiple A-scans are averaged over the x- and y-directions producing a profile of the measured intensity with respect to frequency or depth (post-processed image).

In Fig. 1, we see the experimental data for a three-layer medium with total length 0.7 mm, having two layers (top and bottom) filled with Noa61 ($n_1 = n_3 \approx 1.55$) and a middle one filled with DragonSkin ($n_2 \approx 1.405$). The spectrometer uses a grating with central wavelength 840 nm, going from 700 to 960 nm. On top right, we see an A-scan of the "raw" data (depth information), meaning the intensity of the combined sample and reference fields at a given point on the surface plane. The left picture is the post-processed B-scan (two-dimensional cross-sectional of the volumetric data). The bottom right picture presents the averaged (over lateral dimension) post-processed version of the data on the left. We could say that the post-processed data correspond to the time-domain data and are of interest since there we can see that the form of the intensity pattern is related to the interference happening because of the reflections at the different layer interfaces. The n-th "peak" appears at the position of the n-th interface. We will address later the nature and importance of the other "peaks". All x-axes are in pixel units.

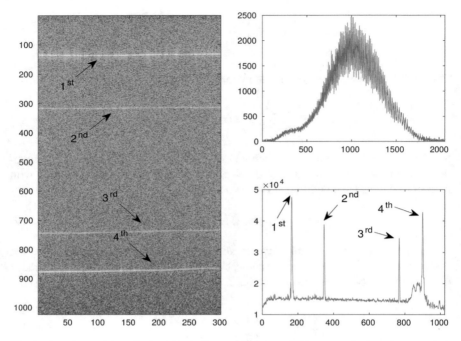

Fig. 1 Experimental data obtained from a frequency-domain OCT system of a three-layer medium with piecewise constant refractive index. Courtesy of Ryan Sentosa and Lisa Krainz, Medical University of Vienna

We refer to [2, 25, 32, 34] for recent works using similar setup and assumptions. Our work differs from previous methods in that we consider a dispersive medium. We deal also with absorbing media, a property that is usually neglected. We address three different cases for the layered medium:

- $\hat{n}_j(\omega) \equiv n_j,\quad j = 1, \ldots, N$ (non-dispersive),
- $\hat{n}_j(\omega) \in \mathbb{R},\quad j = 1, \ldots, N$ (dispersive),
- $\hat{n}_j(\omega) \in \mathbb{C},\quad j = 1, \ldots, N$ (dispersive with absorption).

The paper is organized as follows. In Sect. 2 we present the forward problem, meaning given the medium (location and properties) find the measurement data. We derive formulas that are also needed for the corresponding inverse problem, which we address in Sect. 3. Iterative schemes are presented for dispersive media and a mathematical model is given for the case of absorbing media. In Sect. 4 we give numerical results for simulated data, and we show that the parameter identification problem can be solved under few assumptions.

2 The Forward Problem

We derive mathematical models for the direct problem in OCT for multi-layer media with piecewise inhomogeneous refractive index. We start with a single-layer medium and then we generalize to more layers. The multiple reflections are also taken into account. Most of the formulas presented in this section, like the solutions of the initial value problems or the reflection and transmission operators (analogue to the Fresnel equations) can be found in classic books on partial differential equations [8, 29] and optics [1, 3, 13], respectively. However, we summarize them here, on one hand because we want to derive a rigorous mathematical model in both time and frequency domains and on the other hand because they are needed for the corresponding inverse problems. The easier but essential time-independent case is treated first. Then, we consider the time-dependent case by moving to the frequency-domain for real and complex valued parameters.

2.1 Non-dispersive Medium

Here, we simplify (10), and we consider the following form for the refractive index

$$\hat{n}(z) = \begin{cases} n_0, & z \in \mathbb{R} \setminus \overline{L}, \\ n_j, & z \in L_j, \end{cases} \tag{11}$$

for $j = 1, \ldots, N$. We describe the light propagation using (8) together with (9). Under the above assumption, we obtain

$$\partial_{tt} u(t, z) = \frac{c^2}{\hat{n}^2(z)} \Delta u(t, z), \quad t \in \mathbb{R}, \ z \in \mathbb{R}, \tag{12}$$

the one-dimensional wave equation. In the following, we use $c_j = c/n_j$, $j = 0, \ldots, N$. Let us assume that the initial field is given by

$$u_0(t, z) = f_0(z - c_0 t), \tag{13}$$

together with the assumption that $\operatorname{supp} f_0 \subset (-\infty, z_1)$, where z_1 represents the surface (first boundary point) of the medium L. This assumption on the support of the function reflects the condition that the laser beam does not interact with the probe until time $t = 0$.

We model the single-layer medium as $L = (z_1, z_2)$, for $z_1 < z_2$, but initially we consider the case

$$\hat{n}(z) = \begin{cases} n_0, & z < z_1, \\ n_1, & z > z_1. \end{cases} \tag{14}$$

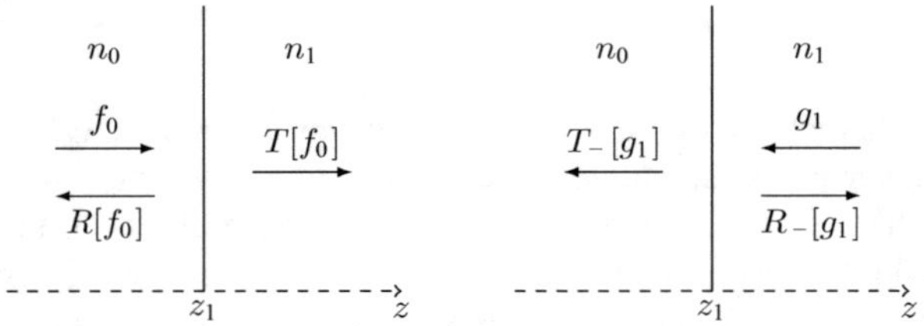

Fig. 2 Wave propagation. The reflection and transmission operators for the sub-problem (15) (left) and the sub-problem (18) (right)

Then, we obtain the system

$$\partial_{tt}u(t,z) = \frac{c^2}{n^2(z)}\Delta u(t,z), \quad t \in \mathbb{R}, \ z \in \mathbb{R},$$
$$u(t,z) = f_0(z - c_0 t), \qquad t < 0, \ z \in \mathbb{R}. \tag{15}$$

The above system of equations describes a wave traveling from the left incident on the interface at $z_1 \in \mathbb{R}$, see the left picture of Fig. 2. It is easy to derive the solution, which is given by

$$u(t,z) = \begin{cases} u_-(t,z) = f_0(z - c_0 t) + g_0(z + c_0 t), & t \geq 0, \ z < z_1, \\ u_+(t,z) = f_1(z - c_1 t), & t \geq 0, \ z > z_1. \end{cases} \tag{16}$$

Here, the function g_0 and f_1 describes the reflected and transmitted field, respectively. Given the continuity condition at $z = z_1$, meaning

$$\lim_{z \uparrow z_1} u_-(t,z) = \lim_{z \downarrow z_1} u_+(t,z), \quad \lim_{z \uparrow z_1} \partial_z u_-(t,z) = \lim_{z \downarrow z_1} \partial_z u_+(t,z),$$

we find a representation of g_0 and f_1 via operators. We denote the reflection operator by R and the transmission operator by T, defined by

$$R : f_0 \mapsto g_0, \quad R[f_0](z + c_0 t) = \frac{c_1 - c_0}{c_1 + c_0} f_0 \left(2z_1 - (z + c_0 t)\right), \tag{17}$$

and

$$T : f_0 \mapsto f_1, \quad T[f_0](z - c_1 t) = \frac{2c_1}{c_1 + c_0} f_0 \left(z_1 + \frac{c_0}{c_1}((z - c_1 t) - z_1)\right).$$

The fact that $\operatorname{supp} f_0 \subset (-\infty, z_1)$ implies that for every $t < 0$, $R[f_0](z + c_0 t) = 0$, in $(-\infty, z_1)$, and $T[f_0](z - c_1 t) = 0$, in (z_1, ∞). This is true, since neither the reflected nor the transmitted wave exists before the interaction of the initial wave with the boundary. Finally, we define the operator $U_1 : f_0 \mapsto u$, mapping the initial function f_0 to the solution u, given by (16), of the problem (15).

Now we consider the following problem

$$\partial_{tt} u(t, z) = \tfrac{c^2}{\hat{n}^2(z)} \Delta u(t, z), \quad t \in \mathbb{R},\ z \in \mathbb{R},$$
$$u(t, z) = g_1(z + c_1 t), \quad t < 0,\ z \in \mathbb{R}, \tag{18}$$

for an initial wave g_1 with $\mathrm{supp} g_1 \subset (z_1, \infty)$, for $\hat{n}$ as in (14).

This problem refers to the case of a wave incident from the right on the boundary $z = z_1$, see the right picture in Fig. 2. Again, we obtain a reflected and a transmitted part of the wave. The solution of this problem is given by

$$u(t, z) = \begin{cases} u_-(t, z) = g_0(z + c_0 t), & t \geq 0,\ z < z_1, \\ u_+(t, z) = f_1(z - c_1 t) + g_1(z + c_1 t), & t \geq 0,\ z > z_1. \end{cases}$$

As previously, we find a representation of the reflected and transmitted waves using operators acting on the initial wave. Here, we denote the reflection operator by R_- and the transmission operator by T_-, having the forms

$$R_- : g_1 \mapsto f_1, \quad R_-[g_1](z - c_1 t) = \frac{c_0 - c_1}{c_1 + c_0} g_1 \left(2z_1 - (z - c_1 t)\right),$$

and

$$T_- : g_1 \mapsto g_0, \quad T_-[g_1](z + c_0 t) = \frac{2c_0}{c_1 + c_0} g_1 \left(z_1 + \frac{c_1}{c_0}((z + c_0 t) - z_1)\right).$$

We define the solution operator $U_2 : g_1 \mapsto u$, mapping the initial g_1 to the solution of the problem (18).

It is trivial to model an operator $U_3 : f_1 \mapsto u$, where f_1 satisfies $supp f_1 \subset (-\infty, z_2)$, for an interface at $z = z_2$, with $\hat{n}(z) = n_2$, for $z > z_2$. This setup models how a transmitted, from the boundary at $z = z_1$, wave propagates for $t \geq 0$. We know that on $(-\infty, z_2)$, $U_3[f_1]$ is of the form

$$U_3[f_1](t, z) = f_1(z - c_1 t) + g_1(z + c_1 t).$$

We define in addition the operator $R_+ : f_1 \mapsto g_1$. These three different cases are combined to produce the following result.

Proposition 1 *Let $L = (z_1, z_2)$ be a single-layer medium, and let the refractive index be given by*

$$\hat{n}(z) = \begin{cases} n_0, & z < z_1, \\ n_1, & z \in (z_1, z_2), \\ n_2, & z > z_2. \end{cases}$$

If the initial wave f_0, given by (13), satisfies $\operatorname{supp} f_0 \subset (-\infty, z_1)$, then for a fixed $y \in (z_1, z_2)$ the solution of (12), together with $u(t < 0, z) = f_0$, is given by

$$u(t, z) = \mathbb{1}_{(-\infty, y)}(z) \left(U_1[f_0](t, z) + \sum_{j=0}^{\infty} U_2 \left[(R_+ R_-)^j R_+ T f_0 \right](t, z) \right)$$

$$+ \mathbb{1}_{(y, \infty)}(z) \sum_{j=0}^{\infty} U_3 \left[(R_- R_+)^j T f_0 \right](t, z), \quad t \geq 0,$$

(19)

with U_1, U_2, and U_3 defined as before.

Proof The function u, given by (19), is by construction a solution to the wave equation problem in both $(-\infty, y)$ and (y, ∞). Thus, we have only to check if both parts coincide in the interval (z_1, z_2). To do so, we recall the definitions

$$U_1[f](t, z) = T[f](z - c_1 t),$$
$$U_2[g](t, z) = R_-[g](z - c_1 t) + g(z + c_1 t),$$
$$U_3[f](t, z) = f(z - c_1 t) + R_+[f](z + c_1 t),$$

for $t \geq 0$, and $z \in (z_1, z_2)$. By plugging these formulas in (19) we get for $(-\infty, y)$ the term

$$T[f_0](z - c_1 t) + \sum_{j=0}^{\infty} R_- \left[(R_+ R_-)^j R_+ T f_0 \right](z - c_1 t) + \sum_{j=0}^{\infty} (R_+ R_-)^j R_+ T [f_0](z + c_1 t),$$

and for (y, ∞) the term

$$\sum_{j=0}^{\infty} (R_- R_+)^j T [f_0](z - c_1 t) + \sum_{j=0}^{\infty} R_+ \left[(R_- R_+)^j T f_0 \right](z + c_1 t).$$

Since all involved operators are bounded, both series converge and we see that the two terms coincide. The last thing to show is that (19) also satisfies the initial condition $u(t, z) = f_0(z - c_0 t)$, for all $t < 0$ and $z \in \mathbb{R}$. This can be seen by considering the supports of the operators U_1, U_2 and U_3, as they are defined previously. $\qquad \square$

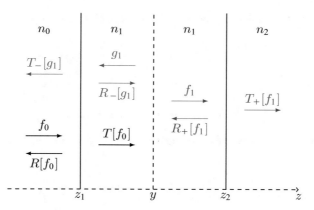

Fig. 3 A graphic representation of (19) for $j = 0$. The fields described by the operator U_1 are in black, the fields in blue are involved in U_3 for $f_1 = T[f_0]$, and in red we see the fields of U_2, where $g_1 = R_+ T[f_0]$

The formula (19) consists of three terms and accounts also for the multiple reflections occurring in the single-layer medium. Each term neglects either the boundary at z_1, or the one at z_2. The operator U_1 maps the initial wave f_0 to the solution u, in the real line. Then, the transmitted wave $T[f_0]$ is traveling back and forth between z_1 and z_2, describing the multiple reflections, given by the field

$$(R_- R_+)^j T[f_0].$$ (20)

The operator U_2 now uses for every $j \in \mathbb{N}$ the reflection of (20) at z_2 as initial and gives back a solution of the sub-problem (18). The last term models the wave interacting with the boundary at $z = z_2$, by application of U_3, which uses (20) as an initial function. In Fig. 3 we see the fields described by each operator in different colors for $j = 0$.

In the following example, we present the forms of the single- and double-reflected wave from the boundary at $z = z_2$, measured in $(-\infty, z_1)$.

Example 1 We know already that the reflected wave from the boundary $z = z_1$, is given by (17). We present now the reflected waves ϱ_{η_r}, $\eta_r = 1, 2$ in the interval $(-\infty, z_1)$, where η_r counts for the numbers of the undergoing reflections, meaning

$$\varrho_1(t, z) = T_- R_+ T[f_0], \quad \text{and} \quad \varrho_2(t, z) = T_- R_+ R_- R_+ T[f_0].$$

For $\eta_r = 1$, using the definition of the operator T applied to f_0 we get

$$\varrho_1(t, z) = \frac{2c_1}{c_1 + c_0} T_- R_+ f_0 \left(z_1 + \frac{c_0}{c_1}((z - c_1 t) - z_1) \right).$$

The argument of f_0 is now a function of $z - c_1 t$, and we can apply R_+, resulting in

$$\varrho_1(t, z) = \frac{c_2 - c_1}{c_2 + c_1} \frac{2c_1}{c_1 + c_0} T_- f_0 \left(z_1 + \frac{c_0}{c_1}(2z_2 - (z + c_1 t) - z_1) \right).$$

Thus, we have

$$R_+ T[f_0](z - c_0 t) = g(z + c_1 t),$$

a function of $z + c_1 t$, where the operator T_- can be to applied to give

$$\varrho_1(t, z) = \frac{2c_0}{c_1 + c_0} \frac{c_2 - c_1}{c_2 + c_1} \frac{2c_1}{c_1 + c_0} f_0 \left(z_1 + \frac{c_0}{c_1}(2(z_2 - z_1) - \frac{c_1}{c_0}((z + c_0 t) - z_1)) \right).$$

Following the same procedure for $\eta_r = 2$ now, we end up with the following form

$$\varrho_2(t, z) = \frac{2c_0}{c_1 + c_0} \frac{c_0 - c_1}{c_1 + c_0} \left(\frac{c_2 - c_1}{c_2 + c_1} \right)^2 \frac{2c_1}{c_1 + c_0}$$

$$\times f_0 \left(z_1 + \frac{c_0}{c_1}(4(z_2 - z_1) - \frac{c_1}{c_0}((z + c_0 t) - z_1)) \right).$$

Now, we move to the case of a multi-layer medium. The solution will be derived using the above formulas and consider the problem layer-by-layer. We define $L_j = (z_j, z_{j+1})$, for $j = 1, \ldots, N$. The refractive index is given by (11) and we define

$$\hat{n}_-(z) = \begin{cases} n_0, & z < z_1, \\ n_1, & z > z_1, \end{cases} \qquad \hat{n}_+(z) = \begin{cases} n_1, & z < z_2, \\ \tilde{\hat{n}}_+(z), & z > z_2. \end{cases} \tag{21}$$

The parameter $\tilde{\hat{n}}_+$ represents the refractive index in the remaining $N - 1$ layers.

Next we want to find a solution of (12) for

$$\hat{n}(z) = \begin{cases} \hat{n}_-(z), & z < y, \\ \hat{n}_+(z), & z > y, \end{cases}$$

and $y \in (z_1, z_2)$. The first case is exactly the same as the problem already discussed for the single-layer case, meaning that the application of the operators U_1 and U_2 is still valid. For the later case, we consider an initial wave f with $\text{supp} f \subset (-\infty, z_2)$ and by $U_3[f]$ we denote the solution of this sub-problem. We know that in $(-\infty, z_2)$, $U_3[f]$ gives

$$u(t, z) = f(z - c_1 t) + g(z + c_1 t).$$

We define an operator R_+ with $R_+[f] = g$, corresponding to the multi-reflected light from the boundaries $z_2, \ldots, z_{N+1}$, if we illuminate by f. Then, we get the following result.

Proposition 2 *Let $\hat{n}$ be defined by (11), and $\hat{n}_-$, $\hat{n}_+$ as in (21). Then for $y \in (z_1, z_2)$ the solution of (12), together with $u(t < 0, z) = f_0$, is given by*

$$u(t, z) = \mathbb{1}_{(-\infty, y)}(z) \left(U_1[f_0](t, z) + \sum_{j=0}^{\infty} U_2 \left[(R_+ R_-)^j R_+ T f_0 \right](t, z) \right)$$

$$+ \mathbb{1}_{(y, \infty)}(z) \sum_{j=0}^{\infty} U_3 \left[(R_- R_+)^j T f_0 \right](t, z), \quad t \geq 0. \tag{22}$$

Remark 1 The formula (22) is analogous to (19), see also Fig. 3. The main and most important difference from the single-layer case is that now we cannot have an explicit representation of the operator U_3.

We know that the solution of (12) in $(-\infty, z_1)$ admits the form

$$u(t, z) = f_0(z - c_0 t) + g(z + c_0 t),$$

and we define an operator $\tilde{R}$, through $\tilde{R}[f_0] = g$.

Proposition 3 *Let the operators R, R_- and T be defined as previously and $\tilde{R}[f_0]$ be given. We define $I[f_0](t, x) = f_0(z - c_0 t)$. Then, the following holds*

$$R_+[\tilde{f}] = \tilde{g}, \tag{23}$$

where

$$\tilde{f} = (T + R_- U_2^{-1}(\tilde{R} - U_1 + I))[f_0], \quad and \quad \tilde{g} = U_2^{-1}(\tilde{R} - U_1 + I)[f_0]. \tag{24}$$

Proof From Proposition 2 we know that

$$u(t, z) = U_1[f_0](t, z) + \sum_{j=0}^{\infty} U_2 \left[(R_+ R_-)^j R_+ T f_0 \right](t, z), \quad t \geq 0, z < z_1,$$

describes a solution. It also holds that

$$u(t, z) = f_0(z - c_0 t) + \tilde{R}[f_0](z + c_0 t), \quad t \geq 0, z < z_1.$$

Then, using the definition of $I[f_0]$, we get

$$U_2^{-1}(\tilde{R} - U_1 + I)[f_0] = (1 - R_+ R_-)^{-1} R_+ T[f_0],$$

which admits the equivalent form

$$U_2^{-1}(\tilde{R} - U_1 + I)[f_0] - R_+ R_- U_2^{-1}(\tilde{R} - U_1 + I)[f_0] = R_+ T[f_0].$$

This results in (23) for $\tilde{f}$ and $\tilde{g}$, as in (24). □

Remark 2 The function $\tilde{g}$ describes the total amount of light which travels back from the remaining $N - 1$ layers, meaning it considers all multiple reflections.

Lemma 1 *Let L be a multi-layer medium consisting of $N \in \mathbb{N}$ layers, and let the refractive index be given by* (11). *Then, the solution of* (12), *together with $u(t < 0, z) = f_0$, can be computed layer-by-layer.*

Proof Starting with the first layer, we use $\hat{n}$ defined in (11) and (21) and we apply Proposition 2. We thus obtain $\tilde{f}$ and $\tilde{g}$, presented in Proposition 3. Then, the function $\tilde{f}$ is the initial wave for the corresponding problem with parameter now given by

$$
\hat{n}(z) = \begin{cases} n_1, & z < z_2, \\ n_2, & z \in (z_2, z_3), \\ \tilde{\hat{n}}(z), & z > z_3, \end{cases}
$$

where $\tilde{\hat{n}}$ represents the refractive index of the next $N - 2$ layers. Repeating the same argument, we use (22), with f_0 replaced by $\tilde{f}$, for the updated operators. We continue this procedure for the new parameters and operators and we end up with the solution for $\hat{n}$ given by (11).   □

After some lengthy but straightforward calculations, we can generalize the formulas of Example 1 for the k-th layer of the medium and $\eta_r \in \mathbb{N}$, resulting in the field

$$
\varrho_{\eta_r}(t, z) = \sum_{q=1}^{\eta_r} \left(\frac{c_{k-1} - c_k}{c_k + c_{k-1}} \right)^{q-1} \left(\frac{c_{k+1} - c_k}{c_{k+1} + c_k} \right)^q \prod_{j=1}^{k} \frac{4c_{j-1}c_j}{(c_j + c_{j-1})^2}
$$

$$
\times f_0 \left(\left(\sum_{j=1}^{k-1} 2z_j(1 - \frac{c_{j-1}}{c_j}) \prod_{l=1}^{j-1} \frac{c_{l-1}}{c_l} + z_k(2 - (2 + 2(q-1)) \frac{c_{k-1}}{c_k}) \prod_{l=1}^{k-1} \frac{c_{l-1}}{c_l} \right. \right.
$$

$$
\left. \left. + z_{k+1} \frac{c_0}{c_k}(2 + 2(q-1)) - (z + c_0 t) \right), \quad t \geq 0, \ z < z_1, \right.
$$

(25)

valid for an initial function f_0, with $\mathrm{supp} f_0 \subset (-\infty, z_1)$.

2.2 Dispersive Medium

In this section, we consider the form (10) for the refractive index and we set $\hat{n}_0(\omega) = \hat{n}_0 > 0$. We assume $\Re\{\hat{n}(\omega)\} > 0$, and $\Im\{\hat{n}(\omega)\} \geq 0$, for all $\omega \in \mathbb{R}$. Unfortunately, an explicit solution, as in Sect. 2.1, cannot be derived here for a time dependent parameter. However, applying the Fourier transform, with respect to time, to (8), we get the Helmholtz equation

$$\Delta \hat{u}(\omega, z) + \frac{\omega^2}{c^2} \hat{n}^2(\omega, z) \hat{u}(\omega, z) = 0, \quad \omega \in \mathbb{R}, \ z \in \mathbb{R}, \tag{26}$$

together with an appropriate radiation condition (equivalent to the initial condition in the time domain) that guarantees uniqueness. For $y \in (z_1, z_2)$, we define

$$\hat{n}(\omega, z) = \begin{cases} \hat{n}_-(\omega, z), & z < y, \\ \hat{n}_+(\omega, z), & z > y, \end{cases}$$

with

$$\hat{n}_-(\omega, z) = \begin{cases} \hat{n}_0, & z < z_1, \\ \hat{n}_1(\omega), & z > z_1, \end{cases} \quad \text{and} \quad \hat{n}_+(\omega, z) = \begin{cases} \hat{n}_1(\omega), & z < z_2, \\ \tilde{n}(\omega, z), & z > z_2. \end{cases}$$

The refractive index $\tilde{n}$ accounts for the parameter of the remaining $N - 1$ layers, meaning $\tilde{n} - \hat{n}_0$ is compactly supported.

Initially, we consider the problem of a right-going incident wave of the form

$$\hat{u}_0(\omega, z) = \alpha_0(\omega) e^{i \frac{\omega}{c} \hat{n}_0 z}, \tag{27}$$

incident at the interface $z = z_1$. Then, the corresponding problem reads

$$\Delta \hat{u}(\omega, z) + \frac{\omega^2}{c^2} \hat{n}_-^2(\omega) \hat{u}(\omega, z) = 0, \quad \omega \in \mathbb{R}, \ z \in \mathbb{R},$$

$$\partial_z \hat{u} - i \frac{\omega}{c} \hat{n}_1(\omega) \hat{u} = 0, \quad \omega \in \mathbb{R}, \ z = z_1^+,$$

for an artificial boundary point $z_1^+ > z_1$. The boundary radiation condition is such that there is no left-going wave at the region $(z_1, +\infty)$.

The solution admits the form

$$\hat{u}(\omega, z) = \begin{cases} \hat{u}_0 + R[\alpha_0](\omega) e^{-i \frac{\omega}{c} \hat{n}_0 z}, & z < z_1, \\ T[\alpha_0](\omega) e^{i \frac{\omega}{c} \hat{n}_1(\omega) z}, & z > z_1, \end{cases}$$

where we define the reflection and transmission operators R and T, respectively, by

$$R : \alpha_0(\omega) \mapsto \frac{\hat{n}_0 - \hat{n}_1(\omega)}{\hat{n}_0 + \hat{n}_1(\omega)} \alpha_0(\omega) e^{2i\frac{\omega}{c}\hat{n}_0 z_1},$$

$$T : \alpha_0(\omega) \mapsto \frac{2\hat{n}_0}{\hat{n}_0 + \hat{n}_1(\omega)} \alpha_0(\omega) e^{i\frac{\omega}{c}(\hat{n}_0 - \hat{n}_1(\omega))z_1}.$$

(28)

The solution operator is then given by $V_1 : \hat{u}_0 \mapsto \hat{u}$. The next sub-problem is described by

$$\Delta \hat{u}(\omega, z) + \frac{\omega^2}{c^2} \hat{n}_-^2(\omega) \hat{u}(\omega, z) = 0, \quad \omega \in \mathbb{R}, \ z \in \mathbb{R},$$

$$\partial_z \hat{u} + i\frac{\omega}{c}\hat{n}_0 \hat{u} = 0, \quad \omega \in \mathbb{R}, \ z = z_1^-.$$

for an incident left-going wave of the form $\hat{u}_0(\omega, z) = \beta_1(\omega) e^{-i\frac{\omega}{c}\hat{n}_1(\omega)z}$, and an artificial boundary point at $z_1^- < z_1$. The boundary radiation condition is that the left-going wave in $(-\infty, z_1)$ is zero. The solution now is given by

$$\hat{u}(\omega, z) = \begin{cases} T_-[\beta_1](\omega) e^{-i\frac{\omega}{c}\hat{n}_0 z}, & z < z_1, \\ \hat{u}_0 + R_-[\beta_1](\omega) e^{i\frac{\omega}{c}\hat{n}_1(\omega)z}, & z > z_1, \end{cases}$$

where R_- and T_- are defined by

$$R_- : \beta_1(\omega) \mapsto \frac{\hat{n}_1(\omega) - \hat{n}_0}{\hat{n}_1(\omega) + \hat{n}_0} \beta_1(\omega) e^{2i\frac{\omega}{c}\hat{n}_1(\omega)z_1},$$

$$T_- : \beta_1(\omega) \mapsto \frac{2\hat{n}_1(\omega)}{\hat{n}_1(\omega) + \hat{n}_0} \beta_1(\omega) e^{i\frac{\omega}{c}(\hat{n}_0 - \hat{n}_1)(\omega)z_1}.$$

Let again $V_2 : \hat{u}_0 \mapsto \hat{u}$ denote the corresponding solution operator.

The final sub-problem deals with the scattering of a right-going wave of the form $\hat{u}_0(\omega, z) = \alpha_1(\omega) e^{i\frac{\omega}{c}\hat{n}_1(\omega)z}$ by a medium supported in $(z_2, +\infty)$ with refractive index $\tilde{n}$. The governing equations are

$$\Delta \hat{u}(\omega, z) + \frac{\omega^2}{c^2} \hat{n}_+^2(\omega) \hat{u}(\omega, z) = 0, \quad \omega \in \mathbb{R}, \ z \in \mathbb{R},$$

$$\lim_{z \to +\infty} \left(\partial_z \hat{u} - i\frac{\omega}{c}\hat{n}_0 \hat{u} \right) = 0, \quad \omega \in \mathbb{R}.$$

The radiation condition now ensures that at infinity exist only right-going waves. The solution is given by

$$\hat{u}(\omega, z) = \alpha_1(\omega) e^{i\frac{\omega}{c}\hat{n}_1(\omega)z} + \beta_1(\omega) e^{-i\frac{\omega}{c}\hat{n}_1(\omega)z}, \quad z < z_2.$$

We define $R_+ : \alpha_1(\omega) \mapsto \beta_1(\omega)$ and the relevant operator $V_3 : \hat{u}_0 \mapsto \hat{u}$ mapping the incident field to the solution of this specific problem. We remark that the operator R_+ cannot be computed explicitly because it contains also the information from the remaining $N - 1$ layers.

Proposition 4 *Let the incident wave be of the form* (27). *We define*

$$\hat{u}_{0,-}^j(\omega, z) = [(R_+ R_-)^j R_+ T \alpha_0](\omega) e^{-i\frac{\omega}{c}\hat{n}_1(\omega)z},$$

$$\hat{u}_{0,+}^j(\omega, z) = [(R_- R_+)^j T \alpha_0](\omega) e^{i\frac{\omega}{c}\hat{n}_1(\omega)z}.$$

Then, the field

$$\hat{u}(\omega, z) = \mathbb{1}_{(-\infty, y)}(z) \left(V_1[\hat{u}_0] + \sum_{j=0}^{\infty} V_2[\hat{u}_{0,-}^j] \right)(\omega, z)$$

$$+ \mathbb{1}_{(y, \infty)}(z) \left(\sum_{j=0}^{\infty} V_3[\hat{u}_{0,+}^j] \right)(\omega, z), \tag{29}$$

for fixed $y \in (z_1, z_2)$, *is the solution of the Helmholtz equation* (26), *for the refractive index defined as above, and satisfies the radiation condition.*

Proof By construction, $\hat{u}$ fulfills (26) in $(-\infty, y)$ and (y, ∞), and the radiation condition. Thus, it remains to show that the two parts coincide in (z_1, z_2). Recalling the definitions of V_1, V_2, and V_3, restricted in (z_1, z_2), we get

$$V_1[\hat{u}_0](\omega, z) + \sum_{j=0}^{\infty} V_2[\hat{u}_{0,-}^j](\omega, z) = T[\alpha_0](\omega) e^{i\frac{\omega}{c}\hat{n}_1(\omega)z}$$

$$+ \sum_{j=0}^{\infty} R_- [(R_+ R_-)^j R_+ T \alpha_0](\omega) e^{i\frac{\omega}{c}\hat{n}_1(\omega)z} + \sum_{j=0}^{\infty} (R_+ R_-)^j R_+ T[\alpha_0](\omega) e^{-i\frac{\omega}{c}\hat{n}_1(\omega)z}$$

and

$$\sum_{j=0}^{\infty} V_3[\hat{u}_{0,+}^j](\omega, z) = \sum_{j=0}^{\infty} (R_- R_+)^j T[\alpha_0](\omega) e^{i\frac{\omega}{c}\hat{n}_1(\omega)z}$$

$$+ \sum_{j=0}^{\infty} R_+ [(R_- R_+)^j T \alpha_0](\omega) e^{-i\frac{\omega}{c}\hat{n}_1(\omega)z}.$$

We reorder the terms and we observe that they coincide in (z_1, z_2). $\qquad\square$

Remark 3 If $L = (z_1, z_2)$ denotes a single-layer medium, with material parameter $\hat{n}$, given by

$$\hat{n}(\omega, z) = \begin{cases} \hat{n}_0, & z < z_1, \\ \hat{n}_1(\omega), & z \in (z_1, z_2), \\ \hat{n}_2(\omega), & z > z_2, \end{cases}$$

then we can compute R_+ explicitly, and also the operator V_3.

Example 2 The amplitude of the j-th reflection in a certain layer is described by the term

$$(R_+ R_-)^j R_+ T[\alpha_0](\omega).$$

The single reflected wave from the most left boundary of L is given by (28). For the k-th layer of the medium, we obtain the back-reflected field

$$\hat{\varrho}(\omega, z) = \sum_{q=1}^{\eta_r} \left(\frac{\hat{n}_k(\omega) - \hat{n}_{k+1}(\omega)}{\hat{n}_{k+1}(\omega) + \hat{n}_k(\omega)} \right)^q \left(\frac{\hat{n}_k(\omega) - \hat{n}_{k-1}(\omega)}{\hat{n}_{k-1}(\omega) + \hat{n}_k(\omega)} \right)^{q-1} \prod_{j=1}^{k} \frac{4\hat{n}_{j-1}(\omega)\hat{n}_j(\omega)}{(\hat{n}_{j-1}(\omega) + \hat{n}_j(\omega))^2}$$

$$\times \alpha_0(\omega) e^{i \frac{\omega}{c} \left(\hat{n}_k(\omega)(2q z_{k+1} - 2(q-1)z_k) + \sum_{l=1}^{k} 2(\hat{n}_{l-1} - \hat{n}_l)(\omega)z_l \right)} e^{-i \frac{\omega}{c} \hat{n}_0 z}, \quad z < z_1,$$
$$\tag{30}$$

where α_0 is the amplitude of the incident wave $\hat{u}_0$.

The solution of (26) in $(-\infty, z_1)$ admits the form

$$\hat{u}(\omega, z) = \alpha_0(\omega) e^{i \frac{\omega}{c} \hat{n}_0 z} + \tilde{\beta}(\omega) e^{-i \frac{\omega}{c} \hat{n}_0 z},$$

and we define $\tilde{R}(\omega) : \alpha_0(\omega) \mapsto \tilde{\beta}(\omega)$.

Lemma 2 *Let the incident wave be of the form (27), and let $\tilde{R}[\alpha_0]$ be known. Then, the following relation holds*

$$R_+[\tilde{\alpha}_0](\omega) = \tilde{\beta}_0(\omega),$$

for

$$\tilde{\alpha}_0 = T[\alpha_0] + R_- \left(T_-^{-1} \left(\tilde{R}[\alpha_0] - R[\alpha_0] \right) \right), \quad \text{and} \quad \tilde{\beta}_0 = T_-^{-1} \left(\tilde{R}[\alpha_0] - R[\alpha_0] \right),$$
$$\tag{31}$$

calculated from the previously defined operators R_-, T and T_-.

Proof We know that in $(-\infty, z_1)$,

$$\alpha_0(\omega)e^{i\frac{\omega}{c}\hat{n}_0 z} + \tilde{R}[\alpha_0](\omega)e^{-i\frac{\omega}{c}\hat{n}_0 z} = V_1[\hat{u}_0](\omega, z) + \sum_{j=0}^{\infty} V_2[\hat{u}_{0,-}^j](\omega, z),$$

for $\hat{u}_{0,-}^j$, defined as in Proposition 4. Using the definitions of V_1 and V_2, we get

$$\tilde{R}[\alpha_0](\omega)e^{-i\frac{\omega}{c}\hat{n}_0 z} - R[\alpha_0](\omega)e^{-i\frac{\omega}{c}\hat{n}_0 z} = T_-\left(\sum_{j=0}^{\infty}(R_+R_-)^j R_+T[\alpha_0]\right)(\omega)e^{-i\frac{\omega}{c}\hat{n}_0 z}.$$

This results in

$$T_-^{-1}\left(\tilde{R}[\alpha_0] - R[\alpha_0]\right) = \sum_{j=0}^{\infty}(R_+R_-)^j R_+T[\alpha_0],$$

which is equivalent to

$$(1 - R_+R_-)\left(T_-^{-1}\left(\tilde{R}[\alpha_0] - R[\alpha_0]\right)\right) = R_+T[\alpha_0].$$

This completes the proof.   □

The amplitudes $\tilde{\alpha}_0$ and $\tilde{\beta}_0$, defined in (31), correspond to the amplitudes of the Fourier transforms of $\tilde{f}$ and $\tilde{g}$, given in Proposition 3. Furthermore, one can derive an analogue of Lemma 1 also for a dispersive medium.

3 The Inverse Problem

We address the inverse problem of recovering the position, the size and the optical properties of a multi-layer medium with piecewise inhomogeneous refractive index. We identify the position by the distance from the detector to the most left boundary of the medium, and the size by reconstructing the constant refractive index n_0 of the background medium. Initially, we discuss the problem of phase retrieval and possible directions to overcome it and then we present reconstruction methods for non-dispersive and dispersive media. We end this section by giving a method, which with the use of the Kramers–Kronig relations, makes the reconstruction of a complex-valued refractive index (absorbing medium) possible. Let $z = z_d$ denote the position of the point detector.

3.1   Phase Retrieval and OCT

The phase retrieval problem, meaning the reconstruction of a function from the magnitude of its Fourier transform, has attracted much attention in the optical imaging community, see [27] for an overview. When dealing with experimental data, additional problems arise, like different types of noise and incomplete data. Mathematically speaking, the problem corresponds to a least squares minimization problem for a non-convex functional. In our case, where we are given one-dimensional data of the form (6) or (7), unique reconstruction of the phase is not possible [15]. However, there exist convergent algorithms that produce satisfactory results under some assumptions on the signal, like bounded support and non-negativity constraints. These algorithms are alternating between time and frequency domains, using usually less coefficients than samples, which makes the exact recovery almost impossible.

In OCT, this problem has been also well studied, see for example [21, 23, 26]. The main idea is either to consider a phase-shifting device in the reference arm or to combine OCT with holographic techniques. The first case, the one we consider here, produces different measurements by placing the mirror at different positions, meaning by changing the path-length difference between the two arms.

As already discussed in Sect. 1, we have measurements of the form

$$\hat{m}(r; \omega) = |(\hat{u} - \hat{u}_0)(\omega, z_d) + (\hat{u}_r - \hat{u}_0)(\omega, z_d)|, \quad \omega \in \mathbb{R},$$

for r fixed, where u_r denotes the y-component of the reference field E_r, and we also acquire the data

$$\hat{m}_s(\omega) = |(\hat{u} - \hat{u}_0)(\omega, z_d)|, \quad \omega \in \mathbb{R}.$$

Since we know the incident field $\hat{u}_0$ explicitly, we can also compute the reference field $\hat{u}_r - \hat{u}_0$ at the point detector. Then, the problem of phase retrieval we address here is to recover $\hat{u} - \hat{u}_0$ from the knowledge of $\hat{m}$ and $\hat{m}_s$ for all $\omega \in \mathbb{R}$. We know, from [18, 19], that if $u_r - u_0$ is compactly supported, then there exist at most two solutions $u - u_0$. See the left picture of Fig. 4, where we visualize graphically the two solutions by plotting in the complex plane the two above equations at specific frequency for the setup of the third example presented later in Sect. 4.

If in addition, there exist a constant $\gamma \in [0, 1)$, such that

$$|\Re\{\hat{u}_r - \hat{u}_0\}| \leq \gamma |\Im\{\hat{u}_r - \hat{u}_0\}|,$$

then, there exists at most one solution in $L^2(\mathbb{R})$ with compact support in $[0, \infty)$. However, it is hard to verify that the reference field fulfills this condition and we observed that, in all numerical examples, this inequality does not hold, for an incident plane wave. Thus, in order to decide which solution of the two is correct, extra information is needed. Motivated by the phase-shifting procedure, we consider

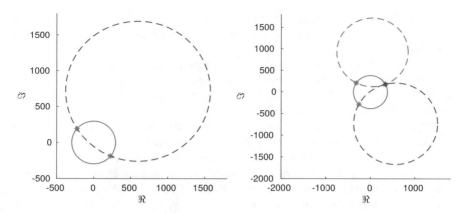

Fig. 4 Left: The intersection points of the circles $\hat{m}(r_1; \omega_1) \in \mathbb{C}$ (red) and $\hat{m}_s(\omega_1) \in \mathbb{C}$ (blue). Right: The intersection points of the circles $\hat{m}(r_1; \omega_2) \in \mathbb{C}$ (red), $\hat{m}(r_2; \omega_2) \in \mathbb{C}$ (green) and $\hat{m}_s(\omega_2) \in \mathbb{C}$ (blue). The red asterisk indicates the unique solution. The setup is the same as the one in the third example in Sect. 4

data for two different positions of the mirror, let us say $r = r_1, r_2$. Then, we get the data

$$\hat{m}(r_1; \omega), \quad \hat{m}(r_2; \omega), \quad \text{and} \quad \hat{m}_s(\omega).$$

Using $\hat{m}(r_1; \cdot)$ and $\hat{m}_s$, we get two possible solutions, and from $\hat{m}(r_2; \cdot)$ and $\hat{m}_s$, other two. But since $\hat{m}_s$ is the same in both cases, we find the unique solution as the common solution of the two pairs. This is illustrated at the right picture of Fig. 4, where we plot the three above relations at two different frequencies. This way, we get unique solutions at every available frequency.

Thus, having measurements for two different positions of the reference mirror, we may consider that frequency-domain OCT provide us with the quantity

$$(\hat{u} - \hat{u}_0)(\omega, z_d), \quad \omega \in \mathbb{R}, \tag{32}$$

the equivalent measurements of a time-domain OCT system, see (5).

3.2 Reconstructions in Time Domain

We consider initially a non-dispersive medium. Then, the refractive index is given by (11) and the time-dependent OCT data admit the form

$$m(t, z_d) = u(t, z_d) - f_0(z_d - c_0 t), \quad t \in [0, T], \tag{33}$$

for $T > 0$. Here, we assume that the initial wave (known explicitly) does not contribute to the measurements. The following presented algorithms are based on a layer-by-layer procedure. At the first step, we reconstruct the parameters for a given layer and then we update the data, to be used for the next layer. Thus, we assume that we have already recovered the boundary point z_{k-1} and the coefficient c_{k-1} of the layer L_{k-1}. We denote by $m^{(k)}(t, z_d)$, the data corresponding to a $(N-k+1)$−layer medium, with the most left layer being the L_k.

As we see in Fig. 1, the (time-dependent) data consist mainly of $N + 1$ major "peaks", for a N−layer medium, and some minor "peaks", related to the light undergoing multiple-reflections in the medium. The experimental data are, of course, also noisy and may show some small "peaks" because of the OCT system. The first two "peaks" correspond, for sure, to the single-reflected light from the first two boundaries. We propose a scheme to neglect minor "peaks" due to multiple reflected light. Then, the first major "peak" in $m^{(k)}$, corresponds to the back-reflected light from the interface at z_k.

Step 1:   We isolate the first "peak" by cutting off around a certain time interval $[T_1, T_2]$, meaning we consider

$$\tilde{m}^{(k)}(t, z_d) = \mathbb{1}_{[T_1, T_2]}(t) m^{(k)}(t, z_d).$$

The time interval can be fixed for all layers and depends on the time support of the initial wave. On the other hand, this wave can be described by (25), if we use $\eta_r = 1$ and replace k by $k - 1$. Therefore, we obtain the equation

$$\tilde{m}^{(k)}(t, z_d) = \frac{c_k - c_{k-1}}{c_k + c_{k-1}} \tilde{f}_0 \left(2z_k \frac{c_0}{c_{k-1}} - (z_d + c_0 t) \right), \quad t \in \mathbb{R}, \tag{34}$$

with $\tilde{f}_0$ given by

$$\tilde{f}_0(z) = \prod_{j=1}^{k-1} \frac{4c_{j-1}c_j}{(c_j + c_{j-1})^2} f_0 \left(\left(\sum_{j=1}^{k-2} 2z_j (1 - \frac{c_{j-1}}{c_j}) \prod_{l=1}^{j-1} \frac{c_{l-1}}{c_l} \right. \right.$$
$$\left. \left. + z_{k-1} (2 - 2\frac{c_{k-2}}{c_{k-1}}) \prod_{l=1}^{k-2} \frac{c_{l-1}}{c_l} \right) + z \right).$$

The supremum of (34), using its shift-invariance property, gives the value of c_k. The position of the interface z_k can then be recovered from (34), by solving the minimization problem

$$\min_{z \in \mathbb{R}} \left| \tilde{m}^{(k)}(t, z_d) - \frac{c_k - c_{k-1}}{c_k + c_{k-1}} \tilde{f}_0 (2z \frac{c_0}{c_{k-1}} - (z_d + c_0 t)) \right|, \quad \text{for all } t \in \mathbb{R}.$$

Since both parameters are time-independent, there exist also other variants for solving this overdetermined problem.

Step 2:   Before moving to the layer L_{k+1}, we have to update the data function. We could just remove the contribution of the current layer, meaning $\tilde{m}^{(k)}$. However, the first "peak" might not correspond to the reflection from z_k, for $k > 1$, but to contributions of multiple reflected wave from previous layers, arriving at the detector before the major wave. Since, we have recovered the properties of L_k, we can compute all future multiple reflections from this layer using (25), let us call them $\mathcal{R}[c_k, z_k]$.
Then, we update the data as

$$m^{(k+1)} = m^{(k)} - \tilde{m}^{(k)} - \mathcal{R}[c_k, z_k].$$

Repeating these steps, we end up with the following result.

Lemma 3 *Let L be a multi-layer medium, with $N \in \mathbb{N}$ layers, characterized by $\hat{n}$, given by (11). Then from the knowledge of n_0, the initial wave f_0, and the measurement data (33), following the above iterative scheme, we can uniquely reconstruct n_j and z_j for $j = 1, \ldots, N + 1$.*

The above scheme can be written in an operator form, by the application of Propositions 2 and 3. For the sake of presentation, we consider the case of the first layer in order to avoid redefining all operators.

Step 1:   Recall the definition of the operator $\tilde{R}$, applied to the initial wave f_0, which describes the total amount of the reflected light. Considering the data (33), we get

$$m(t, z_d) = \tilde{R}[f_0](z_d + c_0 t), \quad t \in [0, T].$$

Following the same procedure, using (17), we can recover c_1 and z_1, from the reduced data equation

$$m^{(1)}(t, z_d) = R[f_0](z_d + c_0 t).$$

Step 2:   We update all operators and from Proposition 3, we obtain $\tilde{f}$ and $\tilde{g}$, given by (24). From the definition of $\tilde{g}$, we see that we obtain the function

$$m^{(2)}(t, z_d) = m(t, z_d) - m^{(1)}(t, z_d),$$

describing the updated data. The advantage here is that we do not need to subtract the multiple reflections term, since they are already included in the updated version of $\tilde{g}$.

3.3   Reconstructions in Frequency Domain

We consider the case of a dispersive medium, with a piecewise inhomogeneous refractive index $\hat{n}(\omega)$, for $\omega \in \mathbb{R}$. Initially, we assume $\hat{n} \in \mathbb{R}$. The derived iterative scheme can be applied also to the simpler non-dispersive case, giving a reconstruction method in the frequency domain for a time-independent parameter.

3.3.1   Dispersive Medium

The refractive index is given by (10) and initially, we restrict ourselves to the case of real-valued $\hat{n}$. The data $\hat{m}$ are given by (32). As before, we present a layer-by-layer scheme. We assume that the boundary point z_{k-1} and the coefficient $\hat{n}_{k-1}$ of the layer L_{k-1} are already recovered. We denote by $\hat{m}^{(k)}(\omega, z_d)$, the data corresponding to the $(N - k + 1)-$layer medium. We choose the time interval $\mathbb{1}_{[T_1,T_2]}$ similarly to Sect. 3.2 but here we have to take into account dispersion. However, this results only to slightly longer time interval, since in the wavelength range, where OCT operates, scattering dominates absorption.

Step 1:   As we have seen in Fig. 1, for example, from the data $\hat{m}^{(k)}$ we cannot distinguish the different "peaks". Thus, we have to switch back to the time domain in order to isolate the first "peak". We apply

$$\tilde{\hat{m}}^{(k)}(\omega) = \mathcal{F}\left(\mathbb{1}_{[T_1,T_2]}\mathcal{F}^{-1}(\hat{m}^{(k)})\right)(\omega), \quad \omega \in \mathbb{R}.$$

We use (30) for $k := k - 1$ and $\eta_r = 1$ to get

$$\tilde{\hat{m}}^{(k)}(\omega) = \frac{\hat{n}_{k-1}(\omega) - \hat{n}_k(\omega)}{\hat{n}_{k-1}(\omega) + \hat{n}_k(\omega)} \tilde{\alpha}_0(\omega) e^{i\frac{\omega}{c}\hat{n}_{k-1}(\omega)2z_k} e^{-i\frac{\omega}{c}\hat{n}_0 z_d}, \tag{35}$$

with

$$\tilde{\alpha}_0(\omega) = \prod_{j=1}^{k-1} \frac{4\hat{n}_{j-1}(\omega)\hat{n}_j(\omega)}{(\hat{n}_{j-1}(\omega) + \hat{n}_j(\omega))^2} \alpha_0(\omega) e^{i\frac{\omega}{c}\sum_{l=1}^{k-1} 2(\hat{n}_{l-1}-\hat{n}_l)(\omega)z_l},$$

describing an already known quantity. The absolute value of (35) gives

$$\hat{n}_k(\omega) = \hat{n}_{k-1}(\omega) \left(\frac{1 + |\tilde{\hat{m}}^{(k)}(\omega)/\tilde{\alpha}_0(\omega)|}{1 - |\tilde{\hat{m}}^{(k)}(\omega)/\tilde{\alpha}_0(\omega)|}\right)^{\pm 1},$$

which together with a suitable condition on $\hat{n}_k$ allow us to recover the refractive index. To reconstruct the position z_k we define the function

$$f_k(\omega) = \frac{\tilde{\hat{m}}^{(k)}(\omega)/\tilde{\alpha}_0(\omega)}{|\tilde{\hat{m}}^{(k)}(\omega)/\tilde{\alpha}_0(\omega)|} sign(\hat{n}_{k-1}(\omega) - \hat{n}_k(\omega)),$$

and we observe that the absolute value of its derivative, together with (35), results in

$$c|f_k'(\omega)| = |2z_k(\hat{n}_{k-1}(\omega) + \omega\hat{n}_{k-1}'(\omega)) - \hat{n}_0 z_d|.$$

Together with a non-negativity constrain, we obtain z_k.

Step 2:   As in the time domain case, we update the data by subtracting (35) and the terms representing the multiple reflections from the already recovered layers, called $\mathcal{R}[\hat{n}_k, z_k]$. We define

$$\hat{m}^{(k+1)} = \hat{m}^{(k)} - \tilde{\hat{m}}^{(k)} - \mathcal{R}[\hat{n}_k, z_k].$$

Repeating the steps, a reconstruction of the properties and the lengths of all the remaining layers is obtained.

Lemma 4 *Let L be a multi-layer medium, with $N \in \mathbb{N}$ layers, characterized by $\hat{n}$, given by the Fourier transform of (10). If we restrict ourselves to the case $\hat{n}(\omega) \in \mathbb{R}$, for all $\omega \in \mathbb{R}$, then the above iterative scheme, allows us to uniquely reconstruct $\hat{n}_j$ and z_j for $j = 1, \ldots, N + 1$, given $\hat{n}_0$, the incident wave $\hat{u}_0$ and the measurement data (32).*

3.3.2   Absorbing Medium

Here, we consider the case of a complex-valued material parameter, $\hat{n}(\omega) \in \mathbb{C}$, for every $\omega \in \mathbb{R}$. The real part describes how the medium reflects the light and the imaginary part (wavelength dependent) determines how the light absorbs in the medium. In [6, 7] we considered the multi-modal PAT/OCT system, meaning that we had additional internal data from PAT, in order to recover both parts of the refractive index. Here, the multi-layer structure allows us to derive an iterative method that requires only OCT data. We decompose $\hat{n}$ as

$$\hat{n}(\omega) = \nu(\omega) + i\kappa(\omega), \quad \nu, \kappa \in \mathbb{R}.$$

The measurements are again given by (32). The above presented iterative scheme, fails in this case. Indeed, recall the formula (35), which we considered for recovering the parameters of the $k-$th layer. Taking the absolute value, for $\hat{n}_{k-1}(\omega)$, $\hat{n}_k(\omega) \in \mathbb{C}$, gives

$$|\tilde{\hat{m}}^{(k)}(\omega)| = \frac{|\hat{n}_{k-1}(\omega) - \hat{n}_k(\omega)|}{|\hat{n}_{k-1}(\omega) + \hat{n}_k(\omega)|}|\tilde{\alpha}_0(\omega)|e^{-\frac{\omega}{c}\kappa_{k-1}(\omega)2z_k}.$$

The last term in the above expression describes how the amplitude of the wave decreases, i.e. attenuation, and prevents us from a step-by-step solution, since z_k still appears. Thus, we propose a different scheme that takes into account also the relations between the parts ν and κ, meaning the Kramers–Kronig relations. We stress here that $\hat{n}$ is holomorphic in the upper complex plane, satisfying $\hat{n}(\omega) = \hat{n}(-\omega)^*$. The parts of the complex-valued refractive index are connected through

$$\nu(\omega) - 1 = \frac{2}{\pi} \int_0^\infty \frac{\omega' \kappa(\omega')}{\omega'^2 - \omega^2} d\omega',$$

$$\kappa(\omega) = -\frac{2\omega}{\pi} \int_0^\infty \frac{\nu(\omega') - 1}{\omega'^2 - \omega^2} d\omega'. \tag{36}$$

In addition, defining the reflection coefficient,

$$\rho_k(\omega) = \frac{\hat{n}_{k-1}(\omega) - \hat{n}_k(\omega)}{\hat{n}_{k-1}(\omega) + \hat{n}_k(\omega)} \in \mathbb{C}, \tag{37}$$

and using its expression in polar coordinates $\rho_k = |\rho_k| e^{i\theta_k}$, we obtain

$$\ln(\rho_k(\omega)) = \ln(|\rho_k(\omega)|) + i\theta_k(\omega),$$

a function that diverges logarithmically as $\omega \to \infty$, and is not square-integrable [22]. However, the following relation holds for the phase of the complex-valued reflectivity

$$\theta_k(\omega) = -\frac{2\omega}{\pi} \int_0^\infty \frac{\ln(|\rho_k(\omega')|)}{\omega'^2 - \omega^2} d\omega'. \tag{38}$$

We define the operator

$$\mathcal{H}(\omega) : f \mapsto -\frac{2\omega}{\pi} \int_0^\infty \frac{f(\omega')}{\omega'^2 - \omega^2} d\omega',$$

and we get the relations in compact form

$$\kappa_k(\omega) = \mathcal{H}(\nu_k - 1)(\omega), \quad \text{and} \quad \theta_k(\omega) = \mathcal{H}(\ln|\rho_k|)(\omega).$$

Of course, when working with the Kramers–Kronig relations (36) and (38), one has to deal with the problem that they assume that information is available for the whole spectrum, something that is not true for experimental data. Another problem could be the existence of zero's of the reflection coefficient in the half plane. There exist generalizations of those formulas that can overcome these problems, like the subtractive relations which require few additional data. We refer to [12, 16, 24, 35] for works dealing with the applicability and variants of the Kramers–Kronig relations. This practical problem is out of the scope of this paper and will be considered in future work, where we will examine numerically, with simulated and real data, the validity of the proposed scheme.

Step 1:   At first, we consider the reconstruction of the interface z_k. We apply once the logarithm to the absolute value of (35) and then we take imaginary part of the logarithm of (35). Using the definition (37), we obtain the system of equations

$$\ln(|\tilde{\hat{m}}^{(k)}(\omega)|) = \ln(|\rho_k(\omega)|) + \ln(\tilde{\alpha}_0(\omega)) - 2\tfrac{\omega}{c}\kappa_{k-1}(\omega)z_k,$$
$$\Im\{\ln(\tilde{\hat{m}}^{(k)}(\omega))\} = \theta_k(\omega) - \tfrac{\omega}{c}\hat{n}_0 z_d + 2\tfrac{\omega}{c}\nu_{k-1}(\omega)z_k. \tag{39}$$

We define the data functions

$$\hat{m}_1^{(k)}(\omega) := \ln(|\tilde{\hat{m}}^{(k)}(\omega)|) - \ln(\tilde{\alpha}_0(\omega)),$$
$$\hat{m}_2^{(k)}(\omega) := \Im\{\ln(\tilde{\hat{m}}^{(k)}(\omega))\} + \tfrac{\omega}{c}\hat{n}_0 z_d,$$

and the system (39) takes the form

$$\ln(|\rho_k(\omega)|) - 2\tfrac{\omega}{c}\kappa_{k-1}(\omega)z_k = \hat{m}_1^{(k)}(\omega),$$
$$\theta_k(\omega) + 2\tfrac{\omega}{c}(\nu_{k-1}(\omega) - 1)z_k + 2\tfrac{\omega}{c}z_k = \hat{m}_2^{(k)}(\omega),$$

to be solved for z_k. We rewrite the last equation using the formulas (36) and (38) and the first equation, to obtain

$$\hat{m}_2^{(k)}(\omega) = 2\tfrac{\omega}{c}z_k - \frac{2\omega}{\pi}\int_0^\infty \frac{\ln(|\rho_k(\omega')|)}{\omega'^2 - \omega^2}d\omega' + \frac{4\omega z_k}{\pi c}\int_0^\infty \frac{\omega'\kappa_{k-1}(\omega')}{\omega'^2 - \omega^2}d\omega'$$
$$= 2\tfrac{\omega}{c}z_k - \frac{2\omega}{\pi}\int_0^\infty \frac{\ln(|\rho_k(\omega')|) - 2\tfrac{\omega'}{c}\kappa_{k-1}(\omega')z_k}{\omega'^2 - \omega^2}d\omega'$$
$$= 2\tfrac{\omega}{c}z_k + \mathcal{H}(\hat{m}_1^{(k)})(\omega).$$

The last equation is solved at given frequency $\omega^* \neq 0$, in order to obtain the location of the interface

$$z_k = \frac{c}{2\omega^*}\left(\hat{m}_2^{(k)}(\omega^*) - \mathcal{H}(\hat{m}_1^{(k)})(\omega^*)\right).$$

We can now recover $\hat{n}_k$ from (35) which admits the from

$$\hat{n}_k(\omega) = \hat{n}_{k-1}(\omega)\frac{\tilde{\alpha}_0(\omega)e^{i\frac{\omega}{c}(\hat{n}_{k-1}(\omega)2z_k - \hat{n}_0 z_d)} - \tilde{\hat{m}}^{(k)}(\omega)}{\tilde{\alpha}_0(\omega)e^{i\frac{\omega}{c}(\hat{n}_{k-1}(\omega)2z_k - \hat{n}_0 z_d)} + \tilde{\hat{m}}^{(k)}(\omega)},$$

by equating the real and imaginary parts.

Step 2:   We update the data as in the non-absorbing case.

4   Numerical Implementation

We solve the direct problem considering two different schemes depending on the properties of the refractive index, see (10) and (11). First, we consider the time-independent refractive index and data from a time-domain OCT system.

4.1   Reconstructions in Time Domain

We model the incident field as a gaussian wave centered around a frequency ω_0 moving in the z-direction of the form

$$u_0(t, z) = e^{-\frac{(z-z_0-ct)^2}{2\sigma^2}} \cos\left(\frac{\omega_0}{c}(z - z_0 - ct)\right), \tag{40}$$

with width σ, where z_0 denotes the source position and $c \approx 3 \times 10^8$ m/s is the speed of light.

The simulated data are created by solving (12) using a finite difference scheme. We restrict $z \in [0, 1.5]$mm and we set $T > 0$, the final time. We consider absorbing boundary conditions at the end points and we set $u(0, z) = u_0(0, z)$ and $\partial_t u(0, z) = 0$ as initial conditions. The left-going wave is ignored. We consider equidistant grid points with step size $\Delta z = \lambda_0/100$, where $\lambda_0 = 2\pi c/\omega_0$, is the central wavelength, and time step Δt such that the CFL condition is satisfied.

The measurement data are given by

$$m(t) = |(u - u_0)(t, z_d)|, \quad t \in (0, T], \tag{41}$$

where z_d denotes the position of the point detector. We add noise with respect to the L^2 norm

$$m_\delta = m + \delta \frac{\|m\|_2}{\|v\|_2} v,$$

where v is a vector with components normally distributed random variables and δ denotes the noise level. We have to stress that the total time is such that the data contain also information from the multiple reflections inside the medium. We define the length of the k-th layer

$$\ell_k = z_{k+1} - z_k, \quad \text{for} \quad k = 1, \ldots, N,$$

and we set $\ell_{-1} = |z_d - z_0|$, and $\ell_0 = z_0 - z_d$. We denote by

$$\rho_k = \frac{n_{k-1} - n_k}{n_{k-1} + n_k}, \quad \text{for} \quad k = 1, \ldots, N,$$

the reflection coefficient at the interface $z = z_k$.

Here, we assume that we know only $n_0 = 1$, and the positions of the source and the detector. Thus, we aim for recovering the position, the size and the optical properties of the medium.

The proposed iterative scheme for a N-layer medium is presented in Algorithm 1, where the output is the reconstructed refractive indices and lengths of the layers. First, we order the observed "peaks" at the image with respect to time, producing the set of data (t_l, p_l), for $l = 1, 2, \ldots, \Lambda$. The number $\Lambda \geq N$ describes the number of single and multiple reflections arrived at the detector before the final time T. In order to obtain a physically compatible solution we impose some bounds $[\underline{n}, \overline{n}]$ on the refractive index. This condition is not necessary for data with phase information. We update the data by neglecting the multiple reflections. To do so, once we have recovered the length and the refractive index of a layer, we neglect the "peaks" appearing later referring to multiple reflections inside this layer. Of course, because of numerical error and noisy data, we give a tolerance depending on the time duration of the wave.

We define the error function

$$\epsilon = \left(\sum_{k=1}^{N} (n_k \ell_k - \tilde{n}_k \tilde{\ell}_k)^2 \right)^{1/2},$$

where (n_k, ℓ_k) and $(\tilde{n}_k, \tilde{\ell}_k)$, for $k = 1, \ldots, N$ are the exact and the reconstructed values, respectively.

In the first example, we consider a three-layer medium positioned at $\ell_0 = 0.5$ mm, with $\ell_{-1} = 0$. We set $(n_1, n_2, n_3) = (1.55, 1.41, 1.48)$ and lengths $(\ell_1, \ell_2, \ell_3) = (0.2, 0.3, 0.1)$mm. The obtained data (41) for this example are given at the left picture in Fig. 5.

The results are presented in Table 1 for $[\underline{n}, \overline{n}] = [1.345, 2]$ and tol $= 0.1$ps. We obtain accurate and stable reconstructions, with $\epsilon = 3.34 \times 10^{-7}$. This algorithm can be easily applied to multi-layer media and it is presented here since it will be the core of the more complicated algorithm in the Fourier domain.

4.2 Reconstructions Having Phase Information in Frequency Domain

We aim to reconstruct the time-dependent refractive index, meaning its frequency-dependent Fourier transform. As discussed already in Sect. 3.1, it is possible from the phase-less OCT data to recover the full information, implying that we consider as measurement data the function

$$\hat{m}(\omega) = (\hat{u} - \hat{u}_0)(\omega, z_d), \quad \omega \in [\underline{\omega}, \overline{\omega}]. \tag{42}$$

Result: $\tilde{n}_k$ and $\tilde{\ell}_{k-1}$, for $k = 1, \ldots, N$.
Input: $k = 0$, $\rho_0 = 0$, $n_0 = 1$, ℓ_{-1}, tol and (t_l, p_l), for $l = 1, \ldots, \Lambda$;
while $k \leq N$ **do**

    /* Step 1: Reconstruction of the refractive index.   */

    $\rho_{k+1} = \dfrac{p_{k+1}}{\prod_{j=1}^{k}(1-\rho_j^2)}$, $\tilde{n}_{k+1} = \tilde{n}_k \dfrac{1-\rho_{k+1}}{1+\rho_{k+1}}$;

    if $\tilde{n}_{k+1} \notin [\underline{n}, \overline{n}]$ **then**

         $\rho_{k+1} = -\rho_{k+1}$, $\tilde{n}_{k+1} = \tilde{n}_k \dfrac{1-\rho_{k+1}}{1+\rho_{k+1}}$;

    end

    /* Step 2: Reconstruction of the length.   */

    $\tilde{\ell}_k = \frac{1}{2}\left(c\left(\frac{t_{k+2}-t_{k+1}}{n_k}\right) - \tilde{\ell}_{k-1}\right)$;

    /* Step 3: Update the data.   */

    for $j = 1 : \lfloor \Lambda/N \rfloor$ **do**

         $\tau_{k+1} = t_{k+1} + j\frac{2\tilde{n}_k\tilde{\ell}_k}{c}$;

         for $\kappa = k : \Lambda$ **do**

             if $|t_\kappa - \tau_{k+1}| < tol$ **then**

                 $p_\kappa = 0$;

             end

         end

    end

    k = k +1;

end

Algorithm 1: Iterative scheme (in time) using time-domain data

Here, the frequency interval $[\underline{\omega}, \overline{\omega}]$, with $\overline{\omega} > \underline{\omega} > 0$, models the OCT data, recorded by a CCD camera placed after a spectrometer with wavelength range $[2\pi c/\overline{\omega}, 2\pi c/\underline{\omega}]$, in a frequency-domain OCT system.

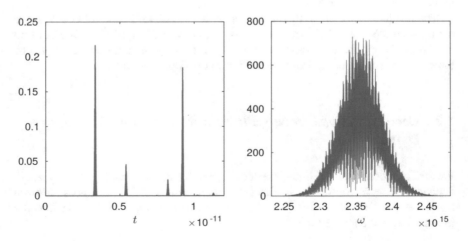

Fig. 5 The simulated data (absolute value) in the time-domain for the first example (left) and in the frequency-domain for the second example (right)

Table 1 Reconstructed values using Algorithm 1 for a three-layer medium

Length (mm)	ℓ_0	ℓ_1	ℓ_2	ℓ_3
Exact	0.50000	0.20000	0.30000	0.10000
Reconstructed (noise free)	0.50000	0.20004	0.29990	0.10006
reconstructed (5% noise)	0.49960	0.20021	0.30030	0.09976
Refractive index	n_1	n_2	n_3	n_4
Exact	1.55000	1.41000	1.48000	1.00000
Reconstructed (noise free)	1.55107	1.41070	1.48087	1.00014
Reconstructed (5% noise)	1.55272	1.40740	1.48170	0.99851

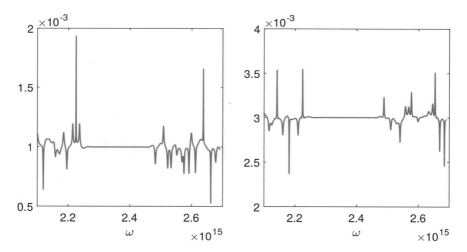

Fig. 6 The function $\phi(\omega)$, for $\omega \in [\underline{\omega}, \overline{\omega}]$, at the first (left) and the last (right) iteration step of Algorithm 2

4.2.1 Non-dispersive Medium

In order to construct the data (42), we consider the time-dependent back-reflected field derived in the previous section, we add noise and we take its Fourier transform with respect to time. Then, we truncate the signal at the interval $[\underline{\omega}, \overline{\omega}]$, see the right picture in Fig. 5.

In Algorithm 2 we present the main steps of the iterative scheme as described in Sect. 3.3. In Step 1, we take advantage of the causality property of the time-dependent signal and we zero-pad $\hat{m}(\omega)$, for all $\omega \in \mathbb{R} \setminus [\underline{\omega}, \overline{\omega}]$, and then we recover the signal as two times the real part of the inverse Fourier transformed field. In the second step, we initially approximated the derivative with respect to frequency using finite differences but it did not produce nice reconstructions due to the highly oscillating signal. We replace the derivative with a high-order differentiator filter taking into account the sampling rate of the signal. In Fig. 6, we plot the function $\phi(\omega)$, $\omega \in [\underline{\omega}, \overline{\omega}]$, and we see that it is constant in a central

Table 2 Reconstructed values using Algorithm 2 for a three-layer medium

Length (mm)	ℓ_0	ℓ_1	ℓ_2	ℓ_3
Exact	0.70000	0.15000	0.40000	0.13000
Reconstructed (noise free)	0.69885	0.15210	0.39628	0.13451
Reconstructed (5% noise)	0.70340	0.15387	0.39337	0.13669
Refractive index	n_1	n_2	n_3	n_4
Exact	1.55000	1.40500	1.55000	1.00000
Reconstructed (noise free)	1.55107	1.40568	1.55107	0.99782
Reconstructed (5% noise)	1.55164	1.40599	1.55139	0.99695

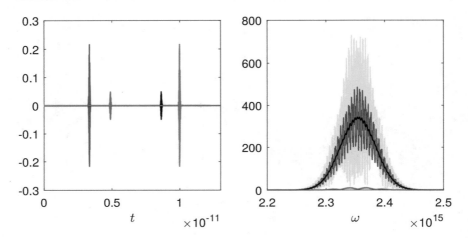

Fig. 7 The step 3 of the Algorithm 2, where we update the data. The (color) curve (right) is the signal in the frequency domain if we neglect the (color) and all the previous "peaks" in the time-domain signal (left)

interval (called trusted) and oscillates close to the end points. Thus, we denote by ω^* either a chosen frequency in the trusted interval or the mean of the frequencies in this trusted interval. We address here that one could also average over the whole spectrum and still get reasonable results.

In the second example, we consider again a three-layer medium with parameters $(n_1, n_2, n_3) = (1.55, 1.405, 1.55)$ and lengths $(\ell_1, \ell_2, \ell_3) = (0.15, 0.5, 0.13)$ mm. Here, $\ell_0 = 0.7$ and $\ell_{-1} = 0.2$ mm. The central frequency is given by $\omega_0 = 2\pi c/\lambda_0$, with $\lambda_0 = 800$ nm. The sampling rate is $f_s = 100c/\lambda_0$. The recovered parameters for noise-free and noisy data are presented in Table 2. The relative error is $\epsilon = 2.164 \times 10^{-5}$. In Fig. 7, we see how the data change as the Algorithm 2 progresses. The picture on the right shows the data in frequency domain with respect to the "peaks" presented in the left picture where we see the time-domain data. The yellow curve (right) represents the full data, the red curve (right) the data if we neglect the red "peak" (left), the blue curve shows the data if we neglect also the blue "peak", and so on. The green curve (right) represents the

Result: $\tilde{n}_k$ and $\tilde{\ell}_{k-1}$, for $k = 1, \dots, N$.
Input: $k = 0$, $\rho_0 = 0$, $n_0 = 1$, ℓ_{-1}, and $\hat{m}(\omega)$, for $\omega \in [\underline{\omega}, \overline{\omega}]$;
while $k \leq N$ **do**

    /* Step 1: Reconstruction of the refractive index.   */
    zero-padding and IFFT of the signal $\hat{m}$;
    Isolate the first peak and FFT the signal to obtain $\hat{m}^{(k)}$;
    $\rho_{k+1} = \frac{\max |\hat{m}^{(k)}|}{\max(\alpha)}, \quad \tilde{n}_{k+1} = \tilde{n}_k \frac{1-\rho_{k+1}}{1+\rho_{k+1}};$
    if $\tilde{n}_{k+1} \notin [\underline{n}, \overline{n}]$ **then**
        | $\rho_{k+1} = -\rho_{k+1}, \quad \tilde{n}_{k+1} = \tilde{n}_k \frac{1-\rho_{k+1}}{1+\rho_{k+1}};$
    end
    /* Step 2: Reconstruction of the length.   */
    define $f_k(\omega) = \frac{\hat{m}^{(k)}}{\alpha} / |\frac{\hat{m}^{(k)}}{\alpha}|$;
    if $k = 0$ **then**
        | $d_k = n_0 \ell_{-1};$
        else
        | | $d_k = n_0 \ell_{-1} - 2\tilde{n}_{k-1}\tilde{\ell}_{k-1};$
        end
    end
    $\phi(\omega) = c|\partial_\omega f_k(\omega)|, \quad \tilde{\ell}_k = -\frac{-\phi(\omega^*)-d_k}{2\tilde{n}_k};$
    if $\tilde{\ell}_k < 0$ **then**
        | $\tilde{\ell}_k = -\frac{\phi(\omega^*)-d_k}{2\tilde{n}_k};$
    end
    /* Step 3: Update the data.   */
    $\hat{m}(\omega) = (\hat{m}(\omega) - \hat{m}^{(k)}(\omega))/(1 - \rho_{k+1}^2);$
    $k = k + 1;$

end

Algorithm 2: Iterative scheme (in frequency) using phase information for non-dispersive medium

signal from the multiple reflections. We observe that the OCT signal maintains the Gaussian form of the incident wave, centered around the central frequency, and the different reflections result in the oscillations of the field.

4.2.2 Dispersive Medium

The incident field in the frequency domain takes the form

$$\hat{u}_0(\omega, z) = \sqrt{2\pi} \frac{\sigma}{2c} e^{-\frac{\sigma^2(\omega-\omega_0)^2}{2c^2}} e^{i\frac{\omega}{c}(z-z_0)}, \quad \omega > 0, \ z \in \mathbb{R},$$

which is the Fourier transform with respect to time of u_0, given by (40), restricted to positive frequencies. This field describes a plane wave moving in the z-direction having a Gaussian profile perpendicular to the incident direction, centered around ω_0. We generate the data considering the formula (30) and then we add noise. The

Algorithm 3 summarizes the steps of the iterative scheme, which for a frequency-independent refractive index simplifies to Algorithm 2.

We model the wavelength-dependent refractive index of the medium using the standard formula, known as Cauchy's equation,

$$n(\lambda) = \beta_1 + \frac{\beta_2}{\lambda^2} + \frac{\beta_3}{\lambda^4},$$

for some fitting coefficients β_j, $j = 1, 2, 3$. In Fig. 8, we see the exact refractive index of the first (left) and the third (right) layer for the medium used in the third example. Afterwards, we consider the refractive index as a function of frequency.

The medium lengths are given by $(\ell_1, \ell_2, \ell_3) = (0.2, 0.3, 0.1)$mm, and we set $\ell_0 = 0.7$mm and $\ell_{-1} = 0$. The second layer has constant refractive index given by $n_2(\omega) = 1.41$. The reconstructions of $n_1(\omega)$ and $n_3(\omega)$ are presented in Fig. 9 for data with 2% noise. In Table 3, we see the recovered lengths and the refractive indices at specific frequencies.

As already discussed, the calculations close to the end points were not stable and since here we are interested in reconstructing the frequency-dependent refractive index, we restrict the computational domain and then we extrapolate the recovered functions in order to update the data. The effect of the frequency domain on the reconstructions is presented in Fig. 10 where we plot the L^2-norm (in semi-logarithmic scale) of the difference between the exact and the computed refractive indices n_1 (blue and green curves) and n_3 (red and purple curves) for seven different computational domains. We start from the most left domain (left top picture) and we move to the most right domain (left bottom picture). The blue and red curves are for $\delta = 1\%$ noise and the green and purple for $\delta = 2\%$ noise. We clearly see that as we

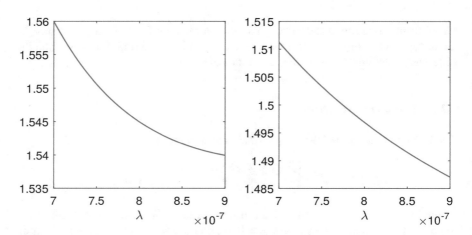

Fig. 8 The behavior of the medium with respect to wavelength (dispersion). The refractive index of the first (left) and the third (right) layer for the third example in the range [700, 900] nm of the spectrometer

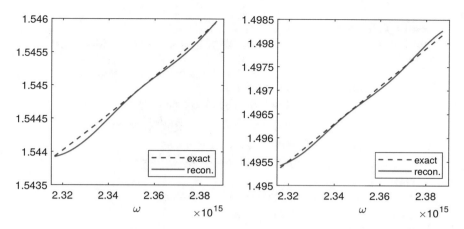

Fig. 9 The exact (dashed blue line) and the reconstructed (red solid line) refractive index of the first (left) and the third (right) layer. These are the results of the Algorithm 3 for noisy data

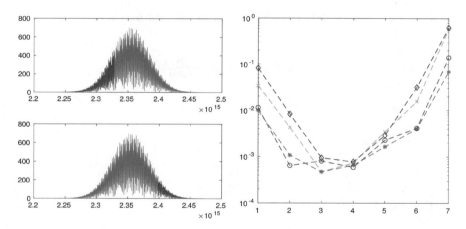

Fig. 10 The performance of the Algorithm 3 for different frequency sub-domains. On the right, we see the L^2−norm (in semi-logarithmic scale) of the difference between the exact and the computed n_1 and n_3 for different noise levels. The case 1 corresponds to the most left (left top) domain and the case 7 to the most right (left bottom) domain

move to the end points and the error increases the calculations become unstable. We chose the domain $\omega \in [2.316, 2.387] \times 10^{15} s^{-1}$, corresponding to the fourth case, for our reconstructions.

Result: $\tilde{n}_k(\omega)$ and $\tilde{\ell}_{k-1}$, for $k = 1, \ldots, N$.
Input: $k = 0$, $\rho_0 = 0$, $n_0 = 1$, ℓ_{-1}, $\mathcal{W} \subset [\underline{\omega}, \overline{\omega}]$ and $\hat{m}(\omega)$, for $\omega \in [\underline{\omega}, \overline{\omega}]$;
while $k \leq N$ **do**

    `/* Step 1: Reconstruction of the refractive index.    */`
    zero-padding and IFFT of the signal $\hat{m}$;
    Isolate the first peak and FFT the signal to obtain $\hat{m}^{(k)}$;

$$\rho_{k+1}(\omega) = \frac{|\hat{m}^{(k)}(\omega)|}{\alpha(\omega)}, \quad \tilde{n}_{k+1}(\omega) = \tilde{n}_k(\omega)\frac{1-\rho_{k+1}(\omega)}{1+\rho_{k+1}(\omega)}, \quad \omega \in \mathcal{W};$$

    if $max\{\tilde{n}_{k+1}(\omega)\} > \overline{n}$ **or** $min\{\tilde{n}_{k+1}(\omega)\} < \underline{n}$ **then**

$$\rho_{k+1}(\omega) = -\rho_{k+1}(\omega), \quad \tilde{n}_{k+1}(\omega) = \tilde{n}_k(\omega)\frac{1-\rho_{k+1}(\omega)}{1+\rho_{k+1}(\omega)};$$

    end
    `/* Step 2: Reconstruction of the length.    */`
    define $f_k(\omega) = \frac{\hat{m}^{(k)}}{\alpha} / |\frac{\hat{m}^{(k)}}{\alpha}|, \quad \omega \in \mathcal{W};$
    if $k = 0$ **then**

        $d_k(\omega) = n_0\ell_{-1};$

        else

            $d_k(\omega) = n_0\ell_{-1} - 2(\tilde{n}_{k-1}(\omega) + \omega\partial_\omega\tilde{n}_{k-1}(\omega))\tilde{\ell}_{k-1};$

        end

    end

$$\phi(\omega) = c|\partial_\omega f_k(\omega)|, \quad \psi_k(\omega) = -\frac{-\phi(\omega^*) - d_k(\omega)}{2(\tilde{n}_k + \omega\partial_\omega\tilde{n}_k)};$$

    if $max\{\psi_k\} < 0$ **then**

$$\psi_k(\omega) = -\frac{\phi(\omega^*) - d_k(\omega)}{2(\tilde{n}_k + \omega\partial_\omega\tilde{n}_k)};$$

    end
    $\tilde{\ell}_k = \psi_k(\omega^*);$
    `/* Step 3: Update the data.    */`
    extrapolate $\rho_{k+1}(\omega)$ from $\mathcal{W}$ to $[\underline{\omega}, \overline{\omega}]$;
    $\hat{m}(\omega) = (\hat{m}(\omega) - \hat{m}^{(k)}(\omega))/(1 - \rho_{k+1}^2(\omega));$
    k = k +1;

end

Algorithm 3: Iterative scheme (in frequency) using phase information for dispersive medium.

Table 3 Reconstructed values using Algorithm 3 for a three-layer medium. The values of the refractive indices are given at $\omega^* = 2.351 \times 10^{15}$

Length (mm)	ℓ_0	ℓ_1	ℓ_2	ℓ_3
Exact	0.70000	0.20000	0.30000	0.10000
Reconstructed (2% noise)	0.69790	0.20139	0.29569	0.10329
Refractive index	$n_1(\omega^*)$	$n_2(\omega^*)$	$n_3(\omega^*)$	n_4
Exact	1.54488	1.41000	1.49675	1.00000
Reconstructed (2% noise)	1.54499	1.41094	1.49884	1.00484

5 Conclusions

In this work we addressed the inverse problem of recovering the optical properties of a multi-layer medium from simulated data modelling a frequency-domain OCT system. We considered the cases of non-dispersive, dispersive and absorbing media. We proposed reconstruction methods and we presented numerical examples justifying the feasibility of the derived schemes. Stable reconstruction with respect to noise were presented. The methods are based on standard equations, equivalent to the Fresnel equations, and to ideas from stripping algorithms. The originality of this work lies in the combination of them into a new iterative method that addresses also the frequency-dependent case, which needs special treatment. As a future work, we plan to examine the applicability of the iterative schemes for experimental data and test numerically the method for absorbing media.

Acknowledgments The work of PE and LV was supported by the Austrian Science Fund (FWF) in the project F6804–N36 (Quantitative Coupled Physics Imaging) within the Special Research Programme SFB F68: "Tomography Across the Scales".

References

1. M. Born, E. Wolf, *Principles of Optics*, 7th edn. (Cambridge University Press, Cambridge, 1999)
2. O. Bruno, J. Chaubell, One-dimensional inverse scattering problem for optical coherence tomography. Inverse Problems **21**, 499–524 (2005)
3. W. Chew, *Waves and Fields in Inhomogeneous Media* (Van Nostrand Reinhold, New York, 1990)
4. W. Drexler, J.G. Fujimoto, *Optical Coherence Tomography: Technology and Applications*, 2nd edn. (Springer, Cham, 2015)
5. P. Elbau, L. Mindrinos, O. Scherzer, Mathematical methods of optical coherence tomography, in *Handbook of Mathematical Methods in Imaging*. ed. by O. Scherzer (Springer New York, 2015), pp. 1169–1204
6. P. Elbau, L. Mindrinos, O. Scherzer, Inverse problems of combined photoacoustic and optical coherence tomography. Math. Methods Appl. Sci. **40**(3), 505–522 (2017)
7. P. Elbau, L. Mindrinos, O. Scherzer, Quantitative reconstructions in multi-modal photoacoustic and optical coherence tomography imaging. Inverse Problems **34**(1), 014006 (2018)
8. L.C. Evans, *Partial Differential Equations*. Graduate Studies in Mathematics, 2nd. edn. (American Mathematical Society, Providence, 2010)
9. A.F. Fercher, Optical coherence tomography. J. Biomed. Opt. **1**(2), 157–173 (1996)
10. A.F. Fercher, Optical coherence tomography - development, principles, applications. Z. Med. Phys. **20**, 251–276 (2010)
11. A.F. Fercher, C.K. Hitzenberger, W. Drexler, G. Kamp, H. Sattmann, In vivo Optical coherence tomography. Am. Shortjournal Ophthalmol. **116**, 113–114 (1993)
12. P. Grosse, V. Offermann, Analysis of reflectance data using the Kramers-Kronig relations. Appl. Phys. A **52**(2), 138–144 (1991)
13. E. Hecht, *Optics*, 4th edn. (Addison Wesley, Boston, 2002)
14. M.R. Hee, J.A. Izatt, E.A. Swanson, D. Huang, J.S. Schuman, C.P. Lin, C.A. Puliafito, J.G. Fujimoto, Optical coherence tomography of the human retina. Arch. Ophthalmol. **113**(3), 325–332 (1995)
15. E. Hofstetter, Construction of time-limited functions with specified autocorrelation functions. IEEE Trans. Inf. Theory **10**(2), 119–126 (1964)

16. S.A.R. Horsley, M. Artoni, G.C. La Rocca, Spatial Kramers–Kronig relations and the reflection of waves. Nat. Photon. **9**(7), 436 (2015)
17. D. Huang, E.A. Swanson, C.P. Lin, J.S. Schuman, G. Stinson, W. Chang, M.R. Hee, T. Flotte, K. Gregory, C.A. Puliafito, J.G. Fujimoto, Optical coherence tomography. Am. Assoc. Adv. Sci. **254**(5035), 1178–1181 (1991)
18. M.V. Klibanov, V.G. Kamburg, Uniqueness of a one-dimensional phase retrieval problem. Inverse Problems **30**(7), 075004 (2014)
19. M.V. Klibanov, P.E. Sacks, A.V. Tikhonravov, The phase retrieval problem. Inverse Problems **11**(1), 1–28 (1995)
20. A. Krishnaswamy, G.V. Baranoski, A Biophysically-based spectral model of light interaction with human skin. Comput. Graph. Forum **23**(3), 331–340 (2004)
21. R.A. Leitgeb, C.K. Hitzenberger, A.F. Fercher, T. Bajraszewski, Phase-shifting algorithm to achieve high-speed long-depth-range probing by frequency-domain optical coherence tomography. Opt. Lett. **28**(22), 2201–2203 (2003)
22. V. Lucarini, J.J. Saarinen, K. Peiponen, E.M. Vartiainen, *Kramers-Kronig Relations in Optical Materials Research*. Springer Series in Optical Sciences (Springer, Berlin 2005)
23. S. Mukherjee, C.S. Seelamantula, An iterative algorithm for phase retrieval with sparsity constraints: application to frequency domain optical coherence tomography, in *2012 IEEE International Conference on Acoustics, Speech and Signal Processing (ICASSP)* (2012), pp. 553–556
24. K.F. Palmer, M.Z. Williams, B.A. Budde, Multiply subtractive Kramers–Kronig analysis of optical data. Appl. Opt. **37**(13), 2660–2673 (1998)
25. C.S. Seelamantula, S. Mulleti, Super-resolution reconstruction in frequency-domain optical-coherence tomography using the finite-rate-of-innovation principle. IEEE Trans. Signal Process. **62**(19), 5020–5029 (2014)
26. C.S. Seelamantula, M.L. Villiger, R.A. Leitgeb, M. Unser, Exact and efficient signal reconstruction in frequency-domain optical-coherence tomography. J. Opt. Soc. Am. A **25**(7), 1762–1771 (2008)
27. Y. Shechtman, Y.C. Eldar, O. Cohen, H.N. Chapman, J. Miao, M. Segev, Phase retrieval with application to optical imaging: a contemporary overview. IEEE Signal Process. Mag. **32**(3), 87–109 (2015)
28. E. Somersalo, Layer stripping for time-harmonic Maxwell's equations with high frequency. Inverse Problems **10**(2), 449–466 (1994)
29. W.A. Strauss, *Partial Differential Equations: An Introduction*, 2nd edn. (Wiley, New York, 2007)
30. E.A. Swanson, J.A. Izatt, M.R. Hee, D. Huang, C.P. Lin, J.S. Schuman, C.A. Puliafito, J.G. Fujimoto, In vivo retinal imaging by optical coherence tomography. Opt. Lett. **18**, 1864–1866 (1993)
31. J. Sylvester, D. Winebrenner, F. Gylys-Colwell, Layer stripping for the Helmholtz equation. SIAM J. Appl. Math. **56**(3), 736–754 (1996)
32. L. Thrane, H.T. Yura, P.E. Andersen, Analysis of optical coherence tomography systems based on the extended Huygens Fresnel principle. J. Opt. Soc. Am. A **17**(3), 484–490 (2000)
33. P.H. Tomlins, R.K. Wang, Theory developments and applications of optical coherence tomography. J. Phys. D Appl. Phys. **38**, 2519–2535 (2005)
34. P.H. Tomlins, R.K. Wang, Matrix approach to quantitative refractive index analysis by Fourier domain optical coherence tomography. J. Opt. Soc. Am. A Opt. Image Sci. Vis. **23**(8), 1897–1907 (2006)
35. R.H. Young, Validity of the Kramers-Kronig transformation used in reflection spectroscopy. J. Opt. Soc. Am. **67**(4), 520–523 (1977)

Review of Image Similarity Measures for Joint Image Reconstruction from Multiple Measurements

Ming Jiang

Abstract It is fundamental in image processing how to measure image similarity quantitatively for tasks such as image quality assessment, image registration, image reconstruction from multiple measurements, etc.. An image similarity measure (ISM) is both task-dependent and feature-dependent, and must be designed according to the characteristics of specific tasks and features. Simply applying distances from the mathematical metric theory or general divergences to spaces of images or spaces of image features usually does not provide appropriate ISMs. In this chapter, we review several ISMs for image reconstruction problems from multiple measurements of various types in recent work. The multiple measurements considered here include multi-modality, multi-spectral, and multi-temporal measurements, with multi-modality tomography, multi-spectral XCT, and dynamic tomography, as the imaging applications, respectively. We focus on motivations and constructions of the ISMs and avoid their general rigorous mathematical presentations to simplify notations for the readability for a general audience. ISMs under review are proposed for image structural similarity and have been successfully applied to image reconstruction from multiple measurements.

1 Introduction

Image similarity measure (ISM) is fundamental for imaging science. It is fundamental in image processing how to measure image similarity quantitatively for image quality assessment, image registration, etc., and recently for joint image reconstruction from multiple measurements. In the common sense, it provides a measure of the similarity of two images. It seems that there a number of mathematical theories that can be applied, such as the various distances from the theory of metric spaces.

M. Jiang (✉)
School of Mathematical Sciences, Peking University, Beijing, China
e-mail: ming-jiang@pku.edu.cn

© Springer Nature Switzerland AG 2021
B. Kaltenbacher et al. (eds.), *Time-dependent Problems in Imaging and Parameter Identification*, https://doi.org/10.1007/978-3-030-57784-1_9

However, a further thinking indicates that ISM is intricate, both task-dependent and feature-dependent. Therefore, simple measures such as the mean absolute error (MAE) or root mean squared error (RMSE) are inappropriate for measuring image similarity because of lacking locational and structural information. Hence, such simple measures do not perform well for specific image processing tasks, such as image retrieval [66], image quality assessment (IQA) [53], image registration [8, 28, 60, 74], colour image processing [56], video processing [26, 41, 54, 85], dynamic tomography [34, 50, 76], and joint multi-modality image reconstruction [13, 25, 35, 39, 40, 44, 51, 63].

Generically, similarity is directional and should not be treated as a symmetric relation[77]. The predicates that *A is similar to B* and that *A and B are similar to each other* are logically inequivalent, where *A* and *B* are stimuli or features of objects under studying. " *The variant is more similar to the prototype than vice versa*" [77]. Please refer to [77] for more discussions and examples. Symmetric ISMs could be appropriate for some image processing tasks such as for IQA. However, symmetric ISMs are vulnerable to ill-transferring of distinct features in joint image reconstruction (JIR) for multi-modality tomography [44]. Nevertheless, the performance of a JIR method depends on other factors in addition to its ISM, no matter it is symmetric or not. Please refer to the discussions in Sect. 2.2.

Similarity is of non-transitive nature in general. The similarity of *A* to *B* and *B* to *C* does not imply a similarity of *A* to *C*, where *A*, *B* and *C* are stimuli or features of objects under studying. This is because they can be similar at different feature components when there are multiple feature components. For example, *A* could be similar to *B* by one feature *x* and *B* to *C* by another different feature *y*, but *A* is not similar to *C* either by feature *x* or by feature *y*. The following example is from [77]. Jamaica is similar to Cuba because of geographical proximity; Cuba is similar to Russia because of their political affinity; but Jamaica is not similar to Russia at all. With the same idea, it is not difficulty to find instances for the transitivity of image similarity to fail for image processing tasks [45]. A kid shares similar facial looking with his/her father and also with his/her mother; but his/her father and mother can have no similar facial looking at all. Please refer to [78] for another convincing example.

The structure of this chapter is as follows. In Sect. 2, the JIR problem from multi-measuement is introduced with a general discussion on lessons for ISMs and JIR in Sect. 2.2. In Sect. 3, a review of ISMs in recent work is reviewed. We focus on motivations and constructions of the ISMs and avoid their general rigorous mathematical presentations to simplify notations for the readability of general readers. In Sect. 4, relevant issues and possible problems for future study are discussed. We conclude this manuscript in Sect. 5.

2   A Framework for Joint Image Reconstruction from Multiple Measurements with Image Similarity Measures

The rapid development of sensing technology in multi-modality, multi-temporal and multi-spectral measurements has enabled a number of new imaging techniques such as multi-modality, dynamic tomography, and multi-spectral XCT. Joint image reconstruction (JIR) from multiple measurements is to estimate jointly images from all measured data, so that data from different modality, different instance and different spectrum can be used to complement each other.

Biomedical imaging is one of the fields where tomographic techniques from multiple measurements are actively under development to provide the visualizing of fused anatomical and/or functional information for biological and pharmaceutical study or clinical diagnosis. In one hand, the physical challenge is how to integrate multiple imaging modalities into one hybrid imaging system, such as PET/XCT, PET/MRI, SPECT/XCT, SPECT/MRI, DOT/MRI and DOT/XCT systems.[1] On the other hand, the mathematical challenge is how to jointly reconstruct images from all the data measured.

For multi-modality tomography, the multi-modality measurement for imaging modalities involved is conducted simultaneously for static objects and functional distributions, and the JIR is to reconstruct jointly images for all the modalities with appropriate cross-modality priors, rather than separately for each modality or sequentially one image reconstruction followed by another [25, 35, 71, 79]. For dynamic tomography, the multi-temporal measurement for certain imaging modalities is conducted sequentially for dynamic object and functional distributions changing during the data acquisition process, and the JIR is to reconstruct images with appropriate cross-temporal priors, with unknown motion [69, 70], or estimated motion from data [67] or from physical motion models [10], or known motion [19, 37]. For multi-spectral XCT, the multi-spectral measurement is conducted by energy-discriminating detectors for predetermined energy channels, and the JIR is to reconstruct jointly multi-channel images with appropriate cross-channel priors [33, 49, 64].

It can be foreseen that there will be imaging systems by combining the multi-modality, multi-temporal and multi-spectral measurements. The aforementioned cross-modality, cross-temporal and cross-channel priors will be necessary for the JIR of such systems. These priors characterize the structural similarity for images from different modalities, instances, and channels, and are in fact kinds of image similarity measures (ISMs). However, it is subtle in choosing or designing ISMs for different applications.

---

[1]These abbreviations are DOT for diffuse optical tomography, MRI for magnetic resonance imaging, PET for positron emission tomography, SPECT for single-photon emission computerized tomography, XCT for x-ray computerized tomography.

In Sect. 2.1 we will present a conceptual framework for the JIR from multiple measurements with ISMs. In Sect. 2.2, we will discuss general lessons in designing ISMs for different imaging applications.

2.1 Joint Image Reconstruction with Image Similarity Measures

A conceptual framework for the JIR from multiple measurements is as follows [35]. For $m = 1, \cdots, M$, let

$$A_m(u_m) = g_m, \tag{1}$$

be the m-th measurement for the m-th forward process A_m, respectively. Depending on the imaging applications, the forward process A_m, image u_m and measurement g_m are interpreted differently. For multi-modality tomography, for the m-th modality, A_m is the forward process for the modality, and u_m is the image of the modality, and g_m is the corresponding measurement. For dynamic tomography, for the m-th instance, A_m is the forward process for an imaging modality, which is the same for $m = 1, \cdots, M$ though subject to the different imaging geometry only, and u_m is the image at the m-th instance, and g_m is the corresponding measurement. For multi-spectral XCT, for the m-th channel, A_m is the energy-dependent attenuated Radon transform in predetermined energy window, u_m is the linear combination of the channel images, g_m is the corresponding measurement. For different channels, the measurement g_m is subject to the energy-dependent intensity flux and detector efficiency for each energy window [49].

For each image reconstruction from one single measurement (1), the conventional regularization approach is to minimize the following penalized reconstruction functional,

$$E_m(u_m) = \|A_m(u_m) - g_m\|^2 + \alpha_m R_m(u_m) \tag{2}$$

where $R_m(u_m)$ is the regularization for u_m and $\alpha_m > 0$ the regularization parameter, respectively, for $m = 1, \cdots, M$.

The JIR from multiple measurements is based on the observation that images of the same object possess partially similar structural features from different modalities, or from different instances if the motion is well sampled by the data acquisition process, or from different channels if the energy windows are well predetermined to capture the K-edge jumps, respectively for multi-modality, or multi-temporal, or multi-spectral measurement. Such similar structural features could be pursued for enhancing image quality by introducing an extra penalty into the reconstruction functional in (2). Assume that T is an image feature operator and

D is an image similarity measure for this image feature. Then we can incorporate the feature similarity into the reconstruction functional as follows, for $m = 1, \cdots, M$,

$$F_m(u_m) = \|A_m(u_m) - g_m\|^2 + \alpha_m R_m(u_m) + \sum_{\tilde{m}} \gamma_{\tilde{m}} D\left[T(u_m), T(u_{\tilde{m}})\right], \qquad (3)$$

where $\tilde{m}$ indicates possible similar images $u_{\tilde{m}}$ other than u_m, i.e., $\tilde{m} \neq m$ and $\gamma_{\tilde{m}} > 0$ the corresponding regularization parameter for this similarity, the sum is over possible $\tilde{m} \in \{1, \cdots, M\} \setminus \{m\}$. For multi-modality tomography, $\tilde{m}$ can be any image modality involved. For dynamic tomography and multi-spectral XCT, $\tilde{m}$ can be nearby instances or channels, such as $m - 1$ or $m + 1$. The JIR from multiple measurements is then to obtain reconstructed images by minimizing the following joint functional

$$F\left[\{u_m\}_{m=1}^M\right] = \sum_{m=1}^M \tau_m F_m(u_m), \qquad (4)$$

where the weights $\tau_m > 0$, for $m = 1, \cdots, M$.

One popular approach is to perform alternative minimization over each $F_m(u_m)$, i.e.,

$$u_m^{k+1} = \underset{u_m}{\arg\min} \; F_m(u_m), \qquad (5)$$

in certain order over $m \in \{1, \cdots, M\}$, for $k = 1, \cdots$, until some stopping criteria are reached. Please note that the effect of the weights τ_m is combined into the parameters α_m and $\gamma_{\tilde{m}}$ after re-parametrization, and hence can be ignored in implementation in this alternative minimization approach.

The above alternative minimization approach is only one of the possible methods for minimizing the joint functional F in (4) for the JIR from multiple measurements. When the objective functional F is not jointly convex with respect to all its arguments $\{u_m\}_{m=1}^M$, the above alternative minimization process will be trapped at local minimizers, and global minimization methods should be explored. The above formulations use the least-squares functional as the data fidelity under the assumption that data noise in the measurement is Gaussian. When the data noise is not Gaussian, other data fidelity functional should be used which will change the joint convexity of the JIR objective functional F. Moreover, the penalty functional $R_m(u_m)$ and especially the ISM D also can change the joint convexity of the JIR objective functional F. In this regard, it should be remarked that a simultaneous minimization algorithm for the JIR of PET/MRI is established in [40], where the objective functional for the JIR is jointly convex. Please refer to [40] for details.

The above formulation for the JIR from multiple measurements can be obtained from a Bayesian approach [5, 35] under the conditions that measurements in (1) are independent. The penalty functional $R_m(u_m)$ and the sum $\sum_{\tilde{m}} \gamma_{\tilde{m}} D\left[T\left(u_{m}\right), T\left(u_{\tilde{m}}\right)\right]$ of ISM penalties in (3) are equivalent to the prior and conditional priors for u_{m} in the joint probability distribution. The JIR from multiple measurements by minimizing the joint functional F in (4) is the maximum *a posterior* (MAP) estimate [2, 3, 82]. For the perspective of Bayesian statistical inference, the MAP estimate is the estimator for the simple 0–1 loss function [82]. Although the MAP estimate has been dominant in image reconstruction, other estimators with different loss functions, such as the minimum mean squares estimate (MMSE) with the squared-error Loss function [2, 82], can be pursued. When sizes of images are big, the computing load for MMSE is increased significantly [80].

2.2 Lessons on Image Similarity Measures for Joint Image Reconstruction

Unlike the image reconstruction for single measurement in (2), the ISM D in (3) is necessary for the cross communication of complementary feature information during a JIR process, and is one of the significant factors for the performance of its corresponding JIR method. Nevertheless, the role of the ISM should not be emphasized to an excessive degree, because the performance of a JIR method is not dominated by its ISM only. The actual performance of a JIR method depends on other factors such as how much similar and dissimilar features contribute to the total ISM in terms of feature strength, how regularization parameters and iteration numbers, and other algorithmic parameters are tuned, and also on its implementation techniques.

In general, ISM is task-dependent and feature-dependent and should be designed according to the characteristics of specific tasks and features accordingly. For multi-modality tomography, difficulties are due to different ranges, different contrasts, and different structural information of different modality images because of different underlying physics. There are features in one modality but not available in others. Hence, images of the same object from different modalities share similar features only partially in some regions. ISM must encourage the reconstruction of features when there is sufficient evidence from measured data and avoid the non-existent features to be reconstructed. The phenomenon of inappropriately reconstruction of non-existent features due to inappropriate ISMs is called '*ill-transfer of features*', and should be addressed in designing ISMs [44]. For dynamic tomography, although the ranges and contrasts are the same among images of all instances, there are new image features due to emerging motion that are in some instances but not available in others. For multi-spectral XCT, although the ranges and contrasts are fixed and about the same among images of all instances, there are new image features due to K-edge jumps that are in some instances but not available in others [49, 65, 68]. Therefore, the ill-transfer of features could happen in a JIR process for dynamic tomography or multi-spectral XCT with an inappropriate ISM. However, the issue of ill-transfer of features with dynamic tomography or multi-spectral XCT is different from that with multi-modality tomography, because it can be resolved by increasing the sampling rate during the data acquisition process and with proper configurations for the reconstruction process as aforementioned at the end of the previous paragraph. On the other side, recent progresses have demonstrated that JIR for multi-modality tomography can obtain images with quality better than single modality image reconstruction [24, 25, 39, 49, 51, 63, 64].

One convenient approach is to apply distances from metric spaces as ISMs. Features are elements of the underlying metric spaces. Usually a function is applied to the distance. It was demonstrated that the function is generically an exponential decay function [72] . However, such an approach will induce similarities with symmetry and transitivity. Symmetry is a conventional assumption for metrics. Transitivity follows from the triangle inequality. Both symmetry and transitivity could be appropriate for certain image processing tasks but inappropriate for other tasks, especially for JIR from multiple measurements, for reasons in the following paragraphs. Hence, this approach is not reported in recent work for JIR from multiple measurements.

From the mathematical perspective of metric spaces, symmetric ISMs, such that $D(p, q) = D(q, p)$, are vulnerable to ill-transfer of features. For symmetric ISMs, features in both images u_m and $u_{\bar{m}}$ must match each other, and then ill-transfer of features is unavoidable without further data evidence, deliberate algorithmic and implementation configurations. It should be remarked that this symmetric issue of ISM should not be over-emphasized because permissible feature transfers during a JIR process is eventually determined by the quality and evidence from the measured data.

As discussed in Sect. 1, ISMs satisfying the triangle inequality are transitive. This will make it vulnerable to ill-transfer of features in a JIR process from multiple measurements. For example, a joint 3-modality image reconstruction for MRI/DOT/XCT. Features in MRI images could be transferred to XCT images via DOT, even though such features does not exist in XCT images. Nevertheless, the discussions on the symmetric vulnerability and practical behaviour of the JIR methods in the first paragraph of this subsection are applicable to the current vulnerability discussion with the triangle inequality. Moreover, it can be proved that if D is a distance, then

$$|D(x, z) - D(y, z)| \leq D(x, y), \tag{6}$$

for any x, y and z. This will induce a uniform similarity structure in the feature space. However, whether this uniform structure is reasonable or inappropriate for a JIR task is unclear to the author at the time of writing this chapter. The triangle inequality and the symmetry of D are independent. But if D satisfies a strong form of the triangle inequality,

$$D(x, y) \leq D(x, z) + D(y, z), \tag{7}$$

and $D(x, x) = 0$, for any x, y and z, then D is symmetric.

Another approach is to use the entropy-like ISMs such as the mutual information (MI) and joint entropy (JE), Kullback–Leibler divergence or more general f-divergences [35, 60–62, 74]. Although MI has been generally applied to image registration, "*It is not an easy measure to understand: the underlying process of how misregistration influences the probability distribution is difficult to envisage. How it influences the relation between joint and marginal distributions is even more mystifying*" [60]. "*The information theoretic functionals are not easily predictable in their behavior.*" [58]. Applying both MI and JE simply to pixel values loses the locational, or spatial, or structural information because shuffling pixels randomly will provide the same MI or JE. Nevertheless, spatial information can be introduced by using structural features at a number of scales [73, 74]. However, the objective functionals will be non-convex and difficult to optimize in implementation [73]. Both MI and JE are symmetric, and hence are prone to the ill-transfer of features. It was found that MI enforces the ill-transferring more than JE does [58], though their effect of ill-transferring depends on the features used. Another issue is how to estimate the joint probability distributions efficiently and effectively. "*Various problems can emerge, like sensitivity to sample size, number of histogram bins or interpolation.*" [48].

It was found that there are f-measures that are able to perform better than MI at the cost of more difficult inference [61, 74]. This kind of ISMs does not in general satisfy the symmetry and triangle inequality [14–18, 38]. However, although several topologies can be induced by them [38], none of the topologies is both numerically stable and computationally tractable because of lacking triangle inequality and efficient computable representation. The topological effect on algorithmic stability due to floating errors cannot be ignored [47].

3   Existing Examples of Image Similarity Measures

In this section, we are to present several ISMs for JIR from multiple measurements in recent works. All the ISMs are designed to capture the structural similarity in terms of image gradients or edges. In the following, both image functions u_1 and u_2 are defined on the same image domain $\Omega \subset \mathbf{R}^N$, for $N = 2$ or 3 unless explicitly indicated otherwise. For notational simplicity and the readability of general readers, we focus on the motivations and constructions of the ISMs and avoid their general rigorous mathematical presentations. Interested readers can refer to the references for further details.

3.1   Vectorial Norms

The vectorial total variation (VTV) was proposed in [4] for colour image denoising and later extended to a number of colour image processing applications in [9], and later for the JIR of multi-modality tomography in [13, 35], and 4D cone-beam CT [55]. It was called joint total variation in [35]. For smooth image functions u_1 and u_2, it is defined as the following

$$\text{VTV}[u_1, u_2] = \int_\Omega \sqrt{|\nabla u_1|^2 + |\nabla u_2|^2}, \tag{8}$$

i.e., the integral of the $L_{2,1}$ norm of the matrix $\nabla u = [\nabla u_1, \nabla u_2]$ where ∇u_i is a column vector with components $\frac{\partial u_i}{\partial x_j}$ for $j = 1, \cdots, N$, for $i = 1$ and 2. For the $L_{p,q}$-norm for matrix, we use the convention as in [81].[2]

The intuition of the VTV can be understood with the following example from [13]. Let the image functions u_1 and u_2 are two 1-dimensional step functions with jumps at x_1 and x_2, respectively. If $x_1 = x_2$, then

$$\text{VTV}[u_1, u_2] = \sqrt{2}. \tag{10}$$

If $x_1 \neq x_2$, then

$$\text{VTV}[u_1, u_2] = 2. \tag{11}$$

Hence, by minimizing the VTV will enforce the alignment of image edges of u_1 and u_2. In other words, the VTV will encourage image edges to be of the same sparse structures with the $L_{2,1}$ norm of the matrix ∇u in terms of the L_1 norm of each column of it [1, 12]. However, when edges become complicated and artificial edges emerges due to the stair-casing effect of the TV regularization [11], the behaviour of the VTV cannot be as simple as in the above intuitive simple example. Please refer to [25] for another explanation in terms of the diffusion process of modalities.

In addition to the $L_{2,1}$ norm in (8), other matrix norms can also be used to formulate vectorial norms. Among them, the nuclear norm, or the Schatten 1-norm, i.e., the sum of the singular values of the Jacobian matrix ∇u, was proposed to promote rank sparsity instead of L_1 sparsity from the VTV [42], because the nuclear norm is the convex envelope of matrix rank [27, Section 5.1.4]. The resultant vector norm is called the total nuclear variation (TNV). It has been proved that the VTN prefers u_1 and u_2 to share parallel gradient directions [42]. The TNV was proposed for colour image processing in [42, 52]. It has been applied to multi-spectral XCT in [33, 49, 64], a registration problem for a dynamic contrast enhanced MRI sequence

---

[2]For an $S \times T$ matrix $A = \left[a_{i,j} \right]_{1 \leq i \leq S, 1 \leq j \leq T}$,

$$\|A\|_{p,q} = \left[\sum_{j=1}^{T} \left(\sum_{1 \leq i \leq S} |a_{i,j}|^p \right)^{\frac{q}{p}} \right]^{\frac{1}{q}}. \tag{9}$$

and histological serial sectioning [8]. For smooth image functions u_1 and u_2, the TNV is equal to

$$\text{TNV}[u_1, u_2] = \int_\Omega \text{trace}\left[(\nabla u)^{\text{tr}} \nabla u\right]. \tag{12}$$

The total generalized variation (TGV) [6, 7], or the higher-order total variation (HOT) [83, 84], was independently proposed as image models of piece-wise polynomials to overcome the stair-casing effect from the TV due to the piece-wise constant model of the TV. Both can be applied as vectorial norms for the joint vector $u = [u_1, u_2]$. The vectorial TGV has been applied to PET/MRI in [40, 51] and multi-modal electron tomography in [43] by promoting joint sparsity of the edge sets of all channels. For smooth image functions $u = [u_1, u_2]$ the vectorial second order TGV is equal to

$$\text{TGV}_\alpha^2[u] = \inf_{w \in C^\infty(\Omega, \mathbf{R}^N)^2} \alpha_1 \left\| |\nabla u - w|_{2,2} \right\|_1 + \alpha_0 \left\| |\mathcal{E}[w]|_{2,2} \right\|_1 \tag{13}$$

where α_1 and α_0 are positive constants, $|\cdot|_{2,2}$ is the $L_{2,2}$ norm or Frobenius norm of matrices, $\mathcal{E}[w] = \frac{1}{2}\left[\nabla w + (\nabla w)^{\text{tr}}\right]$ is the deformation tensor [75] or the symmetrized gradient of w, $\|\cdot\|$ is the L_1 norm. As in the case of the TNV, the nuclear norm can also be used to replace the first $L_{2,2}$ norm in (13) to enforce rank sparsity. For smooth image functions $u = [u_1, u_2]$ the nuclear second order TGV is equal to

$$\text{NTGV}_\alpha^2[u] = \inf_{w \in C^\infty(\Omega, \mathbf{R}^N)^2} \alpha_1 \left\| |\nabla u - w|_{\text{nuc}} \right\|_1 + \alpha_0 \left\| |\mathcal{E}[w]|_{2,2} \right\|_1. \tag{14}$$

For the nuclear second order TGV, it was reported in [51] that "*Overall, the nuclear norm yielded the best results*" and in that "*the performance was comparable to a Frobenius-norm based penalization*".

The ISMs in this subsection are defined as vectorial norms for the joint gradient vector $\nabla u = [\nabla u_1, \nabla u_2]$ in certain function spaces. They are both convex and symmetric in its joint arguments $[\nabla u_1, \nabla u_2]$, and satisfy the triangle inequality.

3.2 Parallel Level Sets

Structural similarity can also be pursued by investigating the parallelism of the level sets of u_1 and u_2 [35, 36]. The parallelism of the level sets was defined as structural similarity in [35]. Because the gradients of u_1 and u_2 are the normals to their level sets, the parallelism of the level sets is equivalent to the parallelism of the gradients of u_1 and u_2, and can be analytically formulated by the cross-product and inner-product of vectors [21, 29–32, 36]

$$\mathbf{a} \times \mathbf{b} = |a| \cdot |b| \cdot \sin \angle(\mathbf{a}, \mathbf{b}), \tag{15}$$

$$\langle \mathbf{a}, \mathbf{b} \rangle = |a| \cdot |b| \cdot \cos \angle(\mathbf{a}, \mathbf{b}), \tag{16}$$

for two vectors $\mathbf{a}$ and $\mathbf{b} \in \mathbf{R}^3$, where $\angle(\mathbf{a}, \mathbf{b})$ is the angle between $\mathbf{a}$ and $\mathbf{b}$, $\times$ and $\langle \cdot, \cdot \rangle$ are the canonical cross product and inner product in $\mathbf{R}^3$, and $|\cdot|$ is the canonical Euclidean norm. The following ISMs, were proposed in [36] for image registration,

$$D^c[u_1, u_2] = \int_\Omega \left\| \frac{\nabla u_1}{\|\nabla u_1\|} \times \frac{\nabla u_2}{\|\nabla u_2\|} \right\|^2, \tag{17}$$

$$D^d[u_1, u_2] = -\int_\Omega \left\langle \frac{\nabla u_1}{\|\nabla u_1\|}, \frac{\nabla u_2}{\|\nabla u_2\|} \right\rangle^2, \tag{18}$$

and the following ISM was applied to the multi-modality JIR in [13, 35]

$$S[u_1, u_2] = \int_\Omega \|\nabla u_1 \times \nabla u_2\|^2. \tag{19}$$

S is separately convex in each of its argument but not convex in its joint argument [22, Example 7.4.3, p.98]. In (17) and (18), we assume that the gradients are not zero at each point of Ω for notational simplicity. One typical numerical technique to replace gradients norms by $\sqrt{\|\nabla u_1\|^2 + \varepsilon}$, for example. The integrand of (19), i.e., the cross product term for structural similarity, was used as a constraint for structural similarity in [29–32]. The parallelism of gradients as in (18) has been used for image registration in addition to mutual information in [59]. The structural similarity by the inner product was proposed for colour image processing in [23]. The following ISM by parallel level sets (PLS) was proposed

$$\text{PLS}[u_1, u_2] = \int_\Omega \varphi \left[\psi \left(|\nabla u_1(x)| \cdot |\nabla u_2(x)| \right) - \psi \left(|\langle \nabla u_1(x), \nabla u_2(x) \rangle| \right) \right], \tag{20}$$

with strictly increasing functions φ and $\psi : [0, \infty] \to [0, \infty]$. PLS has been applied to PET/MRI in [25, 71]. Please refer to [25] for the theoretical study on its properties in terms of the diffusion process of modalities. When φ and ψ are both equal to the identity function, it follows that

$$\text{PLS}[u_1, u_2] = \int_\Omega |\nabla u_1(x)| \cdot |\nabla u_2(x)| - |\langle \nabla u_1(x), \nabla u_2(x) \rangle|, \tag{21}$$

which is referred to as the linear parallel level sets in [25]. The joint convexity of the general PLS has been studied in details in [22]. It has been proved that even the special case in (21) is not jointly convex in its joint argument and that it is not separately convex in each argument if another argument is non-zero.

3.3   Infimal Convolution

For multi-modality tomography, it can happen that the gradients ∇u_1 and ∇u_2 are of different orders of magnitude due to different ranges, different contrasts, and different structural information of different modality images because of different underlying physics. For the VTV in (8), if gradients with unbalanced orders of magnitude appear, the gradient with the high magnitude will dominate the vectorial norm, lead to a regularization by itself and ignorance of the gradient with the small magnitude, rather than alignment of both gradient directions. The same remark is applicable to some ISMs in previous subsections.

To resolve this issue, the infimal convolution of Bregman divergence of the TV is proposed [56, 63]. The Bregman divergence induced by the TV at u is equal to, for smooth image functions u and v,

$$D_{\mathrm{TV}}[v; u] = \int_\Omega \|\nabla v\| \left(1 - \left\langle \frac{\nabla v}{\|\nabla v\|}, \frac{\nabla u}{\|\nabla u\|} \right\rangle \right). \tag{22}$$

Again, in (22) and the following of this subsection, we assume that the gradients are not zero at each point of Ω for notational simplicity. $D_{\mathrm{TV}}[\cdot, \cdot]$ is asymmetric and convex. Although $D_{\mathrm{TV}}[v, u]$ is free of the issue of gradient magnitude, it does not encourage parallel gradients aligned at opposite directions. The infimal convolution of $D_{\mathrm{TV}}[\cdot, \cdot]$ is introduced to allow the alignment of gradients with both same and opposite directions,

$$\mathrm{ICB}_{\mathrm{TV}}[v; u] = \inf_{w \in C^\infty(\Omega)} D_{\mathrm{TV}}[v - w; u] + D_{\mathrm{TV}}[w; -u]. \tag{23}$$

Because

$$D_{\mathrm{TV}}[w; -u] = \int_\Omega \|\nabla w\| \left(1 - \left\langle \frac{\nabla w}{\|\nabla w\|}, \frac{-\nabla u}{\|\nabla u\|} \right\rangle \right), \tag{24}$$

minimizing the infimal convolution in (23) is to decompose the gradient ∇v into two parts intuitively: one is to match the direction of ∇u, another is to match the opposite direction of ∇u. The infimal convolution in (23) has been applied to PET/MRI in [63] and dynamic SPECT in [20]. The infimal convolution of the TGV has been proposed and applied to video decompression in [41] and dynamic MRI in [67].

The Bregman divergence $D_{\mathrm{TV}}[\cdot; \cdot]$ induced by the TV in (22) is only separately convex in its first argument with the second argument being fixed. Hence, although the infimal convolution maintains convexity, $\mathrm{ICB}_{\mathrm{TV}}[\cdot; \cdot]$ is only separately convex in its first argument with the second argument being fixed, and is not jointly convex in both arguments. Please refer to [56, 63] for further theoretical analysis.

3.4 Application of Tversky's Feature Contrast Model

In his prominent paper [77], Tversky proposed his '*feature contrast model*' for the similarity of binary features. Features are binary in the sense that a given feature of an object either is or is not in its set of features A. These features can also be seen as the set of predicates that are true. An example of such binary feature is the edge set of an image. With convincing and entertaining arguments, Tversky demonstrated why the requirements of symmetry and triangle inequality for similarity measures are inappropriate to explain a number of psychological experiments. Tversky proposed a set of axioms about similarity measures of binary features, which includes the axioms of matching, independence, solvability, invariance, and proved mathematically that feature similarity measures must of the following form [77],

$$S(a, b) = p(A \cap B) - \gamma_1 p(A \setminus B) - \gamma_2 p(B \setminus A), \tag{25}$$

where A and B denote the sets of binary features associated with the objects a and b, respectively, γ_1 and γ_2 are non-negative constants. p is an additive function such that $p(A \cup B) = p(A) + p(B)$ whenever $A \cap B = \emptyset$. Please note there the convention in [77] is that the more similar a to b, the bigger the similarity $S(a, b)$ is. Similarity measures thus obtained increase with addition of common features and/or deletion of distinctive features (i.e., features that belong to one object but not to the other) [77], and meets well the demand of ISM for avoiding *ill-transfer of features* in a JIR process from multiple measurements.

For 2D images, a natural choice for p with respect to image edge sets is the 1-dimensional Hausdorff measure $\mathcal{H}$, i.e., the length of image edges. Therefore, we obtain,

$$S(u_1, u_2) = \mathcal{H}(K_1 \cap K_2) - \gamma_1 \mathcal{H}(K_1 \setminus K_2) - \gamma_2 \mathcal{H}(K_2 \setminus K_1), \tag{26}$$

where K_1 and K_2 are the image edge sets of images u_1 and u_2, respectively. Higher-dimensional Hausdorff measures can also be used in (26) for higher-dimensional images [46]. To simplify notations, let us consider the case with 2-measurement in (1), i.e., $M = 2$. By using Mumford-Shah regularization functional for each modality in (2),

$$R_m(u_m) = \int_{\Omega \setminus K_m} |\nabla u_m|^2 + \beta_m \mathcal{H}(K_m), \tag{27}$$

and $D = -S$ as the ISM, we arrive at the following extended Mumford–Shah functionals E_1 and E_2 for JIR from 2-measurement, after re-parametrization with the same notations,

$$E_1(u_1, K_1) = \|A_1(u_1) - g_1\|^2 + \alpha_1 \int_{\Omega \setminus K_1} |\nabla u_1|^2 + \beta_1 \left[\mathcal{H}(K_1 \cap K_2) + \gamma_1 \mathcal{H}(K_1 \setminus K_2) \right],$$

$$(28)$$

$$E_2(u_2, K_2) = \|A_2(u_2) - g_2\|^2 + \alpha_2 \int_{\Omega \setminus K_2} |\nabla u_2|^2 + \beta_2 \left[\mathcal{H}(K_2 \cap K_1) + \gamma_2 \mathcal{H}(K_2 \setminus K_1) \right].$$

$$(29)$$

The above extended Mumford–Shah regularization has been proposed in [57] and applied for DOT/XCT in [39]. The ISM D is non-convex, asymmetric if $\gamma_1 \neq \gamma_2$ and does not satisfy the triangle inequality because of the complexity of the edge space.

4 Discussions

The ISMs reviewed and their corresponding JIR methods are reported to be successfully in improving the image quality compared to image reconstruction from single measurement. As discussed in the beginning of Sect. 2.2, the performance of an ISM and its JIR method depends on many factors. As for image registration [60], the evaluation and comparison of IIR methods is complicated and difficult. The complexities and difficulties in evaluating image registration methods are the same for JIR methods, because of the variants of imaging modalities, datasets, parameter selection methods, implementation techniques, runtime environments, software settings and hardware configurations, etc., and also the lacking of implementation details. Hence, the current author does not try to provide a comparison of the ISMs reviewed their corresponding JIR methods in this manuscript.

Many ISMs reviewed are gradient based, and hence are vulnerable to data noise. However, as discussed in the beginning of Sect. 2.2, the actual performance of a JIR method depends on many other factors. Hence, the performance of a JIR method with a gradient-based ISM is pending to how regularization is enforced to suppress data noise during the reconstruction process. ISMs satisfying the triangle inequality should be of numerical stability because of their uniform continuity property (6), in spite of their shortcoming due to transitivity from the triangle inequality as discussed in the second last paragraph in Sect. 1.

Many ISMs reviewed have its origin in colour image processing by cross-channel structural similarity such as the VTV in [4, 9] and the TNT in [42] or geometry of level sets of image functions by the parallelism or alignments of image gradient such as PLS in [23] and the infimal convolution of Bregman divergence of the TV [56]. It is still unknown how the underlying imaging physics could be exploited for designing ISMs for JIR, in spite of that JIR involves generally different imaging physics. The different imaging physics principles could be cues to pursue for features, feature structures, and formulating new ISMs.

Although the VTV and its variants in Sect. 3.1 could suffer from the issue of gradients with unbalanced orders of magnitude as discussed at the beginning of Sect. 3.3, and the extended Mumford-Shah regularization are free from it, it will be interesting to design ISMs that, in addition to the sole use of edges, could take account of the edge strength, for example, in terms of the gradient magnitude. In this regard, an extension of the Tversky's feature contrast model for non-binary features will be of guidance for constructing such ISMs. There was an extension of the Tversky's feature contrast model for non-binary features in [66]. However, this extension only utilizes the image gradient magnitudes but not their directions. Because of the role of image gradient directions for the image structural similarity as revealed in this review, it is necessary to investigate ISMs taking account of both strength and direction of edges. A more challenging problem is how to formulate image similarity measures when there are multiple features involved. This is necessary because more features will help characterize image similarity. Although an ISM can be formulated for each feature, the difficulty is how to combine them into an ISM when the similarities of independent features are inconsistent. For example, in the cases of two features A and B, for one image pairs, it could happen that the image similarity from feature A is bigger than that from feature B, while the image similarity from feature B is smaller than that from feature B.

Images from different modalities usually have different spatial resolutions. Hence, it is a problem how to measure image similarity at multiple-resolutions. When there are more than two imaging modalities, it is naturally that a certain iteration over the pairs of imaging modalities could be selected, and then the JIR of each pair of them can be performed, so that the JIR could be performed. A question raises also naturally: are there other approaches for JIR in addition to the iteration-over-pair process? When the joint JIR functional is jointly convex, global minimization algorithms could be established as in [40]. Moreover, if the joint JIR functional is jointly convex, the alternative iteration-over-pair process will not be trapped at local minimizers. Hence, ISMs that will induce the joint convexity of the joint JIR functional is highly desired, though they are difficult to design.

5 Conclusions

The quantitative measurement of image similarity is fundamental for tasks such as image quality assessment, image registration, image reconstruction from multiple measurements, and more. An ISM is both task-dependent and feature-dependent and must be designed according to the characteristics of specific tasks and features. Simply applying distances from the mathematical metric theory or general divergences to spaces of images or spaces of image features usually does not provide applicable appropriate ISMs. In this chapter, we have reviewed several ISMs for image reconstruction problems from multiple measurements of various types in recent work.. The multiple measurements considered here include multi-modality, multi-spectral, and multi-temporal measurements, with multi-modality tomography, multi-spectral XCT, and dynamic tomography, as the imaging applications, respectively. We have focused on motivations and constructions of the ISMs, avoided their general rigorous mathematical presentations to simplify notations for the readability for a general audience, and discussed relevant issues and possible problems for future study. The ISMs reviewed and their corresponding JIR methods have been successfully applied to tomographic imaging applications.

Acknowledgments The author is grateful for discussions and comments from Profs. Peter Maaß of University Bremen, Simon Arridge and Bangti Jin of University College London, Alfred K. Louis of Saarland University, Ronny Ramlau of the Johann Radon Institute for Computational and Applied Mathematics (RICAM), the Austrian Academy of Sciences, and their friendly hosting during my sabbatical visits to them. The author thanks the anonymous reviewer for insightful and constructive comments which help improve the presentation and quality of my manuscript, and appreciate the timely help on English usage from Profs. John W. Emerson of Yale University and Todd E. Quinto of Tufts University. This work is partially supported by National Science Foundation of China (61520106004, 11961141007) and Sino-German Center (GZ 1025).

References

1. E.V.D. Berg, M.P. Friedlander, Theoretical and empirical results for recovery from multiple measurements. IEEE Trans. Inf. Theory **56**(5), 2516–2527 (2010)
2. J.O. Berger, *Statistical Decision Theory and Bayesian Analysis*, 2nd edn. (Springer, New York, 1985)
3. J.M. Bernardo, A.F.M. Smith, *Statistical Decision Theory and Bayesian Analysis* (Wiley, New York, 2000)
4. P. Blomgren, T.F. Chan, Color TV: total variation methods for restoration of vector-valued images. IEEE Trans. Image Process. **7**(3), 304–309 (1998)
5. J.E. Bowsher, V.E. Johnson, T.G. Turkington, R.J. Jaszczak, C. Floyd, R.E. Coleman, Bayesian reconstruction and use of anatomical a priori information for emission tomography. IEEE Trans. Med. Imag. **15**(5), 673–686 (1996)
6. K. Bredies, M. Holler, Regularization of linear inverse problems with total generalized variation. J. Inverse Ill-Posed Problems **22**(6), 871–913 (2014)
7. K. Bredies, K. Kunisch, T. Pock, Total generalized variation. SIAM J. Imag. Sci. **3**(3), 492–526 (2010)
8. K. Brehmer, B. Wacker, J. Modersitzki, A novel similarity measure for image sequences, in *8th International Workshop, Biomedical Image Registration. WBIR 2018*, ed. by S. Klein, M. Staring, S. Durrleman, S. Sommer. Lecture Notes in Computer Science (Springer, Cham, 2018), pp. 47–56
9. X. Bresson, T.F. Chan, Fast dual minimization of the vectorial total variation norm and applications to color image processing. Inverse Problems Imag. **2**(4), 455–484 (2008)
10. M. Burger, H. Dirks, L. Frerking, A. Hauptmann, T. Helin, S. Siltanen, A variational reconstruction method for undersampled dynamic x-ray tomography based on physical motion models. Inverse Problems **33**(12), 124008 (2017)
11. T. Chan, S. Esedoglu, F. Park, A. Yip, Total variation image restoration: overview and recent developments, in *Handbook of Mathematical Models in Computer Vision*, ed. by N. Paragios, Y. Chen, O. Faugeras (Springer, Boston, 2006), pp. 17–31
12. S.F. Cotter, B.D. Rao, E. Kjersti, K. Kreutz-Delgado, Sparse solutions to linear inverse problems with multiple measurement vectors. IEEE Trans. Signal Process. **53**(7), 2477–2488 (2005)

13. B. Crestel, G. Stadler, O. Ghattas, A comparative study of structural similarity and regularization for joint inverse problems governed by PDEs. Inverse Problems **35**(2), 024003 (2019)
14. I. Csiszár, Informationstheoretische Konvergenzbegriffe im Raum der Wahrscheinlichkeitsverteilungen. Magyar Tud. Akad. Mat. Kutató Int. Közl. **7**, 137–158 (1962)
15. I. Csiszár, Eine informationstheoretische Ungleichung und Ihre Anwendung auf den Beweis der Ergodizität von Markoffschen Ketten. Magyar Tud. Akad. Mat. Kutató Int. Közl. **8**, 85–108 (1963)
16. I. Csiszár, Über topologische und metrische Eigenschaften der relativen Information der Ordnung p, in *Third Prague Conference Information Theory, Statistical Decision Functions, Random Processes (Liblice, 1962)* (Trans.) (Publishing House Czech Academy of Sciences, Prague, 1964), pp. 63–73
17. I. Csiszár, Information-type measures of difference of probability distributions and indirect observations. Stud. Sci. Math. Hungar. **2**, 299–318 (1967)
18. I. Csiszár, On topology properties of f-divergences. Stud. Sci. Math. Hungar. **2**, 329–339 (1967)
19. L. Desbat, S. Roux, P. Grangeat, Compensation of some time dependent deformations in tomography. IEEE Trans. Med. Imag. **26**(2), 261–269 (2007)
20. Q.Q. Ding, M. Burger, X.Q. Zhang, Dynamic SPECT reconstruction with temporal edge correlation. Inverse Problems **34**(1), 22 (2018)
21. M. Droske, M. Rumpf, A variational approach to nonrigid morphological image registration. SIAM J. Appl. Math. **64**(2), 668–687 (2004)
22. M.J. Ehrhardt, Joint reconstruction for multi-modality imaging with common structure. Ph.D. Thesis, University College London (2015)
23. M.J. Ehrhardt, S.R. Arridge, Vector-valued image processing by parallel level sets. IEEE Trans. Image Process. **23**(1), 9–18 (2014)
24. M.J. Ehrhardt, M.M. Betcke, Multi contrast MRI reconstruction with structure-guided total variation. SIAM J. Imag. Sci. **9**(3), 1084–1106 (2016)
25. M.J. Ehrhardt, K. Thielemans, L. Pizarro, D. Atkinson, S. Ourselin, B.F. Hutton, S.R. Arridge, Joint reconstruction of PET-MRI by exploiting structural similarity. Inverse Problems **31**(1), 015001 (2015)
26. Y.M. Fang, Z.J. Fang, F.N. Yuan, Y. Yang, S.Y. Yang, N.N. Xiong, Optimized multioperator image retargeting based on perceptual similarity measure. IEEE Trans. Syst. Man Cybern.-Syst. **47**(11), 2956–2966 (2017)
27. M. Fazel, Matrix rank minimization with applications. Ph.D. Thesis, Stanford University (2002)
28. E. Ferrante, N. Paragios, Slice-to-volume medical image registration: a survey. Med. Image Anal. **39**, 101–123 (2017)
29. L.A. Gallardo, Multiple cross-gradient joint inversion for geospectral imaging. Geophys. Res. Lett. **34**(19), L19301 (2007)
30. L.A. Gallardo, M.A. Meju, Characterization of heterogeneous near-surface materials by joint 2D inversion of DC resistivity and seismic data. Geophys. Res. Lett. **30**(13), 1 (2003)
31. L.A. Gallardo, M.A. Meju, Joint two-dimensional DC resistivity and seismic travel time inversion with cross-gradients constraints. J. Geophys. Res. Solid Earth **109**(B3), B03311 (2004)
32. L.A. Gallardo, M.A. Meju, Structure-coupled multiphysics imaging in geophysical sciences. Rev. Geophys. **49**(1), RG1003 (2011)
33. H. Gao, H. Yu, S. Osher, G. Wang, Multi-energy CT based on a prior rank, intensity and sparsity model (PRISM). Inverse Problems **27**(11), 115012 (2011)
34. K. Gong, J.X. Cheng-Liao, G.B. Wang, K.T. Chen, C. Catana, J.Y. Qi, Direct Patlak reconstruction from dynamic PET data using the kernel method with MRI information based on structural similarity. IEEE Trans. Med. Imag. **37**(4), 955–965 (2018)
35. E. Haber, M.H. Gazit, Model fusion and joint inversion. Surv. Geophys. **34**(5), 675–695 (2013)
36. E. Haber, J. Modersitzki, Beyond mutual information: a simple and robust alternative, in *Bildverarbeitung für die Medizin* (Springer, Berlin, 2005), pp. 350–354

37. B.N. Hahn, M.L. Kienle Garrido, An efficient reconstruction approach for a class of dynamic imaging operators. Inverse Problems **35**(9), 094005 (2019)
38. P. Harremoës, Information topologies with applications, in *Entropy, Search, Complexity*. Bolyai Society Mathematical Studies (Springer, Berlin, 2007), pp. 113–150
39. D. He, M. Jiang, A.K. Louis, P. Maass, T. Page, Joint bi-modal image reconstruction of DOT and XCT with an extended Mumford-Shah functional, in *2019 IEEE 16th International Symposium on Biomedical Imaging*. (ISBI 2019) (2019), pp. 1463–1466
40. M. Holler, R. Huber, F. Knoll, Coupled regularization with multiple data discrepancies. Inverse Problems **34**(8), 084003 (2018)
41. M. Holler, K. Kunisch, On infimal convolution of TV-type functionals and applications to video and image reconstruction. SIAM J. Imag. Sci. **7**(4), 2258–2300 (2014)
42. K.M. Holt, Total nuclear variation and Jacobian extensions of total variation for vector fields. IEEE Trans. Image Process. **23**(9), 3975–3989 (2014)
43. R. Huber, G. Haberfehlner, M. Holler, G. Kothleitner, K. Bredies, Total generalized variation regularization for multi-modal electron tomography. Nanoscale **11**(12), 5617–5632 (2019)
44. M. Jiang, Perspectives of similarity measures for joint multimodality image reconstruction, in *Tomographic Inverse Problems: Theory and Applications*, Mathematisches Forschungsinstitut Oberwolfach (2019), pp. 37–40. Report No. 4/2019
45. M. Jiang, Perspectives of similarity measures for joint multimodality image reconstruction, in *Wokshop on Tomographic Inverse Problems: Theory and Applications* (2019)
46. M. Jiang, P. Maass, T. Page, Regularizing properties of the Mumford-Shah functional for imaging applications. Inverse Problems **30**(3), 035007 (2014)
47. B. Jin, Personal communications (2019)
48. H. Kalinić, S. Lonĉarić, B. Bijnens, Absolute joint moments: a novel image similarity measure. EURASIP J. Image Video Process. **2013**(1), 24 (2013)
49. D. Kazantsev, J.S. Jorgensen, M.S. Andersen, W.R.B. Lionheart, P.D. Lee, P.J. Withers, Joint image reconstruction method with correlative multi-channel prior for x-ray spectral computed tomography. Inverse Problems **34**(6), 064001 (2018)
50. D. Kazantsev, G. Van Eyndhoven, W.R.B. Lionheart, P.J. Withers, K.J. Dobson, S.A. McDonald, R., Atwood, P.D. Lee, Employing temporal self-similarity across the entire time domain in computed tomography reconstruction. Philos. Trans. R. Soc. A-Math. Phys. Eng. Sci. **373**(2043), 20140389 (2015)
51. F. Knoll, M. Holler, T. Koesters, R. Otazo, K. Bredies, D.K. Sodickson, Joint MR–PET reconstruction using a multi-channel image regularizer. IEEE Trans. Med. Imag. **36**(1), 1–16 (2017)
52. S. Lefkimmiatis, A. Roussos, M. Unser, P. Maragos, Convex generalizations of total variation based on the structure tensor with applications to inverse problems, in *Scale Space and Variational Methods in Computer Vision: SSVM 2013*. Lecture Notes in Computer Science (Springer, Berlin, 2013), pp. 48–60
53. D. Li, T. Jiang, M. Jiang, Recent advances and challenges in video quality assessment. ZTE Commun. **17**(01), 3–11 (2019)
54. Y. Li, L. Hu, K. Xia, J. Luo, Fast distributed video deduplication via locality-sensitive hashing with similarity ranking. EURASIP J. Image Video Process. **2019**, 1–11 (2019)
55. J. Liu, X. Zhang, X. Zhang, H. Zhao, Y. Gao, D. Thomas, D.A. Low, H. Gao, 5D respiratory motion model based image reconstruction algorithm for 4D cone-beam computed tomography. Inverse Problems **31**(11), 115007 (2015)
56. M. Moeller, E.M. Brinkmann, M. Burger, T. Seybold, Color Bregman TV. SIAM J. Imag. Sci. **7**(4), 2771–2806 (2014)
57. T.S. Page, Image reconstruction by Mumford-Shah regularization with a priori edge information. Ph.D. Thesis, Universität Bremen (2015)
58. C. Panagiotou, S. Somayajula, A.P. Gibson, M. Schweiger, R.M. Leahy, S.R. Arridge, Information theoretic regularization in diffuse optical tomography. J. Opt. Soc. Am. A-Opt. Image Sci. Vis. **26**(5), 1277–1290 (2009)

59. J.P.W. Pluim, J.B.A. Maintz, M.A. Viergever, Image registration by maximization of combined mutual information and gradient information. IEEE Transa. Med. Imag. **19**(8), 809–814 (2000)
60. J.P.W. Pluim, J.B.A. Maintz, M.A. Viergever, Mutual-information-based registration of medical images: a survey. IEEE Trans. Med. Imag. **22**(8), 986–1004 (2003)
61. J.P.W. Pluim, J.B.A. Maintz, M.A. Viergever, f-information measures in medical image registration. IEEE Trans. Med. Imag. **23**(12), 1508–1516 (2004)
62. C. Pöschl, O. Scherzer, Distance measures and applications to multimodal variational imaging, in *Handbook of Mathematical Methods in Imaging* (Springer, New York, 2015), pp. 125–155
63. J. Rasch, E.M. Brinkmann, M. Burger, Joint reconstruction via coupled Bregman iterations with applications to PET-MR imaging. Inverse Problems **34**(1), 014001 (2018)
64. D.S. Rigie, P.J. La Riviere, Joint reconstruction of multi-channel, spectral CT data via constrained total nuclear variation minimization. Phys. Med. Biol. **60**(5), 1741–1762 (2015)
65. E. Roessl, R. Proksa, K-edge imaging in x-ray computed tomography using multi-bin photon counting detectors. Phys. Med. Biol. **52**(15), 4679–4696 (2007)
66. S. Santini, R. Jain, Similarity measures. IEEE Trans. Pattern Anal. Mach. Intell. **21**(9), 871–883 (1999)
67. M. Schloegl, M. Holler, A. Schwarzl, K. Bredies, R. Stollberger, Infimal convolution of total generalized variation functionals for dynamic MRI. Magn. Reson. Med. **78**(1), 142–155 (2017)
68. J.P. Schlomka, E. Roessl, R. Dorscheid, S. Dill, G. Martens, T. Istel, C. Baumer, C. Herrmann, R. Steadman, G. Zeitler, A. Livne, R. Proksa, Experimental feasibility of multi-energy photon-counting K-edge imaging in pre-clinical computed tomography. Phys. Med. Biol. **53**(15), 4031–4047 (2008)
69. U. Schmitt, A.K. Louis, Efficient algorithms for the regularization of dynamic inverse problems: I. theory. Inverse Problems **18**(3), 645–658 (2002)
70. U. Schmitt, A.K. Louis, C. Wolters, M. Vauhkonen, Efficient algorithms for the regularization of dynamic inverse problems: II. Applications. Inverse Problems **18**(3), 659–676 (2002)
71. G. Schramm, M. Holler, A. Rezaei, K. Vunckx, F. Knoll, K. Bredies, F. Boada, J. Nuyts, Evaluation of parallel level sets and Bowsher's method as segmentation-free anatomical priors for time-of-flight PET reconstruction. IEEE Trans. Med. Imag. **37**(2), 590–603 (2018)
72. R.N. Shepard, Toward a universal law of generalization for psychological science. Science **237**(4820), 1317–1323 (1987)
73. S. Somayajula, C. Panagiotou, A. Rangarajan, Q.Z. Li, S.R. Arridge, R.M. Leahy, PET image reconstruction using information theoretic anatomical priors. IEEE Trans. Med. Imag. **30**(3), 537–549 (2011)
74. A. Sotiras, C. Davatzikos, N. Paragios, Deformable medical image registration: a survey. IEEE Trans. Med. Imag. **32**(7), 1153–1190 (2013)
75. R. Temam, G. Strang, Functions of bounded deformation. Arch. Ration. Mech. Anal. **75**(1), 7–21 (1980)
76. T. Thireou, G. Kontaxakis, L.G. Strauss, A. Dimitrakopoulou-Strauss, S. Pavlopoulos, A. Santos, Feasibility study of the use of similarity maps in the evaluation of oncological dynamic positron emission tomography images. Med. Biol. Eng. Comput. **43**(1), 23–32 (2005)
77. A. Tversky, Features of similarity. Psychol. Rev. **84**(4), 327–352 (1977)
78. R.C. Veltkamp, M. Hagedoorn, Shape similarity measures, properties and constructions, in *Advances in Visual Information Systems, Proceedings*, ed. by R. Laurini. Lecture Notes in Computer Science (2000), pp. 467–476
79. G. Wang, M. Kalra, V. Murugan, Y. Xi, L. Gjesteby, M. Getzin, Q.S. Yang, W.X. Cong, M. Vannier, Vision 20/20: simultaneous CT-MRI – next chapter of multimodality imaging. Med. Phys. **42**(10), 5879–5889 (2015)
80. Y. Wang, X. Jiang, B. Yu, M. Jiang, A hierarchical Bayesian approach for aerosol retrieval using MISR data. J. Am. Stat. Assoc. **108**(502), 483–493 (2013)
81. Wikipedia contributors: Matrix norm — Wikipedia, the free encyclopedia (2019)
82. G. Wrinkler, *Image Analysis, Random Fields and Dynamic Monte Carlo Methods* (Springer, Berlin, 1995)

83. J.S. Yang, H.Y. Yu, M. Jiang, G. Wang, High-order total variation minimization for interior tomography. Inverse Problems **26**(3), 035013 (2010)
84. J.S. Yang, H.Y. Yu, M. Jiang, G. Wang, High-order total variation minimization for interior SPECT. Inverse Problems **28**(1), 015001 (2012)
85. Y. Zhou, X. Bai, W.Y. Liu, L.J. Latecki, Similarity fusion for visual tracking. Int. J. Comput. Vis. **118**(3), 337–363 (2016)

Holmgren-John Unique Continuation Theorem for Viscoelastic Systems

Maarten V. de Hoop, Ching-Lung Lin, and Gen Nakamura

Abstract We consider Holmgren-John's uniqueness theorem for a partial differential equation with a memory term when the coefficients of the equation are analytic. This is a special case of the general unique continuation property (UCP) for the equation if its coefficients are analytic. As in the case in the absence of a memory term, the Cauchy-Kowalevski theorem is the key to prove this. The UCP is an important tool in the analysis of related inverse problems. A typical partial differential equation with memory term is the equation describing viscoelastic behavior. Here, we prove the UCP for the viscoelastic equation when the relaxation tensor is analytic and allowed to be fully anisotropic.

1 Introduction

Many solids, such as earth materials, are viscoelastic. Viscoelasticity is a manifestation of rheology. The deformation of such solids are described by a viscoelastic equation with memory. In terms of displacement vector $u(x, t), t \in (0, T), x \in \Omega \subset \mathbb{R}^n$ with $n = 1, 2, 3$, this equation is given as

$$\rho(x)\partial_t^2 u(x, t) - \nabla \cdot (C(x)\nabla u(x, t)) + \nabla \cdot \int_0^t G(x, t - \tau)\nabla u(x, \tau)d\tau = 0, \quad (1)$$

M. V. de Hoop
Departments of Computational and Applied Mathematics and Earth, Environmental and Planetary Sciences, Rice University, Houston, TX, USA
e-mail: mvd2@rice.edu

C.-L. Lin
Department of Mathematics, National Cheng Kung University, Tainan, Taiwan
e-mail: cllin2@mail.ncku.edu.tw

G. Nakamura (✉)
Department of Mathematics, Hokkaido University, Sapporo, Japan
e-mail: nakamuragenn@gmail.com

© Springer Nature Switzerland AG 2021
B. Kaltenbacher et al. (eds.), *Time-dependent Problems in Imaging and Parameter Identification*, https://doi.org/10.1007/978-3-030-57784-1_10

287

where Ω is the reference domain for the deformation, $\rho(x) > 0$ is the density, $C(x)$ is the elasticity tensor, $-G(x, t)$ is the t-derivative of the relaxation tensor $H(x, t)$ and the $C(x)$ can be identified as $C(x) = H(x, 0)$. The original viscoelastic equation with memory is given as

$$\rho(x)\partial_t^2 u(x, t) - \nabla \cdot \int_0^t H(x, t - \tau)\nabla\partial_\tau u(x, \tau)d\tau = 0. \tag{2}$$

To derive (1) from (2), we need to invoke the assumption that $u(x, 0) = 0$, $x \in \Omega$. Physically it is natural to assume that $C(x)$ satisfies the strong ellipticity condition:

$$\sum_{i,j,k,l=1}^{n} C_{ijkl}(x)\xi_j\xi_\ell\eta_i\eta_k \geq \delta(\sum_{j=1}^{n} \xi_j^2)(\sum_{l=1}^{n} \eta_l^2), \ \xi_j, \ \eta_l \in \mathbb{R}, \ x \in \overline{\Omega} \tag{3}$$

for some constant $\delta > 0$.

To describe long-time behavior of viscoelastic relaxation, such as in the study of postseismic relaxation [4], the inertia term, $\rho(x)\partial_t^2 u(x, t)$, plays no role and is omitted yielding the quasi-static counterpart of (1).

By assuming that the coefficients of (1) are analytic, we will show that the unique continuation property (UCP) of solutions in the x-direction holds for this equation and also holds for its dual equation. More precisely we have the following

Theorem 1 *If the Cauchy data of a C^2-class solution $u(x, t)$ of (1) is zero on a regular surface in Ω over a large enough time interval $(0, T)$, then $u(x, t)$ is zero in Ω over some sub-time interval $(0, T_1) \subset (0, T)$.*

Remark 1 For the quasi-static case, the proof of the UCP can be made somewhat more concise. More precisely, the Holmgren transformation (9) given in Sect. 2 does not have to include the term ct^2 for z_n which makes the duality argument in Sect. 5 more straighforward.

The UCP plays an important role in the analysis of inverse problems. For example it is used for showing the uniqueness of identifying an unknown obstacle inside an acoustic medium (see [2]). It remains challenging to relax the condition of analyticity also in the case of elastic systems without memory terms. With analytic coefficients (components of the stiffness tensor) and without memory terms, the UCP is proved as a byproduct of the Cauchy-Kowalevski theorem and the Holmgren transformation via a duality argument. In the same spirit, we aim to prove the UCP as a byproduct of the Cauchy-Kowalevski theorem. The condition of causality, $u(x, 0) = 0$ for $x \in \Omega$, seems bothersome in the context of the Cauchy-Kowalevski theorem for (1) as it gives the unique analytic solution for the Cauchy problem only. However, by assuming this condition for the solution $u(t, x)$ of the original viscoelastic equation, we can still obtain the UCP by using analytic solutions to the dual of (1).

The memory term is a very special non-local term. We mention a result pertaining to the Cauchy-Kowalevski theorem for a generalized Camassa-Holm equation with a non-local operator [1]. The Camassa-Holm equation describes a wave in shallow water. One of the pioneers who contributed to an abstract extension of the Cauchy-Kowalevski theorem was Ovsjannikov [7] who was an expert in the theory of shallow water waves. Earlier than Ovsjannikov's work, there was work by Yamanaka [8] who first gave the framework of an abstract version of the Cauchy-Kowalevski theorem. Further progress was made by Nirenberg [5]; his work was completed concisely by Nishida [6] who was again an expert in the theory of shallow water waves. This is a very brief literature overview about the abstract version of the Cauchy-Kowalevski theorem which by no means is complete. Concerning the UCP for the viscoelastic equations, as far as we know, we are unaware of any result on the Cauchy-Kowalewski theorem which is the key to prove the UCP for (1).

We will adopt the argument given in [5] to prove the Cauchy-Kowalewski theorem by using a particular norm which is natural to prove the uniqueness of solution to the Cauchy problem. Then, as we mentioned before, we will show the UCP by the Holmgren transformation and a duality argument. We note that we will hereafter restrict our analysis to the cases $n = 2, 3$ while the analysis of the case $n = 1$ is straightforward.

The remainder of this paper is organized as follows. In the next section we will give some preliminaries. At the end of Sect. 3 we will state the Cauchy-Kowalevski theorem. The proof of this theorem will be given in the succeeding two sections. In Sect. 3, we prove the existence of a solution to the Cauchy problem and in Sect. 4, we establish the uniqueness of the solution to this Cauchy problem. In the final section, we show the duality relation between the solution of the Cauchy problem of (1) and any solution of its dual equation. After that, the UCP follows in a standard fashion.

2   Preliminaries and Cauchy-Kowalevski's Theorem

We subject (1) to a coordinate transformation and then transform it to a first-order system. We let $x_n = \phi(x')$ with $x = (x_1, \cdots, x_{n-1}, x_n) = (x', x_n)$ and $\phi(x')$ be an analytic function in a neighborhood of $x' = 0$ such that $\phi(0) = 0$. We consider a coordinate transformation, Φ,

$$x \mapsto y = \Phi(x), \quad x' = y' = (y_1, \cdots, y_{n-1}), \quad y_n = x_n - \phi(x'), \tag{4}$$

denote $\Psi = \Phi^{-1}$, $J(x) = \det \nabla\Phi(x)$, and assume that $J(x) > 0$ near $x = 0$. Then (1) becomes

$$\tilde{\rho}(y)\partial_t^2\tilde{u}(y, t) = \mathcal{J}(y)^{-1}\big\{\nabla_y\cdot(\tilde{C}(y)\nabla_y\tilde{u}(y, t)) - \nabla_y\cdot\int_0^t \tilde{G}(y, t-\tau)\nabla_y\tilde{u}(y, \tau)\big\}\,d\tau, \tag{5}$$

where we have introduced the following notation,

$$\mathcal{J}(y) := J(\Psi(y))^{-1}, \quad \Psi := \Phi^{-1} \tag{6}$$

and $\tilde{\rho}(y)$, $\tilde{u}(y,t)$, $\tilde{C}(y)$ and so on are given by

$$\tilde{\rho}(y) := \rho(\Psi(y)), \quad \tilde{u}(y,t) := u(\Psi(y),t) \tag{7}$$

and $\tilde{C}(y) = (\tilde{C}_{iqkr}(y))$ with

$$\tilde{C}_{iqkr}(y) := \mathcal{J}(y) \sum_{q,r=1}^{n} \partial_{x_j} y_q(\Psi(y)) \partial_{x_l} y_r(\Psi(y)) C_{ijkl}(\Psi(y)), \quad C(x) = (C_{ijkl}(x)). \tag{8}$$

We further consider the Holmgren transformation given as

$$(y,t) \mapsto (z,t) = (H_t(y),t), \quad y' = z' := (z_1, \cdots, z_{n-1}), \quad z_n = y_n + c|y'|^2 + ct^2, \tag{9}$$

with some constant $c > 0$. Note, here, that H_t is invertible near the origin, and that ∂_{y_j}, $1 \leq j \leq n$, ∂_t in the (y,t)-space takes the form

$$\partial_{y_j} = \partial_{z_j} + 2cz_j\partial_{z_n}, \quad j \neq n, \quad \partial_{y_n} = \partial_{z_n}, \quad \partial_t = 2ct\partial_{z_n} + \partial_t \tag{10}$$

in the (z,t)-space. Then, ignoring the lower order terms, (5) becomes

$$\tilde{\rho}'(4c^2t^2\partial_{z_n}^2 + 4ct\partial_{z_n}\partial_t + \partial_t^2)\tilde{u}' - (\mathcal{J}')^{-1}\big\{\tilde{C}' : \nabla_z^2\tilde{u}'$$
$$- \int_0^t \tilde{G}'(\cdot, t-\tau) : \nabla_z^2\tilde{u}'(\cdot, \tau)\, d\tau\big\} = 0, \tag{11}$$

where we have used the notation $\tilde{C}' := \tilde{C}'(z,t) = (\tilde{C}'_{ijkl}(z,t))$ with

$$\tilde{C}'_{ijkl} := \tilde{C}_{ijkl}(H_t^{-1}(z)), \quad 1 \leq j, l \leq n-1,$$

$$\tilde{C}'_{ijkn} := 2c \sum_{l=1}^{n-1} z_l \tilde{C}_{ijkl}(H_t^{-1}(x)) + \tilde{C}_{ijkn}(H_t^{-1}(z)), \quad 1 \leq j \leq n-1,$$

$$\tilde{C}'_{inkl} := 2c \sum_{j=1}^{n-1} z_j \tilde{C}_{ijkl}(H_t^{-1}(z)) + \tilde{C}_{inkl}(H_t^{-1}(z)), \quad 1 \leq l \leq n-1,$$

$$\tilde{C}'_{inkn} := 4c^2 \sum_{j,l=1}^{n-1} z_j z_l \tilde{C}_{ijkl} + 2c \sum_{l=1}^{n-1} z_l \tilde{C}_{inkl} + 2c \sum_{j=1}^{n-1} z_j \tilde{C}_{ijkn}$$

and

$$(\tilde{\rho}', \mathcal{J}') := (\tilde{\rho}, \mathcal{J})(H_t^{-1}(z)),$$

$$\tilde{u}' = (\tilde{u}'_1, \cdots, \tilde{u}'_n) := \tilde{u}(H_t^{-1}(z), t), \tag{12}$$

$$\tilde{G}'(z, t) := \tilde{G}(H_t(z), t),$$

while

$$\tilde{C}' : \nabla_z^2 \tilde{u}' = \sum_{j,k,l=1}^{n} \tilde{C}'_{ijkl} \partial_{z_j} \partial_{z_l} \tilde{u}'_k.$$

Hence, $\tilde{G}'(z, t)$ in (12) gives $\tilde{G}'(z, t - \tau) = \tilde{G}(H_{t-\tau}(z), t - \tau)$. We will find later that the memory term of (11) is, in fact, a lower order term. Thus, recalling (3), it follows that (11) is non-characteristic with respect to the hyperplane $z_n = 0$ near $(z, t) = (0, 0)$. This means that we can solve the equation with respect to $\partial_{z_n}^2 \tilde{u}'$. It should be remarked here that we ignored the lower order terms to have (11) from (5) only for the sake of simplifying the description and this will not change the argument at all even for the case including the lower order terms.

Next, we transform (11) to a first-order system with memory in terms of extended unknowns

$$U = (U', U'', U'''), \quad U' = \tilde{u}', \quad U'' = \nabla_{z',t}\tilde{u}', \quad U''' = \partial_{z_n}\tilde{u}',$$

where $z = (z_1, \cdots, z_{n-1}, z_n) = (z', z_n)$. Noting that $\partial_{z_n} U' = U''$, $\partial_{z_n} U'' = \nabla_{z',t} U'''$, we find that (11) is equivalent to

$$\partial_{z_n} U = A(z, t, \overline{D^1_{z',t}})U + \int_0^t \left\{ B(z, t-\tau)\partial_{z_n} + D(z, t-\tau, \overline{D^1_{z'}}) \right\} U(\cdot, \tau)\, d\tau, \tag{13}$$

where $D(z, t, \overline{D^1_{z',t}})$ is a linear differential operator with derivatives

$$\overline{D^1_{z',t}} := (\partial^\alpha_{z',t} : |\alpha| \leq 1), \quad \alpha \in Z^n_+, \ Z_+ := \mathbb{N} \cup \{0\}$$

and coefficients analytic in (z, t).

By abuse of notation, we write $x = z$ and define Banach spaces X_s, $0 \leq s < 1$ as follows. Each X_s is the set of vector functions $F(x, t)$ which are holomorphic in polydiscs

$$D_s := \{(x', t) = (x_1, \cdots, x_{n-1}, t) \in \mathbb{C}^n : |x_j| < sR, \ 1 \leq j \leq n-1, \ |t| < sR\} \tag{14}$$

for a fixed $R > 0$ with norm

$$\|F\|_s := \sup\{|F(x', t)| : (x', t) \in D_s\}. \tag{15}$$

Clearly, we have

$$X_s \subset X_{s'}, \quad \| \cdot \|_{s'} \leq \| \cdot \|_s, \quad 0 \leq s' < s. \tag{16}$$

Hence, X_s, $0 \leq s < 1$ is a Banach scale. By the Cauchy integral formula for holomorphic functions, we have

$$\| \partial_{x_j} F \|_{s'} \leq R^{-1} \| F \|_s (s - s')^{-1}, \quad 1 \leq j \leq n - 1,$$
$$\| \partial_t F \|_{s'} \leq R^{-1} \| F \|_s (s - s')^{-1} \tag{17}$$

for $0 \leq s' < s$.

We finally write (13) in the form

$$\frac{\partial u}{\partial x_n}(x_n, t) = \tilde{A}(x_n, t) u(x_n, t) + \int_0^t \tilde{B}(x_n, t - \eta) u(x_n, \eta) d\eta \text{ with } \tilde{B} = B \partial_{x_n} + \tilde{D},$$
$$u(0, t) = u_0(t),$$
$$\tag{18}$$

where x_n, u, $\tilde{A}$, B, $\tilde{D}$ correspond to z_n, U, A, B, D in (13). We have suppressed the other independent variables in the notation. Now we are ready to state the Cauchy-Kowalevski theorem.

Theorem 2 *Assume that the coefficients of the equation in* (18) *are* $\overline{D_2}$*- holomorphic valued continuous functions in an open interval* $(-\delta_0, \delta_0)$. *Then there exist* $0 < \delta < \delta_0$ *depending only on the coefficients and a unique solution* $u(x, t)$ *of the Cauchy problem* (18) *which is a* D_1*-holomorphic valued* C^1 *function in* $x_n \in I := (-\delta, \delta)$ *for any given holomorphic function* $u_0(t)$ *in* D_1. *Here, for example, a* D_1*-holomorphic valued function means that it is a real analytic function in* $D_1 \cap \mathbb{R}^n$ *and it can be extended to a holomorphic function in* D_1, *where* $D_1 = D_s$ *in* (14) *with* $s = 1$.

Remark 2

(i) We note that we took the region for (x', t) in which the coefficients of equation (18) are holomorphic twice as large as that for the solution $u(x, t)$, to handle the memory term of the equation.

(ii) Note the form (26) of solution $u(x, t)$ given in Sect. 3 to show the existence of a solution for Theorem 2 and take into account of the analyticity of coefficients of (18) with respect to x_n. Then, the same argument to prove Theorem 2 gives that the solution $u(x, t)$ of (18) is holomorphic with respect to x_n in a complex neighborhood $\mathcal{I}$ of the real open interval I given in the above theorem. Then by Hartog's theorem (see page 70, [3]), we can say that $u(x, t)$ is holomorphic in $\{(x, t) = (x', x_n, t) : x_n \in \mathcal{I}, (x', t) \in D_1\}$.

3   Existence: Contraction Property

In this section, we prove the existence of a solution of the Cauchy problem (18). To this end, we consider the solution $u(x_n, t)$ of the integral equation

$$u(x_n, t) = u_0(t) + \int_0^{x_n} \left(\tilde{A}(\xi, t)u(\xi, t)d\xi + \int_0^t \tilde{B}(\xi, t - \eta)u(\xi, \eta)d\eta \right) d\xi$$

$$= u_0(t) + \int_0^{x_n} A(\xi, t)u(\xi, t)d\xi + \int_0^{x_n} \int_0^t B(\xi, t - \eta)\partial_\xi u(\xi, \eta)d\eta d\xi$$

$$= u_0(t) + \int_0^{x_n} A(\xi, t)u(\xi, t)d\xi + \int_0^t B(x_n, t - \eta)u(x_n, \eta)d\eta$$

$$- \int_0^t B(0, t - \eta)u(0, \eta)d\eta, \qquad (19)$$

where

$$A(\xi, t)u(\xi, t) = \tilde{A}(\xi, t)u(\xi, t) + \int_0^t \left(\tilde{D}(\xi, t-\eta) - \partial_\xi B(\xi, t-\eta) \right)u(\xi, \eta)d\eta \qquad (20)$$

and the integrations with respect to ξ and η are taken along $\overrightarrow{0\,x_n}$ and $\overrightarrow{0\,t}$, respectively.

We let E be a vector space defined as

$$E = \{u(x_n, t) : u(x_n, \cdot) \in C^0(-a(1 - s), a(1 - s)) \text{ for } 0 \le s < 1, \ M[u] < \infty\} \qquad (21)$$

with

$$M[u] := \sup_{\{|x_n| < a(1-s), 0 \le s < 1\}} \|u(x_n, \cdot)\|_s \left(1 - \frac{|x_n|}{a(1 - s)} \right), \qquad (22)$$

where, for each $0 \le s < 1$, $u(x_n, t)$ is a X_s valued continuous function in $|x_n| < a(1 - s)$. We suppress the variables $x' = (x_1, \cdot, x_{n-1})$, that is, for each fixed $|x_n| < a(1 - s)$ we consider $u(x_n, t) = u(x_n, t; x') \in X_s$ as a function of (x', t). We also note that the definition of $M[u]$ is different from the one given in [5]. This is to make the proof of the uniqueness of solution u to (18) more straightforward.

By (16) and observing that

$$\left(1 - \frac{|x_n|}{a(1 - s')} \right)^{-1} \le \left(1 - \frac{|x_n|}{a(1 - s)} \right)^{-1} \text{ in } |x_n| < a(1 - s) \qquad (23)$$

for $s' < s$, it is not difficult to prove that E is complete with respect to the norm $M[\cdot]$ by using the well-known diagonal sequence trick argument in analysis considering a monotonically increasing sequence $0 \le s_n \to 1$, $n \to \infty$ for the parameter s of X_s. Hence, E is a Banach space.

We define

$$Tu(x_n, t) := u_0(t) + \int_0^{x_n} A(\xi, t)u(\xi, t)d\xi$$

$$+ \int_0^t (B^t u)(x_n, \eta)d\eta - \int_0^t (B^t u)(0, \eta)d\eta \qquad (24)$$

and

$$w(x_n, t) = Su(x_n, t) := \int_0^{x_n} A(\xi, t)u(\xi, t)d\xi$$

$$+ \int_0^t (B^t u)(x_n, \eta)d\eta - \int_0^t (B^t u)(0, \eta)d\eta, \qquad (25)$$

where we have used the abbreviation $(B^t u)(x_n, \eta) = B(x_n, t - \eta)u(x_n, \eta)$. We first show that there exists a solution $u \in E$ such that $Tu = u$ in this section, and then show that this solution, u, is unique in the next section.

The solution u above, which is a fixed point of T, will be obtained as

$$u = \sum_{l=0}^{\infty} S^l u_0. \qquad (26)$$

Hence it is sufficient to prove

$$M[w] \leq \frac{1}{2} M[u]. \qquad (27)$$

For $|x_n| < a(1 - s)$,

$$\|w(x_n, \cdot)\|_s \leq \int_0^{|x_n|} \|Au(\xi, \cdot)\|_s d|\xi| + cR\|u(x_n, \cdot)\|_s + cR\|u(0, \cdot)\|_s, \qquad (28)$$

where $d|\xi|$ denotes the length element along the straight line $\overrightarrow{0\,x_n}$ and $R > 0$ was used to define the polydisc D_s given by (14). Also a direct computation gives

$$\int_0^{|x_n|} \|\tilde{A}u(\xi, \cdot)\|_s d|\xi| \leq c \int_0^{|x_n|} \|u(\xi, \cdot)\|_{s(|\xi|)}(s(|\xi|) - s)^{-1}d|\xi|$$

$$\leq cM[u] \int_0^{|x_n|} (s(|\xi|) - s)^{-1} \left(1 - \frac{\xi}{a - as(|\xi|)}\right)^{-1} d|\xi|. \qquad (29)$$

Here and hereafter c is a positive constant depending only on $\tilde{A}$, B, $\tilde{D}$ of (15) in Sect. 2.

We let $s(|\xi|) = \frac{1}{2}\left(1 + s - \frac{|\xi|}{a}\right)$, then

$$s(|\xi|) - s = \frac{1}{2}\left(1 - s - \frac{|\xi|}{a}\right), \quad a - as(|\xi|) - |\xi| = \frac{a}{2}\left(1 - s - \frac{|\xi|}{a}\right). \quad (30)$$

By (30), we have

$$(s(|\xi|) - s)^{-1}\left(1 - \frac{|\xi|}{a - as(|\xi|)}\right)^{-1} = 2(1 - s)\left(1 - s - \frac{|\xi|}{a}\right)^{-2} + 2|\xi|a^{-1}\left(1 - s - \frac{|\xi|}{a}\right)^{-2}$$

$$\leq 2a^2(1 - s)(a - as - |\xi|)^{-2} + 2a|x_n|(a - as - |\xi|)^{-2}. \quad (31)$$

Combining (29) and (31), we obtain

$$\int_0^{|x_n|} \|\tilde{A}u(\xi, \cdot)\|_s d|\xi|$$

$$\leq cM[u]2a^2(1 - s)(1 + |x_n|/a(1 - s))\int_0^{|x_n|} (a - as - |\xi|)^{-2}d|\xi|$$

$$\leq cM[u]2a^2(1 - s)(1 + |x_n|/a(1 - s))(a - as - |\xi|)^{-1}$$

$$\leq cM[u]4a^2(1 - s)(a - as - |\xi|)^{-1}$$

$$= 4acM[u]\left(1 - \frac{|x_n|}{a(1 - s)}\right)^{-1}. \quad (32)$$

The other term in A satisfies the estimate,

$$\int_0^{|x_n|} \left\|\int_0^t \left(\tilde{D}(\xi, t - \eta) - \partial_\xi B(\xi, t - \eta)\right)u(\xi, \eta)d\eta\right\|_s d|\xi|$$

$$\leq 4acRM[u]\left(1 - \frac{|x_n|}{a(1 - s)}\right)^{-1}. \quad (33)$$

Similar estimates hold for the last two terms of (25). Combining these with (28), (32) and (33), we find that

$$\left(1 - \frac{|x_n|}{a(1 - s)}\right)\|w(x_n, \cdot)\|_s \leq 4acM[u] + 4acRM[u] + cRM[u] + cRM[u]. \quad (34)$$

Thus, we derive from (34) that

$$M[w] \leq 4acM[u] + 4acRM[u] + cRM[u] + cRM[u] \leq \frac{1}{2}M[u], \quad (35)$$

where we have chosen a and R such that $ac < 1/32$ and $cR < \min(1/8, c)$, respectively.

4 Uniqueness

In this section, we prove the uniqueness of solution u to (18). To begin with, we let $v(x_n, t)$ be a solution to (18). Then $w(x_n, t) = u(x_n, t) - v(x_n, t)$ satisfies

$$w(x_n, t) = \int_0^{x_n} A(\xi, t) w(\xi, t) d\xi + \int_0^t (B^t w)(x_n, \eta) d\eta - \int_0^t (B^t w)(0, \eta) d\eta. \tag{36}$$

Hence, by repeating the argument given in Sect. 3, we obtain (27). We recall that

$$M[w] := \sup_{\{|x_n| < a(1-s), 0 \le s < 1\}} \|w(x_n, \cdot)\|_s \left(1 - \frac{|x_n|}{a(1-s)}\right).$$

Since $|x_n| < a(1-s)$ implies $0 < 1 - \frac{|x_n|}{a(1-s)} \le 1$, we have

$$M[w] \le \sup_{\{|x_n| < a(1-s), 0 \le s < 1\}} \|w(x_n, \cdot)\|_s. \tag{37}$$

If the right-hand side of (37) is finite, then we have the uniqueness from (27). However, letting $1 > s \to 1$, we have $x_n \to 0$ and yet we may have $\|w(x_n, \cdot)\|_s \to \infty$. To avoid this difficulty we will modify the definition of $M[w]$ so that we can have $M[w] < \infty$.

We start by fixing $0 < s_0 < 1$ and define

$$M_0[w] := \sup_{\{|x_n| < a(s_0-s),\ 0 \le s < s_0\}} \|w(x_n, \cdot)\|_s \left(1 - \frac{|x_n|}{a(s_0 - s)}\right). \tag{38}$$

Since $w \in E$ and $M_0[w]$ is monotonically non-decreasing as s_0 increases, we have

$$M_0[w] \le \sup_{\{|x_n| < a(s_0-s)\}} \|w(x_n, \cdot)\|_{s_0} < \infty. \tag{39}$$

The uniqueness follows upon proving that

$$M_0[w] \le \frac{1}{2} M_0[w]. \tag{40}$$

We will repeat the argument in Sect. 3 to prove (40). We only show the estimate corresponding to $\tilde{A}$. For $|x_n| < a(s_0 - s)$, we find that

$$\int_0^{|x_n|} \|Aw(\xi, \cdot)\|_s d|\xi| \leq \int_0^{|x_n|} \|w(\xi, \cdot)\|_{s(\xi)} (s(\xi) - s)^{-1} d|\xi|$$

$$\leq c M_0[w] \int_0^{|x_n|} (s(|\xi|) - s)^{-1} \left(1 - \frac{|\xi|}{as_0 - as(|\xi|)}\right)^{-1} d|\xi|. \qquad (41)$$

We let $s(|\xi|) = \frac{1}{2}\left(s_0 + s - \frac{|\xi|}{a}\right)$, then

$$s(|\xi|) - s = \frac{1}{2}\left(s_0 - s - \frac{|\xi|}{a}\right), \quad as_0 - as(|\xi|) - |\xi| = \frac{a}{2}\left(s_0 - s - \frac{|\xi|}{a}\right). \qquad (42)$$

By (42), we have that

$$(s(|\xi|) - s)^{-1} \left(1 - \frac{|\xi|}{as_0 - as(|\xi|)}\right)^{-1}$$

$$= 2(s_0 - s)\left(s_0 - s - \frac{|\xi|}{a}\right)^{-2} + 2\xi a^{-1}\left(s_0 - s - \frac{|\xi|}{a}\right)^{-2}$$

$$\leq 2a^2(s_0 - s)(as_0 - as - |\xi|)^{-2} + 2a|x_n|(as_0 - as - |\xi|)^{-2}. \qquad (43)$$

Combining (41) and (43), we obtain that

$$\int_0^{|x_n|} \|\tilde{A}w(\xi, t)\|_s d|\xi|$$

$$\leq c M_0[w] 2a^2(s_0 - s)\left(1 + \frac{|x_n|}{a(s_0 - s)}\right)\int_0^{|x_n|} (as_0 - as - |\xi|)^{-2} d|\xi|$$

$$\leq c M_0[w] 2a^2(s_0 - s)\left(1 + \frac{x_n}{a(s_0 - s)}\right)(as_0 - as - |\xi|)^{-1}$$

$$\leq c M_0[w] 4a^2(s_0 - s)(as_0 - as - |\xi|)^{-1}$$

$$= 4ac M_0[w]\left(1 - \frac{|x_n|}{a(s_0 - s)}\right)^{-1}. \qquad (44)$$

Then handling the other terms in similar ways, we have

$$M_0[w] \leq 4ac M_0[w] + 4ac R M_0[w] + c R M_0[w] + c R M_0[w] \leq \frac{1}{2} M_0[w], \qquad (45)$$

where we have chosen a and R such that $ac < 1/32$ and $cR < \min(1/8, c)$, respectively. Therefore, $M_0[w] = 0$ which implies $w = 0$.

Combining the results in Sects. 3 and 4, by restricting the solution to the real space, we obtain the Cauchy-Kowalevski theorem for the Cauchy problem to equation (1) with Cauchy data on $x_n = \phi(x')$.

Before closing this section, we note that our Cauchy-Kowlevseki theorem is valid in a neighborhood of any point where the coefficients of the equation are analytic and setting up the Cauchy problem as we did before in a neighborhood of this point.

5 Duality Argument

In this section, we present the duality argument which gives the UCP for the equation given in (18). We let L be the operator defined by

$$Lv(x_n, t) = \partial_{x_n} v(x_n, t) - \tilde{A} v(x_n, t) - \int_0^t (B^t \partial_{x_n} + \tilde{D}^t) v(x_n, \eta) d\eta, \qquad (46)$$

where we have again suppressed the variables other than (x_n, t). For a more detailed form of L, see (18).

We let $H > 0$ be small and define W by

$$W := \{(x, t) = (x', x_n, t) \in \mathbb{R}^{n+1} : x_n \in (0, H), \ t > 0, \ x_n > c(|x'|^2 + t^2)\}. \tag{47}$$

Suppose that $u \in C^1(\overline{W})$ is a solution of (18), that is, $Lu = 0$ in W, and assume that

$$u = 0 \text{ for } \{t \le 0\} \cup \{x_n \le c(|x'|^2 + t^2)\}. \tag{48}$$

Then we will show that $u = 0$ in W. This is the key step in proving the UCP. The remaining steps or arguments showing the UCP are quite standard. For details, see, for instance, [3]. The definition of W is coming from the Holmgren transformation (9), and $u = 0$ in $x_n \le c(|x'|^2 + t^2)$ corresponds to $u = 0$ on $y_n = 0$ (see (4) and (9)) together with extending $u = 0$ to $y_n < 0$, which is possible due to the fact that (5) is non-characteristic with respect to $y_n = 0$ for small t. Furthermore, $u = 0$ for $t \le 0$ is coming from the physical assumption that was already explained just after (2). Hence, u also satisfies

$$Lu = 0 \text{ in } \{0 < x_n < H\} \setminus W. \tag{49}$$

We introduce some further notation. For $0 < h < H$, we let

$$W_h := W \cap \{x_n < h\}, \ \ T_h := \sup\{t > 0 : (x, t) \in W(h)\},$$

$$P_h := \text{the projection of } W_h \text{ to the } x\text{-space}, \tag{50}$$

$$Q_h := P_h \times (0, T_h).$$

From (48) and (49), we have

$$u = 0 \text{ in } (Q_h \setminus W_h) \cup \{t < 0\} \tag{51}$$

and

$$Lu = 0 \text{ in } Q_h, \tag{52}$$

respectively. For $w \in C^1(\overline{Q_h})$ we evaluate

$$\int_{Q_h} Lu(x, t) \cdot w(x, t) dx dt$$

$$= \int_{Q_h} \partial_{x_n} u(x, t) \cdot w(x, t) dx dt - \int_{Q_h} \tilde{A}u(x, t) \cdot w(x, t) dx dt$$

$$- \int_{Q_h} \left(\int_0^t (B^t \partial_{x_n} + \tilde{D}^t) u(x, s) ds \right) \cdot w(x, t) dx dt. \tag{53}$$

We observe that by using (51), we have

$$\int_{Q_h} \left(\int_0^t B(x, t - s) \partial_{x_n} u(x, s) ds \right) \cdot w(x, t) dx dt$$

$$= \int_{P_h} \int_0^{T_h} \int_0^t B(x, t - s) \partial_{x_n} u(x, s) \cdot w(x, t) ds dt dx$$

$$= \int_{P_h} \int_0^{T_h} \int_s^{T_h} B(x, t - s) \partial_{x_n} u(x, s) \cdot w(x, t) dt ds dx$$

$$= \int_{P_h} \int_0^{T_h} \int_t^{T_h} B(x, s - t) \partial_{x_n} u(x, t) \cdot w(x, s) ds dt dx$$

$$= - \int_{Q_h} u(x, t) \cdot \left(\int_t^{T_h} \partial_{x_n} (B^*(x, s - t) w(x, s)) ds \right) dt dx$$

$$+ \int_{Q_h \cap \{x_n = h\}} u((x', h), t) \cdot \left(\int_t^{T_h} B^*((x', h), s - t) w((x', h), s) ds \right) dt dx'$$

$$\tag{54}$$

and

$$\int_{Q_h} M(x, t) \partial u(x, t) \cdot w(x, t) dx dt = - \int_{Q_h} u(x, t) \cdot \partial \left(M^*(x, t) w(x, t) \right) dx dt \tag{55}$$

for any matrix $M(x, t) \in C^\infty(\overline{Q_h})$, where ∂ is either $\partial = \partial_{x_j}$ with $1 \leq j \leq n - 1$ or $\partial = \partial_t$, and the notation * is used to denote the action taking the dual of an operator.

The dual operator, L^*, of L is given by

$$L^*w(x,t) := -\partial_{x_n}w(x,t) - \tilde{A}^*w(x,t)$$

$$+ \int_t^{T_h} \partial_{x_n}(B^*(x,s-t)w(x,s))ds - \int_t^{T_h} \tilde{D}^*(x,s-t)w(x,s)ds. \qquad (56)$$

We write

$$z(x',t) := w((x',h),t) - \int_t^{T_h} B^*((x',h),s-t)w((x',h),s)ds. \qquad (57)$$

Then from (52)–(55), we find that

$$0 = \int_{Q_h} Lu(x,t) \cdot w(x,t)dxdt$$

$$= \int_{Q_h \cap \{x_n=h\}} u \cdot z\,dx'dt + \int_{Q_h} u(x,t) \cdot L^*w(x,t)dxdt$$

$$= \int_{Q_h \cap \{x_n=h\}} u \cdot z\,dx'dt \qquad (58)$$

if $L^*w = 0$ in Q_h.

For any given function $z(x',t)$ analytic in a neighborhood of $\overline{Q_h \cap \{x_n=h\}}$, consider the integral equation (57) with respect to $w((x',h),t)$. Then we can show that by an argument that is similar, and is an easier version of the proof of Theorem 2, that there exists a unique solution $w((x',h),t)$ to this integral equation which is analytic in a neighborhood of $\overline{Q_h \cap \{x_n=h\}}$.

Finally, we consider the Cauchy problem,

$$L^*w = 0 \quad \text{in } Q_h, \quad w = w((x',h),t) \text{ at } x_n = h. \qquad (59)$$

By Theorem 2, this problem admits a unique solution $w \in C^1(\overline{Q_h})$. Then due to the denseness of such Cauchy data in $L^2(Q_h \cap \{x_n=h\})$, we have $u = 0$ at $x_n = h$ for each $h \in (0, H)$. This immediately implies $u = 0$ in W.

Acknowledgments MVdH gratefully acknowledges support from the Simons Foundation under the MATH + X program, the National Science Foundation under grant DMS-1815143, and the corporate members of the Geo-Mathematical Imaging Group at Rice University. GN was supported by Grant-in-Aid for Scientific Research of the Japan Society for the Promotion of Science (No. 15K17555, No. 19K03554) during this study.

References

1. R.F. Barostichi, A.A. Himonas, G. Petronilho, A Cauchy-Kovalevsiky theorem for nonlinear and nonlocal equations. Analy. Geom. PROMS **127**, 59–66 (2015)
2. V. Isakov, On uniqueness of obstacles and boundary conditions from restricted dynamical and scattering data. Inverse Probl. Imaging **2**, 151–165 (2008)

3. F. John, *Partial Differential Equations*, 4th edn. (Springer, Berlin, 1982)
4. E. Klein, L. Fleitout, C. Vignay, J.D. Garaud, Afterslip and viscoelastic relaxation model inferred from the large-scale post-seismic deformation following the 2010 M_w 8.8 Maule earthquake (Chile). Geophys. J. Int. **205**, 1455–1472 (2016)
5. L. Nirenberg, An abstract form of the nonlinear Cauchy-Kowalewski theorem. J. Diff. Geom. **6**, 561–576 (1972)
6. T. Nishida, A note on a theorem of Nirenberg. J. Diff. Geom. **12**, 629–633 (1977)
7. L.V. Ovsyannikov, Singular operator in a scale of Banach spaces. Dokl. Akad. Nauk SSSR **163**, 819–822 (1965)
8. T. Yamanaka, Noe on Kowalevskaja's system of partial differential equations. Comment. Math. Univ. St. Paul **9**, 7–10 (1960)

Tomographic Reconstruction for Single Conjugate Adaptive Optics

Jenny Niebsch and Ronny Ramlau

Abstract Single Conjugate Adaptive Optic systems use the light of one bright guide star and a deformable mirror to correct for the loss of image quality of earthbound astronomical telescopes caused by turbulences in the atmosphere. The system achieves best correction in guide star direction. The imaging quality of the scientific object, which is usually separated from the guide star, can further be improved if the turbulence distribution is known. We propose to use wavefront sensor measurements from the past to recover the turbulence in the atmosphere. Mathematically, a limited angle tomography problem has to be solved. We present a model for the related tomography equations and discuss solvability and uniqueness of the solutions. Based on our analysis we develop an algorithm for the inversion and obtain a first numerical reconstruction.

1 Introduction

The image quality of modern earthbound astronomical telescopes suffers heavily from turbulences in the atmosphere. Patches of warm or cold air, located in layers of the atmosphere, distort the light coming from the scientific objects of interest, resulting in blurred images. This effect is in particular pronounced for the new generation of Extremely Large Telescopes (ELT) which are currently under construction. A remedy is the use of *Adaptive Optics (AO)*: These systems correct the aberrations of the incoming wavefronts of the scientific object by means of one or more deformable mirrors (DM). Based on measurements of the incoming

J. Niebsch
Johann Radon Institute Linz, Linz, Austria
e-mail: jenny.niebsch@ricam.oeaw.ac.at

R. Ramlau (✉)
Institute of Industrial Mathematics, Johannes Kepler University Linz, Linz, Austria

Johann Radon Institute Linz, Linz, Austria
e-mail: ronny.ramlau@jku.at

© Springer Nature Switzerland AG 2021
B. Kaltenbacher et al. (eds.), *Time-dependent Problems in Imaging and Parameter Identification*, https://doi.org/10.1007/978-3-030-57784-1_11

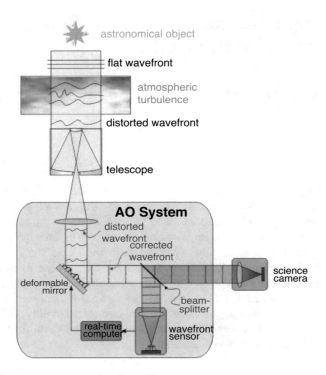

Fig. 1 Sketch of SCAO system, taken from [4]

wavefronts from one or several guide stars, the mirror shape is chosen such that the distortions from the incoming wavefront are corrected in the reflected wavefront. For a detailed description of the principles of Adaptive Optics we refer to [6, 26, 27].

There are different modes of operation for AO. The first—and simplest—is *Single Conjugate Adaptive Optics (SCAO)*, see Fig. 1. It is used if the scientific object is close to a bright star that acts as a Natural Guide Star (NGS). As a point source that is far away, the incoming light from the NGS resembles a plane wave that is distorted by the turbulences in different layers of the atmosphere. A wavefront sensor (WFS) measures the incoming wavefront, and an Real Time Computing (RTC) system determines a shape of the DM that flattens the wavefront of the NGS, see Fig. 2, which is now recorded as a sharp image. As the light from the nearby scientific object passes through nearly the same part of the atmosphere the DM also corrects its image.

For objects that have no NGS nearby, tomography based AO systems are suitable. These systems use multiple guide stars—both natural and artificial—each equipped with a wavefront sensor, and deformable mirrors for correction. Artificial guide stars are created by laser beams and therefore are called Laser Guide Stars (LGS). They are used whenever no suitable NGS is close by. Modern AO systems, e.g. for the ELT, will be equipped with up to 6 LGS. The incoming

Fig. 2 Correction of a wavefront by a deformable mirror [2]

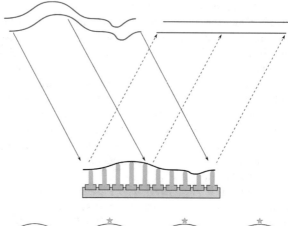

Fig. 3 Different operating modes of AO systems. Light blue marks the area of corrected imaging quality [2]

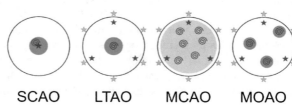

SCAO   LTAO   MCAO   MOAO

wavefronts from the different Guide Stars are used for a tomography of the atmosphere, i.e., the turbulence distribution of the atmosphere above the telescope is reconstructed. The underlying mathematical problem is a limited angle tomography and therefore severely ill-posed [3, 17]. However, as we only strive to reconstruct a layered atmosphere composed of a finite number of layers, the ill-posedness of the problem is somewhat mitigated, see [18]. Three different AO systems are based on atmospheric tomography: *Multi Conjugated Adaptive Optics (MCAO), Laser Tomography (LTAO)* and *Multi Object Adaptive Object (MOAO)*. LTAO uses the reconstructed atmosphere and one DM which is deformed s.t. the scientific object of interest is optimally sharpened. MOAO is based on the same concept, but uses several mirrors that are optimized to sharpen separated objects at the same time. In contrast, MCAO uses up to three different mirrors, conjugated to different heights, to achieve a high imaging quality on a large connected patch of the sky, thus allowing to observe larger structures. See also Fig. 3 for the different systems. Let us finally remark that all computations required for the control of the DMs need to be done in real time, and have to be repeated about every 2 ms for the full observation process. For further information on the systems we refer to [1, 5, 14, 20, 24] and for the mathematics of atmospheric tomography we refer to [7–12, 15, 21–23, 25, 29–34].

As mentioned above, the imaging quality for a SCAO system decreases the further the scientific object is away from the Guide Star. A knowledge of the turbulence distribution of the atmospheric layers would therefore also be useful in the SCAO case, as it would allow for a better estimate of the wavefront aberration in the direction of the observed object. Of course, a tomography of the atmosphere based on measurements from one direction is impossible. Our idea of a tomography-like reconstruction of the atmosphere for SCAO is based on the fact that the layers are blown over the telescope by a—at least on small time scales—constant wind velocity while the layers themselves are not changing (frozen flow assumption, see, e.g., [27]). Please note that wind speed and direction might be different for each layer. Basically, the measurements from the wavefront sensor at different time steps are created by shifted turbulent layers. As we will see, the connection of the data for a certain number of time steps and the turbulence of the layers can be described by a system of equations that is very similar to the atmospheric tomography. We may add that frozen flow assumption has been used previously for a better estimate of the measured wavefront [19].

The paper is organized as follows: In Sect. 2.1 we give a short overview about wavefront reconstruction from sensor measurements in SCAO. In Sect. 2.2 the tomography operator for SCAO will be derived, whereas Sect. 2.3 focuses on solvability and uniqueness of the resulting operator equation. Section 2.4 decomposes the tomography operator on a rectangular domain using the Fourier basis of $L_2([-R, R]^2)$. Section 3 contains the description of the reconstruction algorithm, the test setting and numerical results of a simulation to test the feasibility of our approach. We close with a short summary and an outlook to future work.

2   A Mathematical Approach for Tomography for SCAO

In this section we will present the general idea of SCAO—Tomography as well as the related tomography equations. The Fourier representation of this operator on a rectangular domain allows to derive conditions for the (unique) solvability of the problem. As a consequence, conditions on the model of the atmosphere and the number of time steps used for the reconstruction can be derived.

2.1   Wavefront Reconstruction for SCAO

As already mentioned in the introduction, SCAO is the simplest AO system, utilising one deformable mirror, optically conjugated to the ground layer of the atmosphere, and one wavefront sensor. The telescope is positioned such that the NGS is located in the centre of the field of view and the light emitted from the NGS approximately propagates in direction of the zenith to the telescope aperture. The WFS measures the incoming wavefront of the NGS, i.e., it sees summed turbulence contributions of the layers in that direction. Based on the sensed incoming wavefront, the deformable mirror is adapted in such a way that it compensates for this wavefront. Unfortunately, the sensor cannot measure the wavefronts directly. E.g., a Shack-Hartmann sensor, Fig. 4, measures averaged gradients in x- and y-direction over

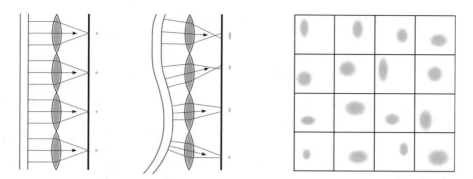

Fig. 4 Sketch of a Shack-Hartmann Sensor with 16 subapertures. The deviation of the center of the recorded spots from the center of each subaperture Ω_i in x and y direction are the measurements (s_x^i, s_y^i) [2]

sub-apertures of the sensor. Measurements **s** and the related WF φ are connected via the WFS operator Γ. For the Shack-Hartmann sensor, the WFS operator is given by

$$\mathbf{s} = (s_x^i, s_y^i)_{i \in I} = \Gamma \varphi, \tag{1}$$

$$s_x^i := \int_{\Omega_i} \frac{\partial}{\partial x} \varphi(x, y) \, dx, \tag{2}$$

$$s_y^i := \int_{\Omega_i} \frac{\partial}{\partial y} \varphi(x, y) \, dy. \tag{3}$$

The shape of the mirror can be derived directly from the wavefront. However, in order to determine the wavefront φ from wavefront measurements, Eq. (1) has to be solved about every 1–2 ms due to the fast changing atmosphere. Several methods were developed to compute φ from the data **s**. MVM methods connect the correction commands for the DM and the sensor data via a single control matrix which requires matrix-vector-multiplication (MVM) to solve the problem, leading to a numerical complexity of $\mathcal{O}(N^2)$. As the number of measurements grows with the size of the telescopes, faster algorithms have been developed to guarantee real time reconstruction. A matrix free approach for the reconstruction of wavefronts from sensor data in real time is the CuReD algorithm (Cumulated Reconstructor with Domain decomposition) [28, 35].

Our approach is based on the wavefronts instead of the measurements, thus we assume that the wavefronts have already been reconstructed by a suitable reconstruction method.

2.2   Derivation of the Tomography Equations

In classical SCAO as described in Sect. 2.1, the best correction is in direction of the NGS. The quality of the image of the observed scientific object decreases rapidly with increasing angular distance from the NGS, as the WF emitted by that object takes a different path through the atmosphere than the NGS and therefore is not optimally compensated by the DM. Incorporating a reconstruction of the atmosphere in the SCAO mode would enable us to correct the incoming WF from the scientific object. The observation quality, i.e., the Strehl ratio (which is closely related to the L_2 error of the reconstruction) of the object is expected to improve while the Strehl ratio in direction of the NGS will probably decrease.

In modelling the effect of the turbulence we will use the assumption that the atmosphere has a layered structure, and that the effect of the layered turbulence distribution on a planar wavefront can be expressed by the summation of the turbulent layers in the appropriate directions, see, e.g., [8]. More specific, the incoming wavefront $\varphi(\mathbf{r}, t)$ at the telescope aperture Ω_A can be written as the sum of the turbulence contributions of the layer functions $\Phi^{(l)}$, i.e.,

$$\varphi(\mathbf{r}, t) = \sum_{l=1}^{L} \Phi^{(l)}(\mathbf{r}, t), \tag{4}$$

where $r \in \Omega_A$ represents the 2D spatial coordinates in the telescope pupil and t indicates the time. Based on the Taylor frozen flow assumption, which states that each layer propagates with its own speed and direction, represented by the wind shift vector $\mathbf{v}_l \in \mathbb{R}^2$, the temporal evolution of the single layers from time $t - \tau$ to t can be attributed to a spatial shift with displacement $\tau \mathbf{v}_l$, i.e.,

$$\Phi^{(l)}(\mathbf{r}, t - \tau) = \Phi^{(l)}(\mathbf{r} + \tau \mathbf{v}_l, t). \tag{5}$$

Assuming equidistant time steps Δ_T and choosing $\tau = k \Delta_T$ we have

$$\Phi^{(l)}(\mathbf{r}, t - k \Delta_T) = \Phi^{(l)}(\mathbf{r} + k \Delta_T \mathbf{v}_l, t), \tag{6}$$

see also [19]. We can use (6) to compute the layers $\Phi^{(l)}$ at the actual time t from data of the previous time steps $(t - k \Delta_T), k = 0, \cdots, K$. Using (4), (6), the incoming wavefronts at different time steps can be computed as

$$\varphi_k := \varphi(\mathbf{r}, t - k \Delta_T) = \sum_{l=1}^{L} \Phi^{(l)}(\mathbf{r} + k \Delta_T \mathbf{v}_l, t), \tag{7}$$

see Fig. 5 for an illustration.

Please note that we assume that the wind vectors $\mathbf{v}_l$ are explicitly known. We wish, however, to remark that techniques are available to either estimate the wind

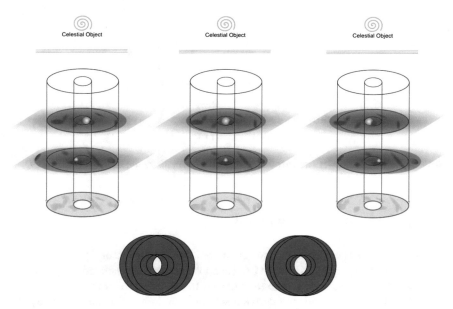

Fig. 5 Illustration of the turbulence contributions of the layers $\Phi^{(1)}$ and $\Phi^{(2)}$ to the incoming wavefronts $\varphi_3, \varphi_2, \varphi_1$. The telescope aperture is restricted by the red circles. The dark blue areas on the turbulent layers belongs to those atmospheric cutouts which contribute to the incoming wavefronts of the evolution equations. The picture below shows the support of the layer functions

vectors from the wavefront sensor measurements or to measure them directly with an additional instrument [19].

Equation (7) forms the basis for our tomography equation. We define by $\Omega_A(k\Delta_T\mathbf{v}_l) = \{\mathbf{r} \in \mathbb{R}^2 : \mathbf{r} - k\Delta_T\mathbf{v}_l \in \Omega_A\}$ the area of layer l seen by the telescope at time step k and $\Omega_l = \bigcup_{k=0}^{K} \Omega_A(k\Delta_T\mathbf{v}_l)$. Since we also want to consider the Fourier transform of the tomography equation we define the rectangular area $\bar{\Omega} = [-R, R]^2$ where R is the smallest number such that $\bigcup_{l=1}^{L} \Omega_l \subset [-R, R]^2$.

Setting $X = L^2(\bar{\Omega})$, and, for a fixed time t, $\Phi^{(l)}(\mathbf{r}, t) = \Phi^{(l)}(\mathbf{r}) \in X$ for $l = 1, \cdots, L$, we define the operator $\mathbf{A}_k : \prod_{l=1}^{L} X \to X$ as

$$(\mathbf{A}_k\boldsymbol{\Phi})(\mathbf{r}) := \sum_{l=1}^{L} \Phi^{(l)}(\mathbf{r} + \mathbf{v}_l k\Delta_T), \quad \mathbf{r} \in \bar{\Omega}, \tag{8}$$

where $\boldsymbol{\Phi} = (\Phi^{(1)}, \cdots, \Phi^{(L)})$ represents the full layered atmosphere. With definition (8) and (7) the tomography-like operator equation at time t is then given as

$$\mathbf{A}\boldsymbol{\Phi} := \begin{pmatrix} \mathbf{A}_0\boldsymbol{\Phi} \\ \mathbf{A}_1\boldsymbol{\Phi} \\ \vdots \\ \mathbf{A}_K\boldsymbol{\Phi} \end{pmatrix} = \begin{pmatrix} \varphi_0 \\ \varphi_1 \\ \vdots \\ \varphi_K \end{pmatrix} =: \boldsymbol{\varphi}, \tag{9}$$

$$\text{with} \quad \mathbf{A} : \prod_{l=1}^{L} X \to \prod_{j=1}^{K+1} X.$$

Now the computation of the turbulence profile reduces to the solution of Eq. (9).

2.3 Solvability and Uniqueness

In this section we focus on the existence of solutions of our tomography equation as well as on the uniqueness. As a shift in a function is nicely represented by the Fourier transform, the shift in the layers in (8) suggests to consider the tomography equation in the Fourier space in order to derive necessary conditions for the solvability and uniqueness of the tomography equation (9).

In the following, we use the 2-D Fourier transform on $\mathbb{R}^2$ defined as

$$(\mathcal{F}f)(\mathbf{s}) = \int_{\mathbb{R}^2} f(\mathbf{r})e^{-2\pi i\langle\mathbf{r},\mathbf{s}\rangle}d\mathbf{r},$$

where $\langle\cdot,\cdot\rangle$ is the usual inner product on $\mathbb{R}^2$.

For our theoretical analysis we assume that the atmosphere layer functions and the wavefronts are defined on $\mathbb{R}^2$. Specifically, we require that

Assumption 1 *For the layers* $\boldsymbol{\Phi} = (\Phi^{(1)}, \ldots, \Phi^{(L)})$ *holds* $\Phi^{(l)} \in L^1(\mathbb{R}^2) \cap L^2(\mathbb{R}^2)$.

Remark 1 Turbulences in the atmosphere can be modelled either by a Kolmogrov or a van Karman statistics [27]. Thus, in average, the turbulent layers belong to the Sobolev space $H^{11/6}(\mathbb{R}^2)$ and thus also to $L^2(\mathbb{R}^2)$. As there is anyway no hope to reconstruct the atmosphere outside the area where there are measurements available, we can extend the layers outside this area by zero or simply assume that the layers are compactly supported. In this case the layer will also belong to $L^1(\mathbb{R}^2)$.

Equation (9) in the Fourier domain is given by

$$(\mathcal{F}\mathbf{A}\boldsymbol{\Phi})(\mathbf{s}) = \begin{pmatrix} (\mathcal{F}\mathbf{A}_0\boldsymbol{\Phi})(\mathbf{s}) \\ (\mathcal{F}\mathbf{A}_1\boldsymbol{\Phi})(\mathbf{s}) \\ \vdots \\ (\mathcal{F}\mathbf{A}_K\boldsymbol{\Phi})(\mathbf{s}) \end{pmatrix} = \begin{pmatrix} (\mathcal{F}\varphi_0)(\mathbf{s}) \\ (\mathcal{F}\varphi_1)(\mathbf{s}) \\ \vdots \\ (\mathcal{F}\varphi_K)(\mathbf{s}) \end{pmatrix},$$

where the Fourier transform is applied componentwise. Because of the linearity and the time shifting property of the Fourier transform,

$$(\mathcal{F}f(\mathbf{r} - \mathbf{r}_0))(\mathbf{s}) = e^{-2\pi i \langle \mathbf{r}_0, \mathbf{s} \rangle}(\mathcal{F}f(\mathbf{r}))(\mathbf{s}),$$

we have for each $k = 0, \cdots, K$,

$$(\mathcal{F}\mathbf{A}_k\boldsymbol{\Phi})(\mathbf{s}) = \sum_{l=1}^{L} e^{2\pi i \langle \mathbf{v}_l k \Delta_T, \mathbf{s} \rangle}(\mathcal{F}\Phi^{(l)}(\mathbf{r}))(\mathbf{s}) = (\mathcal{F}\varphi_k)(\mathbf{s}).$$

Hence the Fourier transform of (9) can be expressed in matrix form as

$$\begin{pmatrix} (\mathcal{F}\varphi_0)(\mathbf{s}) \\ (\mathcal{F}\varphi_1)(\mathbf{s}) \\ \vdots \\ (\mathcal{F}\varphi_K)(\mathbf{s}) \end{pmatrix} = \underbrace{\begin{pmatrix} 1 & \cdots & 1 \\ e^{\Delta_T 2\pi i \langle \mathbf{v}_1, \mathbf{s} \rangle} & \cdots & e^{\Delta_T 2\pi i \langle \mathbf{v}_L, \mathbf{s} \rangle} \\ \vdots & \ddots & \vdots \\ e^{K\Delta_T 2\pi i \langle \mathbf{v}_1, \mathbf{s} \rangle} & \cdots & e^{K\Delta_T 2\pi i \langle \mathbf{v}_L, \mathbf{s} \rangle} \end{pmatrix}}_{=:\mathbf{F}(\mathbf{s})} \begin{pmatrix} (\mathcal{F}\Phi^{(1)})(\mathbf{s}) \\ (\mathcal{F}\Phi^{(2)})(\mathbf{s}) \\ \vdots \\ (\mathcal{F}\Phi^{(L)})(\mathbf{s}) \end{pmatrix}. \quad (10)$$

Representation (10) allows to link the solvability of the tomography equation to the invertibility of the matrix $\mathbf{F}(\mathbf{s})$.

Proposition 1 *Assume that for $l, m = 1, \cdots, L$ and $l \neq m$ holds $\mathbf{v}_l \neq \mathbf{v}_m$.*

(1) The matrix $\mathbf{F}(\mathbf{0})$ has rank 1.
(2) Assume that for $\mathbf{s} \neq \mathbf{0}$ holds for all $n \in \mathbb{Z}$

$$\langle \mathbf{v}_l - \mathbf{v}_m, \mathbf{s} \rangle \neq \frac{n}{\Delta_T}. \quad (11)$$

Then the columns of $\mathbf{F}(\mathbf{s})$ in (10) are linearly independent if $L \leq K + 1$ and linearly dependent if $L > K + 1$.

Proof

(1) If $\mathbf{s} = \mathbf{0}$, then all entries in $\mathbf{F}(\mathbf{0})$ are equal to 1, and therefore $\mathbf{F}(\mathbf{0})$ has rank 1.
(2) Let $x_l := e^{\Delta_T 2\pi i \langle \mathbf{v}_l, \mathbf{s} \rangle}$, $l = 1, \cdots, L$. Because of (28), $x_l \neq x_m$ holds for all $l \neq m, l, m = 1, \cdots, L$. With $e^{k\Delta_T 2\pi i \langle \mathbf{v}_l, \mathbf{s} \rangle} =: x_l^k$ for $k = 1, \cdots, K$, the matrix $\mathbf{F}(\mathbf{s})$ is a Vandermonde matrix

$$\mathbf{F} = \begin{pmatrix} 1 & \cdots & 1 \\ x_1 & \cdots & x_L \\ x_1^2 & \cdots & x_L^2 \\ \vdots & \ddots & \vdots \\ x_1^K & \cdots & x_L^K \end{pmatrix}. \tag{12}$$

In case $L = K + 1$, its determinant can be computed as $\prod_{l=2}^{L} \prod_{m=1}^{l-1}(x_l - x_m)$, [16], and it is nonzero under our assumptions and therefore $\mathbf{F}$ has linearly independent columns.

In case $L < K + 1$, the matrix (12) can be expanded with $N = K + 1 - L$ columns $(1, x_{L+j}, x_{L+j}^2, \cdots, x_{L+j}^K)^T$ for $j = 1, \cdots, N$, where the x_{L+j} are chosen distinct from all x_l and from each other. The expanded matrix is a square Vandermonde matrix hence all columns are linearly independent which implies that the first L columns are as well.

In case $L > K + 1$, the first $K + 1$ columns are linearly independent with the same argument as in above and form a basis of $\mathbb{C}^{K+1}$. It follows that the remaining column vectors are linearly dependent on the first $K + 1$ column vectors. □

Whenever $\mathbf{v}_l \neq \mathbf{v}_m$ and condition (28) hold, then $\mathbf{F}(\mathbf{s})$ is invertible, and thus the Fourier transform of the turbulent layers is uniquely reconstructable for those $\mathbf{s}$. For the remaining $\mathbf{s}$ we have the following result:

Proposition 2 *Assume that for $l, m = 1, \cdots, L$ and $l \neq m$ holds $\mathbf{v}_l \neq \mathbf{v}_m$. Further, assume that either $\mathbf{s} = 0$ or*

$$\langle \mathbf{v}_l - \mathbf{v}_m, \mathbf{s} \rangle = \frac{n}{\Delta_T}. \tag{13}$$

holds for some $l \neq m$, $l, m = 1, \ldots, L$, $n \in \mathbb{Z}$ and $\mathbf{s} \neq 0$. Then there exists a sequence $\mathbf{s}_k \to \mathbf{s}$ as $k \to \infty$, and $\mathbf{F}(\mathbf{s}_k)$ is invertible.

Proof We start with the case $\mathbf{s} \neq 0$. If (13) holds for some (l, m) and $n \neq 0$, we set $\mathbf{s}_k := (1 - \frac{1}{k})\mathbf{s} \neq 0$ for $k \in \mathbb{N}$ and $k > 1$. Clearly, $\mathbf{s}_k \to \mathbf{s}$ as $k \to \infty$, and

$$\langle \mathbf{v}_l - \mathbf{v}_m, \mathbf{s}_k \rangle = (1 - \frac{1}{k})\frac{n}{\Delta_T} \to \frac{n}{\Delta_T}.$$

As of course also $\langle \mathbf{v}_l - \mathbf{v}_m, \mathbf{s}_k \rangle \neq \frac{n}{\Delta_T}$, $\mathbf{F}(\mathbf{s}_k)$ is invertible at least for large k.

Now assume that

$$\langle \mathbf{v}_l - \mathbf{v}_m, \mathbf{s} \rangle = 0$$

for some (l, m). We set $\mathbf{s}_k := (1 - \frac{1}{k})\mathbf{s} + \frac{1}{k}(\mathbf{v}_l - \mathbf{v}_m)$ and obtain

$$\langle \mathbf{v}_{\tilde{l}} - \mathbf{v}_{\tilde{m}}, \mathbf{s}_k \rangle = \langle \mathbf{v}_{\tilde{l}} - \mathbf{v}_{\tilde{m}}, \mathbf{s} \rangle + \frac{1}{k} \langle \mathbf{v}_{\tilde{l}} - \mathbf{v}_{\tilde{m}}, \mathbf{v}_l - \mathbf{v}_m \rangle \tag{14}$$

$$= \begin{cases} \frac{c_{\tilde{l},\tilde{m}}}{k} \|\mathbf{v}_l - \mathbf{v}_m\|^2 \text{ if } (\mathbf{v}_l - \mathbf{v}_m) \parallel (\mathbf{v}_{\tilde{l}} - \mathbf{v}_{\tilde{m}}) \\ \\ (1 - \frac{1}{k}) \langle \mathbf{v}_{\tilde{l}} - \mathbf{v}_{\tilde{m}}, \mathbf{s} \rangle + \frac{1}{k} \langle \mathbf{v}_{\tilde{l}} - \mathbf{v}_{\tilde{m}}, \mathbf{v}_l - \mathbf{v}_m \rangle \end{cases} \tag{15}$$

$$\neq \frac{n}{\Delta_T}, \tag{16}$$

$n \in \mathbb{Z}$, at least for a subsequence of a $\mathbf{s}_k$, and thus $\mathbf{F}$ is invertible on the subsequence. Please note that the above argument only holds as the wind vectors are elements in $\mathbb{R}^2$. It remains to consider the case $\mathbf{s} = \mathbf{0}$. Now we choose a vector $\mathbf{v} \in \mathbb{R}^2$ such that $\langle \mathbf{v}_l - \mathbf{v}_m, \mathbf{v} \rangle \neq 0$ for all (l, m) and set $\mathbf{s}_k := \frac{1}{k} v$. It follows again that there exists at least a subsequence of $\mathbf{s}_k$ such that

$$\langle \mathbf{v}_l - \mathbf{v}_m, \mathbf{s}_k \rangle \neq \frac{n}{\Delta_T},$$

and $\mathbf{F}(\mathbf{s}_k)$ is again invertible, which concludes the proof. □

Now we are able to give a result on the unique solvability.

Proposition 3 *Assume that $\Phi^{(l)} \in L^1(\mathbb{R}^2) \cap L^2(\mathbb{R}^2)$, $l = 1, \ldots, L$ and that the wind speed vectors $\mathbf{v}_l$ fulfill the condition $\mathbf{v}_l - \mathbf{v}_m \neq 0$ for $m \neq l$ and $l, m = 1, \ldots, L$. Then $\boldsymbol{\Phi} = (\Phi^{(1)}, \ldots, \Phi^{(L)})$ is uniquely reconstructable if $L \leq K + 1$.*

Proof If $L \leq K + 1$ then $\mathbf{F}(\mathbf{s})$ is invertible as long as (28) holds. According to (10), $\mathcal{F}(\Phi^{(l)})$ is thus well defined by the measurements. Now assume (28) is violated, i.e., (13) holds. According to Proposition 2 for those $\mathbf{s}$ there exists a sequence $\mathbf{s}_k \to \mathbf{s}$ where $\mathbf{F}(\mathbf{s}_k)$ is invertible and therefore the related values of $\mathcal{F}(\Phi^{(l)})(\mathbf{s}_k)$ are uniquely defined by the measurements. As $\Phi^{(l)} \in L^1$, its Fourier transform $\mathcal{F}(\Phi^{(l)})$ is continuous and therefore $\mathcal{F}(\Phi^{(l)})(s) := \lim_{k \to \infty} \mathcal{F}(\Phi^{(l)})(\mathbf{s}_k)$ is uniquely determined. □

From Proposition 1 it follows that the assumption that the wind shifts of the layers are distinct is vital. It is also quite natural: Assume that we are given an atmosphere composed of two layers that move with the same speed and in the same direction. It follows immediately from (8) that the wavefront φ_k at time step $k\Delta_T$ is just the shifted version of φ_0 and thus contains no additional information of the layered atmosphere. More general, all layers that move with the same wind vector behave in the data like a single layer that contains the sum of the turbulence contributions from those layers and can therefore not be reconstructed uniquely.

The matrix $\mathbf{F}$ is invertible if the number of time steps K is chosen as $L - 1$, i.e., including the data from time frame i the number of data equals the number of layers. If the wind shifts are not distinct or if $L > K + 1$, then a least squares solution can be obtained by applying the generalized inverse $\mathbf{F}^\dagger = (\mathbf{F}^*\mathbf{F})^{-1}\mathbf{F}^*$ [13]. This is summarised in

Corollary 1 *Under the assumptions of Proposition 1 the solution of the tomography equation on $L^2(\mathbb{R}^2)$ is given by*

1.

$$\Phi(\mathbf{r}) = \mathcal{F}^{-1}(\mathbf{F}^{-1}(\mathbf{s})(\mathcal{F}\boldsymbol{\varphi})(\mathbf{s}))(\mathbf{r}), \quad if \quad K = L - 1,$$

2.

$$\Phi(\mathbf{r}) = \mathcal{F}^{-1}(\mathbf{F}^+(\mathbf{s})(\mathcal{F}\boldsymbol{\varphi})(\mathbf{s}))(\mathbf{r}) \quad otherwise$$

$$with \quad \mathbf{F}^\dagger = (\mathbf{F}^*\mathbf{F})^{-1}\mathbf{F}^*.$$

Assuming a certain number of layers in the atmosphere, the time steps taken into consideration must be at least of the same number or more. This is confirmed by numerical test computations. The assumption that the wind shift vectors of the layers are not equal (although they can be parallel) is reasonable. In case the vector $(\langle \mathbf{v}_1, \mathbf{s}\rangle, \cdots, \langle \mathbf{v}_L, \mathbf{s}\rangle)$ is zero or close to it the matrix $\mathbf{F}$ is rank deficient.

2.4 A Fourier Series Representation of the Tomography Operator

In a real life situation wavefronts can only be determined on the telescope aperture. In a first approximation of the real data situation we assume that the layer functions as well as the wavefronts are defined on $\bar{\Omega} = [-R, R]^2$. As in (9), it makes now sense to use the Fourier series representation of the layer and wavefront functions instead of the Fourier transform on $\mathbb{R}^2$. We assume that the functions in $L^2(\bar{\Omega})$ are complex (complex unit i) valued and define an orthonormal basis of $L^2(\bar{\Omega})$ by

$$\omega_{jm}(x, y) := \frac{1}{2R}e^{i\frac{\pi}{R}(jx+my)}, \qquad\qquad j, m \in \mathbb{Z}. \qquad (17)$$

Next, we decompose the projection operators $\mathbf{A}_k$ in (8) with respect to this basis. Each $\Phi^{(l)}(x, y) \in L^2(\bar{\Omega})$, $l = 1, \cdots, L$, has the representation

$$\Phi^{(l)}(x, y) = \sum_{j,m\in\mathbb{Z}} \Phi^{(l)}_{jm} \omega_{jm}(x, y) \qquad\qquad with$$

$$\Phi^{(l)}_{jm} = \langle \Phi^{(l)}, \omega_{jm}\rangle_{L^2(\bar{\Omega})}.$$

With the notation

$$\boldsymbol{\Phi}_{jm} = (\Phi^{(1)}_{jm}, \cdots, \Phi^{(L)}_{jm})^T,$$

we can represent the full atmosphere as

$$\mathbf{\Phi} = \sum_{j,m\in\mathbb{Z}} \mathbf{\Phi}_{jm}\,\omega_{jm},\tag{18}$$

where the summation of a vector has to be understood componentwise. Now (8) can be written as

$$(\mathbf{A}\mathbf{\Phi})_k(x,y) = \sum_{l=1}^{L}\sum_{j,m\in\mathbb{Z}} \Phi_{jm}^{(l)}\,\omega_{jm}(x+v_x^{(l)}k\Delta_T,\,y+v_y^{(l)}k\Delta_T)$$

$$= \sum_{j,m\in\mathbb{Z}}\left(\sum_{l=1}^{L}\Phi_{jm}^{(l)}\underbrace{2R\cdot\omega_{jm}(j\,v_x^{(l)}k\Delta_T,\,m\,v_y^{(l)}k\Delta_T)}_{=:(A_{jm})_{k,l}}\right)\omega_{jm}(x,y).$$

Denoting by φ_{jm}^k, $k = 0,\cdots,K$, the Fourier coefficients of φ_k and by

$$\boldsymbol{\varphi}_{jm} = (\varphi_{jm}^0,\cdots,\varphi_{jm}^K)^T,\tag{19}$$

we have

$$(\mathbf{A}\mathbf{\Phi})(x,y) = \sum_{j,m\in\mathbb{Z}}(A_{jm}\mathbf{\Phi}_{jm})\,\omega_{jm}(x,y) = \sum_{j,m\in\mathbb{Z}}\boldsymbol{\varphi}_{jm}\,\omega_{jm}(x,y),\tag{20}$$

and therefore the Fourier coefficients $\boldsymbol{\varphi}_{jm}$ of the wavefront can be computed from the Fourier coefficients of the atmosphere layers $\mathbf{\Phi}_{jm}$ for each j and m by

$$\boldsymbol{\varphi}_{jm} = A_{jm}\mathbf{\Phi}_{jm},\tag{21}$$

i.e., the computation of the wavefronts (and therefore also the inverse operation) *decouples* for each (j,m). This is also reflected in

Proposition 4 *Let A_{jm}, $\boldsymbol{\varphi}_{jm}$ and $\mathbf{\Phi}_{jm}$ be defined as above. Then operator A can be described in terms of its action on the Fourier coefficients of the turbulent layers and the wavefronts as*

$$\begin{pmatrix}\vdots\\\boldsymbol{\varphi}_{jm}\\\vdots\end{pmatrix} = diag(A_{jm})\begin{pmatrix}\vdots\\\mathbf{\Phi}_{jm}\\\vdots\end{pmatrix}\tag{22}$$

where $diag(A_{jm})$ denotes a block diagonal matrix with the matrices A_{jm} on the diagonal and zeros outside.

Proof Follows directly from (21).   □

Please note that we are now in a similar situation as in [18], where the authors consider a singular value decomposition of the standard atmospheric tomography operator. Specifically, the operator A relates directly to the matrix $diag(A_{jm})$ and

$$\boldsymbol{\varphi}_{jm} \in \mathbb{R}^K \leftrightarrow \boldsymbol{\varphi}_{jm} \cdot \omega_{jm}(x, y) \in L_2(\bar{\Omega})^K$$

$$\boldsymbol{\Phi}_{jm} \in \mathbb{R}^L \leftrightarrow \boldsymbol{\Phi}_{jm} \cdot \omega_{jm}(x, y) \in L_2(\bar{\Omega})^L.$$

For each of the matrices A_{jm} there exists a singular system, i.e., vectors $v_{jm,n} \in \mathbb{C}^L$, $u_{jm,n} \in \mathbb{C}^K$, and numbers $\sigma_{jm,n}, n = 1, \ldots, r_{jm} \leq min\{L, K\}$ that satisfy

$$A_{jm}\boldsymbol{\Phi}_{jm} = \sum_{n=1}^{r_{jm}} \sigma_{jm,n}\langle v_{jm,n}, \boldsymbol{\Phi}_{jm}\rangle u_{jm,n} \tag{23}$$

$$\langle v_{jm,l}, v_{jm,n}\rangle = \delta_{ln}, \qquad \langle u_{jm,l}, u_{jm,n}\rangle = \delta_{ln}$$

$$\sigma_{jm,1} \geq \ldots \geq \sigma_{jm,r_{jm}} > 0.$$

Here, r_{jm} is the rank of the matrix A_{jm} and the $\sigma_{jm,n}^2$ are the positive eigenvalues of the matrices $A_{jm}^H A_{jm}$ and $A_{jm} A_{jm}^H$, respectively. We obtain

Proposition 5 *The operator A admits a singular value type decomposition*

$$A\boldsymbol{\Phi} = \sum_{j,m\in\mathbb{Z}} \left(\sum_{n=1}^{r_{jm}} \sigma_{jm,n}\langle v_{jm,n}, \boldsymbol{\Phi}_{jm}\rangle u_{jm,n} \right) \omega_{jm} \tag{24}$$

and the sequence $\{\sigma_{jm,n} : j, m \in \mathbb{Z}, n = 1, \ldots, r_{jm}\}$ are the singular values of A. Further on,

$$A^\dagger\boldsymbol{\varphi} = \sum_{j,m\in\mathbb{Z}} \left(\sum_{n=1}^{r_{jm}} \frac{\langle u_{jm,n}, \boldsymbol{\varphi}_{jm}\rangle}{\sigma_{jm,n}} v_{jm,n} \right) \omega_{jm} \tag{25}$$

and

$$\mathcal{N}(A)^\perp = span\{v_{jm,n} \cdot \omega_{jm} \mid j, m \in \mathbb{Z}, 1 \leq n \leq r_{jk}\}. \tag{26}$$

If the vectors $\hat{v}_{jm,n} \in \mathbb{C}^L, 1 \leq n \leq L - r_{jk}$, are a basis of $\mathcal{N}(A_{jm})$, then the nullspace of A is given as

$$\mathcal{N}(A) = span\{\hat{v}_{jm,n}\omega_{jm} \mid j, m \in \mathbb{Z}, 1 \leq n \leq L - r_{jk}\}. \tag{27}$$

For a proof we refer to [18, pp. 844].

Proposition 5 links the nullspace of the operator A to the nullspaces of the matrices A_{jm}. Specifically, each rank deficient matrix A_{jm} contributes to the nullspace of A. Please note that the matrices A_{jm} are closely related to the matrices $F(s)$ defined in (10). Specifically, replacing in $F(s)$ in each matrix element s by (j, m) and 2π by $\frac{\pi}{R}$ gives the matrix A_{jm}. In particular, A_{jm} also is a Vandermonde matrix, and Proposition 1 holds accordingly for A_{jm} if s is replaced by (j, m) and the right hand side of (28) is replaced by $n \cdot \frac{2R}{\Delta_T}$. Additionally, (28) immediately yields that **all** matrices A_{jm} are rank deficient if $v_l = v_m$ for some $m \neq l$, i.e., if at least two of the layers move with the same speed and direction. We also conclude from Proposition 1 that the matrix A_{00} has always rank 1, i.e., constant functions on the layers cannot be reconstructed. Figure 7, plotting the rank of the A_{jm}, shows that A has a nontrivial nullspace in the considered setting. In such a case, it is only possible to reconstruct the atmosphere in a least squares sense. Please note that this is not in contradiction to Proposition 3, as we are now in a periodic setting. We summarize these results in

Proposition 6 *Assume that for $l, m = 1, \cdots, L$ and $l \neq m$ holds $\mathbf{v}_l \neq \mathbf{v}_m$.*

(1) The matrix A_{00} has rank 1.
(2) Assume that for $(j, m) \neq (0, 0)$ holds for all $n \in \mathbb{Z}$

$$\langle \mathbf{v}_l - \mathbf{v}_m, (j, m) \rangle \neq n \cdot \frac{2R}{\Delta_T}. \tag{28}$$

Then the columns of A_{jm} are linearly independent if $L \leq K + 1$ and linearly dependent if $L > K + 1$.

If $\mathbf{v}_l = \mathbf{v}_m$ holds for some $l \neq m$, then all matrices A_{jm} are rank deficient.

3 Numerical Realization

3.1 Algorithm

In this paper we are only concerned with the reconstruction of the atmosphere, and neglect the reconstruction of the wavefronts from sensor data as well as the computation of the commands for the deformable mirrors.

We assume therefore that the wavefronts φ_k for times $t_0, t_0 - 1\Delta_T, t_0 - 2\Delta_T, \cdots, t_0 - K\Delta_T$ are given. The reconstruction algorithm is based on the Fourier coefficients φ_{jm}^k of the wavefront (19) and its shifted versions, solving the subsystems (21) and computing the atmosphere via (18). We recall the definition of the matrices $A_{jm} \in \mathbb{R}^{K+1, L}$ as

$$(A_{jm})_{k,l} := 2R \cdot \omega_{jm}(j \, v_x^{(l)} k \Delta_T, m \, v_y^{(l)} k \Delta_T) \tag{29}$$

Clearly, the (A_{jm}) can be computed in advance for each $j, m \in \{-N_0, \cdots, N_0 - 1\}$ where N_0 is a cut-off index for the Fourier series. As mentioned above, the reconstruction process reduces to the solution of a sequence of small matrix vector systems, which can be done efficiently with standard solvers. For our numerical tests, we used either the generalized inverse of A_{jm} or the conjugate gradient method to compute a solution of each equation (21). Please note that whenever a matrix A_{jm} is rank deficient, there is no hope to recover the original solution. Instead, we compute a least squares solution.

Algorithm 1 Reconstruction algorithm for the atmosphere Φ at time t_0

Choose cut-off index N_0
Precompute A_{jm} for $j, m \in \{-N_0, \cdots, N_0 - 1\}$
Compute φ_{jm}^k and cut off according to $j, m \in \{-N_0, \cdots, N_0 - 1\}$
for $j, m = -N_0 \ldots N_0 - 1$ **do**
    $\Phi_{jm} = \mathbf{solve}(A_{j,m}, \varphi_{jm})$
end for
Compute $\Phi = \sum\limits_{j,m=-N_0}^{N_0} \Phi_{jm} \, \omega_{jm}$.

3.2 Test Setting

For our test computations we used the in-house developed software package *MOST* to create an atmosphere, which produces a realistic atmosphere using a viable C_n^2 profile modeling the strength of turbulence. For a first reconstruction, we use a 3-layers atmosphere, see Table 1 for specifications. The telescope models the ELT of the European Southern Observatory with radius $R = 21m$, whereas the atmosphere is created on the square $[-23.5; 23.5]^2$.

Proposition 3 suggests that we need at least $K = L + 1$ wavefronts as input in order to reconstruct L layers. Thus we have chosen $K = 4$, i.e. 5 time steps of length $\Delta_T = 0.002$ s as input data. The wavefronts φ_k, i.e., the data, are computed by summation of the shifted atmosphere layers at time $(t - k\Delta_T), k = 0, \cdots, K$, on the larger area. The quality of the reconstructed atmosphere will be estimated on $[-21; 21]^2$. Thus we avoid errors at the edges that arise from the periodic boundary

Table 1 Characteristics of the turbulent layers of the simulated 3-layers atmosphere. The C_n^2 profile is a measure characterizing the amount of turbulence located on a specific layer

	Height	Speed	Direction (x, y)	C_n^2-profile
1-Layer	0 m	40 m/s	$(-1,0)$	50%
2-Layer	8000 m	15 m/s	$(1,-1)$	25%
3-Layer	12,000 m	30 m/s	$(1,0)$	25%

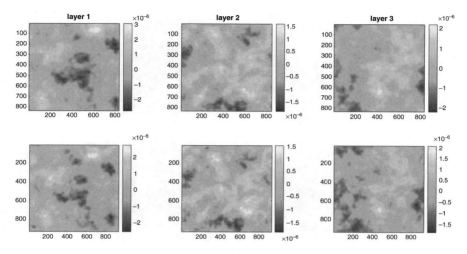

Fig. 6 Reconstruction of the 3-layers atmosphere from exact data: the upper row shows the original layer function, the lower row shows the reconstructed layer functions

conditions of wavefronts and layers. The cut-off index for the Fourier series for all test cases was chosen as $N_0 = 80$, which also resembles the physical fact that the currently available wavefront sensors cannot resolve higher frequencies.

3.3 Reconstruction of a 3-Layers Atmosphere from Exact Data

As a proof of concept we only aim at the reconstruction of the 3-layers atmosphere from undisturbed data $\varphi_0, \cdots, \varphi_4$ according to Algorithm 1. Figure 6 displays the reconstructed atmosphere as well as the original atmosphere. A first visual inspection suggests a good reconstruction quality. However, please note that there is an almost constant offset between the two reconstructions. The relative errors between the original layers $\Phi^{(l)}$ and the reconstructed layers $\Phi_{rec}^{(l)}$ on $[-21; 21]^2$ are $(8.8\%, 2.1\%, 12.9\%)$ for $l = 1, 2, 3$, which is large given that we used exact data. Again, the errors are largely due to the offset. A close inspection shows that all the errors are created at indices (j, m) where the matrix A_{jm} is rank deficient: in these cases, the method computes solutions that are perpendicular to $\mathcal{N}(A_{jm})$, a property that is in general not shared by the coefficients from the underlying original atmosphere. Figure 7 displays the coefficients where A_{jm} is (numerically) rank deficient.

To verify that the reconstruction of all the coefficients which are related to matrices A_{jm} with full rank is correct we compared the reconstruction quality only on the related set of Fourier coefficients: If we drop all coefficients (j, m) where A_{jm} is rank deficient both in the original and the reconstructed layers, we obtain

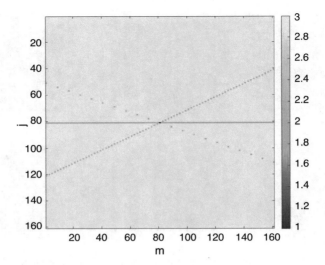

Fig. 7 Plot of the rank of the matrices A_{jm} vs indices (j, m). Yellow shows full rank (3), other colors indicate rank deficiencies of the associated matrix

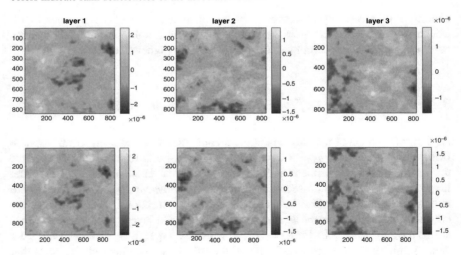

Fig. 8 Three-layers atmosphere from exact data restricted to Fourier coefficients for full rank A_{jm}: The upper row shows the restricted original layer function, the lower row shows the restricted reconstructed layer functions

a perfect reconstruction, see Fig. 8. In this case, the relative error between original and reconstructed atmosphere is $\approx 10^{-23}$ and thus within the numerical accuracy. This confirms that our proposed method is able to reconstruct a solution to the atmospheric tomography problem at least on $\mathcal{N}(A)^{\perp}$.

4 Summary and Future Work

In the previous sections we have developed and analyzed a method that reconstructs the turbulence in the atmosphere above an earthbound telescope based on measurements of incoming wavefronts at different time steps from a Natural Guide Star. The underlying mathematical problem has been analyzed and a reconstruction method has been developed. In future work, several important questions have to be answered: First, appropriate regularization strategies for the inversion of our tomography operator have to be developed. An option is, e.g., to regularize each subsystem with matrix A_{jm} separately. In a second step, the reconstructions have to be carried out using wavefront sensor measurements instead of wavefronts. Finally, the whole algorithm has to be included into a more realistic simulation environment like OCTOPUS from ESO in order to quantify the gain in imaging quality.

Acknowledgments The work of the authors was partially supported by the Austrian Science Fund (FWF), project F 6805-N36: SFB *Tomography Across the Scales* and by the Austrian Ministry of Research (Hochschulraumstrukturmittel) in the project *Observation oriented Astrophysics in the ELT era*. Both authors would like to thank Roland Wagner and Markus Pöttinger for fruitful discussions.

References

1. D.R. Andersen, S.S. Eikenberry, M. Fletcher, B.L. William Gardhuose, J.-P. Veran, D. Gavel, R. Clare, R.G.L. Jolissaint, R. Julian, W. Rambold, The MOAO system of the IRMOS near-infrared Multi-Object Spectrograph for TMT, in *Proceedings of the SPIE*, vol. 6269 (2006)
2. G. Auzinger, New reconstruction approaches in adaptive optics for extremely large telescopes. PhD thesis, Johannes Kepler University Linz, 2017
3. M. Davison, The ill-conditioned nature of the limited angle tomography problem. SIAM J. Appl. Math. **43**, 428–448 (1983)
4. V. Dhillion, Adaptive optics. http://www.vikdhillon.staff.shef.ac.uk/teaching/phy217/telescopes/phy217/_tel/_adaptive.html
5. E. Diolaiti, A. Baruffolo, M. Bellazzini, V. Biliotti, G. Bregoli, C. Butler, P. Ciliegi, J.-M. Conan, G. Cosentino, S. D'Odorico, B. Delabre, H. Foppiani, T. Fusco, N. Hubin, M. Lombini, E. Marchetti, S. Meimon, C. Petit, C. Robert, P. Rossettini, L. Schreiber, R. Tomelleri, MAORY: a multi-conjugate adaptive optics relay for the E-ELT. Messenger **140**, 28–29 (2010)
6. B. Ellerbroek, C. Vogel, Inverse problems in astronomical adaptive optics. Inverse Probl. **25**, 063001 (2009)
7. B. Ellerbroek, L. Gilles, C. Vogel, A computationally efficient wavefront reconstructor for simulation or multi-conjugate adaptive optics on giant telescopes. Proc. SPIE **4839** (2002)
8. T. Fusco, J.-M. Conan, G. Rousset, L. Mugnier, V. Michau, Optimal wave-front reconstruction strategies for multi conjugate adaptive optics. J. Opt. Soc. Am. A **18**, 2527–2538 (2001)
9. L. Gilles, B. Ellerbroek, Split atmospheric tomography using laser and natural guide stars. J. Opt. Soc. Am. **25**, 2427–2435 (2008)
10. L. Gilles, B. Ellerbroek, C. Vogel, Layer-oriented multigrid wavefront reconstruction algorithms for multi-conjugate adaptive optics. Proc. SPIE **4839** (2002). https://doi.org/10.1117/12.459347
11. L. Gilles, B. Ellerbroek, C. Vogel, Preconditioned conjugate gradient wave-front reconstructors for multiconjugate adaptive optics. Appl. Opt. **42**, 5233–5250 (2003)

12. L. Gilles, B. Ellerbroek, C. Vogel, A comparison of multigrid V-cycle versus Fourier domain preconditioning for laser guide star atmospheric tomography, in *Adaptive Optics: Analysis and Methods/Computational Optical Sensing and Imaging/Information Photonics/Signal Recovery and Synthesis Topical Meetings on CD-ROM, OSA Technical Digest (CD)* (Optical Society of America, Washington, 2007)

13. G.H. Golub, C.F. Van Loan, *Matrix Computations* (The Johns Hopkins University Press, Baltimore, 2013)

14. F. Hammer, F. Sayède, E. Gendron, T. Fusco, D. Burgarella, V. Cayatte, J.-M. Conan, F. Courbin, H. Flores, I. Guinouard, et al., The FALCON concept: multi-object spectroscopy combined with MCAO in near-IR, in *Scientific Drivers for ESO Future VLT/VLTI Instrumentation ESO Astrophysics Symposia* (2002), pp. 139–148

15. T. Helin, M. Yudytskiy, Wavelet methods in multi-conjugate adaptive optics. Inverse Probl. **29**, 085003 (2013)

16. R.A. Horn, C.R. Johnson, *Topics in Matrix Analysis* (Cambridge University Press, Cambridge, 1991)

17. F. Natterer, *The Mathematics of Computerized Tomography* (Wiley, New York, 1986)

18. A. Neubauer, R. Ramlau, A singular-value-type decomposition for the atmospheric tomography operator. SIAM J. Appl. Math. **77**, 838–853 (2017)

19. M. Pöttinger, R. Ramlau, G. Auzinger, A new temporal control approach for SCAO systems. Inverse Probl. **36**, 015002 (2019)

20. M. Puech, H. Flores, M. Lehnert, B. Neichel, T. Fusco, P. Rosati, J.-G. Cuby, G. Rousset, Coupling MOAO with integral field spectroscopy: specifications for the VLT and the E-ELT. Mon. Not. R. Astron. Soc. **390**, 1089–1104 (2008)

21. S. Raffetseder, R. Ramlau, M. Yudytskiy, Optimal mirror deformation for multi conjugate adaptive optics systems. Inverse Probl. **32**, 025009 (2016)

22. R. Ramlau, M. Rosensteiner, An efficient solution to the atmospheric turbulence tomography problem using Kaczmarz iteration. Inverse Probl. **28**, 095004 (2012)

23. R. Ramlau, A. Obereder, M. Rosensteiner, D. Saxenhuber, Efficient iterative tip/tilt reconstruction for atmospheric tomography. Inverse Probl. Sci. Eng. **22**, 1345–1366 (2014)

24. F. Rigaut, B. Ellerbroek, R. Flicker, Principles, limitations and performance of multiconjugate adaptive optics. Proc. SPIE **4007**, 1022–1031 (2000)

25. C. Robert, J.-M. Conan, D. Gratadour, L. Schreiber, T. Fusco, Tomographic wavefront error using multi-LGS constellation sensed with Shack-Hartmann wavefront sensors. J. Opt. Soc. Am. A **27**, A201–A215 (2010)

26. F. Roddier, *Adaptive Optics in Astronomy* (Cambridge University Press, Cambridge, 1999)

27. M.C. Roggemann, B. Welsh, *Imaging Through Turbulence*. Laser and Optical Science and Technology Series (CRC Press, New York, 1996)

28. M. Rosensteiner, Wavefront reconstruction for extremely large telescopes via CuRe with domain decomposition. J. Opt. Soc. Am. A **29**, 2328–2336 (2012)

29. M. Rosensteiner, R. Ramlau, The Kaczmarz algorithm for multi-conjugate adaptive optics with laser guide stars. J. Opt. Soc. Am. A **30**, 1680–1686 (2013)

30. D. Saxenhuber, R. Ramlau, A gradient-based method for atmospheric tomography. Inverse Probl. Imaging **10**, 781–805 (2016)

31. M. Tallon, I. Tallon-Bosc, C. Béchet, F. Momey, M. Fradin, E. Thiébaut, Fractal iterative method for fast atmospheric tomography on extremely large telescopes, in *Proc. SPIE 7736, Adaptive Optics Systems II* (2010), pp. 77360X–77360X–10

32. E. Thiébaut, M. Tallon, Fast minimum variance wavefront reconstruction for extremely large telescopes. J. Opt. Soc. Am. A **27**, 1046–1059 (2010)

33. Q. Yang, C. Vogel, B. Ellerbroek, Fourier domain preconditioned conjugate gradient algorithm for atmospheric tomography. Appl. Opt. **45**, 5281–5293 (2006)

34. M. Yudytskiy, T. Helin, R. Ramlau, Finite element-wavelet hybrid algorithm for atmospheric tomography. J. Opt. Soc. Am. A **31**, 550–560 (2014)

35. M. Zhariy, A. Neubauer, M. Rosensteiner, R. Ramlau, Cumulative wavefront reconstructor for the Shack-Hartman sensor. Inverse Probl. Imaging **5**, 893–913 (2011)

Inverse Problems of Single Molecule Localization Microscopy

Montse Lopez-Martinez, Gwenael Mercier, Kamran Sadiq, Otmar Scherzer, Magdalena Schneider, John C. Schotland, Gerhard J. Schütz, and Roger Telschow

Abstract Single molecule localization microscopy is a recently developed super-resolution imaging technique to visualize structural properties of single cells. The basic principle consists in chemically attaching fluorescent dyes to the molecules, which after excitation with a strong laser may emit light. To achieve superresolution, signals of individual fluorophores are separated in time. In this paper we follow the physical and chemical literature and derive mathematical models describing the propagation of light emitted from dyes in single molecule localization microscopy experiments via Maxwell's equations. This forms the basis of formulating inverse problems related to single molecule localization microscopy. We also show that the current status of reconstruction methods is a simplification of more general inverse problems for Maxwell's equations as discussed here.

M. Lopez-Martinez · M. Schneider · G. J. Schütz
Institute of Applied Physics, TU-Wien, Vienna, Austria
e-mail: lopez-martinez@iap.tuwien.ac.at; schneider@iap.tuwien.ac.at; schuetz@iap.tuwien.ac.at

G. Mercier · R. Telschow
Faculty of Mathematics, University of Vienna, Vienna, Austria
e-mail: gwenael.mercier@univie.ac.at; roger.telschow@univie.ac.at

K. Sadiq (✉)
Johann Radon Institute for Computational and Applied Mathematics (RICAM), Linz, Austria
e-mail: kamran.sadiq@ricam.oeaw.ac.at

O. Scherzer
Faculty of Mathematics, University of Vienna, Vienna, Austria

Johann Radon Institute for Computational and Applied Mathematics (RICAM), Linz, Austria
e-mail: otmar.scherzer@univie.ac.at

J. C. Schotland
Department of Mathematics and Department of Physics, University of Michigan, Ann Arbor, MI, USA
e-mail: schotland@umich.edu

© Springer Nature Switzerland AG 2021
B. Kaltenbacher et al. (eds.), *Time-dependent Problems in Imaging and Parameter Identification*, https://doi.org/10.1007/978-3-030-57784-1_12

1 Introduction

The structure and organization of proteins in cells relate directly to their biological function. Many proteins associate with each other and form functional supramolecular arrangements known as oligomers. Protein oligomers appear in a wide range of crucial biological processes, such as signal transduction, ion transport or immune reactions. The accurate characterization of the supramolecular organization of proteins, including oligomer stoichiometry and its spatial distribution, is fundamental to fully understand these biological processes.

Several tools address the study of the structure of small biological units, most popular ones being x-ray crystallography and, most recently, cryo-electron microscopy, which have been used to characterize the structure of individual isolated proteins with a high level of detail [27, 31]. However, currently these tools cannot be applied for studying quaternary protein assemblies in their native cellular environment, due to a lack in chemical contrast: it is impossible to single out the molecular structures of interest within the plethora of other molecular species. A solution is provided by fluorescence microscopy, where a single protein species is addressed by specific fluorescence labelling directly in the cell. While fluorescence microscopy allows for imaging these labelled structures at a high signal to noise ratio, its resolution is limited to around 200 nm due to the diffraction of light. This prohibits a characterization of oligomeric arrangements with conventional light microscopy, since these structures are smaller than the resolution limit. In summary, the current life sciences are limited by a resolution gap, the upper limit of which is set by the diffraction limit of fluorescence microscopy, the lower limit by the difficulty to interpret crystallography experiments of oligomeric protein complexes.

In principle, the arrival of superresolution microscopy techniques allows to overcome this gap. Virtually all superresolution techniques are based on fluorescence microscopy, and as such have to overcome or circumvent the problem of optical diffraction. A fluorescent label emits light that is imaged by the microscopy system as a blurry dot. This dot of diffracted light is known as the point spread function (PSF). Its size d (the diameter of the essential support of the PSF) is determined by the light wavelength λ and by the numerical aperture (NA) of the objective. The angle θ_{max} is one half of the angular aperture (A). Neglecting lens aberrations, it can be described analytically by an Airy function, where the distance between the maximum and its first minimum is given by (see Eq. (96))

$$d = \frac{\lambda}{2n \sin(\theta_{max})} = \frac{\lambda}{2\,NA} \text{ with n the refractive index of the medium.} \quad (1)$$

The size of the PSF determines the limit of resolution of conventional light microscopy. It was first described by Abbe [1], and it is known as Abbe's limit of diffraction: If two fluorescent labels are closer than the distance d, their PSFs overlap, and they cannot be distinguished from each other. For fluorescence

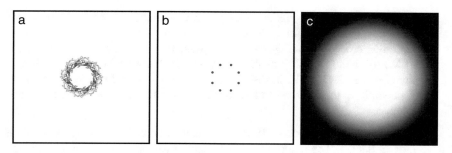

Fig. 1 Illustration of the limit of light diffraction. (**a**) Crystallographic structure of the human NPC (pdb: 5A9Q) [34] viewed with NGL viewer [28]. (**b**) Simplification of the NPC structure showing its eight symmetric units. (**c**) Representation of an ideal, diffraction-limited image of the structure in (**b**)

microscopy, with wavelengths in the visible spectrum and objectives with numerical apertures generally lower than 1.3, this resolution limit is in the order of 200 nm.

For oligomeric protein structures, the distance between their subunits is typically in the range of a few nanometers, far smaller than the diffraction limit. The signals from the individual subunits overlap, and cannot be resolved by conventional light microscopy. A prominent example for such a structure is the nuclear pore complex (NPC), which is a large protein complex located in the nuclear membrane of eukaryotic cells. Its structure is well characterized through electron microscopy [34]. NPCs are composed of around 30 proteins arranged in an 8-fold symmetry forming a pore that regulates the transport across the nuclear membrane. The overall size ranges approximately between 80 to 120 nm depending on the species [22]. As we see in Fig. 1, even if only one protein in each symmetrical subunit is labelled, diffraction leads to one blurry dot as the image of the complex, where we can neither identify the number of subunits nor their spatial arrangement.

Great efforts have been made to overcome this barrier, but it was not until the advent of superresolution microscopy that images with a resolution below the diffraction limit could be obtained. As a key asset, superresolution microscopy techniques circumvent Abbe's limit of diffraction by utilizing photophysical properties of the fluorescent labels: they keep adjacent molecules at different fluorescence states, making it possible to differentiate them from each other. This is achieved using different techniques that can be combined into two general approaches:

1. Techniques that use patterned illumination to control the fluorescence state of the labels, selecting which of them emit at a given moment. This approach includes, among others, Stimulated Emission Depletion (STED) [23, 24, 35], Reversible Saturable Optical Fluorescence Transitions (RESOLFT) [18], Minimal Photon Fluxes (MINFLUX) [4] or Saturated Structured Illumination Microscopy (SSIM) [16] methods.

2. Techniques that use properties of the fluorescent labels to stochastically switch their fluorescent state, so that neighbouring labels do not emit at the same

time. These techniques are commonly termed Single Molecule Localization Microscopy (SMLM) and include, among others, Stochastic Optical Reconstruction Microscopy (STORM) [29], Photoactivated Localization Microscopy (PALM) [6], and DNA- Points Accumulation for Imaging in Nanoscale Topography (DNA-PAINT) [21]. In SMLM, the signals of the individual fluorophores are sequentially localized and used to reconstruct an image with subdiffraction resolution.

In this work, we focus on SMLM techniques, where the working principle is described in Sect. 2. The objective of this paper is to derive mathematical models of light propagation through the imaging device and to formulate associated inverse problems. This sets the base for the formulation of the inverse problems of SMLM, which concerns the localization of the fluorescent labels with high localization precision and the reliable reconstruction of the imaged structures. We show that the currently used imaging workflow in SMLM can be viewed as solving an inverse problem for Maxwell's equations (see Sect. 7). The inverse problem of SMLM has been previously investigated. In [9], a model for light propagation based on Maxwell's equations is proposed and used to localize the positions and strengths of fluorescent dipoles. The model also accounts for the effects of the detection optics and employed a maximum likelihood reconstruction method. The inverse scattering problem with internal sources was investigated in [13], as a means of achieving sub wavelength resolution in SMLM. A local inversion formula was derived and the inverse problem was shown to be well-posed.

2 Single Molecule Localization Microscopy (SMLM)

Principle of SMLM

An SMLM experiment starts with the labelling of the proteins of interest with a fluorophore. There are different strategies for labelling, depending on the type of fluorescent probe, the molecule of interest, and its location in the cell. It should be taken into account that no labelling strategy is perfect, and labelling efficiency will likely be below 100%. In addition, the size of the probe or of the attachment of the linker molecule, in the cases were an intermediate is necessary, can affect the accuracy of the measurement. The influence of these aspects will be addressed in later sections in more detail.

During an SMLM measurement, the experimental conditions are tuned such that most of the fluorophores are in their dark state, and in each frame, a small subset of them is stochastically activated. The active fluorophores are sufficiently isolated from each other so that their PSFs do not overlap. After some time these fluorophores switch to a dark state, and a new subset of fluorophores is stochastically activated. This is repeated thousands of times, to ensure the collection of signals from enough fluorophores. We can see a scheme of an ideal experiment

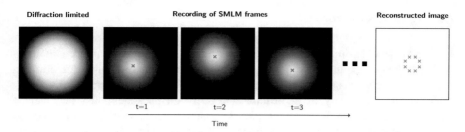

Fig. 2 Scheme of an ideal SMLM experiment. In a classical diffraction-limited image, all the fluorophores are active, and the structure underneath—in this example an NPC—is unresolvable. In contrast, in an SMLM experiment only a sparse subset of fluorophores is active per image. In the first frame ($t = 1$) of this exemplary SMLM movie, only one fluorophore is active, while the others remain in their dark state. The PSF of this fluorophore, can be fitted mathematically, which yields the fluorophore localization. In $t = 2$, the first fluorophore returns to its dark state, and another fluorophore is activated and can now be localized. This is repeated until all fluorophores have been localized. All localizations are collected in a final reconstructed image, which corresponds to the structure shown in Fig. 1b

in Fig. 2. The necessity for sparse labels per image and for enough localizations to reconstruct the structure results in movies with tens of thousands of frames. After data collection, all signals in all frames are fitted individually to obtain the coordinates of the fluorescent probes. All localizations are then collected and used to reconstruct a superresolution image.

The fluorophores used in SMLM are able to spontaneously change their fluorescence state. This property is commonly known as photoswitching or blinking. One dark-bright-dark cycle is usually called a blink. Photoswitching mechanisms are different for different kinds of probes and can be the result of conformational changes in the dye molecule, chemical changes, or binding events. Typically, a combination of light illumination and the choice of special chemical conditions is used for deactivation, i.e. the transitions to a long-lived dark state. The activation, i.e. the transition back from this state, is usually light-induced, although other phenomena may apply (for example, binding events in the case of DNA-PAINT microscopy). An extensive review of available SMLM fluorophores and their properties can be found in [26].

In an ideal experiment, each fluorophore undergoes exactly one blink, in which it emits a high number of photons, and it remains in a dark state for the rest of the measurement. Commonly, however, fluorophores undergo multiple blinking events, or remain inactive during the whole imaging procedure. These non-ideal behaviours directly influence the quality of the final image, and they should be considered when analysing the data, as will be detailed in the next sections. The emission behaviour of a fluorescent label, and therefore the quality of the collected data, will depend largely on the kind of fluorophore used, the labelling strategy followed, and the environmental conditions of the fluorophore [25].

Fitting of Localizations

In an SMLM experiment, thousands of individual frames are recorded. Obtaining the final image requires post-processing of the recorded raw data. All blinking events are analyzed and the positions of the molecules are determined by fitting their signals. A variety of algorithms and software packages exist that can be applied to analyze the data [30]. Often, a Gaussian function is fitted to the detected intensity data using a maximum likelihood or least squares method. The coordinates of the center of the Gaussian peak are then taken as the position of the molecule. Finally, the localizations obtained from all recorded frames are combined to yield the reconstructed image.

Localization Error and Bias

The achievable resolution in SMLM depends on how well the position of a molecule can be estimated by fitting its PSF. The fitting procedure is influenced by various factors of signal quality, including brightness, background noise and the pixel size of the detector. The error in the estimation of the molecule position follows a normal distribution. Its standard deviation is referred to as localization precision σ_{loc}. The mean of the error distribution is the localization accuracy μ_{loc}. In the optimal case, it holds that $\mu_{loc} = 0$, i.e. the estimation is unbiased. However, in practice a bias in the localization procedure may be present, e.g. due to distortions of the PSF. A bias may also arise from the labeling procedure. The size of some labels itself can be rather large, which displaces the position of the fluorophore from the actual molecule of interest by up to tens of nanometers. Various formulas for the estimation of the localization precision σ_{loc} have been proposed in the literature [10]. The theoretical limit for the best achievable localization precision is given by the Cramér-Rao lower bound (CRLB), which is critically dependent on the collected number of photons [32].

Blinking and Overcounting

In SMLM, fluorophores switch between a fluorescent on-state and a non-fluorescent off-state. The transitions between the two states occur stochastically. Ideally, each fluorophore is detected exactly once during the whole imaging procedure, i.e. it is in the on-state in exactly one frame.

However, this is unlikely in a real experimental situation. Due to the stochastic nature of transitions between the states, fluorescent dyes can stay in the on-state for several consecutive frames and, moreover, repeatedly switch between the on- and the off-state. Thus, a single molecule may be detected multiple times.

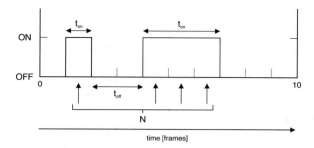

Fig. 3 Exemplary time trace for a fluorophore. The fluorophore can switch between a dark off-state and a bright on-state. Indicated are the on- and off-time (t_{on}, t_{off}), representing the number of consecutive frames the molecule is in its bright or dark state, respectively, and the number of detections N

However, the position coordinates assigned to each detection slightly differ due to localization errors. Hence, it is not possible to distinguish whether localizations belong to one blinking molecule or to different molecules. Overcounting of single protein molecules may also occur as a consequence of non-stoichiometric labeling: Depending on the labeling procedure, a single molecule of interest does not necessarily carry one fluorescent dye only, but may be linked to multiple dyes.

The problem of overcounting is depicted in Fig. 5. Here, individual molecules of the NPC are assumed to be detected multiple times during the imaging procedure, leading to a misrepresentation of the actual structure.

Blinking statistics can be determined experimentally by labeling at sufficiently low concentrations of the dye, so that localizations from individual molecules of interest can be well separated. Analysis of the acquired localization data allows to determine statistics for the number of detections of individual molecules of interest, the duration of emission bursts (t_{on}) and the duration of dark times (t_{off}). In Fig. 3, a schematic of a time trace of occupied states for an individual molecule is shown. An exemplary result for the blinking statistics of a fluorescent dye is depicted in Fig. 4.

A simple approach to account for multiple detections of the same molecule is to merge localizations that occur in close spatial and temporal proximity [2]. However, the results of this method highly depend on the chosen thresholds. Moreover, it cannot account for long-lived dark states. Other post-processing algorithms rely on experimentally derived blinking statistics in order to correct for overcounting. However, care must be taken here, because photophysics of fluorophores, in particular blinking, depends on the local environment of the dye and may likely vary under different experimental conditions [25]. An overview over different methods for correcting overcounting artifacts is given in [5].

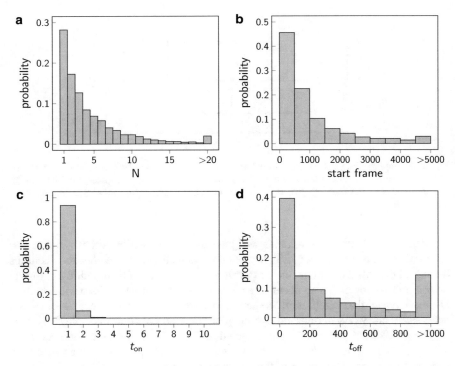

Fig. 4 Experimentally derived blinking statistics for Alexa Fluor 647, a commonly used fluorophore for SMLM. Shown are histograms for the number of detections N of a single fluorophore (**a**), the on-time t_{on} (**b**), and the off-time t_{off} (**c**)

Forward Simulation of SMLM Localization Maps

In the following, we describe the main steps in the simulation of localization maps obtained by a 2D SMLM experiment. Figure 5 shows simulation results of the spatial arrangement for the example of NPCs.

The actual question of interest is the structural arrangement of molecules in a cell membrane. The first step in the simulation is therefore to spread the position of molecules on the region of interest according to the desired distribution. For example, the molecules can be spread randomly, in clusters, or as oligomers of a certain shape. The assigned positions represent ground truth.

As a second step, the simulated molecules are fluorescently labeled. In real experimental conditions, not all molecules of interest are detected: some proteins are not bound to a dye, or the dye is never detected during the imaging time. In the simulations this is accounted for by adjusting the mean labeling efficiency, a parameter in the interval [0, 1] that determines the mean fraction of molecules observed in the experiment. Labeled molecules are selected randomly from all simulated molecules according to the chosen distribution.

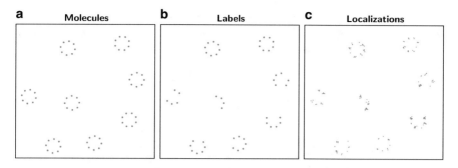

Fig. 5 Simulation of a SMLM experiment for the nuclear pore complex (NPC). (**a**) True spatial arrangement of molecules. The distance between two neighboring molecules of the NPC was set to 40 nm. The molecules are labeled with fluorescent probes. (**b**) with a labeling efficiency of 80%. The SMLM experiment was simulated with a localization precision of $\sigma_{loc} = 5$ nm, and the blinking statistics from Fig. 4. (**c**) Obtained localization map. Due to overcounting and the finite localization precision, individual molecules are observed multiple times

Next, overcounting has to be included in the simulation. As described above, a protein molecule can be detected multiple times during the whole imaging procedure. To account for this in the simulations, the number of detections of each molecule of interest, the frame of its first appearance and the duration of on- and off-times are included. For each labeled molecule, these variables are drawn randomly either from experimentally acquired blinking statistics or from specified theoretical distributions. This allows to assign to each molecule a list of those frames, in which it is detected.

The last step in the simulation is to account for measurement errors. For each detection of a molecule, its true simulated position is displaced by adding a localization error, which is drawn randomly from a normal distribution. The mean and the standard deviation of the error distribution correspond to the localization accuracy and localization precision, respectively. Ideally, the mean value is zero, i.e. the localization is accurate. However, inaccuracy may occur, e.g. due to certain properties of the labeling procedure [10]. Localization precision depends mainly on the collected number of photons and background noise. Typical values that are achieved in SMLM experiments are commonly around 10 nm, but precisions of 1 nm have been claimed.

The final result of the simulation is the localization map, i.e. a list of localization coordinates with the according frame numbers of detection. An exemplary simulated localization map for the NPC is shown in Fig. 5. The obtained localizations are the basis for further analysis.

In the next section we model the experiment mathematically. In order to do so we summarize essential notation first in Table 1.

Table 1 Physical parameters used in the paper and dimensions

Symbol	Description	Reference	Relations	Units
d_0	Maximum thickness of lens			m
d	Thickness of lens (function of height)	Fig. 8		m
f_L	Focal length of the tube lens	Fig. 8		m
f_{obj}	Focal length of the objective	Fig. 8		m
λ	Wavelength	Eq. (1)		m
$n = 1$	Refractive index in vacuum	Eq. (1)		–
n_l	Refractive index of lens	Eq. (88)		–
d	Resolution limit	Eq. (1)		m
NA	Numerical aperture	Eq. (1)	$d = \frac{\lambda}{2\text{NA}}$	–
$\theta_{\max} \in [0, \pi/2)$	Angle of aperture	Eq. (1)	$\text{NA} = \text{n} \sin(\theta_{\max})$	–
ϵ_0	Electric permittivity (vac.)	Eq. (12)		F/m
μ_0	Magnetic permeability (vac.)	Eq. (13)		H/m Henries per m
ω	Wave frequency			$Hz = 1/s$
c	Light speed (vac.)	Eq. (22)		m/s
$\kappa, \kappa_\varepsilon$	Wave number	Eq. (22)	$\kappa = \frac{\omega}{c} = \frac{2\pi}{\lambda}$	$1/m$
χ	Susceptibility	Eq. (21)		–
Ψ	Dipole	Eq. (36)		
Ψ_p, Ψ_s	Dipole components	Eq. (58)		

3 Mathematical Prerequisites

In what follows we summarize some basic mathematical framework:

3.1 Distributions

In order to define distributions (generalized functions) we need to introduce appropriate function spaces first:

Definition 1 The *Schwartz-space* of functions from $\mathbb{R}^n$ to $\mathbb{C}$ is defined as

$$\mathcal{S}(\mathbb{R}^n; \mathbb{C}) := \left\{ \phi \in C^\infty(\mathbb{R}^n; \mathbb{C}) : \text{ for all } \alpha, \beta \in \mathbb{N}_0^n, \|\phi\|_{\alpha,\beta} := \sup_{\mathbf{x} \in \mathbb{R}^n} \left| \mathbf{x}^\alpha \partial^\beta \phi(\mathbf{x}) \right| < \infty \right\}.$$
(2)

Accordingly the *Schwartz-space* of *vector valued functions* is defined by

$$\mathcal{S}(\mathbb{R}^n; \mathbb{C}^m) := \left\{ \mathbf{\Phi} \in C^\infty(\mathbb{R}^n; \mathbb{C}^m) : \Phi_i \in \mathcal{S}(\mathbb{R}^n; \mathbb{C}), i = 1, \dots, m \right\}.$$
(3)

The space of linear functionals $T : S(\mathbb{R}^n; \mathbb{C}^m) \to \mathbb{C}$ for which there exist $k, l \in \mathbb{N}_0$ and some $C > 0$ such that for all $\Phi \in S(\mathbb{R}^n; \mathbb{C}^m)$ the following inequality holds

$$|\langle T, \Phi \rangle| := |T\Phi| \leq C \sum_{i=1}^{m} \sum_{|\alpha| \leq k, |\beta| \leq l} \|\Phi_i\|_{\alpha, \beta} \tag{4}$$

is called *space of tempered distributions* and is denoted by $S'(\mathbb{R}^n; \mathbb{C}^m)$.

Definition 2 (Causal Distribution) A tempered distribution $T \in S'(\mathbb{R}^{n-1} \times \mathbb{R}; \mathbb{C})$ is called *causal* if its support in time is included in $[0, +\infty)$. That is T is causal if and only if for all test functions $\phi \in S(\mathbb{R} \times \mathbb{R}^{n-1}; \mathbb{C})$ which satisfy

$$\phi(\mathbf{x}, t) = 0 \text{ for all } t \geq 0, \ \mathbf{x} \in \mathbb{R}^{n-1},$$

we have

$$\langle T, \phi \rangle = 0.$$

For causal distributions, the quantity $|\langle T, \phi \rangle|$ can be estimated as follows.

Lemma 1 *Let $T \in S'(\mathbb{R}^{n-1} \times \mathbb{R}; \mathbb{C})$ be causal. Then, there exists a constant $C > 0$ (which depends only on l) such that for all test functions $\phi \in S(\mathbb{R}^{n-1} \times \mathbb{R}; \mathbb{C})$ the following estimate holds:*

$$|\langle T, \phi \rangle| \leq C \sup_{\substack{|\alpha| \leq k, |\beta| \leq l \\ \mathbf{x} \in \mathbb{R}^{n-1}}} \sup_{t \geq -1} |(\mathbf{x}, t)^\alpha \partial^\beta \phi(\mathbf{x}, t)|. \tag{5}$$

Proof Let $\varkappa$ be a $C^\infty(\mathbb{R}; \mathbb{R})$ cut-off function, that satisfies $\varkappa(t) = 1$ for $t \geq 0$ and $\varkappa(t) = 0$ for $t \leq -1$. Then, for all test functions $\phi \in S(\mathbb{R}^{n-1} \times \mathbb{R}; \mathbb{C})$, we define $\psi = \varkappa\phi$. For $t \geq 0$, we have $\psi - \phi = 0$, which means, since T is causal, that $\langle T, \psi - \phi \rangle = 0$, or in other words,

$$\langle T, \phi \rangle = \langle T, \psi \rangle. \tag{6}$$

Now, let us use the definition of tempered distribution for T: there exist $k, l \in \mathbb{N}_0$ and $\tilde{C} > 0$ such that

$$|\langle T, \psi \rangle| \leq \tilde{C} \sup_{|\alpha| \leq k, |\beta| \leq l} \sup_{(\mathbf{x}, t) \in \mathbb{R}^{n-1} \times \mathbb{R}} |(\mathbf{x}, t)^\alpha \partial^\beta \psi(\mathbf{x}, t)|. \tag{7}$$

Now, the function ψ has support in $\mathbb{R}^{n-1} \times [-1, +\infty)$, which means that the last inequality can be rewritten as

$$|\langle T, \psi \rangle| \leq C \sup_{|\alpha| \leq k, |\beta| \leq l} \sup_{(\mathbf{x}, t) \in \mathbb{R}^{n-1} \times [-1, +\infty)} |(\mathbf{x}, t)^\alpha \partial^\beta \psi(\mathbf{x}, t)|.$$

Finally, denoting $\beta = (\beta_t, \beta_{x_1}, \cdots, \beta_{x_{n-1}})$, one can expand

$$\partial^\beta \psi = \sum_{i=0}^{\beta_t} \binom{\beta_t}{i} \varkappa^{(i)}(t) \partial^{(\beta_t - i, \beta_{x_1}, \cdots, \beta_{x_{n-1}})} \phi(\mathbf{x}, t).$$

Since the function $\varkappa$ is fixed (independent of ϕ), the quantity

$$C_l = \sup_{i \leqslant l} \sup_{t \in \mathbb{R}} |\varkappa^{(i)}(t)|$$

is finite and independent of ϕ and we have

$$|\partial^\beta \psi(\mathbf{x}, t)| \leqslant 2^l C_l \sup_{|\beta| \leqslant l} |\partial^\beta \phi(\mathbf{x}, t)|$$

and we can finally conclude, taking the supremum on $t \geqslant -1$, that

$$\sup_{|\alpha| \leqslant k, |\beta| \leqslant l} \sup_{(\mathbf{x}, t) \in \mathbb{R}^{n-1} \times [-1, +\infty)} |(\mathbf{x}, t)^\alpha \partial^\beta \psi(\mathbf{x}, t)|$$

$$\leqslant 2^l C_l \sup_{|\alpha| \leqslant k, |\beta| \leqslant l} \sup_{(\mathbf{x}, t) \in \mathbb{R}^{n-1} \times [-1, +\infty)} |(\mathbf{x}, t)^\alpha \partial^\beta \phi(\mathbf{x}, t)|.$$

Since ψ has a support included in $[-1, +\infty)$, one can in the left-hand side of the inequality take the supremum over $t \in \mathbb{R}$. Plugging this inequality into Eq. (7) and recalling Eq. (6), we get Eq. (5). □

We need to notationally differ between δ-distributions in different dimensions:

Definition 3 (δ-Distributions) $\delta : \mathbb{R}^3 \rightarrow \mathbb{R}$ denotes the three-dimensional δ-distribution. $\tilde{\delta} : \mathbb{R} \rightarrow \mathbb{R}$ denotes the one-dimensional δ-distribution. For $r_0 \in \mathbb{R}$, $\tilde{\delta}_{r_0} : \mathbb{R} \rightarrow \mathbb{R}$ is defined by $\tilde{\delta}_{r_0}(r) = \tilde{\delta}(r - r_0)$ for all $r \in \mathbb{R}$. $\tilde{\delta}' : \mathbb{R} \rightarrow \mathbb{R}$ denotes the derivative of the one-dimensional δ-distribution.

3.2   Fourier- and k-Transform

The most important mathematical tool in this paper is the Fourier-transform:

Definition 4 (Temporal Fourier-Transform) Let $T \in \mathcal{S}'(\mathbb{R} \times \mathbb{R}^{n-1}, \mathbb{C})$. We define its *Fourier-transform* $\widehat{T}$ by its action on a test function $\phi \in \mathcal{S}(\mathbb{R} \times \mathbb{R}^{n-1}, \mathbb{C})$

$$\langle \widehat{T}, \phi \rangle := \langle T, \check{\phi} \rangle \tag{8}$$

where

$$\check{\phi}(t) := \frac{1}{\sqrt{2\pi}} \int_{\xi=-\infty}^{\infty} e^{i\xi t} \phi(\xi) \, d\xi.$$

As defined, the operator $T \mapsto \widehat{T}$ is well defined and continuous from $S'(\mathbb{R} \times \mathbb{R}^{n-1}, \mathbb{C})$ into itself [15] with inverse $T \mapsto \check{T}$ with for all test functions ϕ,

$$\left\langle \check{T}, \phi \right\rangle := \left\langle T, \widehat{\phi} \right\rangle$$

where

$$\widehat{\phi}(\xi) := \frac{1}{\sqrt{2\pi}} \int_{t=-\infty}^{\infty} e^{-i\xi t} \phi(t) \, dt$$

is the Fourier-transform on the Schwartz space (it coincides with the one given in Eq. (8)).

The Fourier-transform in spatial variables is called the k-transform:

Definition 5 (k-Transform) Let $\mathbf{r} = \begin{pmatrix} r_1 \\ r_2 \\ r_3 \end{pmatrix} \in \mathbb{R}^3$, and $\mathbf{k} = \begin{pmatrix} k_1 \\ k_2 \\ k_3 \end{pmatrix} \in \mathbb{R}^3$.

- For $i \in \{1, 2, 3\}$, let us denote $\tilde{\mathbf{k}}^{(i)} = \begin{pmatrix} \tilde{k}_1 \\ \tilde{k}_2 \\ \tilde{k}_3 \end{pmatrix} \in \mathbb{R}^3$, where $\tilde{k}_j = \begin{cases} r_j & \text{if } j \neq i \\ k_i & \text{if } j = i \end{cases}$, for $j \in \{1, 2, 3\}$.

The k-transform of the Fourier-transform of $\mathbf{V} : \mathbb{R}^3 \to \mathbb{C}^3$ in direction x_i, is defined by

$$\boxed{\mathcal{F}_i[\mathbf{V}](\tilde{\mathbf{k}}^{(i)}) := \frac{1}{\sqrt{2\pi}} \int_{\mathbb{R}} e^{-ik_i r_i} \mathbf{V}(\mathbf{r}) \, dr_i.}$$

- For $i, \hat{i} \in \{1, 2, 3\}$, let $\tilde{\mathbf{k}}^{(i,\hat{i})} = \begin{pmatrix} \tilde{k}_1 \\ \tilde{k}_2 \\ \tilde{k}_3 \end{pmatrix} \in \mathbb{R}^3$, where $\tilde{k}_j = \begin{cases} r_j & \text{if } j \neq i \text{ and } j \neq \hat{i} \\ k_i & \text{if } j = i \text{ or } j = \hat{i} \end{cases}$, for $j \in \{1, 2, 3\}$.

The k-transform of the Fourier-transform of $\mathbf{V} : \mathbb{R}^3 \to \mathbb{C}^3$ in direction (x_i, x_j) is defined by

$$\boxed{\mathcal{F}_{ij}[\mathbf{V}](\tilde{\mathbf{k}}^{(i,j)}) := \frac{1}{2\pi} \int_{\mathbb{R}} \int_{\mathbb{R}} e^{-i(k_i r_i + k_j r_j)} \mathbf{V}(\mathbf{r}) \, dr_i \, dr_j.}$$

- The k-transform of the Fourier-transform of $\mathbf{V} : \mathbb{R}^3 \to \mathbb{C}^3$ in all three directions is defined by

$$\mathcal{F}[\mathbf{V}](\mathbf{k}) := \mathcal{F}_1[\mathcal{F}_2[\mathcal{F}_3[\mathbf{V}]]](\mathbf{k}) = \frac{1}{(2\pi)^{\frac{3}{2}}} \int_{\mathbb{R}^3} e^{-i\mathbf{k}\cdot\mathbf{r}} \mathbf{V}(\mathbf{r}) d\mathbf{r}.$$

Remark 1 From Definition 3 it follows that for $\mathbf{r}^0 \in \mathbb{R}^3$ fixed

$$\mathcal{F}[\mathbf{r} \to \delta(\mathbf{r} - \mathbf{r}^0)](\mathbf{k}) = \frac{1}{(2\pi)^{\frac{3}{2}}} e^{-i\mathbf{k}\cdot\mathbf{r}^0}. \tag{9}$$

3.3 Coordinate Systems

Definition 6 (Spherical Coordinates) Associated to $\mathbf{r} = \begin{pmatrix} r_1 \\ r_2 \\ r_3 \end{pmatrix} \in \mathbb{R}^3$ is the polar

coordinate representation $(r = |\mathbf{r}|, \theta, \varphi) \in [0, \infty) \times [0, \pi] \times [0, 2\pi)$ such that

$$\mathbf{r} = r \begin{pmatrix} \sin(\theta)\cos(\varphi) \\ \sin(\theta)\sin(\varphi) \\ \cos(\theta) \end{pmatrix}. \tag{10}$$

4 Mathematical Modeling of Light Propagation

We consider an optical single molecule localization microscopy experiment. Therefore a mathematical modeling of the light propagation via Maxwell's equations is appropriate: We consider macroscopic Maxwell's equations (in SI units), in order to model the interaction of the incoming light with the sample. These equations describe the time evolution of the *electric field* $\mathbf{E} : \mathbb{R}^3 \times \mathbb{R} \to \mathbb{R}^3$ and the *magnetic field* $\mathbf{B} : \mathbb{R}^3 \times \mathbb{R} \to \mathbb{R}^3$ for a given *charge density* $\rho : \mathbb{R}^3 \times \mathbb{R} \to \mathbb{R}$ and an *electric current* $\mathbf{J} : \mathbb{R}^3 \times \mathbb{R} \to \mathbb{R}^3$:

$$\nabla_{\mathbf{r}} \cdot \mathbf{D}(\mathbf{r}; t) = \rho(\mathbf{r}, t), \qquad\qquad \mathbf{r} \in \mathbb{R}^3, t \in \mathbb{R}, \tag{11a}$$

$$\nabla_{\mathbf{r}} \cdot \mathbf{B}(\mathbf{r}; t) = 0, \qquad\qquad \mathbf{r} \in \mathbb{R}^3, t \in \mathbb{R}, \tag{11b}$$

$$\nabla_{\mathbf{r}} \times \mathbf{E}(\mathbf{r}, t) = -\partial_t \mathbf{B}(\mathbf{r}; t), \qquad\qquad \mathbf{r} \in \mathbb{R}^3, t \in \mathbb{R} \tag{11c}$$

$$\nabla_{\mathbf{r}} \times \mathbf{H}(\mathbf{r}; t) = \partial_t \mathbf{D}(\mathbf{r}; t) + \mathbf{J}(\mathbf{r}; t), \qquad\qquad \mathbf{r} \in \mathbb{R}^3, t \in \mathbb{R}. \tag{11d}$$

Here

$$\mathbf{D} \equiv \epsilon_0 \mathbf{E} + \mathbf{P} \qquad (12)$$

denotes the *electric displacement* and

$$\mathbf{H} \equiv \frac{1}{\mu_0} \mathbf{B} - \mathbf{M} \qquad (13)$$

denotes the *effective magnetic field*, related to the *electric* and *magnetic polarization fields* $\mathbf{P}$ and $\mathbf{M}$, respectively. All along this paper the differential operators $\nabla_{\mathbf{r}}$, $\nabla_{\mathbf{r}}\cdot$, $\nabla_{\mathbf{r}}\times$, Δ are meant with respect to the variables $\mathbf{r}$. More background on modeling of electromagnetic wave propagation can be found in [20].

In the following we make a series of assumptions for simplifying Maxwell's equations:

4.1 Material Properties

Biological specimens as we are considering in single molecule localization microscopy experiments can be assumed to be non-magnetizable:

Assumption (Non-Magnetizeable Medium) A medium is *non-magnetizable* if

$$\mathbf{M}(\mathbf{r}; t) = 0 \text{ for all } \mathbf{r} \in \mathbb{R}^3, t \in \mathbb{R}. \qquad (14)$$

□

Remark 2 In single molecule localization microscopy experiments, fluorescent dyes are attached to molecules of interest and upon excitation of the probe with a strong laser impulse they emit light. The mathematical modeling of this process is omitted and we are considering only the influence on a macroscopic level, meaning that charge density and currents are induced. A detailed mathematical modeling of the chemical processes would require a modeling with *microsopic* Maxwell's equations, which is omitted here for the sake of simplicity. In a similar context microscopic Maxwell's equations have been considered in Optical Coherence Imaging in [11].

On a macroscopic level, from Eqs. (11a)–(11d) it follows from Assumption 4.1 that

$$\partial_t \rho(\mathbf{r}; t) = - \nabla_{\mathbf{r}} \cdot \mathbf{J}(\mathbf{r}; t) \text{ for all } \mathbf{r} \in \mathbb{R}^3, t \in \mathbb{R}. \qquad (15)$$

Taking into account Assumption 4.1 and combining Eqs. (11c) and (11d) we obtain the *vector Helmholtz equation* for the *electric field* $\mathbf{E}$:

$$\nabla_{\mathbf{r}} \times \nabla_{\mathbf{r}} \times \mathbf{E}(\mathbf{r}; t) + \frac{1}{c^2} \partial_{tt} \mathbf{E}(\mathbf{r}; t) = -\frac{1}{\epsilon_0 c^2} \partial_{tt} \mathbf{P}(\mathbf{r}; t) - \frac{1}{\epsilon_0 c^2} \partial_t \mathbf{J}(\mathbf{r}; t) \text{ for all } \mathbf{r} \in \mathbb{R}^3, t \in \mathbb{R}$$

(16)

where $\mu_0 \epsilon_0 = 1/c^2$, with c being the speed of light in vacuum.

Remark 3 If the right-hand side of Eq. (16) vanishes then $\mathbf{E}$ describes the propagation of the electric field in vacuum. The right-hand side models the interaction of light and matter and the effect of the external charges.

Equation (16) is understood in a distributional sense. That means that for every $\mathbf{\Phi} \in \mathcal{S}(\mathbb{R}^3; \mathbb{R}^3)$ and $\mathbf{\Psi} \in \mathcal{S}(\mathbb{R}; \mathbb{R}^3)$, and with $\mathbf{\Phi} \otimes \mathbf{\Psi} \in \mathcal{S}(\mathbb{R}^3 \times \mathbb{R}; \mathbb{R}^3)$ denoting the vector valued function consisting of componentwise multiplication,

$$\langle \mathbf{E}, (\nabla_{\mathbf{r}} \times \nabla_{\mathbf{r}} \times \mathbf{\Phi}) \otimes \mathbf{\Psi} \rangle + \left\langle \mathbf{E}, \frac{1}{c^2} \mathbf{\Phi} \otimes \partial_{tt} \mathbf{\Psi} \right\rangle = -\left\langle \mathbf{P}, \frac{1}{\epsilon_0 c^2} \mathbf{\Phi} \otimes \partial_{tt} \mathbf{\Psi} \right\rangle + \left\langle \mathbf{J}, \frac{1}{\epsilon_0 c^2} \mathbf{\Phi} \otimes \partial_t \mathbf{\Psi} \right\rangle.$$

(17)

4.2 Linear Optics

In linear optics one assumes a linear relation between the electric polarization $\mathbf{P}$ and the electric field $\mathbf{E}$.

Assumption (Polarization Response Function in Linear Optics) $\mathbf{P}$ and $\mathbf{E}$ satisfy the linear relation,

$$\mathbf{P}(\mathbf{r}; t) = \epsilon_0 \int_{\tau=-\infty}^{\infty} \mathcal{T}(\mathbf{r}; t, \tau) \mathbf{E}(\mathbf{r}, \tau) d\tau,$$

(18)

where $(t; \tau) \to \mathcal{T}(\mathbf{r}; t, \tau) \in \mathbb{R}^{3 \times 3}$ is a matrix valued function that averages the electric field over time. $\mathcal{T}$ is called the (linear) *polarization response function*. For fixed $\mathbf{r}$ the matrix valued function $(t; \tau) \in \mathbb{R}^2 \to \mathcal{T}(\mathbf{r}; t, \tau) \in \mathbb{R}^{3 \times 3}$ is supposed to satisfy the following assumptions:

Causality No polarization is observed before the field is induced, i.e.

$$\mathcal{T}(\mathbf{r}; t, \tau) = 0, \quad \text{for all } t \leq \tau.$$

Time invariance means that $(t; \tau) \to \mathcal{T}(\mathbf{r}; t, \tau)$ is just a function of $t - \tau$. That is, we can write

$$\mathcal{T}(\mathbf{r}; t - \tau) = \mathcal{T}(\mathbf{r}; t, \tau), \quad \text{for all } t, \tau \in \mathbb{R}.$$

Here we use a slight abuse of notation and identify notationally the two functions $\mathcal{T}$ on the left-hand side and right-hand side.

□

Remark 4 Let Assumption 4.2 hold, then $\mathcal{T}(\mathbf{r}; t - \tau) = 0$ for $t \le \tau$.

We now move on to the Fourier-Laplace domain. In order to do so we postulate causality assumptions, which we assume to hold all along the remaining paper:

Assumption (Causality) The functions $\mathbf{J}, \mathbf{P}, \mathbf{E}$ (and thus in turn ρ, $\mathbf{D}$, $\mathbf{H}$) are meaning that

$$\mathbf{J}(t; \mathbf{r}) = \mathbf{P}(t; \mathbf{r}) = \mathbf{E}(t; \mathbf{r}) = 0 \text{ for all } t < 0, \mathbf{r} \in \mathbb{R}^3. \tag{19}$$

□

Let Assumption 4.2 hold (in particular we assume that $\mathcal{T}$ is time invariant and causal), and assume that $\mathbf{J}, \mathbf{P}, \mathbf{E}$ are causal, then from the Fourier convolution theorem it follows that

$$\widehat{\mathbf{P}}(\mathbf{r}; \omega) = \epsilon_0 \chi(\mathbf{r}; \omega) \widehat{\mathbf{E}}(\mathbf{r}; \omega), \quad \text{for all } \mathbf{r} \in \mathbb{R}^3, \omega \in \mathbb{R}, \tag{20}$$

where

$$\chi(\mathbf{r}; \omega) = \int_{\tau=-\infty}^{\infty} \mathcal{T}(\mathbf{r}; \tau) e^{-i\omega\tau} d\tau = \sqrt{2\pi} \widehat{\mathcal{T}}(\mathbf{r}; \omega) \in \mathbb{C}^{3\times3} \text{ for all } \mathbf{r} \in \mathbb{R}^3, \omega \in \mathbb{R}, \tag{21}$$

is called the linear *electric dipolar susceptibility*.

We denote the wave number by

$$\kappa(\omega) := \frac{\omega}{c} \text{ and more general } \kappa_\varepsilon := \kappa_\varepsilon(\omega) = \frac{\omega + i\varepsilon}{c} \text{ for all } \varepsilon > 0. \tag{22}$$

The application of the Fourier-transform to the vector Helmholtz Equation (16) gives the following equation for the Fourier-transform $\widehat{\mathbf{E}} : \mathbb{R}^3 \times \mathbb{R} \to \mathbb{C}^3$ of the electric field:

$$\nabla_{\mathbf{r}} \times \nabla_{\mathbf{r}} \times \widehat{\mathbf{E}}(\mathbf{r}; \omega) - \kappa^2(\omega) \widehat{\mathbf{E}}(\mathbf{r}; \omega) = \frac{1}{\epsilon_0} \kappa^2(\omega) \widehat{\mathbf{P}}(\mathbf{r}; \omega) - \frac{i\omega}{\epsilon_0 c^2} \widehat{\mathbf{J}}(\mathbf{r}; \omega), \quad \text{for all } \mathbf{r} \in \mathbb{R}^3, \omega \in \mathbb{R}$$

and consequently by using Eq. (20) we get

$$\nabla_{\mathbf{r}} \times \nabla_{\mathbf{r}} \times \widehat{\mathbf{E}}(\mathbf{r}; \omega) - \kappa^2(\omega)(\mathbb{I} + \chi(\mathbf{r}; \omega)) \widehat{\mathbf{E}}(\mathbf{r}; \omega) = -\frac{i\omega}{\epsilon_0 c^2} \widehat{\mathbf{J}}(\mathbf{r}; \omega) \quad \text{for all } \mathbf{r} \in \mathbb{R}^3, \omega \in \mathbb{R}, \tag{23}$$

where $\mathbb{I} \in \mathbb{R}^{3\times3}$ is the identity matrix.

4.3   Isotropic Media

Additional simplifications of Maxwell's equations can be made when the medium is assumed to be *isotropic*:

Assumption (Isotropic Medium) Let Assumptions 4.1 and 4.2 hold. The medium is *isotropic* if the susceptibility is a multiple of the identity, that is it can be written as $\chi(\mathbf{r}; t)\mathbb{I} \in \mathbb{C}^{3\times 3}$ with $\chi(\mathbf{r}; t) \in \mathbb{C}$. With a slight abuse of notation, we identify the diagonal matrix and the diagonal entry.   □

4.4   Homogeneous Material

We consider an isotropic, non magnetizable material with a linear polarization response (that is, Assumptions 4.1, 4.2, and 4.3 are satisfied), which in addition is *homogeneous*:

Assumption (Homogeneous Material) An isotropic, non magnetizable material with a linear polarization response is *homogeneous* if $\chi \equiv 0$.   □

For a homogeneous material (that is $\chi \equiv 0$) it follows from Eq. (23) that

$$-\frac{i\omega}{\epsilon_0 c^2}\widehat{\mathbf{J}}(\mathbf{r}; \omega) = \nabla_{\mathbf{r}} \times \nabla_{\mathbf{r}} \times \widehat{\mathbf{E}}(\mathbf{r}; \omega) - \kappa^2(\omega)\widehat{\mathbf{E}}(\mathbf{r}; \omega). \tag{24}$$

Thus, by using the vector identity

$$\nabla_{\mathbf{r}} \times \nabla_{\mathbf{r}} \times \widehat{\mathbf{E}} = \nabla_{\mathbf{r}} \nabla_{\mathbf{r}} \cdot \widehat{\mathbf{E}} - \Delta_{\mathbf{r}} \widehat{\mathbf{E}},$$

we get from Eq. (24)

$$-\frac{i\omega}{\epsilon_0 c^2}\widehat{\mathbf{J}}(\mathbf{r}; \omega) = \nabla_{\mathbf{r}} \nabla_{\mathbf{r}} \cdot \widehat{\mathbf{E}}(\mathbf{r}; \omega) - \Delta_{\mathbf{r}} \widehat{\mathbf{E}}(\mathbf{r}; \omega) - \kappa^2(\omega)\widehat{\mathbf{E}}(\mathbf{r}; \omega). \tag{25}$$

Now, by using Eq. (12) and the assumption on homogeneity, $\chi \equiv 0$, which together with Eq. (20) implies that $\mathbf{P} \equiv 0$, we get

$$\mathbf{D} = \epsilon_0 \mathbf{E} + \mathbf{P} = \epsilon_0 \mathbf{E}.$$

This, together with Eq. (25) shows that

$$-\frac{i\omega}{\epsilon_0 c^2}\widehat{\mathbf{J}}(\mathbf{r}; \omega) = \frac{1}{\epsilon_0} \nabla_{\mathbf{r}} \nabla_{\mathbf{r}} \cdot \widehat{\mathbf{D}}(\mathbf{r}; \omega) - \Delta_{\mathbf{r}} \widehat{\mathbf{E}}(\mathbf{r}; \omega) - \kappa^2(\omega)\widehat{\mathbf{E}}(\mathbf{r}; \omega). \tag{26}$$

Now, by using Eq. (11a) in Fourier domain we get from Eq. (26)

$$-\frac{i\omega}{\epsilon_0 c^2}\widehat{\mathbf{J}}(\mathbf{r};\omega) = \frac{1}{\epsilon_0}\nabla_{\mathbf{r}}\,\widehat{\rho}(\mathbf{r};\omega) - \Delta_{\mathbf{r}}\widehat{\mathbf{E}}(\mathbf{r};\omega) - \kappa^2(\omega)\widehat{\mathbf{E}}(\mathbf{r};\omega). \tag{27}$$

Finally, by using Eq. (15) in Fourier domain,

$$i\omega\widehat{\rho} = -\nabla_{\mathbf{r}}\cdot\widehat{\mathbf{J}}(\mathbf{r};\omega) \tag{28}$$

in Eq. (27) we get

$$-\frac{i\omega}{\epsilon_0 c^2}\widehat{\mathbf{J}}(\mathbf{r};\omega) = -\frac{1}{i\omega\epsilon_0}\nabla_{\mathbf{r}}\nabla_{\mathbf{r}}\cdot\widehat{\mathbf{J}}(\mathbf{r};\omega) - \Delta_{\mathbf{r}}\widehat{\mathbf{E}}(\mathbf{r};\omega) - \kappa^2(\omega)\widehat{\mathbf{E}}(\mathbf{r};\omega).$$

In other words, we have for every $\mathbf{r}\in\mathbb{R}^3$, $\omega\in\mathbb{R}$

$$
\begin{aligned}
\Delta_{\mathbf{r}}\widehat{\mathbf{E}}(\mathbf{r};\omega) + \kappa^2(\omega)\widehat{\mathbf{E}}(\mathbf{r};\omega) &= \frac{i}{\epsilon_0}\left(\frac{\omega}{c^2} + \frac{1}{\omega}\nabla_{\mathbf{r}}\nabla_{\mathbf{r}}\cdot\right)\widehat{\mathbf{J}}(\mathbf{r};\omega) \\
&= \frac{i\omega}{\epsilon_0 c^2}\widehat{\mathbf{J}}(\mathbf{r};\omega) + \frac{1}{\epsilon_0}\nabla_{\mathbf{r}}\,\widehat{\rho}(\mathbf{r};\omega).
\end{aligned}
\tag{29}
$$

For any $\tau\in\mathbb{R}$, a solution of the nonhomogeneous Eq. (29) is given by (see [33]):

$$\widehat{\mathbf{E}}(\mathbf{r};\omega) = \tau\widehat{\mathbf{E}}^+(\mathbf{r};\omega) + (1-\tau)\widehat{\mathbf{E}}^-(\mathbf{r};\omega) \quad \text{for all} \quad \mathbf{r}\in\mathbb{R}^3,\ \omega\in\mathbb{R}, \quad \text{where}$$

$$\widehat{\mathbf{E}}^\pm(\mathbf{r};\omega) := -\int_{\mathbb{R}^3}\mathcal{G}_\omega^\pm(\mathbf{r},\mathbf{r}')\left(\frac{i\omega}{\epsilon_0 c^2}\widehat{\mathbf{J}}(\mathbf{r}';\omega) + \frac{1}{\epsilon_0}\nabla_{\mathbf{r}}\,\widehat{\rho}(\mathbf{r}';\omega)\right)d\mathbf{r}' \tag{30}$$

with Green's functions:

$$\mathcal{G}_\omega^\pm(\mathbf{r},\mathbf{r}') = \frac{e^{\pm i\kappa(\omega)|\mathbf{r}-\mathbf{r}'|}}{4\pi\,|\mathbf{r}-\mathbf{r}'|}. \tag{31}$$

The physically meaningful solution is, as we motivate below, a convolution with the *retarded Green's* function $\mathcal{G}_\omega^+$: That is, the retarded solution of the Helmholtz Equation (23) is given by Eq. (30) with $\tau = 1$ (see [33]):

$$\boxed{\widehat{\mathbf{E}}(\mathbf{r};\omega) = -\int_{\mathbb{R}^3}\mathcal{G}_\omega^+(\mathbf{r},\mathbf{r}')\left(\frac{i\omega}{\epsilon_0 c^2}\widehat{\mathbf{J}}(\mathbf{r}';\omega) + \frac{1}{\epsilon_0}\nabla_{\mathbf{r}}\,\widehat{\rho}(\mathbf{r}';\omega)\right)d\mathbf{r}'.} \tag{32}$$

Remark 5 With a slight abuse of notation we identify $\mathcal{G}_\omega^+$ with $\mathcal{G}_\omega$ and $\widehat{\mathbf{E}}_\omega^+$ with $\widehat{\mathbf{E}}_\omega$, since we are only interested in the retarded solutions.

5   Attenuating Solution and Initial Conditions

Definition 7 (Attenuating and Causal Solution of Eq. (23)) Let $\varepsilon > 0$ and $\kappa_\varepsilon(\omega) = \frac{\omega + i\varepsilon}{c}$ as defined in Eq. (22).

- Then, we call $\widehat{\mathbf{E}}_\varepsilon$ the *approximate attenuating solution* of Eq. (23) if it satisfies the equation

$$\nabla_{\mathbf{r}} \times \nabla_{\mathbf{r}} \times \widehat{\mathbf{E}}_\varepsilon(\mathbf{r}; \omega) - \kappa_\varepsilon^2(\omega)(\mathbb{I} + \chi(\mathbf{r}; \omega))\widehat{\mathbf{E}}_\varepsilon(\mathbf{r}; \omega) = -\frac{i\omega - \varepsilon}{\epsilon_0 c^2}\widehat{\mathbf{J}}_\varepsilon(\mathbf{r}; \omega),$$

(33)

where non-attenuating solution and attenuating solution is related by Eq. (35).
- We call $\widehat{\mathbf{E}}_\varepsilon$ a *causal attenuating solution* of Eq. (23) if $\mathbf{E}_\varepsilon$ (the inverse Fourier-transform of $\widehat{\mathbf{E}}_\varepsilon$) is a *causal* distribution.

In the following we show that $\widehat{\mathbf{E}}_\varepsilon$ approximates the retarded solution of the vector-Helmholtz Equation (32) in a distributional sense:

Theorem 1 *For every $\varepsilon > 0$, let $\widehat{\mathbf{E}}_\varepsilon$ be the solution of Eq. (33), the causal attenuating wave equation, and let $\widehat{\mathbf{E}}$ be the retarded solution of Eq. (23), which is given by Eq. (32), then*

$$\widehat{\mathbf{E}}_\varepsilon \xrightarrow[\varepsilon \to 0]{S'} \widehat{\mathbf{E}}.$$

(34)

Proof We define for all $t \in \mathbb{R}$, $\mathbf{r} \in \mathbb{R}^3$,

$$\mathbf{E}_\varepsilon(\mathbf{r}; t) = \alpha_\varepsilon(t)\mathbf{E}(\mathbf{r}; t) \text{ where } \alpha_\varepsilon(t) := e^{-\varepsilon t}.$$

(35)

Because $\mathbf{E}$ is causal, $\mathbf{E}_\varepsilon$ is a tempered distribution and since $\widehat{\mathbf{E}}$ is a solution of Eq. (23), it follows that for all $\varepsilon > 0$, $\widehat{\mathbf{E}}_\varepsilon$ is a solution of Eq. (33) and in particular it is also causal. We show that $\mathbf{E}_\varepsilon \xrightarrow[\varepsilon \to 0]{S'} \mathbf{E}$ and because the Fourier transform (see Eq. (8)) is a bounded operator on $S'(\mathbb{R}^3 \times \mathbb{R}; \mathbb{R}^3)$ (see [15, Theorem 5.17]), the assertion, Eq. (34), then follows.

To prove that $\mathbf{E}_\varepsilon \xrightarrow[\varepsilon \to 0]{S'} \mathbf{E}$, we need to show that for all $\mathbf{\Phi} \in S(\mathbb{R}^3 \times \mathbb{R}; \mathbb{R}^3)$, $\langle \mathbf{E}_\varepsilon, \mathbf{\Phi} \rangle \to \langle \mathbf{E}, \mathbf{\Phi} \rangle$. Noting that $\langle \mathbf{E}_\varepsilon, \mathbf{\Phi} \rangle = \langle \mathbf{E}, \alpha_\varepsilon \mathbf{\Phi} \rangle$, we therefore need to show that $\langle \mathbf{E}, \mathbf{\Phi} - \alpha_\varepsilon \mathbf{\Phi} \rangle \to 0$. Lemma 1 shows that, because $\mathbf{E}$ is causal, one can write

$$|\langle \mathbf{E}, \mathbf{\Phi} - \alpha_\varepsilon \mathbf{\Phi} \rangle| \leqslant C \sup_{\alpha \leqslant k, \beta \leqslant l} \sup_{t \geqslant -1} |t^\alpha \partial_t^\beta (\mathbf{\Phi} - \alpha_\varepsilon \mathbf{\Phi})(t)|.$$

Now, note that for all $\beta \in \mathbb{N}_0$ and all $t \in \mathbb{R}$,

$$\partial_t^\beta \left[(e^{-\varepsilon t} - 1)\Phi(t) \right] = -\partial_t^\beta \Phi(t) + \sum_{i=0}^{\beta} \binom{\beta}{i} (-\varepsilon)^i e^{-\varepsilon t} \partial_t^{\beta-i} \Phi(t)$$

$$= (e^{-\varepsilon t} - 1)\partial_t^\beta \Phi(t) + \varepsilon A_\varepsilon(t) e^{-\varepsilon t}.$$

where A_ε is a polynomial (with coefficients uniformly bounded with ε) in the derivatives of Φ up to the order $\beta - 1$. Since the derivatives of Φ are Schwartz functions, $\sup_{t \geq -1} |t^\alpha A_\varepsilon(t)|$ is then uniformly bounded in ε, which implies

$$\lim_{\varepsilon \to 0} \sup_{t \geq -1} \left| t^\alpha \varepsilon A_\varepsilon(t) e^{-\varepsilon t} \right| = 0.$$

Now, $B(t) := \partial_t^\beta \Phi$ is also a Schwartz function, which means that for every $k \in \mathbb{N}_0$ there exists C_k such that $\sup_t |(t^{k+2} + 1)B(t)| \leq C_k$. It then follows that for all $t \geq -1$,

$$\left| t^\alpha (e^{-\varepsilon t} - 1)B(t) \right|$$

$$= \left| \frac{t^\alpha}{t^{\alpha+2} + 1} (e^{-\varepsilon t} - 1)(t^{\alpha+2} + 1)B(t) \right| \leq C_\alpha \sup_{t \geq -1} \left| \frac{t^\alpha(e^{-\varepsilon t} - 1)}{t^{\alpha+2} + 1} \right|,$$

where the last supremum converges to zero with $\varepsilon \to 0$. Therefore we conclude that

$$\lim_{\varepsilon \to 0} \sup_{t \geq -1} \left| t^\alpha \partial_t^\beta ((e^{-\varepsilon t} - 1)\Phi(t)) \right| = 0,$$

which means $\langle \mathbf{E}, \Phi - \alpha_\varepsilon \Phi \rangle \to 0$. $\square$

5.1 Dipoles

The emission of fluorescent dyes will be modeled as dipoles.

Definition 8 (Emitting Dipole) An *emitting dipole* is a vector $\Psi = \begin{pmatrix} \Psi_1 & \Psi_2 & \Psi_3 \end{pmatrix}^T$, which is associated to a point $\mathbf{r}^\Psi$ in space; $|\Psi|$ is called *charge intensity* and $\frac{\Psi}{|\Psi|}$ can be represented in spherical coordinates $(\theta_m, \varphi_m) \in \mathbb{S}^2$. Both notations are used synonymously and called the *orientation* of the emitting dipole. That is

$$\Psi = \begin{pmatrix} \Psi_1 \\ \Psi_2 \\ \Psi_3 \end{pmatrix} = \begin{pmatrix} |\Psi| \sin(\theta_m) \cos(\varphi_m) \\ |\Psi| \sin(\theta_m) \sin(\varphi_m) \\ |\Psi| \cos(\theta_m) \end{pmatrix}. \tag{36}$$

The *limiting* density of a dipole at position $\mathbf{r}^{\Psi} = \begin{pmatrix} 0 \\ 0 \\ r_3^{\Psi} \end{pmatrix} \in \mathbb{R}^3$ is defined as a

generalized function in space

$$\widehat{\rho}(\mathbf{r}) := |\Psi| \lim_{s \to 0^+} \frac{\delta_{\mathbf{r}^{\Psi} - s \frac{\Psi}{|\Psi|}}(\mathbf{r}) - \delta_{\mathbf{r}^{\Psi} + s \frac{\Psi}{|\Psi|}}(\mathbf{r})}{2s} \text{ for all } \mathbf{r} \in \mathbb{R}^3. \tag{37}$$

That is, in mathematical terms, the dipole charge is the directional derivative of a three-dimensional δ-distribution in direction $-\frac{\Psi}{|\Psi|}$. Moreover, we denote by

$$\widehat{\mathbf{J}}(\mathbf{r}; \omega) := i\omega\Psi\delta(\mathbf{r} - \mathbf{r}^{\Psi}) \tag{38}$$

the *dipole current* (which is frequency dependent).

In what follows we assume that the emitting dipole is a unit-vector (that is $|\Psi| = 1$), which simplifies the considerations and the notation.

Lemma 2 *Let $\widehat{\mathbf{J}}$ and $\widehat{\rho}$ be as defined in Eqs. (38) and (37), respectively and satisfy Eq. (28). Then*

$$\widehat{\mathbf{R}}(\mathbf{r}; \omega) := \frac{i\omega}{c^2}\widehat{\mathbf{J}}(\mathbf{r}; \omega) + \nabla_{\mathbf{r}}\widehat{\rho}(\mathbf{r}; \omega) \tag{39}$$

satisfies

$$\widehat{\mathbf{R}}(\mathbf{r}; \omega) = -\frac{\omega^2\Psi}{c^2}\delta(\mathbf{r} - \mathbf{r}^{\Psi})$$

$$- \begin{pmatrix} \Psi_1\tilde{\delta}''(x_1)\tilde{\delta}(x_2)\tilde{\delta}(x_3) + \Psi_2\tilde{\delta}'(x_1)\tilde{\delta}'(x_2)\tilde{\delta}(x_3) + \Psi_3\tilde{\delta}'(x_1)\tilde{\delta}(x_2)\tilde{\delta}'(x_3) \\ \Psi_1\tilde{\delta}'(x_1)\tilde{\delta}'(x_2)\tilde{\delta}(x_3) + \Psi_2\tilde{\delta}(x_1)\tilde{\delta}''(x_2)\tilde{\delta}(x_3) + \Psi_3\tilde{\delta}(x_1)\tilde{\delta}'(x_2)\tilde{\delta}'(x_3) \\ \Psi_1\tilde{\delta}'(x_1)\tilde{\delta}(x_2)\tilde{\delta}'(x_3) + \Psi_2\tilde{\delta}(x_1)\tilde{\delta}'(x_2)\tilde{\delta}'(x_3) + \Psi_3\tilde{\delta}(x_1)\tilde{\delta}(x_2)\tilde{\delta}''(x_3) \end{pmatrix},$$
$$\tag{40}$$

where $(\mathbf{x}, x_3)^T := \mathbf{r} - \mathbf{r}^{\Psi}$, where $\mathbf{r}^{\Psi}$ denotes the dipole position.

Proof Taking into account that the three-dimensional δ-distribution can be written as

$$\delta_{\mathbf{r}^{\Psi} \pm s\Psi}(\mathbf{r}) = \prod_{j=1}^{3} \tilde{\delta}_{(\mathbf{r}^{\Psi})_j \pm s\Psi_j}(\mathbf{r}_j) = \prod_{j=1}^{3} \tilde{\delta}(\mathbf{r}_j - (\mathbf{r}^{\Psi})_j \mp s\Psi_j)$$

we find

$$\widehat{\rho}(\mathbf{r}; \omega) = -\sum_{i=1}^{3} \Psi_i(\tilde{\delta}_{(\mathbf{r}^{\Psi})_i})'(\mathbf{r}_i) \prod_{j \neq i} \tilde{\delta}_{(\mathbf{r}^{\Psi})_j}(\mathbf{r}_j) = -\sum_{i=1}^{3} \Psi_i\tilde{\delta}'((\mathbf{r} - \mathbf{r}^{\Psi})_i) \prod_{j \neq i} \tilde{\delta}((\mathbf{r} - \mathbf{r}^{\Psi})_j)$$

and we get

$$- \nabla_{\mathbf{r}} \widehat{\rho}(\mathbf{r}; \omega)$$

$$= \begin{pmatrix} \Psi_1 \tilde{\delta}''(x_1) \tilde{\delta}(x_2) \tilde{\delta}(x_3) + \Psi_2 \tilde{\delta}'(x_1) \tilde{\delta}'(x_2) \tilde{\delta}(x_3) + \Psi_3 \tilde{\delta}'(x_1) \tilde{\delta}(x_2) \tilde{\delta}'(x_3) \\ \Psi_1 \tilde{\delta}'(x_1) \tilde{\delta}'(x_2) \tilde{\delta}(x_3) + \Psi_2 \tilde{\delta}(x_1) \tilde{\delta}''(x_2) \tilde{\delta}(x_3) + \Psi_3 \tilde{\delta}(x_1) \tilde{\delta}'(x_2) \tilde{\delta}'(x_3) \\ \Psi_1 \tilde{\delta}'(x_1) \tilde{\delta}(x_2) \tilde{\delta}'(x_3) + \Psi_2 \tilde{\delta}(x_1) \tilde{\delta}'(x_2) \tilde{\delta}'(x_3) + \Psi_3 \tilde{\delta}(x_1) \tilde{\delta}(x_2) \tilde{\delta}''(x_3) \end{pmatrix}.$$

$$(41)$$

On the other hand

$$-\nabla \cdot \widehat{\mathbf{J}}(\mathbf{r}; \omega) = -i\omega \nabla \cdot (\boldsymbol{\Psi}\delta(\mathbf{r} - \mathbf{r}^{\boldsymbol{\Psi}}))$$

$$= -i\omega \sum_{i=1}^{3} \Psi_i \tilde{\delta}'((\mathbf{r} - \mathbf{r}^{\boldsymbol{\Psi}})_i) \prod_{j \neq i} \tilde{\delta}((\mathbf{r} - \mathbf{r}^{\boldsymbol{\Psi}})_j) = i\omega \widehat{\rho}(\mathbf{r}),$$

and thus Eq. (28) is satisfied.

Moreover, using Eq. (41) in Eq. (28) gives Eq. (40). $\qquad\square$

In the following we calculate the solution $\widehat{\mathbf{E}}$ of Eq. (30), similar as in [12].

The following lemma and its proof are based on [12].

Lemma 3 *Let $\widehat{\mathbf{E}}$ as in Eq. (32) be the retarded solution of Eq. (33) at fixed frequency ω. In what follows we omit therefore the dependency of ω and write $\widehat{\mathbf{E}}(\mathbf{r}) := \widehat{\mathbf{E}}(\mathbf{r}; \omega)$.*

Moreover, let the medium be isotropic, non magnetizable, homogeneous and have a linear polarization response (that is, $\chi \equiv 0$).

As above we assume that a dipole $\boldsymbol{\Psi} \in \mathbb{R}^3$ is located at position $\mathbf{r}^{\boldsymbol{\Psi}} = (0, 0, r_3^{\boldsymbol{\Psi}})^T$.

Moreover, for all $\varepsilon > 0$ let κ_ε be as in Eq. (22) and we define for fixed $k_1, k_2 \in \mathbb{R}$

$$q := \lim_{\varepsilon \to 0^+} q_\varepsilon \text{ where } q_\varepsilon := a_\varepsilon + ib_\varepsilon := \sqrt{\kappa_\varepsilon^2 - k_1^2 - k_2^2} \quad \text{with} \quad b_\varepsilon > 0 \qquad (42)$$

(that is q_ε is the complex root with positive imaginary part). Let now $\mathbf{r} \in \mathbb{R}^3$ be such that $r_3 - r_3^{\boldsymbol{\Psi}} \geq 0$, then

$$\widehat{\mathbf{E}}(\mathbf{r}) = -\frac{1}{2\pi} \frac{1}{\epsilon_0} \mathcal{F}_{12}^{-1} \left[(k_1, k_2) \mapsto \Psi_3 \mathbf{e}_3 \tilde{\delta}(r_3 - r_3^{\boldsymbol{\Psi}}) + \frac{i e^{iq(r_3 - r_3^{\boldsymbol{\Psi}})}}{2q} (\boldsymbol{\Psi} \times \mathbf{k}_q) \times \mathbf{k}_q \right] (r_1, r_2),$$

$$(43)$$

where $\mathbf{k}_q = (k_1, k_2, q)^T$.

Proof First let $\varepsilon > 0$, and we prove an identity of the form Eq. (43) for $\widehat{\mathbf{E}}_\varepsilon$. We note that

$$\mathcal{F}[\nabla_{\mathbf{r}} \times \nabla_{\mathbf{r}} \times \widehat{\mathbf{E}}_\varepsilon](\mathbf{k}) = -(\mathcal{F}[\widehat{\mathbf{E}}_\varepsilon](\mathbf{k}) \times \mathbf{k}) \times \mathbf{k} \text{ for all } \mathbf{k} \in \mathbb{R}^3.$$

Thus from Eq. (23) with $\chi \equiv 0$ it follows by applying the k-transform, and by using Eqs. (38), (22) and (9) that

$$- (\mathcal{F}[\widehat{\mathbf{E}}_\varepsilon](\mathbf{k}) \times \mathbf{k}) \times \mathbf{k} - \kappa_\varepsilon^2 \mathcal{F}[\widehat{\mathbf{E}}_\varepsilon](\mathbf{k}) = -\frac{i\omega - \varepsilon}{\epsilon_0 c^2} \mathcal{F}[\widehat{\mathbf{J}}_\varepsilon](\mathbf{k}) = \frac{\kappa_\varepsilon^2}{(2\pi)^{\frac{3}{2}} \epsilon_0} \mathbf{\Psi} e^{-ik_3 r_3^\Psi}.$$

(44)

Elementary calculation rules for $\times$ provide that

$$(\mathbf{v} \times \mathbf{k}) \times \mathbf{k} = (\mathbf{k} \cdot \mathbf{v})\mathbf{k} - |\mathbf{k}|^2 \mathbf{v} \text{ for all } \mathbf{v}, \mathbf{k} \in \mathbb{R}^3,$$

(45)

which, by application to $\mathbf{v} = \mathcal{F}[\widehat{\mathbf{E}}_\varepsilon](\mathbf{k})$ and $\mathbf{v} = \mathbf{\Psi}$, respectively, shows that

$$|\mathbf{k}|^2 \mathcal{F}[\widehat{\mathbf{E}}_\varepsilon](\mathbf{k}) = -(\mathcal{F}[\widehat{\mathbf{E}}_\varepsilon](\mathbf{k}) \times \mathbf{k}) \times \mathbf{k} + (\mathbf{k} \cdot \mathcal{F}[\widehat{\mathbf{E}}_\varepsilon](\mathbf{k}))\mathbf{k} \text{ and}$$
$$|\mathbf{k}|^2 \mathbf{\Psi} = -(\mathbf{\Psi} \times \mathbf{k}) \times \mathbf{k} + (\mathbf{k} \cdot \mathbf{\Psi})\mathbf{k}.$$

(46)

Therefore, by multiplying Eq. (44) with $|\mathbf{k}|^2$ and using Eq. (46), it follows that

$$(\kappa_\varepsilon^2 - |\mathbf{k}|^2)(\mathcal{F}[\widehat{\mathbf{E}}_\varepsilon](\mathbf{k}) \times \mathbf{k}) \times \mathbf{k} - \kappa_\varepsilon^2 (\mathbf{k} \cdot \mathcal{F}[\widehat{\mathbf{E}}_\varepsilon](\mathbf{k}))\mathbf{k}$$
$$= \frac{\kappa_\varepsilon^2}{(2\pi)^{\frac{3}{2}} \epsilon_0} e^{-ik_3 r_3^\Psi} [-(\mathbf{\Psi} \times \mathbf{k}) \times \mathbf{k} + (\mathbf{k} \cdot \mathbf{\Psi})\mathbf{k}].$$

(47)

Since $\mathbf{k}$ and $(\mathbf{v} \times \mathbf{k}) \times \mathbf{k}$ are orthogonal, it follows from Eq. (47) that:

$$(\mathcal{F}[\widehat{\mathbf{E}}_\varepsilon](\mathbf{k}) \cdot \mathbf{k})\mathbf{k} = -\frac{1}{(2\pi)^{\frac{3}{2}}} \frac{1}{\epsilon_0} e^{-ik_3 r_3^\Psi} (\mathbf{\Psi} \cdot \mathbf{k})\mathbf{k}$$
$$- \frac{1}{(2\pi)^{\frac{3}{2}}} \frac{1}{\epsilon_0} e^{-ik_3 r_3^\Psi} \left(|\mathbf{k}|^2 \mathbf{\Psi} + (\mathbf{\Psi} \times \mathbf{k}) \times \mathbf{k}\right),$$
$$(|\mathbf{k}|^2 - \kappa_\varepsilon^2)(\mathcal{F}[\widehat{\mathbf{E}}_\varepsilon](\mathbf{k}) \times \mathbf{k}) \times \mathbf{k} = \frac{\kappa_\varepsilon^2}{(2\pi)^{\frac{3}{2}} \epsilon_0} e^{-ik_3 r_3^\Psi} (\mathbf{\Psi} \times \mathbf{k}) \times \mathbf{k}.$$

Inserting these two identities into Eq. (46) and noting that since κ_ε is not real, one can divide by $|\mathbf{k}|^2 - \kappa_\varepsilon^2$, yields

$$|\mathbf{k}|^2 \mathcal{F}[\widehat{\mathbf{E}}_\varepsilon](\mathbf{k}) = -\frac{1}{(2\pi)^{\frac{3}{2}}} \frac{1}{\epsilon_0} e^{-ik_3 r_3^\Psi} \left(|\mathbf{k}|^2 \mathbf{\Psi} + (\mathbf{\Psi} \times \mathbf{k}) \times \mathbf{k} + \frac{\kappa_\varepsilon^2}{|\mathbf{k}|^2 - \kappa_\varepsilon^2} (\mathbf{\Psi} \times \mathbf{k}) \times \mathbf{k}\right)$$
$$= -\frac{1}{(2\pi)^{\frac{3}{2}}} \frac{1}{\epsilon_0} e^{-ik_3 r_3^\Psi} \left(|\mathbf{k}|^2 \mathbf{\Psi} + \frac{|\mathbf{k}|^2}{|\mathbf{k}|^2 - \kappa_\varepsilon^2} (\mathbf{\Psi} \times \mathbf{k}) \times \mathbf{k}\right),$$

such that

$$\mathcal{F}[\widehat{\mathbf{E}}_\varepsilon](\mathbf{k}) = -\frac{1}{(2\pi)^{\frac{3}{2}}} \frac{1}{\epsilon_0} e^{-ik_3 r_3^\Psi} \left(\mathbf{\Psi} + \frac{(\mathbf{\Psi} \times \mathbf{k}) \times \mathbf{k}}{|\mathbf{k}|^2 - \kappa_\varepsilon^2}\right).$$

Therefore

$$\widehat{\mathbf{E}}_\varepsilon(\mathbf{r}) = -\frac{1}{(2\pi)^{\frac{3}{2}}}\frac{1}{\epsilon_0}\mathcal{F}_{12}^{-1}\left[\mathcal{F}_3^{-1}\left[\left(\boldsymbol{\Psi} + \frac{(\boldsymbol{\Psi}\times\mathbf{k})\times\mathbf{k}}{|\mathbf{k}|^2 - \kappa_\varepsilon^2}\right)e^{-ik_3r_3^{\boldsymbol{\Psi}}}\right](r_3)\right](r_1, r_2).$$

In order to prove Eq. (43) for $\widehat{\mathbf{E}}_\varepsilon$, it remains to show that

$$\frac{1}{\sqrt{2\pi}}\mathcal{F}_3^{-1}\left[k_3 \to \left(\boldsymbol{\Psi} + \frac{(\boldsymbol{\Psi}\times\mathbf{k})\times\mathbf{k}}{|\mathbf{k}|^2 - \kappa_\varepsilon^2}\right)e^{-ik_3r_3^{\boldsymbol{\Psi}}}\right](r_3) = \Psi_3\mathbf{e}_3\tilde{\delta}(r_3 - r_3^{\boldsymbol{\Psi}})$$

$$+\frac{ie^{iq_\varepsilon(r_3 - r_3^{\boldsymbol{\Psi}})}}{2q_\varepsilon}(\boldsymbol{\Psi}\times\mathbf{k}_{q_\varepsilon})\times\mathbf{k}_{q_\varepsilon},$$

which is done by standard, but quite lengthy computations, which are presented in Appendix 1.

Now, we consider $\varepsilon \to 0$. Theorem 1 combined with the continuity of the inverse Fourier transform $\mathcal{F}_{12}^{-1}$ in $\mathcal{S}'(\mathbb{R}^2, \mathbb{R}^2)$ which implies that

$$\widehat{\mathbf{E}}(\mathbf{r}) = -\frac{1}{2\pi}\frac{1}{\epsilon_0}\mathcal{F}_{12}^{-1}\left[(k_1, k_2) \to \Psi_3\mathbf{e}_3\tilde{\delta}(r_3 - r_3^{\boldsymbol{\Psi}}) + \lim_{\varepsilon\to 0}\frac{ie^{iq_\varepsilon(r_3 - r_3^{\boldsymbol{\Psi}})}}{2q_\varepsilon}(\boldsymbol{\Psi}\times\mathbf{k}_{q_\varepsilon})\times\mathbf{k}_{q_\varepsilon}\right](r_1, r_2).$$

To prove the assertion, we simply need to check that, in $\mathcal{S}'$

$$\lim_{\varepsilon\to 0}\frac{ie^{iq_\varepsilon(r_3 - r_3^{\boldsymbol{\Psi}})}}{2q_\varepsilon}(\boldsymbol{\Psi}\times\mathbf{k}_{q_\varepsilon})\times\mathbf{k}_{q_\varepsilon} = \frac{ie^{iq(r_3 - r_3^{\boldsymbol{\Psi}})}}{2q}(\boldsymbol{\Psi}\times\mathbf{k}_q)\times\mathbf{k}_q.$$

These two quantities being L_{loc}^1 functions, it is enough to show that the limit holds in $L_{\text{loc}}^1(\mathbb{R}\times(\mathbb{R}^2\times\mathbb{R}))$. The L_{loc}^1 convergence is then obtained noticing that

$$e^{iq_\varepsilon(r_3 - r_3^{\boldsymbol{\Psi}})}(\boldsymbol{\Psi}\times\mathbf{k}_{q_\varepsilon})\times\mathbf{k}_{q_\varepsilon} - e^{iq(r_3 - r_3^{\boldsymbol{\Psi}})}(\boldsymbol{\Psi}\times\mathbf{k}_q)\times\mathbf{k}_q \xrightarrow{L^\infty} 0$$

and that

$$\frac{1}{q_\varepsilon} - \frac{1}{q} = \frac{\kappa_\varepsilon^2 - \kappa^2}{(\kappa_\varepsilon^2 - k_1^2 - k_2^2)\sqrt{\kappa^2 - k_1^2 - k_2^2} + (\kappa^2 - k_1^2 - k_2^2)\sqrt{\kappa_\varepsilon^2 - k_1^2 - k_2^2}}$$

converges to zero in L_{loc}^1. Note that this would imply only a convergence in $\mathcal{D}'$, but the two functions are actually uniformly L^∞ outside the compact set $\{k_1^2 + k_2^2 \geq |\kappa|^2 + 1\}$, so the convergence holds in $\mathcal{S}'$ as well. □

Moreover, we make the assumption that the dipole can be *rotating*.

Fig. 6 The axis of cone has angular coordinates θ_m and φ_m in the coordinate system. A general orientation within the cone has coordinates θ and φ in the coordinate system, and axial coordinate β, and azimuthal coordinate η with respect to the cone axis. The outer limit of motion in the cone is given by $\beta = \alpha_m$

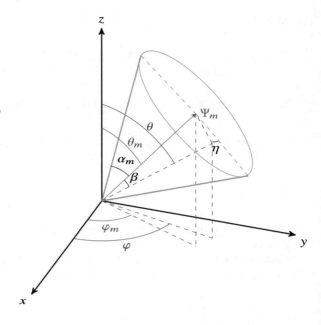

Definition 9 (Rotating Dipole) The emitting dipole is considered wobbling uniformly distributed around the dipole orientation $\frac{\Psi_m}{|\Psi_m|} = (\theta_m, \varphi_m) \in \mathbb{S}^2$ in a cone of semi-angle α_m (see Fig. 6). Assuming a dipole-emission from an oscillating source we get after averaging, a *source* represented as the indicator function

$$\mathbb{1}_m = \frac{1}{|C(\Psi_m, \alpha_m)|} \mathbb{1}_{C(\Psi_m, \alpha_m)}, \tag{48}$$

where

$$C(\Psi_m, \alpha_m) = \left\{ \tau \Phi \in \mathbb{S}^2 : |\angle \Phi \Psi_m| \leq \alpha_m, 0 \leq \tau \leq |\Psi_m| \right\}. \tag{49}$$

Note that $|C(\Psi_m, \alpha_m)| = \frac{1}{3}\pi |\Psi_m|^3 \tan^2(\alpha_m)$. Taking into account Eqs. (38) and (37) the according charge density and current of dye m are given by

$$\widehat{\mathbf{J}}_m(\mathbf{r}; \omega) = i\omega \mathbb{1}_m, \quad \widehat{\rho}_m(\mathbf{r}; \omega) = \frac{i}{\omega} \nabla_{\mathbf{r}} \cdot \widehat{\mathbf{J}}_m(\mathbf{r}; \omega) \text{ and}$$

$$\widehat{\mathbf{R}}_m(\mathbf{r}; \omega) := \frac{i\omega}{c^2} \widehat{\mathbf{J}}_m(\mathbf{r}; \omega) + \nabla_{\mathbf{r}} \widehat{\rho}_m(\mathbf{r}; \omega). \tag{50}$$

6 The Forward Problem

In the following we present mathematical models describing the emission and propagation of light caused by dyes, which are exposed to strong laser light illumination. See Fig. 7 for a schematic representation of the experiment. In single molecule localization microscopy two-dimensional images are recorded after exposing the probe subsequently to strong laser illuminations, such that the dyes appear in dark ("off") and light ("on") state. This allows to separate the fluorescent emission of individual dyes in time, allowing for high resolution images. In order to minimize the notational effort we consider recording of a single image frame first. The mathematical model of consecutive recordings of multiple frames is analogous and requires one additional parameter representing numbering of frames (a virtual time).

In the following we state a series of assumptions, which are used throughout the remainder of the paper:

Assumption (Medium, Monochromatic Source and Response) In the following we assume that

- The incident light is a *monochromatic plane wave* of frequency ω_{inc} and orientation **v**.
- The medium is assumed to be isotropic, non magnetizable, homogeneous and has a linear polarization response.
- Moreover, we assume that a dye can be modeled as an *absorbing dipole* Ψ_a, which emits monochromatic waves of frequency $\omega \neq \omega_{inc}$ resulting in an *emitting dipole*

$$\Psi = (\mathbf{v} \cdot \Psi_a)\mathbf{v}. \tag{51}$$

Indeed what we will measure is the electric field at frequency ω, which is not affected by the incident field at frequency ω_{inc}. As a consequence we only have to consider the electric field at the frequency $\omega \in \mathbb{R}$.

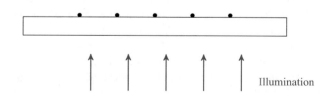

Fig. 7 Illustration of the experiment: Biomolecular structures are placed on the glass surface at position $r_3^{\Psi} \leqslant 0$ and illuminated from the bottom. The glass plate has a thickness r_3^{Ψ}

- In what follows we assume that the considered dipole $\Psi = \begin{pmatrix} \Psi_1 \\ \Psi_2 \\ \Psi_3 \end{pmatrix}$ is located

  at position $\mathbf{r}^{\Psi} = \left(0 \; 0 \; r_3^{\Psi} \right)^T$ with $r_3^{\Psi} \leqslant 0$. Unless stated otherwise $\mathbf{r} \in \mathbb{R}^3$ with
  $r_3 > 0$. The sign assumptions on r_3 and r_3^{Ψ} are in accordance with the experiment:
  the object is assumed left of the lens system (see Fig. 8) and $\mathbf{r}$ is a point of the
  measurement system.
- The dyes absorb light, which can result in fluoresence emission. We describe the
  states of an absorbing dye with index m via a time indicator function: The *on-off*
  indicator

$$\mathbf{I}_m \in \{0, 1\}, \tag{52}$$

  tells us whether the m-th dye is an emitting state or not.

  $\square$

The complete experimental setup of the optical experiment of single molecule
localization microscopy is represented in Fig. 8. For the mathematical modeling we
are considering the propagation of light at different locations of the optical system.
The dyes are considered at positions $\mathbf{r}^{\Psi_m}$ with $r_3^{\Psi_m} \leqslant 0$ and the *focal plane* (which
contains the focal point of the objective) corresponds to the bottom of the glass plate,
which is not mathematically modeled, that is the focal plane is at position $r_3 = 0$.
Note that in particular the dipole *is not located at the focal plane*, unless if $r_3^{\Psi_m} = 0$.
For the sake of simplicity of presentation we consider only a single dye, and leave
the subscript m whenever appropriate.

The mathematical modeling of the experimental setup follows [3], however it is
adapted to our notation:

- In Sect. 6.1 we describe the propagation of the electric field in the medium, that
  is from the bottom of the cell (the assumption is that only molecules labeled with
  a dye at the bottom of the cell emit light) up to the objective (see Fig. 8). This
  domain will be denoted by Ω. Since the objective is far away from the molecule
  (relative to the size of the molecule) the electric field can be approximated well
  by its far field, which is calculated below. The three-dimensional k-transformed
  coordinate system is denoted by $\mathbf{k} \in \mathbb{R}^3$ (see Sect. 6.1).
- In Sect. 6.2 we present in mathematical terms the propagation of the emitted light
  when it passes through the objective; that is after passing through the medium.
  In fact the light rays are aligned parallel by the objective in r_3 direction. The
  objective has a focal length $\mathtt{f}_{obj}$ and it is positioned orthogonal to the r_3 axis
  with left distance to the focal plane (glass plate) $r_3^{obj} = \mathtt{f}_{obj}$ (see Fig. 10).
  Indeed the lens system is complicated and a detailed mathematical modeling is
  not possible. A simplified model assumes that the objective is big compared to the
  wavelength, such that the intensity law of *Geometric Optics* applies (see Fig. 9
  and [7]), and phase shifts due to the curvature of the lenses can be neglected.

- In Sect. 6.3 we calculate the propagation of the light after passing through the back focal plane of the objective, that is in between r_3^{bfp} and $r_3^{\mathrm{tl}i}$ (see Fig. 8), from knowledge of the field at the plane with third coordinate r_3^{bfp}. This is achieved by solving the Helmholtz equation in air between the back focal plane of the objective and the incident plane of the tube lens.
- We assume that the lens is a circular plano-convex *tube lens* with maximal thickness d_0. The thickness is described as a function d. Moreover, we assume that the lens has a *focal length* f_L and that its pupil function is given by $P_L : \mathbb{R}^2 \to \mathbb{R}$,

$$P_L(\mathbf{x}) = \begin{cases} 1 & \text{for } |\mathbf{x}| \le R \\ 0 & \text{for } |\mathbf{x}| > R \end{cases}, \tag{53}$$

in the plane $r_3 = r_3^{\mathrm{tl}}$ (see Fig. 8). The lens is assumed to be *converging*, such that the *paraxial approximation* holds, that is we can assume that the wave vector of the wave is almost aligned with the optical axis [14, Sec. 4.2.3]. The adequate formulas are derived in Sect. 6.4.
- Finally the light is bundled to the image plane, which provides an image described by coordinates $\mathbf{x}_f \in \mathbb{R}^2$ (see Sect. 6.5).

We summarize the different coordinate systems used below in a table:

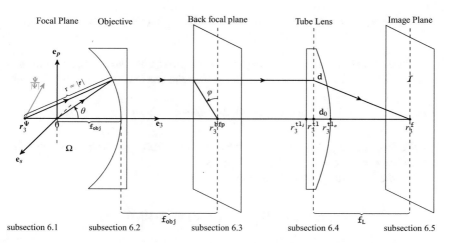

Fig. 8 The *plane of observation* is defined as the plane containing the dipole $\boldsymbol{\Psi}$, the $\mathbf{e}_3$-axis, and the path of a particular ray through the objective, the *back focal plane* and the tube lens (with focal length f_{obj})

Table 2 Some abbreviation used in Lemma 3 and its proof, as well as in Appendix 1 and 2

Position	Coordinates	Fourier
Medium Ω	$\mathbf{r} \in \mathbb{R}^3$, $(r, \theta, \varphi) \in \mathbb{R}_+ \times \mathbb{S}^2$	$\mathbf{k} \in \mathbb{R}^3$
Back focal plane (BFP)	$\mathbf{x} \in \mathbb{R}^2$, $(\rho, \varphi) \in \mathbb{R}_+ \times [0, 2\pi)$	$\mathbf{u} \in \mathbb{R}^2$, $(\xi, \nu) \in \mathbb{R}_+ \times [0, 2\pi)$
Tube Lens	$\mathbf{y} \in \mathbb{R}^2$, $(\varrho, \sigma) \in \mathbb{R}_+ \times [0, 2\pi)$	$\mathbf{v} \in \mathbb{R}^2$, $(\varkappa, \vartheta) \in \mathbb{R}_+ \times [0, 2\pi)$
Image plane (IP) $\mathcal{I}$	$\mathbf{x}_f \in \mathbb{R}^2$	$\mathbf{u}_f \in \mathbb{R}^2$
Between BFP and IP	$(\mathbf{x}, r_3) \in \mathbb{R}^3$	
General notation	$\mathbf{k}_{12} = (k_1, k_2)^T \in \mathbb{R}^2$, $\mathbf{k} = (k_1, k_2, k_3)^T \in \mathbb{R}^3$	
	$\mathbf{k}_z = (k_1, k_2, z)^T$ $z \in \mathbb{C}$, $k_1, k_2 \in \mathbb{R}$	

6.1 Far Field Approximation in the Medium

In this subsection we derive the far field approximation of the Fourier-transform of the electric field, $\widehat{\mathbf{E}}$, *in the medium*. The derivation expands [12].

First, we give the definition of the far field:

Definition 10 The *far field* $F_\infty : \mathbb{S}^2 \to \mathbb{C}^3$ of a function $F : \mathbb{R}^3 \to \mathbb{C}^3$ satisfies: There exists $\hat{C} > 0$ and a function $C : [0, \infty) \to [0, \infty)$ such that

$$\lim_{r \to \infty} |F(r, \theta, \varphi) - C(r) F_\infty(\theta, \varphi)| = 0 \quad \text{with} \quad |r C(r)| \le \hat{C} \text{ for all } r \in [0, \infty).$$
(54)

Lemma 4 *Let the medium be isotropic, non magnetizable, homogeneous and have a linear polarization response. We assume that the considered dipole Ψ is located at position* $\mathbf{r}^{\Psi} = \begin{pmatrix} 0 \\ 0 \\ r_3^{\Psi} \end{pmatrix}$ *with $r_3^{\Psi} < 0$. Moreover, let* $\mathbf{r} = \begin{pmatrix} r_1 \\ r_2 \\ r_3 \end{pmatrix} \in \mathbb{R}^3$ *with $r_3 > r_3^{\Psi}$; The later assumption means that we are considering only light rays, which are propagating into the lens system (see Fig. 8).*

Then the far field *of $\widehat{\mathbf{E}}$ in the medium is given by*

$$e^{i\kappa r_3^{\Psi} \cos(\theta)} \widehat{\mathbf{E}}_\infty(\theta, \varphi) = \cos(\theta) \left(-\Psi_p \cos(\theta) + \Psi_3 \sin(\theta) \right) \mathbf{e}_p - \Psi_s \mathbf{e}_s$$
$$+ \sin(\theta) \left(\Psi_p \cos(\theta) - \Psi_3 \sin(\theta) \right) \mathbf{e}_3$$
(55)

and

$$\boxed{C(r) = \frac{\kappa^2}{4\pi \varepsilon_0} \frac{e^{i\kappa r}}{r}.}$$
(56)

where $\Psi_j = \langle \Psi, \mathbf{e}_j \rangle$, $j = p, s, 3$ are the coefficients of Ψ with respect to the orthonormal basis

$$\mathbf{e}_p := \begin{pmatrix} \cos(\varphi) & \sin(\varphi) & 0 \end{pmatrix}^T, \quad \mathbf{e}_s := \begin{pmatrix} -\sin(\varphi) & \cos(\varphi) & 0 \end{pmatrix}^T, \quad \mathbf{e}_3, \tag{57}$$

that is

$$\Psi = \Psi_p \mathbf{e}_p + \Psi_s \mathbf{e}_s + \Psi_3 \mathbf{e}_3. \tag{58}$$

Proof Taking into account the assumption that $r_3 - r_3^\Psi > 0$, and by representing the vector $\mathbf{r} \in \mathbb{R}^3$ as

$$\mathbf{r} = r_3 \mathbf{e}_3 + r_3 \begin{pmatrix} \mathbf{v} & 0 \end{pmatrix}^T = r_3 \begin{pmatrix} v_1 & v_2 & 1 \end{pmatrix}^T, \tag{59}$$

with a (non-unit) vector $\begin{pmatrix} v_1 & v_2 \end{pmatrix}^T = \mathbf{v} \in \mathbb{R}^2$ in the plane spanned by $\mathbf{e}_1$ and $\mathbf{e}_2$, it follows from Eq. (43) that

$$\widehat{\mathbf{E}}(\mathbf{r}) = -\frac{1}{8\pi^2} \frac{1}{\epsilon_0} \int_{\mathbf{k}_{12} \in \mathbb{R}^2} e^{i r_3 \mathbf{v} \cdot \mathbf{k}_{12}} \left(\frac{i e^{iq(r_3 - r_3^\Psi)}}{q} (\Psi \times \mathbf{k}_q) \times \mathbf{k}_q \right) d\mathbf{k}_{12}, \tag{60}$$

where q and $\mathbf{k}_q$ are as defined in Eq. (42). Note that in Eq. (60) $q = q(\mathbf{k}_{12})$ is defined as in Eq. (115), and therefore the integral on the right-hand side is of the form (neglecting the factor $-\frac{i}{8\pi^2} \frac{1}{\epsilon_0}$)

$$\int_{\mathbf{k}_{12} \in \mathbb{R}^2} e^{i r_3 \zeta(\mathbf{k}_{12})} \beta(\mathbf{k}_{12}) \, d\mathbf{k}_{12}$$

with

$$\zeta(\mathbf{k}_{12}) = \mathbf{k}_{12} \cdot \mathbf{v} + q \quad \text{and} \quad \beta(\mathbf{k}_{12}) = \frac{e^{-i r_3^\Psi q}}{q} (\Psi \times \mathbf{k}_q) \times \mathbf{k}_q. \tag{61}$$

The *stationary phase method*, [19, Th. 7.7.5], states that if $\hat{\mathbf{k}}$ is a critical point of ζ, which has been calculated in Eq. (114), then

$$\int_{\mathbf{k}_{12} \in \mathbb{R}^2} e^{i r_3 \zeta(\mathbf{k}_{12})} \beta(\mathbf{k}_{12}) \, d\mathbf{k}_{12} = e^{i r_3 \zeta(\hat{\mathbf{k}})} \left(\det \left(r_3 H(\zeta)(\hat{\mathbf{k}})/(2\pi i) \right) \right)^{-1/2} \beta(\hat{\mathbf{k}}) + o\left(\frac{1}{r_3} \right).$$

Taking into account Eq. (113) in Lemma 11, and $\hat{\mathbf{k}}_{12}$ of ζ as defined in Eq. (114), and being aware that $q = q(\mathbf{k}_{12})$ (that is q is a function of $\mathbf{k}_{12}$), we apply Eqs. (116), and (115) and get

$$\int_{\mathbf{k}_{12}\in\mathbb{R}^2} \frac{e^{ir_3(\mathbf{k}_{12}\cdot\mathbf{v}+q)}}{q} e^{-iqr_3^{\mathbf{\Psi}}} (\mathbf{\Psi}\times\mathbf{k}_q)\times\mathbf{k}_q\,d\mathbf{k}_{12}$$

$$=2i\pi\kappa^2\frac{e^{ir_3\kappa\sqrt{1+|\mathbf{v}|^2}}}{r_3} e^{-i\frac{\kappa r_3^{\mathbf{\Psi}}}{\sqrt{1+|\mathbf{v}|^2}}} \left(\mathbf{\Psi}\times\frac{\mathbf{r}}{|\mathbf{r}|}\right)\times\frac{\mathbf{r}}{|\mathbf{r}|}+o\left(\frac{1}{r_3}\right)\quad\text{for } r_3\to\infty.$$

Now, we recall Eqs. (10) and (59), which imply that

$$\sqrt{1+|\mathbf{v}|^2}=\frac{1}{\cos(\theta)},\quad \frac{\mathbf{r}}{|\mathbf{r}|}=\begin{pmatrix}\sin(\theta)\cos(\phi)\\\sin(\theta)\sin(\phi)\\\cos(\theta)\end{pmatrix}=\sin(\theta)\mathbf{e}_p+\cos(\theta)\mathbf{e}_3$$

and $|\mathbf{r}|\cos(\theta)=r_3$,

such that we get

$$\int_{\mathbf{k}_{12}\in\mathbb{R}^2}\frac{e^{ir_3(\mathbf{k}_{12}\cdot\mathbf{v}+q)}}{q}e^{-iqr_3^{\mathbf{\Psi}}}(\mathbf{\Psi}\times\mathbf{k}_q)\times\mathbf{k}_q\,d\mathbf{k}_{12}$$

$$=2i\pi\kappa^2 e^{-i\kappa r_3^{\mathbf{\Psi}}\cos(\theta)}\frac{e^{i\kappa|\mathbf{r}|}}{|\mathbf{r}|}\left(\mathbf{\Psi}\times\frac{\mathbf{r}}{|\mathbf{r}|}\right)\times\frac{\mathbf{r}}{|\mathbf{r}|}+o\left(\frac{1}{r_3}\right).$$

This shows that

$$\widehat{\mathbf{E}}(\mathbf{r})=e^{-i\kappa r_3^{\mathbf{\Psi}}\cos(\theta)}\frac{\kappa^2}{4\pi\epsilon_0}\frac{e^{i\kappa|\mathbf{r}|}}{|\mathbf{r}|}\left(\mathbf{\Psi}\times\frac{\mathbf{r}}{|\mathbf{r}|}\right)\times\frac{\mathbf{r}}{|\mathbf{r}|}+o\left(\frac{1}{r_3}\right)=C(r)\widehat{\mathbf{E}}_\infty(\mathbf{r})+o\left(\frac{1}{r_3}\right).$$

$$(62)$$

It remains to compute the second identity of Eq. (55). Expressing $\dfrac{\mathbf{r}}{|\mathbf{r}|}$ and $\mathbf{\Psi}$ in terms of the associated basis $\mathbf{e}_p$, $\mathbf{e}_s$, $\mathbf{e}_3$ from Definition 6, and using Eq. (45), we get from Eq. (58)

$$\left(\mathbf{\Psi}\times\frac{\mathbf{r}}{|\mathbf{r}|}\right)\times\frac{\mathbf{r}}{|\mathbf{r}|}=\left(\mathbf{\Psi}\cdot\frac{\mathbf{r}}{|\mathbf{r}|}\right)\frac{\mathbf{r}}{|\mathbf{r}|}-\mathbf{\Psi}$$

$$=\Big((\sin(\theta)\mathbf{e}_p+\cos(\theta)\mathbf{e}_3)\cdot(\Psi_p\mathbf{e}_p+\Psi_s\mathbf{e}_s+\Psi_3\mathbf{e}_3)\Big)$$

$$(\sin(\theta)\mathbf{e}_p+\cos(\theta)\mathbf{e}_3)-\Psi_p\mathbf{e}_p-\Psi_s\mathbf{e}_s-\Psi_3\mathbf{e}_3$$

$$=\Big(\sin(\theta)\Psi_p+\cos(\theta)\Psi_3\Big)(\sin(\theta)\mathbf{e}_p+\cos(\theta)\mathbf{e}_3)-\Psi_p\mathbf{e}_p$$

$$-\Psi_s\mathbf{e}_s-\Psi_3\mathbf{e}_3,$$

which after rearrangement proves the second identity.   □

In the imaging system the calculation of the electric field is not done at once but in different sections (Sects. 6.1, 6.2, 6.3, 6.4, and 6.5). In each of these sections the electric field is calculated by transmission from the electric field computed at the previous section. In addition, we assume that the light which hits the objective from Ω can be approximated by its far field expansion $C(\mathtt{f}_{\mathtt{obj}})\widehat{\mathbf{E}}_\infty$, which we will use instead of $\widehat{\mathbf{E}}$.

6.2   Propagation of the Electric Field through the Objective

In the following we calculate the electric field in the objective. Assuming that the electric field (light) emitted from the dipoles travels along straight lines in the medium to the objective, the objective aligns the emitted rays from the dipole parallel to the r_3-axis in such a way that the electric field between the incidence surface of the objective and the back focal plane undergoes a phase shift that does not depend on the distance to the optical axis. In the ideal situation, where the wavelength is assumed to be infinitely small compared to the length parameters of the optical system, the electric field can be computed via the *intensity law of geometrical optics* (see Fig. 9 and [7, Sec. 3.1.2] for a derivation).

Assumption The objective consists of a set of optical elements (lenses and mirrors) which are not modelled here (see some examples in [7, Sec. 6.6]). Its aim is to transform spherical waves originated at its focal point into waves which propagate along the optical axis. In what follows, the computations are made ignoring a constant (independent on the point in the back focal plane) phase shift which is underwent by the wave through the objective.   □

Lemma 5 *Let* $\mathbf{r}$ *be a point at the back focal plane of the objective, that is with* r_3 *coordinate* $r_3^{\mathtt{bfp}}$ *and with spherical coordinates* (r, θ, φ). *We define the* radial length *on the propagation plane (planes with constant* r_3 *coordinate) (see Fig. 10), by*

Fig. 9 Intensity law of Geometrical optics: the energy carried along a ray must remain constant. The power transported by a ray is proportional to $|\mathbf{E}|^2\,dA$, where dA is an infinitesimal cross-section perpendicular to the ray propagation. Thus, the fields must satisfy
$|\mathbf{E}_2| = |\mathbf{E}_1|\,\frac{1}{\sqrt{\cos(\theta)}}$

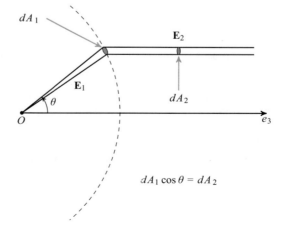

Fig. 10 Approximation used: We assume that the cell is fixed to the glass, and the distance of the dipole $\left| r_3^\Psi \right|$ from the focal plane is sufficiently smaller than $r = |\mathbf{r}|$, such that $\mathtt{f}_{\mathrm{obj}} \approx r$, and $\theta' \approx \theta$

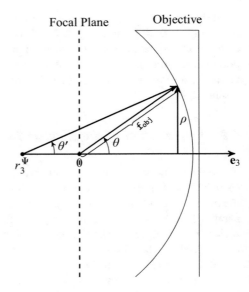

$$\rho := \rho(\theta) := \mathtt{f}_{\mathrm{obj}} \sin(\theta). \tag{63}$$

Then

$$\widehat{\mathbf{E}}^{\mathtt{bfp}}(\rho, \varphi) :=$$

$$\begin{cases} \dfrac{C(\mathtt{f}_{\mathrm{obj}})}{\sqrt{\cos(\theta)}} e^{-i\kappa r_3^\Psi \cos(\theta)} \begin{pmatrix} \cos(\varphi)\left(-\Psi_p \cos(\theta) + \Psi_3 \sin(\theta)\right) - \sin(\varphi)\Psi_s \\ \sin(\varphi)\left(-\Psi_p \cos(\theta) + \Psi_3 \sin(\theta)\right) + \cos(\varphi)\Psi_s \\ 0 \end{pmatrix} & \theta \leqslant \theta_{\max} \\[2em] 0 & \theta > \theta_{\max} \end{cases}$$

$$\tag{64}$$

where $C(\mathtt{f}_{\mathrm{obj}})$ is as defined in Eq. (56), and

$$\theta_{\max} := \arcsin(\mathrm{NA})$$

is the maximal angle θ for rays to enter the objective (the other rays simply do not enter the optical system). Note that the refractive index in air is assumed one.

Proof The electric field is transmitted according to the law of geometrical optics [3, Eq. 16] into the objective at the points

$$\mathbf{r} = \begin{pmatrix} 0 \\ 0 \\ r_3^\Psi \end{pmatrix} + \mathtt{f}_{\mathrm{obj}} \mathbb{S}^2, \tag{65}$$

that is the electric field simply undergoes a rotation of axis $\mathbf{e}_s$ and angle θ as well as a magnification of $\frac{1}{\sqrt{\cos\theta}}$ (see Assumption 6.2).

The rotation with angle θ around the axis $\mathbf{e}_s$ changes the unit vectors as follows:

$$\begin{aligned}
\mathbf{e}_p &\to \sin(\theta)\mathbf{e}_3 + \cos(\theta)\mathbf{e}_p \\
\mathbf{e}_3 &\to \cos(\theta)\mathbf{e}_3 - \sin(\theta)\mathbf{e}_p \\
\mathbf{e}_s &\to \qquad \mathbf{e}_s.
\end{aligned} \tag{66}$$

Now, Eq. (66) shows that the expression of the electric field in the back focal plane will be simpler using coordinates $(\mathbf{e}_p, \mathbf{e}_s, \mathbf{e}_3)$. Equation (55) leads to

$$\widehat{\mathbf{E}}^{\mathrm{bfp}}(\rho,\varphi) = \frac{C(\mathbf{f}_{\mathrm{obj}})}{\sqrt{\cos(\theta)}} e^{-i\kappa r_3^{\Psi}\cos(\theta)} \left\{ \left(-\Psi_p\cos(\theta) + \Psi_3\sin(\theta)\right)\mathbf{e}_p - \Psi_s\mathbf{e}_s \right\},$$

where $C(\mathbf{f}_{\mathrm{obj}})$ is as defined in Eq. (56). Writing the unit vectors $\mathbf{e}_p$ and $\mathbf{e}_s$ in the fixed system of coordinates (x_1, x_2, x_3) gives Eq. (64). $\qquad\square$

6.3 Between the Objective and the Lens

Behind the objective, the light propagates through air until it reaches the tube lens. Denoting by κ^2 the wave number in air (see Eq. (22)) the electric field satisfies the homogeneous Helmholtz equation in the tube lens:

$$\Delta\widehat{\mathbf{E}}(\mathbf{r}) + \kappa^2\widehat{\mathbf{E}}(\mathbf{r}) = 0 \quad \text{in} \quad \mathcal{H} := \left\{ \mathbf{r} \in \mathbb{R}^3 : r_3^{\mathrm{bfp}} < r_3 < r_3^{\mathrm{tl}_i} \right\} \tag{67}$$

together with the boundary condition

$$\widehat{\mathbf{E}}(r_1, r_2, r_3^{\mathrm{bfp}}) = \widehat{\mathbf{E}}^{\mathrm{bfp}}(r_1, r_2) \text{ for all } (r_1, r_2) \in \mathbb{R}^2. \tag{68}$$

The solution of Eq. (67) can actually be calculated by applying a phase shift to $\widehat{\mathbf{E}}^{\mathrm{bfp}}$ as the following lemma shows.

Lemma 6 *Representing* $\mathbf{x} = (\mathbf{f}_{\mathrm{obj}}\sin(\theta)\cos(\varphi), \mathbf{f}_{\mathrm{obj}}\sin(\theta)\sin(\varphi)) \in \mathbb{R}^2$, *then the Fourier transforms of* $\widehat{\mathbf{E}}$ *in the transverse plane of* $(\mathbf{x}, r_3)$ *can be calculated from* $\widehat{\mathbf{E}}(\mathbf{x}, r_3^{\mathrm{bfp}})$ *in the following way*

$$\mathcal{F}_{12}(\widehat{\mathbf{E}})(k_1, k_2, r_3) = \mathcal{F}_{12}[\widehat{\mathbf{E}}^{\mathrm{bfp}}](k_1, k_2)e^{(r_3 - r_3^{\mathrm{bfp}})\sqrt{-\kappa^2 + k_1^2 + k_2^2}}. \tag{69}$$

where the square root can denote both of the complex square roots.

Proof First, we notice that since $\widehat{\mathbf{E}}^{\mathrm{bfp}}$ is bounded with compact support, it is a L^2 function in the plane $\{r_3 = r_3^{\mathrm{bfp}}\}$. Taking the Fourier transform in these two variables, Eqs. (67) and (68) are equivalent to

$$\partial_{r_3}^2 \mathcal{F}_{12}(\widehat{\mathbf{E}})(k_1, k_2, r_3) + (\kappa^2 - k_1^2 - k_2^2)\mathcal{F}_{12}(\widehat{\mathbf{E}})(k_1, k_2, r_3)$$
$$= 0 \text{ for all } \mathbf{r} \in \mathbb{R}^3, \ r_3^{\mathrm{bfp}} < r_3 < r_3^{\mathrm{tl}_i} \tag{70}$$

with the boundary condition

$$\mathcal{F}_{12}(\widehat{\mathbf{E}})(k_1, k_2, r_3^{\mathrm{bfp}}) = \mathcal{F}_{12}[\widehat{\mathbf{E}}^{\mathrm{bfp}}](k_1, k_2) \text{ for all } (k_1, k_2) \in \mathbb{R}^2. \tag{71}$$

Now, Eq. (70) is a simple ODE whose solution writes (for $\kappa^2 - k_1^2 - k_2^2 \neq 0$)

$$\mathcal{F}_{12}(\widehat{\mathbf{E}})(k_1, k_2, r_3) = \mathcal{F}_{12}[\widehat{\mathbf{E}}^{\mathrm{bfp}}](k_1, k_2)e^{(r_3 - r_3^{\mathrm{bfp}})\sqrt{-\kappa^2 + k_1^2 + k_2^2}}.$$

$$\square$$

Among the fields computed in Eq. (69), several are not physical or will not be observed:

- Having $\kappa^2 < k_1^2 + k_2^2$ leads to either a real positive square root which corresponds to a wave exploding as r_3 increases and is therefore not physical or a real negative root, which yields an exponentially decreasing wave (*evanescent*) which exist but, since $(r_3 - r_3^{\mathrm{bfp}})$ is several orders of magnitude bigger than the wave length, will be damped by the time it hits the tube lens. Therefore we also do not consider it.
- When, $\kappa^2 > k_1^2 + k_2^2$, we get two imaginary roots, namely $\pm i\sqrt{\kappa^2 - k_1^2 - k_2^2}$, which corresponds to the two Green functions Eq. (31). For the same reason as above, we will only consider the positive sign.

This can be summerized in the following assumption, that will hold in what follows.

Assumption We only consider $\kappa^2 \geqslant k_1^2 + k_2^2$ and we obtain

$$\mathcal{F}_{12}(\widehat{\mathbf{E}})(k_1, k_2, r_3) = \mathcal{F}_{12}[\widehat{\mathbf{E}}^{\mathrm{bfp}}](k_1, k_2)e^{i(r_3 - r_3^{\mathrm{bfp}})\sqrt{\kappa^2 - k_1^2 - k_2^2}}. \tag{72}$$

$$\square$$

In the following we calculate $\mathcal{F}_{12}[\widehat{\mathbf{E}}^{\mathrm{tl}_i}]$, where $\widehat{\mathbf{E}}^{\mathrm{tl}_i} = \widehat{\mathbf{E}}(\mathbf{x}, r_3^{\mathrm{tl}_i})$.

Lemma 7 *Let $\widehat{\mathbf{E}}^{\mathrm{tl}_i}(\mathbf{x}) = \widehat{\mathbf{E}}(\mathbf{x}, r_3^{\mathrm{tl}_i})$ be the electric field at the indicent plane of the tube thin lens (at $r_3^{\mathrm{tl}_i}$) as defined in Eq. (69), then the Fourier transform of $\widehat{\mathbf{E}}^{\mathrm{tl}_i}$ in this plane in polar coordinates (ξ, ν) of (k_1, k_2) is given, for $\xi^2 \leqslant \kappa^2$, by*

$$\mathcal{F}_{12}[\widehat{\mathbf{E}}^{\mathrm{tl}i}](\xi, v) = \mathcal{F}_{12}[\widehat{\mathbf{E}}^{\mathrm{bfp}}](\xi, v)e^{i(r_3^{\mathrm{tl}i} - r_3^{\mathrm{bfp}})\sqrt{\kappa^2 - \xi^2}}$$

$$= \frac{1}{2}(\mathtt{f}_{\mathrm{obj}})^2 C(\mathtt{f}_{\mathrm{obj}})e^{i(r_3^{\mathrm{tl}i} - r_3^{\mathrm{bfp}})\sqrt{\kappa^2 - \xi^2}}.$$

$$\cdot \begin{pmatrix} -\Psi_1[I_{1,0}(\xi) + I_{2,0}(\xi)] + \Psi_1\cos(2v)[I_{1,2}(\xi) + I_{2,2}(\xi)] + \Psi_2\sin(2v)[I_{1,2}(\xi) + I_{2,2}(\xi)] - 2i\Psi_3\cos(v)I_{2,1}(\xi) \\ -\Psi_2[I_{1,0}(\xi) + I_{2,0}(\xi)] + \Psi_2\cos(2v)[I_{1,2}(\xi) + I_{2,2}(\xi)] + \Psi_1\sin(2v)[I_{1,2}(\xi) + I_{2,2}(\xi)] - 2i\Psi_3\sin(v)I_{2,1}(\xi) \\ 0 \end{pmatrix},$$

$$\tag{73}$$

where

$$I_{1,0}(\xi) = \int_0^{\theta_{\max}} \sqrt{\cos(\theta)}\, \sin(\theta)\, e^{-i\kappa r_3^{\Psi}\cos(\theta)} J_0\left(\mathtt{f}_{\mathrm{obj}}\xi\sin(\theta)\right) d\theta$$

$$I_{1,2}(\xi) = \int_0^{\theta_{\max}} \sqrt{\cos(\theta)}\, \sin(\theta)\, e^{-i\kappa r_3^{\Psi}\cos(\theta)} J_2\left(\mathtt{f}_{\mathrm{obj}}\xi\sin(\theta)\right) d\theta$$

$$I_{2,1}(\xi) = \int_0^{\theta_{\max}} (\cos(\theta))^{3/2}\frac{1 - \cos(2\theta)}{2}e^{-i\kappa r_3^{\Psi}\cos(\theta)} J_1\left(\mathtt{f}_{\mathrm{obj}}\xi\sin(\theta)\right) d\theta \tag{74}$$

$$I_{2,0}(\xi) = \int_0^{\theta_{\max}} \sqrt{\cos(\theta)}\,\frac{\sin(2\theta)}{2}e^{-i\kappa r_3^{\Psi}\cos(\theta)} J_0\left(\mathtt{f}_{\mathrm{obj}}\xi\sin(\theta)\right) d\theta$$

$$I_{2,2}(\xi) = \int_0^{\theta_{\max}} \sqrt{\cos(\theta)}\,\frac{\sin(2\theta)}{2}e^{-i\kappa r_3^{\Psi}\cos(\theta)} J_2\left(\mathtt{f}_{\mathrm{obj}}\xi\sin(\theta)\right) d\theta,$$

J_m denotes the Bessel function of the first kind of order m, and $\theta_{\max}$ is the angle of aperture as defined in Eq. (1).

Proof We use the following notation

$$\mathbf{u} = \begin{pmatrix} u_1 \\ u_2 \end{pmatrix} = \xi\begin{pmatrix} \cos(v) \\ \sin(v) \end{pmatrix} \text{ and } \mathbf{x} = \begin{pmatrix} x_1 \\ x_2 \end{pmatrix} = \rho\begin{pmatrix} \cos(\varphi) \\ \sin(\varphi) \end{pmatrix},$$

where $\rho = \rho(\theta)$ (see Eq. (63)) is the radial length on the back focal plane.

The two-dimensional Fourier transform of the $\widehat{\mathbf{E}}^{\mathrm{obj}}$ (defined in Eq. (64)) reads as follows:

$$\mathcal{F}_{12}[\widehat{\mathbf{E}}^{\mathrm{obj}}](\mathbf{u}) = \frac{1}{2\pi}\int_{\mathbf{x}\in\mathbb{R}^2} \widehat{\mathbf{E}}^{\mathrm{obj}}(\mathbf{x})e^{-i\mathbf{u}\cdot\mathbf{x}}d\mathbf{x}$$

$$= \frac{1}{2\pi}(\mathtt{f}_{\mathrm{obj}})^2\int_0^{\theta_{\max}}\int_0^{2\pi} \widehat{\mathbf{E}}^{\mathrm{obj}}(\rho(\theta), \varphi)e^{-i\xi\rho(\theta)\cos(\varphi - v)}\cos(\theta)\sin(\theta)d\varphi d\theta$$

$$= \frac{(\mathtt{f}_{\mathrm{obj}})^2}{2\pi}C(\mathtt{f}_{\mathrm{obj}})\int_0^{\theta_{\max}}\int_0^{2\pi}\left\{\left(-\Psi_p\cos(\theta) + \Psi_3\sin(\theta)\right)\mathbf{e}_p - \Psi_s\mathbf{e}_s\right\}$$

$$\cdot e^{-i\kappa r_3^{\Psi}\cos(\theta)}e^{-i\xi\rho(\theta)\cos(\varphi - v)}\sqrt{\cos(\theta)}\sin(\theta)d\varphi d\theta. \tag{75}$$

Next we calculate the integral on the right-hand side of Eq. (75):

$$
\int_0^{\theta_{max}} \int_0^{2\pi} \left\{ \left(-\Psi_p \cos(\theta) + \Psi_3 \sin(\theta) \right) \mathbf{e}_p - \Psi_s \mathbf{e}_s \right\} \cdot
$$
$$
\cdot\, e^{-i\kappa r_3^\Psi \cos(\theta)} e^{-i\xi\rho(\theta)\cos(\varphi-\nu)} \sqrt{\cos(\theta)} \sin(\theta) d\varphi d\theta
$$
$$
= - \int_0^{\theta_{max}} e^{-i\kappa r_3^\Psi \cos(\theta)} \sqrt{\cos(\theta)} \frac{\sin(2\theta)}{2} \left(\int_0^{2\pi} \Psi_p \mathbf{e}_p e^{-i\xi\rho(\theta)\cos(\varphi-\nu)} d\varphi \right) d\theta
$$
$$
- \int_0^{\theta_{max}} e^{-i\kappa r_3^\Psi \cos(\theta)} \sqrt{\cos(\theta)} \sin(\theta) \left(\int_0^{2\pi} \Psi_s \mathbf{e}_s e^{-i\xi\rho(\theta)\cos(\varphi-\nu)} d\varphi \right) d\theta
$$
$$
+ \int_0^{\theta_{max}} e^{-i\kappa r_3^\Psi \cos(\theta)} (\cos(\theta))^{3/2} \frac{1-\cos(2\theta)}{2} \left(\int_0^{2\pi} \Psi_3 \mathbf{e}_p e^{-i\xi\rho(\theta)\cos(\varphi-\nu)} d\varphi \right) d\theta,
$$
$$(76)$$

where we use sin and cos summation formulas.

We proceed by first evaluating the inner integrals (involving the φ variable) on the right-hand side of Eq. (76), by transforming the $(\mathbf{e}_p, \mathbf{e}_s, \mathbf{e}_3)$ system to the $(\mathbf{e}_1, \mathbf{e}_2, \mathbf{e}_3)$ system, and then using the Bessel identities Eq. (117), to evaluate the integrals.

Using Eq. (57) it follows from Eq. (36) that

$$
\Psi_p = |\mathbf{\Psi}| \sin(\theta_m) \cos(\varphi_m - \varphi), \quad \Psi_s = |\mathbf{\Psi}| \sin(\theta_m) \sin(\varphi_m - \varphi). \tag{77}
$$

Again by application of Eq. (36) and sin and cos summation formulas we get

$$
\Psi_p \cos(\varphi) = \Psi_1 \frac{1+\cos(2\varphi)}{2} + \Psi_2 \frac{\sin(2\varphi)}{2}, \; \Psi_p \sin(\varphi)
$$
$$
= \Psi_1 \frac{\sin(2\varphi)}{2} + \Psi_2 \frac{1-\cos(2\varphi)}{2},
$$
$$
\Psi_s \cos(\varphi) = -\Psi_1 \frac{\sin(2\varphi)}{2} + \Psi_2 \frac{1+\cos(2\varphi)}{2}, \quad \Psi_s \sin(\varphi)
$$
$$
= -\Psi_1 \frac{1-\cos(2\varphi)}{2} + \Psi_2 \frac{\sin(2\varphi)}{2}.
$$
$$(78)$$

Using Eq. (57), we express the first inner integral on the right-hand side of Eq. (76):

$$
\int_0^{2\pi} \Psi_p e^{-i\xi\rho(\theta)\cos(\varphi-\nu)} d\varphi \mathbf{e}_p = \int_0^{2\pi} \Psi_p \cos(\varphi) e^{-i\xi\rho(\theta)\cos(\varphi-\nu)} d\varphi \mathbf{e}_1
$$
$$
+ \int_0^{2\pi} \Psi_p \sin(\varphi) e^{-i\xi\rho(\theta)\cos(\varphi-\nu)} d\varphi \mathbf{e}_2,
$$
$$(79)$$

and to evelute the integral we use the Bessel identities Eqs. (117), and (78).

We use Eq. (117) for $m = 0$ and $m = 2$, and Eq. (63) to evaluate the first integral in Eq. (79):

$$
\int_0^{2\pi} \Psi_p \cos(\varphi) e^{-i\eta \cos(\varphi - v)} d\varphi = \Psi_1 \int_0^{2\pi} \frac{1 + \cos(2\varphi)}{2} e^{-i\eta \cos(\varphi - v)} d\varphi
$$

$$
+ \Psi_2 \int_0^{2\pi} \frac{\sin(2\varphi)}{2} e^{-i\eta \cos(\varphi - v)} d\varphi
$$

$$
= \frac{\Psi_1}{2} \int_0^{2\pi} e^{-i\eta \cos(\varphi - v)} d\varphi + \frac{\Psi_1}{2} \int_0^{2\pi} \cos(2\varphi) e^{-i\eta \cos(\varphi - v)} d\varphi \tag{80}
$$

$$
+ \frac{\Psi_2}{2} \int_0^{2\pi} \sin(2\varphi) e^{-i\eta \cos(\varphi - v)} d\varphi
$$

$$
= \pi \Psi_1 J_0(\eta) - \pi \Psi_1 \cos(2v) J_2(\eta) - \pi \Psi_2 \sin(2v) J_2(\eta),
$$

where $\eta = f_{obj} \xi \sin(\theta)$, and a calculation similar to Eq. (80) yields

$$
\frac{1}{\pi} \int_0^{2\pi} \Psi_p \sin(\varphi) e^{-i\eta \cos(\varphi - v)} d\varphi = \Psi_2 J_0(\eta) - \Psi_2 \cos(2v) J_2(\eta) - \Psi_1 \sin(2v) J_2(\eta). \tag{81}
$$

Thus, using Eqs. (80) and (81), in Eq. (79), the first integral expression on the right-hand side of Eq. (76), becomes

$$
-\frac{1}{\pi} \int_0^{\theta_{max}} e^{-i\kappa r_3^\Psi \cos(\theta)} \sqrt{\cos(\theta)} \frac{\sin(2\theta)}{2} \left(\int_0^{2\pi} \Psi_p e_p e^{-i\xi \rho(\theta) \cos(\varphi - v)} d\varphi \right) d\theta
$$

$$
= -(\Psi_1 e_1 + \Psi_2 e_2) \int_0^{\theta_{max}} \sqrt{\cos(\theta)} \frac{\sin(2\theta)}{2} e^{-i\kappa r_3^\Psi \cos(\theta)} J_0 \left(f_{obj} \xi \sin(\theta) \right) d\theta
$$

$$
+ (\Psi_1 e_1 + \Psi_2 e_2) \cos(2v) \int_0^{\theta_{max}} \sqrt{\cos(\theta)} \frac{\sin(2\theta)}{2} e^{-i\kappa r_3^\Psi \cos(\theta)} J_2 \left(f_{obj} \xi \sin(\theta) \right) d\theta
$$

$$
+ (\Psi_2 e_1 + \Psi_1 e_2) \sin(2v) \int_0^{\theta_{max}} \sqrt{\cos(\theta)} \frac{\sin(2\theta)}{2} e^{-i\kappa r_3^\Psi \cos(\theta)} J_2 \left(f_{obj} \xi \sin(\theta) \right) d\theta
$$

$$
= -(\Psi_1 e_1 + \Psi_2 e_2) I_{2,0}(\xi, r_3^\Psi) + (\Psi_1 e_1 + \Psi_2 e_2) \cos(2v) I_{2,2}(\xi, r_3^\Psi)
$$

$$
+ (\Psi_2 e_1 + \Psi_1 e_2) \sin(2v) I_{2,2}(\xi, r_3^\Psi), \tag{82}
$$

where integrals $I_{p,q}(\xi, r_3^\Psi)$ are as in Eq. (74).

Similar calculation to Eq. (80) yields

$$
\frac{1}{\pi} \int_0^{2\pi} \Psi_s \sin(\varphi) e^{-i\eta \cos(\varphi - v)} d\varphi = -\Psi_1 J_0(\eta) - \Psi_1 \cos(2v) J_2(\eta) - \Psi_2 \sin(2v) J_2(\eta),
$$

$$
\frac{1}{\pi} \int_0^{2\pi} \Psi_s \cos(\varphi) e^{-i\eta \cos(\varphi - v)} d\varphi = \Psi_2 J_0(\eta) + \Psi_2 \cos(2v) J_2(\eta) + \Psi_1 \sin(2v) J_2(\eta). \tag{83}
$$

Next, using Eqs. (57) and (83), we compute the second integral term on the right-hand side of Eq. (76):

$$-\frac{1}{\pi}\int_0^{\theta_{\max}} e^{-i\kappa r_3^{\Psi}\cos(\theta)}\sqrt{\cos(\theta)}\sin(\theta)\left(\int_0^{2\pi}\Psi_s \mathbf{e}_s e^{-i\xi\rho(\theta)\cos(\varphi-\nu)}d\varphi\right)d\theta$$

$$= -(\Psi_1\mathbf{e}_1 + \Psi_2\mathbf{e}_2)\int_0^{\theta_{\max}}\sqrt{\cos(\theta)}\sin(\theta)e^{-i\kappa r_3^{\Psi}\cos(\theta)}J_0\left(\mathtt{f}_{\mathrm{obj}}\xi\sin(\theta)\right)d\theta$$

$$+ (\Psi_1\mathbf{e}_1 + \Psi_2\mathbf{e}_2)\cos(2\nu)\int_0^{\theta_{\max}}\sqrt{\cos(\theta)}\sin(\theta)e^{-i\kappa r_3^{\Psi}\cos(\theta)}J_2\left(\mathtt{f}_{\mathrm{obj}}\xi\sin(\theta)\right)d\theta$$

$$+ (\Psi_2\mathbf{e}_1 + \Psi_1\mathbf{e}_2)\sin(2\nu)\int_0^{\theta_{\max}}\sqrt{\cos(\theta)}\sin(\theta)e^{-i\kappa r_3^{\Psi}\cos(\theta)}J_2\left(\mathtt{f}_{\mathrm{obj}}\xi\sin(\theta)\right)d\theta$$

$$= -(\Psi_1\mathbf{e}_1 + \Psi_2\mathbf{e}_2)I_{1,0}(\xi, r_3^{\Psi}) + (\Psi_1\mathbf{e}_1 + \Psi_2\mathbf{e}_2)\cos(2\nu)I_{1,2}(\xi, r_3^{\Psi})$$

$$+ (\Psi_2\mathbf{e}_1 + \Psi_1\mathbf{e}_2)\sin(2\nu)I_{1,2}(\xi, r_3^{\Psi}),$$

$$(84)$$

where integrals $I_{p,q}(\xi, r_3^{\Psi})$ are as in Eq. (74).

Next, we compute the last integral term on the right-hand side of Eq. (76):

$$\frac{1}{\pi}\int_0^{\theta_{\max}} e^{-i\kappa r_3^{\Psi}\cos(\theta)}(\cos(\theta))^{3/2}\frac{1-\cos(2\theta)}{2}\left(\int_0^{2\pi}\Psi_3\mathbf{e}_p e^{-i\xi\rho(\theta)\cos(\varphi-\nu)}d\varphi\right)d\theta$$

$$= \frac{1}{\pi}\Psi_3\int_0^{\theta_{\max}} e^{-i\kappa r_3^{\Psi}\cos(\theta)}(\cos(\theta))^{3/2}\frac{1-\cos(2\theta)}{2}$$

$$\left(\mathbf{e}_1\int_0^{2\pi}\cos(\varphi)e^{-i\xi\rho(\theta)\cos(\varphi-\nu)}d\varphi\right)d\theta$$

$$+ \frac{1}{\pi}\Psi_3\int_0^{\theta_{\max}} e^{-i\kappa r_3^{\Psi}\cos(\theta)}(\cos(\theta))^{3/2}\frac{1-\cos(2\theta)}{2}$$

$$\left(\mathbf{e}_2\int_0^{2\pi}\sin(\varphi)e^{-i\xi\rho(\theta)\cos(\varphi-\nu)}d\varphi\right)d\theta$$

$$= -2i\Psi_3\left(\cos(\nu)\mathbf{e}_1 + \sin(\nu)\mathbf{e}_2\right)\int_0^{\theta_{\max}}(\cos(\theta))^{3/2}$$

$$\frac{1-\cos(2\theta)}{2}e^{-i\kappa r_3^{\Psi}\cos(\theta)}J_1\left(\mathtt{f}_{\mathrm{obj}}\xi\sin(\theta)\right)d\theta$$

$$= -2i\Psi_3\left(\cos(\nu)\mathbf{e}_1 + \sin(\nu)\mathbf{e}_2\right)I_{2,1}(\xi, r_3^{\Psi}),$$

$$(85)$$

where in the second equality we use Eq. (117) for $m = 1$, and in the last equality we use the integral $I_{p,q}(\xi, r_3^{\Psi})$ as in Eq. (74).

Using Eqs. (82), (84), and (85), the expression of Eq. (76) becomes

$$\frac{1}{\pi} \int_0^{\theta_{\max}} \int_0^{2\pi} \left\{ \left(-\Psi_p \cos(\theta) + \Psi_3 \sin(\theta) \right) \mathbf{e}_p - \Psi_s \mathbf{e}_s \right\} \cdot$$

$$\cdot e^{-i\kappa r_3^{\Psi} \cos(\theta)} e^{-i\xi \rho(\theta) \cos(\varphi - v)} \sqrt{\cos(\theta)} \sin(\theta) d\varphi d\theta$$

$$= -(\Psi_1 \mathbf{e}_1 + \Psi_2 \mathbf{e}_2)[I_{1,0}(\xi, r_3^{\Psi}) + I_{2,0}(\xi, r_3^{\Psi})] \tag{86}$$

$$+ (\Psi_1 \mathbf{e}_1 + \Psi_2 \mathbf{e}_2) \cos(2v)[I_{1,2}(\xi, r_3^{\Psi}) + I_{2,2}(\xi, r_3^{\Psi})]$$

$$+ (\Psi_2 \mathbf{e}_1 + \Psi_1 \mathbf{e}_2) \sin(2v)[I_{1,2}(\xi, r_3^{\Psi}) + I_{2,2}(\xi, r_3^{\Psi})]$$

$$- 2i\Psi_3 (\cos(v)\mathbf{e}_1 + \sin \mathbf{e}_2) I_{2,1}(\xi),$$

where the integrals $I_{p,q}(\xi)$ are defined in Eq. (74). $\qquad\qquad\square$

6.4   Electric Field Approximation in the Lens

After the light ray has passed through the objective and the back focal plane, a tube lens is placed to focus the light rays onto the image plane.

Definition 11 (Tube Lens Parameters) For the tube lens, we assume that it is a plano-convex *converging lens* with focal length $f_L > 0$, placed with one side at $r_3^{tl_i}$ and the other side at $r_3^{tl_o}$ (see Fig. 8). Moreover the lens has a thickness which is measured orthogonal to $\mathbf{e}_3$ by the function d.

The incoming field at the tube lens $\widehat{\mathbf{E}}^{tl_i}$ (as defined in Eq. (69)) and the outgoing wave field $\widehat{\mathbf{E}}^{tl_o}$ immediately after the lens aperture are related by (we use the same polar coordinates (ρ, φ) in both planes $\{r_3 = r_3^{tl_i}\}$ and $\{r_3 = r_3^{tl_o}\}$)

$$\widehat{\mathbf{E}}^{tl_o}(\rho, \varphi) = e^{i\mu(\rho)} P_L(\rho) \widehat{\mathbf{E}}^{tl_i}(\rho, \varphi), \tag{87}$$

where P_L is the pupil function associated with the tube lens as in Eq. (53), μ is the phase shift experienced by the field through the tube lens (note that it does not depend on φ):

$$\mu(\rho) = \underbrace{\kappa n_l d(\rho)}_{\text{phase delay by lens}} + \underbrace{\kappa (d_0 - d(\rho))}_{\text{phase delay by vacuum}}, \tag{88}$$

where d_0 is the maximum thickness of the lens, $d(\rho)$ is the thickness of the lens at distance ρ from the optical axis, n_l is the refractive index of the lens material, and κ as defined in Eq. (22). The phase delay induced by the lens, under the assumption of a *paraxial approximation* reads as follows (see Table 1 for the summary of all physical parameters below):

$$\mu(\rho) \approx \kappa n_l d_0 - \frac{\kappa}{2f_L}\rho^2 \text{ for all } \rho \in \mathbb{R}. \tag{89}$$

6.5 Electric Field Approximation in the Image Plane

Definition 12 The *electric field at the focal plane*,

$$\mathcal{I} := \left\{ (\mathbf{x}_f, r_3^f) : \mathbf{x}_f \in \mathbb{R}^2 \right\}, \tag{90}$$

is denoted by $\widehat{\mathbf{E}}^f(\mathbf{x}_f) := \widehat{\mathbf{E}}(\mathbf{x}_f, r_3^f)$, where $\widehat{\mathbf{E}}$ solves the boundary value problem

$$\Delta\widehat{\mathbf{E}}(\mathbf{x}, r_3) + \kappa^2(\mathbf{x}, r_3)\widehat{\mathbf{E}}(\mathbf{x}, r_3) = 0 \text{ for all } \mathbf{x} \in \mathbb{R}^2, \ r_3^{tl_o} < r_3 < r_3^f \tag{91}$$

with boundary data

$$\widehat{\mathbf{E}}(\mathbf{x}, r_3^{tl_o}) = \widehat{\mathbf{E}}^{tl_o}(\mathbf{x}) \text{ for all } \mathbf{x} \in \mathbb{R}^2. \tag{92}$$

Note that $\widehat{\mathbf{E}}^{tl_o}$ as defined in Eq. (64) is already an approximation of the electric field outside of the lens system.

Following [14, Eqs 5–14], we can calculate the field $\widehat{\mathbf{E}}$ in the *image plane*.

Lemma 8 *At a point $\mathbf{x}_f$ in the image plane,*

$$\boxed{\widehat{\mathbf{E}}^f(\mathbf{x}_f) = \frac{1}{i\lambda f_L} e^{i\frac{2\pi}{\lambda}(f_L + n_l d_0)} e^{i\frac{\pi}{\lambda f_L}|\mathbf{x}_f|^2} \int_{\mathbf{x}\in\mathbb{R}^2} P_L(\mathbf{x})\widehat{\mathbf{E}}^{tl_i}(\mathbf{x}) e^{-i\frac{2\pi}{\lambda f_L}\langle \mathbf{x}_f, \mathbf{x}\rangle} d\mathbf{x} \text{ for all } \mathbf{x}_f \in \mathbb{R}^2.}$$
$$\tag{93}$$

Proof We apply the Huygens-Fresnel principle (see [14, Eqs 4–17]) to compute the field in the *image plane*:

$$\widehat{\mathbf{E}}^f(\mathbf{x}_f) = \frac{1}{i\lambda d} e^{i\kappa d} e^{i\frac{\kappa}{2d}|\mathbf{x}_f|^2} \int_{\mathbf{x}\in\mathbb{R}^2} \widehat{\mathbf{E}}^{tl_o}(\mathbf{x}) e^{i\frac{\kappa}{2d}|\mathbf{x}|^2} e^{-i\frac{\kappa}{d}\langle \mathbf{x}_f, \mathbf{x}\rangle} d\mathbf{x}$$

$$= \frac{1}{i\lambda d} e^{i\kappa d} e^{i\frac{\kappa}{2d}|\mathbf{x}_f|^2} \int_{\mathbf{x}\in\mathbb{R}^2} P_L(\mathbf{x})\widehat{\mathbf{E}}^{tl_i}(\mathbf{x}) e^{i\mu(\mathbf{x})} e^{i\frac{\kappa}{2d}|\mathbf{x}|^2} e^{-i\frac{\kappa}{d}\langle \mathbf{x}_f, \mathbf{x}\rangle} d\mathbf{x}$$

$$= \frac{1}{i\lambda d} e^{i\kappa d} e^{i\frac{\kappa}{2d}|\mathbf{x}_f|^2} \int_{\mathbf{x}\in\mathbb{R}^2} P_L(\mathbf{x})\widehat{\mathbf{E}}^{tl_i}(\mathbf{x}) e^{i\kappa n_l d_0} e^{-i\frac{\kappa}{2f_L}|\mathbf{x}|^2} e^{i\frac{\kappa}{2d}|\mathbf{x}|^2} e^{-i\frac{\kappa}{d}\langle \mathbf{x}_f, \mathbf{x}\rangle} d\mathbf{x}$$

$$= \frac{1}{i\lambda d} e^{i\frac{2\pi}{\lambda}(d+n_l d_0)} e^{i\frac{\pi}{\lambda d}|\mathbf{x}_f|^2} \int_{\mathbf{x}\in\mathbb{R}^2} P_L(\mathbf{x})\widehat{\mathbf{E}}^{tl_i}(\mathbf{x}) e^{-i\frac{\pi}{\lambda f_L}(1-\frac{f_L}{d})|\mathbf{x}|^2} e^{-i\frac{2\pi}{\lambda d}\langle \mathbf{x}_f, \mathbf{x}\rangle} d\mathbf{x},$$

where in the second equality we use Eq. (87), in the third equality we use the paraxial approximation Eq. (89), and in the last equality we use $\kappa = \frac{2\pi}{\lambda}$. Now, if the image plane is at the distance $d = f_L$, the quadratic phase factor terms within the integrand exactly cancel, leaving

$$
\widehat{\mathbf{E}}^f(\mathbf{x_f}) = \frac{1}{i\lambda f_L} e^{i\frac{2\pi}{\lambda}(f_L + n_l d_0)} e^{i\frac{\pi}{\lambda f_L}|\mathbf{x_f}|^2} \int_{\mathbf{x}\in\mathbb{R}^2} P_L(\mathbf{x})\widehat{\mathbf{E}}^{t1_i}(\mathbf{x}) e^{-i\frac{2\pi}{\lambda f_L}\langle \mathbf{x_f}, \mathbf{x}\rangle} d\mathbf{x},
$$

where the term $e^{i\frac{2\pi}{\lambda}(f_L + n_l d_0)}$ is a constant amplitude, and the term $e^{i\frac{\pi}{\lambda f_L}|\mathbf{x_f}|^2}$ describes a spherical phase curvature in the focal plane. $\qquad\square$

Remark 6 From Eq. (93) it follows by the convolution theorem for the Fourier transform that

$$
\begin{aligned}
\widehat{\mathbf{E}}^f(\mathbf{x_f}) &= \frac{2\pi}{\lambda f_L} e^{i\pi\left(-\frac{1}{2}+\frac{2}{\lambda}(f_L + n_l d_0)+\frac{1}{\lambda f_L}|\mathbf{x_f}|^2\right)} \mathcal{F}_{12}[P_L\widehat{\mathbf{E}}^{t1_i}]\left(2\pi\frac{\mathbf{x_f}}{\lambda f_L}\right) \\
&= C_f \Phi_f \left(\mathcal{F}_{12}[P_L] * \mathcal{F}_{12}[\widehat{\mathbf{E}}^{t1_i}]\right)\left(\frac{2\pi}{\lambda f_L}\mathbf{x_f}\right) \quad \text{for all } \mathbf{x_f} \in \mathbb{R}^2,
\end{aligned}
\tag{94}
$$

where

$$
C_f = \frac{1}{\lambda f_L} \quad \text{and} \quad \Phi_f = e^{i\pi\left(-\frac{1}{2}+\frac{2}{\lambda}(f_L + n_l d_0)+\frac{1}{\lambda f_L}|\mathbf{x_f}|^2\right)}.
\tag{95}
$$

In the next step we calculate the Fourier-transform of a circular pupil function:

Lemma 9 *Let $P : \mathbb{R}^2 \to \mathbb{R}$ be the* circular pupil function *with radius R, as defined in Eq. (53), then*

$$
\mathcal{F}_{12}[P](\mathbf{v}) = R^2 \frac{J_1(R|\mathbf{v}|)}{R|\mathbf{v}|} \text{ for all } \mathbf{v} \in \mathbb{R}^2.
\tag{96}
$$

Proof We use the following polar coordinates:

$$
\mathbf{v} = \varkappa \begin{pmatrix} \cos(\vartheta) \\ \sin(\vartheta) \end{pmatrix} \text{ and } \mathbf{y} = \varrho \begin{pmatrix} \cos(\sigma) \\ \sin(\sigma) \end{pmatrix},
$$

then the Fourier transform of the pupil function

$$
\begin{aligned}
\mathcal{F}_{12}[P](\mathbf{v}) &= \frac{1}{2\pi} \int_{\mathbf{y}\in\mathbb{R}^2} P(\mathbf{y}) e^{-i\mathbf{y}\cdot\mathbf{v}} d\mathbf{y} = \frac{1}{2\pi} \int_0^R \varrho \int_0^{2\pi} e^{-i\varkappa\varrho\cos(\sigma-\vartheta)} d\sigma d\varrho \\
&= \int_0^R \varrho J_0(\varkappa\varrho) d\varrho = \frac{R}{\varkappa} J_1(R\varkappa) = R^2 \frac{J_1(R|\mathbf{v}|)}{R|\mathbf{v}|},
\end{aligned}
\tag{97}
$$

where we use Eq. (117) for $m = 0$ in the third equality. $\qquad\square$

Now, we calculate the k-transform of $\widehat{\mathbf{E}}^{bfp}$ in this approximation.

6.6 Small Aperature

In what follows, we are interested in a small aperture. Of course small apertures have the problem that only little light of the emitted dipoles passes through the lens and thus these considerations are more of theoretical nature.

In case the numerical aperture NA is small, that is if $\theta_{\max}$ is small, it follows from Eq. (64) that

$$\widehat{\mathbf{E}}_{\text{small}}^{\text{bfp}}(\rho, \varphi) = C(\mathtt{f}_{\text{obj}}) \begin{pmatrix} \Psi_1 \\ \Psi_2 \\ 0 \end{pmatrix} \chi_{\rho \leqslant \text{NA}}. \tag{98}$$

In what follows, we denote by P_{NA} the pupil function $\chi_{\rho \leqslant \text{NA}}$. We emphasize that the left-hand side of Eq. (98) is actually an approximation of the right-hand side of Eq. (64). Next we calculate the k-transform of $\widehat{\mathbf{E}}^{\text{bfp}}$ in this approximation. Noting that the objective acts again as a pupil function with disc radius NA we get analogously to Lemma 9 and by using Eq. (73)

$$\mathcal{F}_{12}[\widehat{\mathbf{E}}_{\text{small}}^{\text{tl}i}](\xi, \nu) = C(\mathtt{f}_{\text{obj}}) e^{i(r_3^{\text{tl}i} - r_3^{\text{bfp}})\sqrt{\kappa^2 - \xi^2}}$$
$$\begin{pmatrix} \Psi_1 \\ \Psi_2 \\ 0 \end{pmatrix} \mathcal{F}_{12}[P_{\text{NA}}](\xi, \nu), \quad \xi^2 \leq \kappa^2, \nu \in [0, 2\pi). \tag{99}$$

7 Single Molecule Localization Microscopy Experiments and Inverse Problems

We consider two experiments in two different settings:

Experiment

For *setting* 1 we assume n static emitting dipoles: Several monochromatic plane waves of the same frequency ω_{inc} but with different orientations $\mathbf{v}^{(j)}$, $j = 1, 2, \ldots, M$ with $M > 1$ are used to illuminate the cell. Every emitting dipole emits light in an orientation $\mathbf{\Psi}^{(j)}(k)$, $k = 1, 2, \ldots, n$, $j = 1, 2, \ldots, M$ (depending on the incident field according to Eq. (51)).

Therefore, for each experiment $j = 1, 2, \ldots, M$ the current

$$\widehat{\mathbf{J}}^{(j)}(\mathbf{r}) = i\omega \sum_{k=1}^{n} \mathbf{\Psi}^{(j)}(k)\delta(\mathbf{r} - \mathbf{r}^{\mathbf{\Psi}(k)})$$

and the density

$$\widehat{\rho}^{(j)}(\mathbf{r}) = -\sum_{k=1}^{n} \left|\Psi^{(j)}(k)\right| \delta'(\mathbf{r} - \mathbf{r}^{\Psi(k)}) = -\sum_{k=1}^{n} \delta'(\mathbf{r} - \mathbf{r}^{\Psi(k)})$$

are given via Eqs. (38) and (37) as the superposition of all dipoles. Note, that we assume that all dipoles are unit vectors.

Consequently, the electric field $\widehat{\mathbf{E}} := \widehat{\mathbf{E}}^{(j)}$ solves Eq. (29),

$$\Delta_{\mathbf{r}}\widehat{\mathbf{E}}(\mathbf{r}; \omega) + \kappa^2(\omega)\widehat{\mathbf{E}}(\mathbf{r}; \omega) = \frac{i\omega}{\epsilon_0 c^2}\widehat{\mathbf{J}}^{(j)}(\mathbf{r}) + \frac{1}{\epsilon_0}\nabla_{\mathbf{r}}\widehat{\rho}^{(j)}(\mathbf{r}) \text{ for all } \mathbf{r} \in \Omega.$$

1. The measurements recorded in a *static experiment* are the energies of the electric field in the image plane, after the light has passed through the imaging system. That is, for each experiment $j = 1, 2, \ldots, M$ the data

$$m_i^{(j)}(\mathbf{x}_{\mathrm{f}}; t) = \left|(\mathbf{E}_i^{\mathrm{f}})^{(j)}\right|^2 (\mathbf{x}_{\mathrm{f}}; t) \text{ for all } \mathbf{x}_{\mathrm{f}} \in \mathbb{R}^2, t > 0 \text{ and for } i = 1, 2$$

(100)

are recorded.

2. In the *dynamic experiment* setting the static experiment is repeated. We denote the experiment repetitions with the parameter s: This experimental setup is used in practice because it makes use of blinking dyes, which allows for better localization of the dyes, and thus molecules. That is, the measurements are

$$m^{(j)}(\mathbf{x}_{\mathrm{f}}; t; s) = \left|(\mathbf{E}_i^{\mathrm{f}})^{(j)}\right|^2 (\mathbf{x}_{\mathrm{f}}; t; s) \text{ for all } \mathbf{x}_{\mathrm{f}} \in \mathbb{R}^2, t, s > 0 \text{ for } i = 1, 2.$$

(101)

For *setting* 2 we assume rotating dipoles, which are modelled via Eq. (50)—here in particular we assume that the cone becomes the ball. That is, $\theta = \pi$. Note that in this case several monochromatic excitations do not provide an asset, so we can constrain ourselves to the case $M = 1$. As a consequence of Eq. (50), the emitted currents of all rotating dipoles of a single molecule localization microscopy experiment are given by

$$\widehat{\mathbf{J}}(\mathbf{r}; \omega) = i\omega \sum_{k=1}^{n} \mathbf{I}_k \mathbb{1}_k(\mathbf{r}),$$

(102)

and according to Eq. (29) the electric field satisfies the equation

$$\Delta_{\mathbf{r}}\widehat{\mathbf{E}}(\mathbf{r};\omega) + \kappa^2(\omega)\widehat{\mathbf{E}}(\mathbf{r};\omega) = \frac{1}{\epsilon_0}\sum_m \mathbf{I}_m\widehat{\mathbf{R}}_m(\mathbf{r};\omega) \text{ for all } \mathbf{r} \in \Omega. \tag{103}$$

1. The measurements recorded in a single molecule localization microscopy are the energies of the electric field in the focal plane (see Eq. (93)), after the light has passed through the imaging system. That is, the data

$$m(\mathbf{x}_{\mathrm{f}}; t) = \left|\mathbf{E}^{\mathrm{f}}\right|^2 (\mathbf{x}_{\mathrm{f}}; t) \text{ for all } \mathbf{x}_{\mathrm{f}} \in \mathbb{R}^2, \quad t > 0 \tag{104}$$

are recorded.

2. In the *dynamic experiment* setting the static experiment is repeated, and we denote every repetition experiment with the parameter s: That is the measurements are

$$m(\mathbf{x}_{\mathrm{f}}; t; s) = \left|\mathbf{E}^{\mathrm{f}}\right|^2 (\mathbf{x}_{\mathrm{f}}; t; s) \text{ for all } \mathbf{x}_{\mathrm{f}} \in \mathbb{R}^2, \quad t, s > 0. \tag{105}$$

Note the difference between setting 1 and 2. In the former it is much easier to identify dipoles because the orientation can be resolved and is not changing over time.

7.1 The Limit

Using the small aperture limit of Sect. 6.6 in combination with the formula for the electric field on the image plane Eq. (94), we can compute the electric field in the image plane from Eq. (99). We first make use of a linear approximation of the function $[0, \kappa] \ni \rho \to \sqrt{\kappa^2 - \rho^2} \simeq \kappa$, that is we assume that between the back focal plane of the objective and the lens, the electric field only undergoes a phase shift which does not depend on the distance to the optical axis. It follows then from Eq. (99) that

$$\mathcal{F}_{12}[\widehat{\mathbf{E}}_{\mathrm{small}}^{\mathrm{tl}_i}](\xi, \nu) \simeq C(\mathbf{f}_{\mathrm{obj}})\mathrm{e}^{\mathrm{i}(r_3^{\mathrm{tl}_i} - r_3^{\mathrm{bfp}})\kappa} \begin{pmatrix} \Psi_1 \\ \Psi_2 \\ 0 \end{pmatrix} \mathcal{F}_{12}[P_{\mathrm{NA}}](\xi, \nu). \tag{106}$$

Applying Eq. (94) where we replace $\mathcal{F}_{12}[\widehat{\mathbf{E}}^{\mathrm{tl}_i}]$ by $\mathcal{F}_{12}[\widehat{\mathbf{E}}_{\mathrm{small}}^{\mathrm{tl}_i}]$ and inserting Eq. (106), we obtain

$$\widehat{\mathbf{E}}^{\mathrm{f}}_{\mathrm{small}}(\mathbf{x}_{\mathrm{f}}) = C_{\mathrm{f}} \Phi_{\mathrm{f}} \left(\mathcal{F}_{12}[P_{\mathrm{L}}] * \mathcal{F}_{12}[\widehat{\mathbf{E}}^{\mathrm{t1}_i}_{\mathrm{small}}] \right) \left(\frac{2\pi}{\lambda \mathrm{f}_{\mathrm{L}}} \mathbf{x}_{\mathrm{f}} \right)$$

$$= \begin{pmatrix} \Psi_1 \\ \Psi_2 \\ 0 \end{pmatrix} (\mathcal{F}_{12}[P_{\mathrm{L}}] * \mathcal{F}_{12}[P_{\mathrm{NA}}]) \, (\xi, \nu)$$

$$= \begin{pmatrix} \Psi_1 \\ \Psi_2 \\ 0 \end{pmatrix} \mathcal{F}_{12}[P_{\mathrm{L}} P_{\mathrm{NA}}](\xi, \nu),$$

where we use the following polar coordinates: $\frac{2\pi}{\lambda \mathrm{f}_{\mathrm{L}}} \mathbf{x}_{\mathrm{f}} = \xi \begin{pmatrix} \cos(\nu) \\ \sin(\nu) \end{pmatrix}$.

Now, assuming that $R \geqslant \mathrm{NA}$ (the lens is bigger than the objective), we have $P_{\mathrm{L}} P_{\mathrm{NA}} = P_{\mathrm{NA}}$ and Eq. (97) provides,

$$\widehat{\mathbf{E}}^{\mathrm{f}}_{\mathrm{small}}(\mathbf{x}_{\mathrm{f}}) = C(\mathrm{f}_{\mathrm{obj}}) C_{\mathrm{f}} \Phi_{\mathrm{f}} e^{i(r_3^{\mathrm{t1}_i} - r_3^{\mathrm{bfp}})\kappa}$$

$$\begin{pmatrix} \Psi_1 \\ \Psi_2 \\ 0 \end{pmatrix} \mathrm{NA}^2 \frac{J_1(\mathrm{NA} \frac{2\pi}{\lambda \mathrm{f}_{\mathrm{L}}} |\mathbf{x}_{\mathrm{f}}|)}{\mathrm{NA} \frac{2\pi}{\lambda \mathrm{f}_{\mathrm{L}}} |\mathbf{x}_{\mathrm{f}}|} =: \Gamma(\omega) \begin{pmatrix} \Psi_1 \\ \Psi_2 \\ 0 \end{pmatrix} \mathrm{NA}^2 \frac{J_1(\mathrm{NA} \frac{2\pi}{\lambda \mathrm{f}_{\mathrm{L}}} |\mathbf{x}_{\mathrm{f}}|)}{\mathrm{NA} \frac{2\pi}{\lambda \mathrm{f}_{\mathrm{L}}} |\mathbf{x}_{\mathrm{f}}|}.$$

Finally, what is actually measured in experiments is the intensity $m(\mathbf{x}_{\mathrm{f}}, t) = |\mathbf{E}^{\mathrm{f}}(\mathbf{x}_{\mathrm{f}}, t)|^2$ averaged in time. If we assume that the signal $\mathbf{E}$ is compactly supported in time and that what is measured by the detector contains this whole support, we can use that the time Fourier transform is a unitary operator and write

$$\bar{m}(\mathbf{x}_{\mathrm{f}}) = \int_{t \in \mathbb{R}} |\mathbf{E}^{\mathrm{f}}(\mathbf{x}_{\mathrm{f}})|^2 \, dt$$

$$= \int_{\omega \in \mathbb{R}} |\widehat{\mathbf{E}}^{\mathrm{f}}(\mathbf{x}_{\mathrm{f}})|^2 \, d\omega \qquad (107)$$

$$= \int_{\omega \in \mathbb{R}} |\Gamma(\omega)|^2 \left(\mathrm{NA}^2 \frac{J_1(\mathrm{NA} \frac{2\pi}{\lambda \mathrm{f}_{\mathrm{L}}} |\mathbf{x}_{\mathrm{f}}|)}{\mathrm{NA} \frac{2\pi}{\lambda \mathrm{f}_{\mathrm{L}}} |\mathbf{x}_{\mathrm{f}}|} \right)^2 \left(\Psi_1^2 + \Psi_2^2 \right) d\omega.$$

So, under these approximations (small aperture), what is observed by the detector is an *Airy pattern* (see Fig. 11) whose intensity depends on the optical system but is also proportional to the squared norm of the component of the dipole which lies orthogonally to the optical axis. Making use of the notation introduced before for the dipole orientation, we obtain that *the measured intensity m is proportional to* $\cos^2(\theta_m)$. Since the Airy function can be well approximated by a Gaussian function, the measured signals of the emitted dyes very much looks like a superposition of Gaussian functions.

Fig. 11 Airy pattern

Now, having summarized the mathematical modeling of single molecule localization microscopy we can state the associated inverse problems:

Definition 13 (Inverse Problem) The inverse problem of single molecule localization microscopy in the two different settings for the two experiments consists in calculating

$$(\Psi_m, \mathbf{r}_m, \chi_m(s))_{m=1,\dots,M}, \text{ from the measurements } m.$$

Indeed the inverse problem could also be generalized to reconstruct Eq. (103) in addition, which would result in an *inverse scattering problem* [8]. However, this complex problem is not considered here further.

In current practice of single molecule localization microscopy the simplified formulas Eq. (107) are used for reconstruction of the center of gravities (r_m^1, r_m^2) in the measurement data m induced by the point spread function PSF_ω.

8 Conclusion

The main objective of this work was to model mathematically the propagation of light emitted from dyes in a superresolution imaging experiment. We formulated basic inverse problems related to single molecule superresolution microscopy, with the goal to have a basis for computational and quantitative single molecule superresolution imaging. The derivation of the according equations follows the physical and chemical literature of superresolution microscopy, in particular [3, 17], which are combined with the mathematical theory of distributions to translate physical and chemical terminologies into a mathematical framework.

Appendix 1: Derivation of Particular Fourier Transforms

Before reading through the appendix it might be useful to recall the notation of Table 2. The derivation in Lemma 3 uses the *residual theorem*.

Theorem 2 (Residual Theorem) *Let the function* $z \in \mathbb{C} \to \tilde{f}(z) := f(z)e^{iaz}$ *with* $a > 0$ *satisfy the following properties:*

1. f *is analytic with at most finitely many poles* p_i, $i = 1, \ldots, m$, *which do not lie on the real axis.*
2. *There exists* $M, R > 0$ *such that for every* $z \in \mathbb{C}$ *satisfying* $\Im(z) \geq 0$ *and* $|z| \geq R$

$$|f(z)| \leq \frac{M}{|z|}. \tag{108}$$

Then

$$\int_{\mathbb{R}} \tilde{f}(x)dx = 2\pi i \sum_{i=1}^{m} Res(\tilde{f}; p_i).$$

Using this lemma we are able to prove the following result used in Lemma 3:

Lemma 10 *Let the assumptions and notation of Lemma 3 hold (in particular this means that* $r_3, r_3^{\Psi} \in \mathbb{R}$ *satisfy* $r_3 - r_3^{\Psi} > 0$), *then*

$$\frac{1}{\sqrt{2\pi}} \mathcal{F}_3^{-1} \left[k_3 \to \left(\Psi + \frac{(\Psi \times \mathbf{k}) \times \mathbf{k}}{|\mathbf{k}|^2 - \kappa_\varepsilon^2} \right) e^{-ik_3 r_3^{\Psi}} \right] (r_3)$$

$$= \Psi_3 \mathbf{e}_3 \tilde{\delta}(r_3 - r_3^{\Psi}) + \frac{ie^{iq_\varepsilon(r_3 - r_3^{\Psi})}}{2q_\varepsilon} (\Psi \times \mathbf{k}_{q_\varepsilon}) \times \mathbf{k}_{q_\varepsilon}. \tag{109}$$

First, we note that

$$\Psi + \frac{(\Psi \times \mathbf{k}) \times \mathbf{k}}{k_3^2 - q_\varepsilon^2} = \left(\Psi - \Psi_3 \mathbf{e}_3 + \frac{(\Psi \times \mathbf{k}) \times \mathbf{k}}{k_3^2 - q_\varepsilon^2} \right) + \Psi_3 \mathbf{e}_3.$$

Now, we calculate the Fourier-transform of each of the two terms on the right-hand side. The second term can be calculated from Eq. (9) and is given by

$$\mathcal{F}_3^{-1}[k_3 \mapsto e^{-ik_3 r_3^{\Psi}}](r_3) = \sqrt{2\pi} \tilde{\delta}(r_3 - r_3^{\Psi}). \tag{110}$$

For the calculation of the first term, let

$$z \in \mathbb{C} \to f(z) := \frac{1}{z^2 - q_\varepsilon^2} \left(\Psi \times \begin{pmatrix} k_1 \\ k_2 \\ z \end{pmatrix} \right) \times \begin{pmatrix} k_1 \\ k_2 \\ z \end{pmatrix} + \Psi - \Psi_3 \mathbf{e}_3, \tag{111}$$

and we use the residual theorem:

- Clearly f is analytic with potential poles at $k_3 = \pm q_\varepsilon$.
- To verify Eq. (108) we use the elementary calculation rules for $\nabla_\mathbf{r} \times$, summarized in Eq. (45) and get

$$
\frac{(\boldsymbol{\Psi} \times \mathbf{k}) \times \mathbf{k}}{k_3^2 - q_\varepsilon^2} = -\frac{1}{k_3^2 - q_\varepsilon^2}
$$

$$
\left(k_3^2(\boldsymbol{\Psi} - \Psi_3 \mathbf{e}_3) - k_3 \begin{pmatrix} \Psi_3 k_1 \\ \Psi_3 k_2 \\ \Psi_1 k_1 + \Psi_2 k_2 \end{pmatrix} + (k_1^2 + k_2^2)\boldsymbol{\Psi} - \begin{pmatrix} (\Psi_1 k_1 + \Psi_2 k_2)k_1 \\ (\Psi_1 k_1 + \Psi_2 k_2)k_2 \\ 0 \end{pmatrix} \right)
$$

$$
= -\frac{1}{k_3^2 - q_\varepsilon^2}\left(k_3^2 (\boldsymbol{\Psi} - \Psi_3 \mathbf{e}_3) - k_3 \begin{pmatrix} \Psi_3 k_1 \\ \Psi_3 k_2 \\ \Psi_1 k_1 + \Psi_2 k_2 \end{pmatrix} \right.
$$

$$
\left. + (k_1^2 + k_2^2)(\boldsymbol{\Psi} - \Psi_3 \mathbf{e}_3) + \begin{pmatrix} -\Psi_1 k_1^2 - \Psi_2 k_2 k_1 \\ -\Psi_2 k_2^2 - \Psi_1 k_1 k_2 \\ (k_1^2 + k_2^2)\Psi_3 \end{pmatrix} \right)
$$

$$
= -\frac{1}{k_3^2 - q_\varepsilon^2}\left((k_3^2 - q_\varepsilon^2)(\boldsymbol{\Psi} - \Psi_3 \mathbf{e}_3) - k_3 \begin{pmatrix} \Psi_3 k_1 \\ \Psi_3 k_2 \\ \Psi_1 k_1 + \Psi_2 k_2 \end{pmatrix} + \kappa^2 (\boldsymbol{\Psi} - \Psi_3 \mathbf{e}_3) \right.
$$

$$
\left. + \begin{pmatrix} -\Psi_1 k_1^2 - \Psi_2 k_2 k_1 \\ -\Psi_2 k_2^2 - \Psi_1 k_1 k_2 \\ (k_1^2 + k_2^2)\Psi_3 \end{pmatrix} \right) = -\boldsymbol{\Psi} + \Psi_3 \mathbf{e}_3 + \frac{k_3}{k_3^2 - q_\varepsilon^2}\mathbf{a} - \frac{1}{k_3^2 - q_\varepsilon^2}\mathbf{b}
$$

where

$$
\mathbf{a} = \begin{pmatrix} \Psi_3 k_1 \\ \Psi_3 k_2 \\ \Psi_1 k_1 + \Psi_2 k_2 \end{pmatrix} \quad \text{and } \mathbf{b} = \begin{pmatrix} (\kappa^2 - k_1^2)\Psi_1 - \Psi_2 k_1 k_2 \\ (\kappa^2 - k_2^2)\Psi_2 - \Psi_1 k_1 k_2 \\ (k_1^2 + k_2^2)\Psi_3 \end{pmatrix}.
$$

Using that

$$
\frac{k_3}{k_3^2 - q_\varepsilon^2} = \frac{1}{2}\left(\frac{1}{k_3 - q_\varepsilon} + \frac{1}{k_3 + q_\varepsilon} \right) \quad \text{and} \quad \frac{1}{k_3^2 - q_\varepsilon^2} = \frac{1}{2q_\varepsilon}\left(\frac{1}{k_3 - q_\varepsilon} - \frac{1}{k_3 + q_\varepsilon} \right)
$$

we get

$$
\frac{(\boldsymbol{\Psi} \times \mathbf{k}) \times \mathbf{k}}{k_3^2 - q_\varepsilon^2} + \boldsymbol{\Psi} - \Psi_3 \mathbf{e}_3 = \frac{1}{k_3 - q_\varepsilon}\left(\frac{1}{2}\mathbf{a} - \frac{1}{2q_\varepsilon}\mathbf{b} \right) + \frac{1}{k_3 + q_\varepsilon}\left(\frac{1}{2}\mathbf{a} + \frac{1}{2q_\varepsilon}\mathbf{b} \right).
$$

$$(112)$$

This shows Eq. (108).

Therefore we can apply the residual Theorem 2 for f defined in (111). Then, since q_ε is complex with positive imaginary part, f has only one pole in the upper half plane, and we get

$$\int_{\mathbb{R}} f(k_3) e^{ik_3(r_3 - r_3^{\Psi})} dk_3 = 2\pi i \operatorname{Res}\left(f(z) e^{iz(r_3 - r_3^{\Psi})}; z = q_\varepsilon \right)$$

$$= 2\pi i \left(\frac{1}{2q_\varepsilon} \left(\Psi \times \begin{pmatrix} k_1 \\ k_2 \\ q_\varepsilon \end{pmatrix} \right) \times \begin{pmatrix} k_1 \\ k_2 \\ q_\varepsilon \end{pmatrix} e^{iq_\varepsilon(r_3 - r_3^{\Psi})} \right).$$

This implies

$$\frac{1}{\sqrt{2\pi}} \mathscr{F}_3^{-1} \left[k_3 \to \left(\Psi + \frac{(\Psi \times \mathbf{k}) \times \mathbf{k}}{|\mathbf{k}|^2 - \kappa_\varepsilon^2} \right) e^{-ik_3 r_3^{\Psi}} \right](r_3)$$

$$= \Psi_3 \mathbf{e}_3 \tilde{\delta}(r_3 - r_3^{\Psi}) + \frac{i e^{iq_\varepsilon(r_3 - r_3^{\Psi})}}{2q_\varepsilon} (\Psi \times \mathbf{k}_{q_\varepsilon}) \times \mathbf{k}_{q_\varepsilon}.$$

Appendix 2: Derivation of the Far Field Approximation for the Electric Field

Lemma 11 *Let ζ be the function defined in Eq. (61). That is, for all $\mathbf{k}_{12} \in \mathbb{R}^2$ $\zeta(\mathbf{k}_{12}) = \mathbf{k}_{12} \cdot \mathbf{v} + q(\mathbf{k}_{12})$, with $q = \sqrt{\kappa^2 - k_1^2 - k_2^2}$; note the formal definition of q in Eq. (42) is via the limit $\varepsilon \to 0$. Then the gradient of ζ is given by*

$$\nabla \zeta(\mathbf{k}_{12}) = \mathbf{v} - \frac{\mathbf{k}_{12}}{\sqrt{\kappa^2 - |\mathbf{k}_{12}|^2}}, \tag{113}$$

which vanishes for

$$\hat{\mathbf{k}}_{12} := \frac{\kappa}{\sqrt{1 + |\mathbf{v}|^2}} \mathbf{v}. \tag{114}$$

Consequently $\mathbf{r} = r_3 \mathbf{e}_3 + r_3 \left(\mathbf{v} \ 1 \right)^T$ as in Eq. (59) satisfies

$$\frac{\mathbf{r}}{|\mathbf{r}|} = \frac{1}{\sqrt{1 + |\mathbf{v}|^2}} \begin{pmatrix} \mathbf{v} \\ 1 \end{pmatrix}. \tag{115}$$

Moreover, with q as defined in Eq. (42), we get the following identities

$$\zeta(\hat{\mathbf{k}}) = \hat{\mathbf{k}}_{12} \cdot \mathbf{v} + q = \kappa\sqrt{1 + |\mathbf{v}|^2} \quad and$$

$$q = q(\hat{\mathbf{k}}) = \sqrt{\kappa^2 - \hat{k}_1^2 - \hat{k}_2^2} = \frac{\kappa}{\sqrt{1 + |\mathbf{v}|^2}}. \tag{116}$$

The Hessian of ζ is given by

$$H(\zeta)(\mathbf{k}_{12}) = \begin{pmatrix} -\frac{1}{\sqrt{\kappa^2 - |\mathbf{k}_{12}|^2}} - \frac{k_1^2}{(\kappa^2 - |\mathbf{k}_{12}|^2)^{3/2}} & -\frac{k_1 k_2}{(\kappa^2 - |\mathbf{k}_{12}|^2)^{3/2}} \\ -\frac{k_1 k_2}{(\kappa^2 - |\mathbf{k}_{12}|^2)^{3/2}} & -\frac{1}{\sqrt{\kappa^2 - |\mathbf{k}_{12}|^2}} - \frac{k_2^2}{(\kappa^2 - |\mathbf{k}_{12}|^2)^{3/2}} \end{pmatrix},$$

which evaluated at $\hat{\mathbf{k}}$ gives

$$H(\zeta)(\hat{\mathbf{k}}) =$$

$$\begin{pmatrix} -\frac{c}{\omega}\sqrt{1 + |\mathbf{v}|^2} - \frac{c^3}{\omega^3}(1 + |\mathbf{v}|^2)^{3/2} \cdot \frac{v_1^2 \omega^2}{c^2(1 + |\mathbf{v}|^2)} & -\frac{c^3}{\omega^3}(1 + |\mathbf{v}|^2)^{3/2} \cdot \frac{v_1 v_2 \omega^2}{c^2(1 + |\mathbf{v}|^2)} \\ -\frac{c^3}{\omega^3}(1 + |\mathbf{v}|^2)^{3/2} \cdot \frac{v_1 v_2 \omega^2}{c^2(1 + |\mathbf{v}|^2)} & -\frac{c}{\omega}\sqrt{1 + |\mathbf{v}|^2} - \frac{c^3}{\omega^3}(1 + |\mathbf{v}|^2)^{3/2} \cdot \frac{v_2^2 \omega^2}{c^2(1 + |\mathbf{v}|^2)} \end{pmatrix},$$

$$= -\frac{c}{\omega}\sqrt{1 + |\mathbf{v}|^2} \begin{pmatrix} 1 + v_1^2 & v_1 v_2 \\ v_1 v_2 & 1 + v_2^2 \end{pmatrix},$$

and the determinant satisfies

$$= \frac{c^2(1 + |\mathbf{v}|^2)^2}{\omega^2} > 0.$$

Appendix 3: Bessel Identities

For $x \in \mathbb{R}$ and $\varphi_0 \in [0, 2\pi)$ the Bessel-identities hold:

$$2\pi(-1)^m \mathrm{i}^m J_m(x) \cos(m\varphi_0) = \int_0^{2\pi} e^{-\mathrm{i}x\cos(\varphi - \varphi_0)} \cos(m\varphi) d\varphi,$$

$$2\pi(-1)^m \mathrm{i}^m J_m(x) \sin(m\varphi_0) = \int_0^{2\pi} e^{-\mathrm{i}x\cos(\varphi - \varphi_0)} \sin(m\varphi) d\varphi, \tag{117}$$

where J_m is the Bessel function of the first kind of order m.

Acknowledgments OS is supported by the Austrian Science Fund (FWF), within SFB F68 (Tomography across the scales), project F6807-N36 (Tomography with Uncertainties) and I3661-N27 (Novel Error Measures and Source Conditions of Regularization Methods for Inverse Problems). KS is supported by the Austrian Science Fund (FWF), project P31053-N32 (Weighted

X-ray transform and applications). MLM, MS and GS are also supported by the SFB F68, project F6809-N36 (Ultra-high Resolution Microscopy). JS is supported in part by the NSF grant DMS-1912821 and the AFOSR grant FA9550-19-1-0320.

References

1. E. Abbe, Beiträge zur Theorie des Mikroskops und der mikroskopischen Wahrnehmung. Arch. Mikrosk. Anat. **9**(1), 413–468 (1873)
2. P. Annibale, S. Vanni, M. Scarselli, U. Rothlisberger, A. Radenovic, Quantitative photo activated localization microscopy: unraveling the effects of photoblinking. PLoS One **6**(7), e22678 (2011)
3. D. Axelrod, Fluorescence excitation and imaging of single molecules near dielectric-coated and bare surfaces: a theoretical study. J. Microsc. **247**(2), 147–160 (2012)
4. F. Balzarotti, Y. Eilers, K.C. Gwosch, A.H. Gynnå, V. Westphal, F.D. Stefani, J. Elf, S.W. Hell, Nanometer resolution imaging and tracking of fluorescent molecules with minimal photon fluxes. Science **355**(6325), 606–612 (2017)
5. F. Baumgart, A.M. Arnold, B.K. Rossboth, M. Brameshuber, G.J. Schutz, What we talk about when we talk about nanoclusters. Methods Appl. Fluores. **7**(1), 013001 (2019)
6. E. Betzig, G.H. Patterson, R. Sougrat, O.W. Lindwasser, S. Olenych, J.S. Bonifacino, M.W. Davidson, J. Lippincott-Schwartz, H.F. Hess, Imaging intracellular fluorescent proteins at nanometer resolution. Science **313**(5793), 1642–1645 (2006)
7. M. Born, E. Wolf, *Principles of Optics*, 7th edn. (Cambridge University, Cambridge, 1999)
8. F. Cakoni, D. Colton, *A Qualitative Approach to Inverse Scattering Theory* (Springer, New York, 2014)
9. B.J. Davis, A.K. Swan, M.S. Ünlü, W.C. Karl, B.B. Goldberg, J.C. Schotland, P.S. Carney, Spectral self-interference microscopy for low-signal nanoscale axial imaging. J. Opt. Soc. Am. A **24**(11), 3587–3599 (2007)
10. H. Deschout, F.C. Zanacchi, M. Mlodzianoski, A. Diaspro, J. Bewersdorf, S.T. Hess, K. Braeckmans, Precisely and accurately localizing single emitters in fluorescence microscopy. Nat. Meth. **11**(3), 253–266 (2014)
11. P. Elbau, L. Mindrinos, O. Scherzer, Mathematical methods of optical coherence tomography, in *Handbook of Mathematical Methods in Imaging*, ed. by O. Scherzer (Springer, New York, 2015), pp. 1169–1204
12. G.W. Ford, W.H. Weber, Electromagnetic interactions of molecules with metal surfaces. Phys. Rep. **113**(4), 195–287 (1984)
13. A.C. Gilbert, H.W. Levinson, J.C. Schotland, Imaging from the inside out: inverse scattering with photoactivated internal sources. Opt. Lett. **43**(12), 3005–3008 (2018)
14. J.W. Goodman, *Introduction to Fourier Optics* (Roberts, New York, 2005)
15. G. Grubb, Distributions and operators, in *Graduate Texts in Mathematics* (Springer, New York, 2009)
16. M.G.L. Gustafsson, Nonlinear structured-illumination microscopy: Wide-field fluorescence imaging with theoretically unlimited resolution. Proc. Nat. Acad. Sci. USA **102**(37), 13081–13086 (2005)
17. E.H. Hellen, D. Axelrod, Fluorescence emission at dielectric and metal-film interfaces. J. Opt. Soc. Am. B **4**(3), 337–350 (1987)
18. M. Hofmann, C. Eggeling, S. Jakobs, S.W. Hell, Breaking the diffraction barrier in fluorescence microscopy at low light intensities by using reversibly photoswitchable proteins. Proc. Nat. Acad. Sci. U.S.A. **102**(49), 17565–17569 (2005)
19. L. Hörmander, *The Analysis of Linear Partial Differential Operators I*, 2nd edn. (Springer, Berlin, 2003)
20. J.D. Jackson, *Classical Electrodynamics*, 3rd edn. (Wiley, New York, 1998)

21. R. Jungmann, C. Steinhauer, M. Scheible, A. Kuzyk, P. Tinnefeld, F.C. Simmel, Single-molecule kinetics and super-resolution microscopy by fluorescence imaging of transient binding on DNA origami. Nano Lett. **10**(11), 4756–4761 (2010)
22. G. Kabachinski, T.U. Schwartz, The nuclear pore complex—structure and function at a glance. J. Cell Sci. **128**(3), 423–429 (2015)
23. T.A. Klar, S. Jakobs, M. Dyba, A. Egner, S.W. Hell, Fluorescence microscopy with diffraction resolution barrier broken by stimulated emission. Proc. Nat. Acad. Sci. USA **97**(15), 8206–8210 (2000)
24. T.A. Klar, E. Engel, S.W. Hell, Breaking Abbe's diffraction resolution limit in fluorescence microscopy with stimulated emission depletion beams of various shapes. Phys. Rev. E **64**(6), 066613 (2001)
25. M. Levitus, S. Ranjit, Cyanine dyes in biophysical research: the photophysics of polymethine fluorescent dyes in biomolecular environments. Quart. Rev. Biophys. **44**(1), 123–151 (2011)
26. H.L. Li, J.C. Vaughan, Switchable fluorophores for single-molecule localization microscopy. Chem. Rev. **118**(18), 9412–9454 (2018)
27. I. Orlov, et al.: The integrative role of cryo electron microscopy in molecular and cellular structural biology. Biol. Cell **109**(2), 81–93 (2017)
28. A.S. Rose, A.R. Bradley, Y. Valasatava, J.M. Duarte, A. Prlic, P.W. Rose, Ngl viewer: web-based molecular graphics for large complexes. Bioinformatics **34**(21), 3755–3758 (2018)
29. M.J. Rust, M. Bates, X.W. Zhuang, Sub-diffraction-limit imaging by stochastic optical reconstruction microscopy (storm). Nat. Meth. **3**(10), 793–795 (2006)
30. D. Sage, et al., Super-resolution fight club: assessment of 2d and 3d single-molecule localization microscopy software. Nat. Methods **16**(5), 387–395 (2019)
31. Y.G. Shi, A glimpse of structural biology through x-ray crystallography. Cell **159**(5), 995–1014 (2014)
32. C.S. Smith, N. Joseph, B. Rieger, K.A. Lidke, Fast, single-molecule localization that achieves theoretically minimum uncertainty. Nat. Meth. **7**(5), 373–375 (2010)
33. C.-T. Tai, *Dyadic Green's Functions in Electromagnetic Theory* (IEEE, New York, 1971)
34. A. von Appen, J. Kosinski, L. Sparks, et al., In situ structural analysis of the human nuclear pore complex. Nature **526**, 140–143 (2015)
35. V. Westphal, S.W. Hell, Nanoscale resolution in the focal plane of an optical microscope. Phys. Rev. Lett. **94**(14), 143903 (2005)

Parameter Identification for the Landau–Lifshitz–Gilbert Equation in Magnetic Particle Imaging

Barbara Kaltenbacher, Tram Thi Ngoc Nguyen, Anne Wald, and Thomas Schuster

Abstract Magnetic particle imaging (MPI) is a tracer-based technique for medical imaging where the tracer consists of ironoxide nanoparticles. The key idea is to measure the particle response to a temporally changing external magnetic field to compute the spatial concentration of the tracer inside the object. A decent mathematical model demands for a data-driven computation of the system function which does not only describe the measurement geometry but also encodes the interaction of the particles with the external magnetic field. The physical model of this interaction is given by the Landau–Lifshitz–Gilbert (LLG) equation. The determination of the system function can be seen as an inverse problem of its own which can be interpreted as a calibration problem for MPI. In this contribution the calibration problem is formulated as an inverse parameter identification problem for the LLG equation. We give a detailed analysis of the direct as well as the inverse problem in an all-at-once as well as in a reduced setting. The analytical results yield a deeper understanding of inverse problems connected to the LLG equation and provide a starting point for the development of robust numerical solution methods in MPI.

1 Introduction

Magnetic particle imaging (MPI) is a dynamic imaging modality for medical applications that has first been introduced in 2005 by B. Gleich and J. Weizenecker [10]. Magnetic nanoparticles, consisting of a magnetic iron oxide core and a nonmagnetic

B. Kaltenbacher
Department of Mathematics, Alpen-Adria-Universität Klagenfurt, Klagenfurt, Austria
e-mail: barbara.kaltenbacher@aau.at

T. T. N. Nguyen
Institute of Mathematics and Scientific Computing, University of Graz, Graz, Austria
e-mail: tram.nguyen@uni-graz.at

A. Wald (✉) · T. Schuster
Department of Mathematics, Saarland University, Saarbrücken, Saarland, Germany
e-mail: anne.wald@num.uni-sb.de; thomas.schuster@num.uni-sb.de

© Springer Nature Switzerland AG 2021
B. Kaltenbacher et al. (eds.), *Time-dependent Problems in Imaging and Parameter Identification*, https://doi.org/10.1007/978-3-030-57784-1_13

coating, are inserted into the body to serve as a tracer. The key idea is to measure the nonlinear response of the nanoparticles to a temporally changing external magnetic field in order to draw conclusions on the spatial concentration of the particles inside the body. Since the particles are distributed along the bloodstream of a patient, the particle concentration yields information on the blood flow and is thus suitable for cardiovascular diagnosis or cancer detection [23, 24]. An overview of MPI basics is given in [23]. Since MPI requires the nanoparticles as a tracer, it mostly yields quantitative information on their distribution, but does not image the morphology of the body, such as the tissue density. The latter can be visualized using computerized tomography (CT) [29] or magnetic resonance imaging (MRI) [15]. These do not require a tracer, but involve ionizing radiation in the case of CT or, in the case of MRI, a strong magnetic field and a potentially high acquisition time. Other tracer-based methods are, e.g., single photon emission computerized tomography (SPECT) and positron emission tomography (PET) [8, 30, 36], which both involve radioactive radiation. The magnetic nanoparticles that are used in MPI, on the other hand, are not harmful for organisms. For a more detailed comparison of these methods, we would like to refer the reader to [23].

At this point there have been promising preclinical studies on the performance of MPI, showing that this imaging modality has a great potential for medical diagnosis since it is highly sensitive with a good spatial and temporal resolution, and the data acquisition is very fast [24]. However, particularly in view of an application to image the human body, there remain some obstacles. One obstacle is the time-consuming calibration process. In this work, we assume that the concentration of the nanoparticles inside the body remains static throughout both the calibration process and the actual image acquisition. Mathematically, the forward problem of MPI then can essentially be formulated as an integral equation of the first kind for the particle concentration (or distribution) c,

$$u(t) = \int_\Omega c(x)s(x, t)\, dx,$$

where the integration kernel s is called the *system function*. The system function encodes some geometrical aspects of the MPI scanner, such as the coil sensitivities of the receive coils in which the particle signal u is measured, but mostly it is determined by the particle behavior in response to the applied external magnetic field.

The actual inverse problem in MPI is to reconstruct the concentration c under the knowledge of the system function s from the measured data u. To this end, the system function has to be determined prior to the scanning procedure. This is usually done by evaluating a series of full scans of the field of view, where in each scan a delta sample is placed in a different pixel until the entire field of view is covered [23]. Another option is a model-based approach for s (see for example [22, 28]), which basically involves a model for the particle magnetization. Since this model often depends on unknown parameters, the model-based determination of the system function itself can again be formulated as an inverse problem. This article now addresses this latter type of inverse problem, i.e., the identification of the system function for a known set of concentrations from calibration measurements. More precisely, our goal is to find a decent model for the time-derivative of the particle magnetization $\mathbf{m}$, which is proportional to s.

So far, in model-based approaches for the system function, the particle magnetization $\mathbf{m}$ is not modeled directly. Instead, one describes the mean magnetization $\overline{\mathbf{m}}$ of the particles via the Langevin function, i.e., the response of the particles is modeled on the mesoscopic scale [21, 23]. This approach is based on the assumption that the particles are in thermodynamic equilibrium and respond directly to the external field. For this reason, the mean magnetization is assumed to be a function of the external field, such that the mean magnetization is always aligned with the external field. The momentum of the mean magnetization is calculated via the Langevin function. This model, however, neglects some properties of the particle behavior. In particular, the magnetic moments of the particles do not align instantly with the external field [4].

In this work, we thus address an approach from micromagnetics, which models the time-dependent behavior of the magnetic material inside the particles' cores on the micro scale and allows to take into account various additional physical properties such as particle-particle interaction. For an overview, see for example [25]. Since the core material is iron oxide, which is a ferrimagnetic material that shows a similar behavior as ferromagnets [5, 6], we use the *Landau–Lifshitz–Gilbert (LLG) equation*

$$\frac{\partial}{\partial t}\mathbf{m} = -\widetilde{\alpha}_1 \mathbf{m} \times (\mathbf{m} \times \mathbf{H}_{\mathrm{eff}}) + \widetilde{\alpha}_2 \mathbf{m} \times \mathbf{H}_{\mathrm{eff}},$$

see also [9, 26], for the evolution of the magnetization $\mathbf{m}$ of the core material. The field $\mathbf{H}_{\mathrm{eff}}$ incorporates the external magnetic field together with other relevant physical effects. According to the LLG equation, the magnetization $\mathbf{m}$ performs a damped precession around the field vector of the external field, which leads to a relaxation effect. The LLG equation has been widely applied to describe the time evolution in micromagnetics [2, 7, 11].

In contrast to the imaging problem of MPI, the inverse problem of determining the magnetization $\mathbf{m}$ along with the constants $\widetilde{\alpha}_1, \widetilde{\alpha}_2$ turns out to be a nonlinear inverse problem, which is typical for parameter identification problems for partial differential equations, for example electrical impedance tomography [1], terahertz tomography [38], ultrasound imaging [3] and other applications from imaging and nondestructive testing [20].

We use the *all-at-once* as well as the *reduced* formulation of this inverse problem in a Hilbert space setting, see also [16, 17, 31], and analyze both cases including well-definedness of the forward mapping, continuity, and Fréchet differentiability and calculate the adjoint mappings for the Fréchet derivatives. By consequence, iterative methods such as the Landweber method [14, 27], also in combination with Kaczmarz' method [12, 13], Newton methods (see, e.g., [33]), or subspace techniques [37] can be applied for the numerical solution. An overview of suitable regularization techniques is given in [18, 19].

We begin with a detailed introduction to the modelling in MPI. In particular, we describe the full forward problem and present the initial boundary value problem for the LLG equation that we use to describe the magnetization evolution. In Sect. 3, we formulate the inverse problem of calibration both in the all-at-once and in the reduced setting to obtain the final operator equation that is analyzed in the subsequent section. First, in Sect. 4.1, we present an analysis for the all-at-once setting. The inverse problem in the reduced setting is then addressed in Sect. 4.2. Finally, we conclude our findings in Sect. 5 and give an outlook on further research.

Throughout the article, we make use of the following notation: The differential operators $-\Delta$ and ∇ are applied by components to a vector field. In particular this means that by $\nabla \mathbf{u}$ we denote the transpose of the Jacobian of $\mathbf{u}$. Moreover, $\langle \mathbf{a}, \mathbf{b} \rangle$ or $\mathbf{a} \cdot \mathbf{b}$ denotes the Euclidean inner product between two vectors and $A : B$ the Frobenius inner product between two matrices.

2   The Underlying Physical Model for MPI

The basic physical principle that is exploited in MPI is Faraday's law of induction, which states that whenever the magnetic flux density $\mathbf{B}$ through a coil changes in time, this change induces an electric current in the coil. This current, or rather the respective voltage, can be measured. In MPI, the magnetic flux density $\mathbf{B}$ consists of the external applied magnetic field $\mathbf{H}_{\text{ext}}$ and the particle magnetization $\mathbf{M}^{\text{P}}$, i.e.,

$$\mathbf{B} = \mu_0 \left(\mathbf{H}_{\text{ext}} + \mathbf{M}^{\text{P}} \right),$$

where μ_0 is the magnetic permeability in vacuum. The particle magnetization $\mathbf{M}^{\text{P}}(x, t)$ in $x \in \Omega \subseteq \mathbb{R}^3$ depends linearly on the concentration $c(x)$ of magnetic material, which corresponds to the particle concentration, in $x \in \Omega$ and on the magnetization $\mathbf{m}(x, t)$ of the magnetic material. We thus have

$$\mathbf{M}^{\text{P}}(x, t) = c(x)\mathbf{m}(x, t),$$

where $|\mathbf{m}| = m_S > 0$, i.e., the vector $\mathbf{m}$ has the fixed length m_S that depends on the magnetic core material inside the particles. At this point it is important to remark that we use a slightly different approach to separate the particle concentration, which carries the spatial information on the particles, from the magnetization behavior of the magnetic material and the measuring process. In our approach, the concentration is a dimensionless quantity, whereas in most models, it is defined as the number of particles per unit volume (see, e.g. [23]).

A detailed derivation of the forward model in MPI, based on the equilibrium model for the magnetization, can be found in [23]. The steps that are related to the measuring process can be adapted to our approach. For the reader's convenience, we want to give a short overview and introduce the parameters related to the scanner setup.

If the receive coil is a simple conductor loop, which encloses a surface S, the voltage that is induced can be expressed by

$$u(t) = -\frac{\mathrm{d}}{\mathrm{d}t} \int_S \mathbf{B}(x, t) \cdot \mathrm{d}\mathbf{A} = -\mu_0 \frac{\mathrm{d}}{\mathrm{d}t} \int_S \left(\mathbf{H}_{\text{ext}} + \mathbf{M}^P \right) \cdot \mathrm{d}\mathbf{A}. \tag{1}$$

The signal that is recorded in the receive coil thus originates from temporal changes of the external magnetic field $\mathbf{H}$ as well as of the particle magnetization $\mathbf{M}^P$,

$$u(t) = -\mu_0 \left(\int_\Omega \mathbf{p}^R(x) \cdot \frac{\partial}{\partial t} \mathbf{H}_{\text{ext}}(x, t) \, \mathrm{d}x + \int_\Omega \mathbf{p}^R(x) \cdot \frac{\partial}{\partial t} \mathbf{M}^P(x, t) \, \mathrm{d}x \right) \tag{2}$$

$$=: u^E(t) + u^P(t) \tag{3}$$

For the signal that is caused by the change in the particle magnetization we obtain

$$u^P(t) = -\mu_0 \frac{\mathrm{d}}{\mathrm{d}t} \int_\Omega \mathbf{p}^R(x) \cdot \mathbf{M}^P(x, t) \, \mathrm{d}x$$

$$= -\mu_0 \int_\Omega \mathbf{p}^R(x) \cdot \frac{\partial}{\partial t} \mathbf{M}^P(x, t) \, \mathrm{d}x$$

$$= -\mu_0 \int_\Omega c(x) \mathbf{p}^R(x) \cdot \frac{\partial}{\partial t} \mathbf{m}(x, t) \, \mathrm{d}x$$

$$= -\mu_0 \int_\Omega c(x) s(x, t) \, \mathrm{d}x.$$

The function

$$s(x, t) := \mathbf{p}^R(x) \cdot \frac{\partial}{\partial t} \mathbf{m}(x, t) = \left\langle \mathbf{p}^R(x), \frac{\partial}{\partial t} \mathbf{m}(x, t) \right\rangle_{\mathbb{R}^3} \tag{4}$$

is called the *system function* and can be interpreted as a potential to induce a signal in the receive coil. The function $\mathbf{p}^R$ is called the coil sensitivity and is determined by the architecture of the respective receive coil. For our purposes, we assume that $\mathbf{p}^R$ is known. The measured signal that originates from the magnetic particles can thus essentially be calculated via an integral equation of the first kind with a time-dependent integration kernel s.

The particle magnetization, however, changes in time in response to changes of the external field. It is thus an important objective to encode the interplay of the external field and the particles in a sufficiently accurate physical model. The magnetization of the magnetic particles that are used in MPI can be considered on different scales. The following characterization from ferromagnetism has been taken from [25]:

On the *atomic level*, one can describe the behavior of a magnetic material as a spin system and take into account stochastic effects that arise, for example, from Brownian motion.

In the *microscopic scale*, continuum physics is applied to work with deterministic equations describing the magnetization of the magnetic material.

In the *mesoscopic scale*, we can describe the magnetization behavior via a mean magnetization, which is an average particle magnetic moment.

Finally, on a *macroscopic scale*, all aspects that arise from the microstructure are neglected and the magnetization is described by phenomenological constitutive laws.

In this work, we intend to use a model from micromagnetism, allowing us to work with a deterministic equation to describe the magnetization of the magnetic material. The core material of the nanoparticles consists of iron-oxide or magnetite, which is a ferrimagnetic material. The magnetization curve of ferrimagnetic materials is similar to the curve that is observed for ferromagnets, but with a lower saturation magnetization (see, e.g., [5, 6]). This approach has also been suggested in [32]. The evolution of the magnetization in time is described by the *Landau–Lifshitz–Gilbert (LLG) equation*

$$\mathbf{m}_t := \frac{\partial}{\partial t}\mathbf{m} = -\widetilde{\alpha}_1 \mathbf{m} \times (\mathbf{m} \times \mathbf{H}_{\text{eff}}) + \widetilde{\alpha}_2 \mathbf{m} \times \mathbf{H}_{\text{eff}}, \tag{5}$$

see [9, 25] and the therein cited literature. The coefficients

$$\widetilde{\alpha}_1 := \frac{\gamma \alpha_D}{m_S(1 + \alpha_D^2)} > 0, \quad \widetilde{\alpha}_2 := \frac{\gamma}{(1 + \alpha_D^2)} > 0$$

are material parameters that contain the gyromagnetic constant γ, the saturation magnetization m_S of the core material and a damping parameter α_D. The vector field $\mathbf{H}_{\text{eff}}$ is called the *effective magnetic field*. It is defined as the negative gradient $-D\mathcal{E}(\mathbf{m})$ of the *Landau energy* $\mathcal{E}(\mathbf{m})$ of a ferromagnet, see, e.g., [25]. Taking into account only the interaction with the external magnetic field $\mathbf{H}$ and particle-particle interactions, this energy is given by

$$\mathcal{E}_A(\mathbf{m}) = A \int_\Omega |\nabla \mathbf{m}|^2 \, dx - \mu_0 m_S \int_\Omega \langle \mathbf{H}, \mathbf{m}\rangle_{\mathbb{R}^3} \, dx,$$

where $A \geq 0$ is a scalar parameter (the exchange stiffness constant [9]). We thus have

$$\mathbf{H}_{\text{eff}} = 2A\,\Delta\mathbf{m} + \mu_0 m_S \mathbf{H}_{\text{ext}}. \tag{6}$$

Together with Neumann boundary conditions and a suitable initial condition our model for the magnetization thus reads

$$\mathbf{m}_t = -\alpha_1 \mathbf{m} \times (\mathbf{m} \times (\Delta\mathbf{m} + \mathbf{h}_{\text{ext}})) + \alpha_2 \mathbf{m} \times (\Delta\mathbf{m} + \mathbf{h}_{\text{ext}}) \quad \text{in } [0, T] \times \Omega, \tag{7}$$

$$0 = \partial_\nu \mathbf{m} \qquad\qquad\qquad\qquad\qquad\qquad\qquad\quad \text{on } [0, T] \times \partial\Omega, \tag{8}$$

$$\mathbf{m}_0 = \mathbf{m}(t = 0), \; |\mathbf{m}_0| = m_S \qquad\qquad\qquad\qquad \text{in } \Omega, \tag{9}$$

where $\mathbf{h}_{\text{ext}} = \frac{\mu_0 m_S}{2A} \mathbf{H}_{\text{ext}}$ and $\alpha_1 := 2A\widetilde{\alpha}_1, \alpha_2 := 2A\widetilde{\alpha}_2 > 0$. The initial value $\mathbf{m}_0 = \mathbf{m}(t = 0)$ corresponds to the magnetization of the magnetic material in the beginning of the measurement. To obtain a reasonable value for $\mathbf{m}_0$, we take into account that the external magnetic field is switched on before the measuring process starts, i.e., $\mathbf{m}_0$ is the state of the magnetization that is acquired when the external field is static. This allows us to precompute $\mathbf{m}_0$ as the solution of the stationary problem

$$\alpha_1 \mathbf{m}_0 \times (\mathbf{m}_0 \times (\Delta\mathbf{m}_0 + \mathbf{h}_{\text{ext}}(t = 0))) = \alpha_2 \mathbf{m}_0 \times (\Delta\mathbf{m}_0 + \mathbf{h}_{\text{ext}}(t = 0)) \tag{10}$$

with Neumann boundary conditions.

Remark 1 In the stationary case, damping does not play a role, and if we additionally neglect particle-particle interactions, we obtain the approximative equation

$$\hat{\mathbf{m}}_0 \times \left(\hat{\mathbf{m}}_0 \times \mathbf{h}_{\text{ext}}(t = 0)\right) = 0$$

with an approximation $\hat{\mathbf{m}}_0$ to $\hat{\mathbf{m}}$, since $\alpha_2 \approx 0$ and $\mathbf{H}_{\text{eff}} \approx \mu_0 m_S \mathbf{H}_{\text{ext}}$. The above equation yields $\hat{\mathbf{m}}_0 \parallel \mathbf{h}_{\text{ext}}(t = 0)$. Together with $|\hat{\mathbf{m}}_0| = m_S$ this yields

$$\hat{\mathbf{m}}_0 = m_S \frac{\mathbf{h}_{\text{ext}}(t = 0)}{|\mathbf{h}_{\text{ext}}(t = 0)|}.$$

This represents a good approximation to $\mathbf{m}_0$ where $\mathbf{h}_{\text{ext}}$ is strong at the time point $t = 0$:

$$\mathbf{m}_0 \approx \hat{\mathbf{m}}_0 = m_S \frac{\mathbf{h}_{\text{ext}}(t = 0)}{|\mathbf{h}_{\text{ext}}(t = 0)|}.$$

2.1 The Observation Operator in MPI

Faraday's law states that a temporally changing magnetic field induces an electric current in a conductor loop or coil, which yields the relation (1). By consequence, not only the change in the particle magnetization contributes to the induced current, but also the dynamic external magnetic field $\mathbf{H}_{\text{ext}}$. Since we need the particle signal for the determination of the particle magnetization, we need to separate the particle signal from the excitation signal due to the external field. This is realized by processing the signal in a suitable way using filters.

MPI scanners usually use multiple receive coils to measure the induced particle signal at different positions in the scanner. We assume that we have $L \in \mathbb{N}$ receive coils with coil sensitivities $\mathbf{p}_\ell^R$, $\ell = 1, \ldots, L$, and the measured signal is given by

$$\widetilde{v}_\ell(t) = -\mu_0 \int_0^T \widetilde{a}_\ell(t - \tau) \int_\Omega c(x) \mathbf{p}_\ell^R(x) \cdot \frac{\partial}{\partial \tau} \mathbf{m}(x, \tau) \, dx \, d\tau, \tag{11}$$

where T is the repetition time of the acquisition process, i.e., the time that is needed for one full scan of the object, and $a_\ell : [0, T] \to \mathbb{R}$ is the transfer function with periodic continuation $\widetilde{a}_\ell : \mathbb{R} \to \mathbb{R}$. The transfer function serves as a filter to separate particle and excitation signal, i.e., it is chosen such that

$$\widetilde{v}_\ell^E(t) := \left(\widetilde{a}_\ell * u_\ell^E\right)(t) = -\mu_0 \int_0^T \widetilde{a}_\ell(t - \tau) \int_\Omega \mathbf{p}_\ell^R(x) \cdot \frac{\partial}{\partial t} \mathbf{H}_{\text{ext}}(x, t) \, dx \, dt \approx 0.$$

In practice, $\widetilde{a}_\ell$ is often a band pass filter. For a more detailed discussion of the transfer function, see also [23]. In this work, the transfer function is known analytically.

We define

$$\mathbf{K}_\ell(t, \tau, x) := -\mu_0 \widetilde{a}_\ell(t - \tau) c(x) \mathbf{p}_\ell^R(x),$$

such that the measured particle signals are given by

$$v_\ell(t) = \int_0^T \int_\Omega \mathbf{K}_\ell(t, \tau, x) \cdot \frac{\partial}{\partial \tau} \mathbf{m}(x, \tau) \, d\tau \, dx, \tag{12}$$

where $\mathbf{m}$ fulfills (7), (8), (9).

To determine $\mathbf{m}$ in $\Omega \times (0, T)$, we use the data $v_{k\ell}(t)$, $k = 1, \ldots, K$, $\ell = 1, \ldots, L$, from the scans that we obtain for different particle concentrations c_k, $k = 1, \ldots, K$, $K \in \mathbb{N}$. The forward operator thus reads

$$v_{k\ell}(t) = \int_0^T \int_\Omega \mathbf{K}_{k\ell}(t, \tau, x) \cdot \frac{\partial}{\partial \tau} \mathbf{m}(x, \tau) \, dx \, d\tau,$$

$$\mathbf{K}_{k\ell}(t, \tau, x) := -\mu_0 \widetilde{a}_\ell(t - \tau) c_k(x) \mathbf{p}_\ell^R(x). \tag{13}$$

2.2   Equivalent Formulations of the LLG Equation

In this section, we derive additional formulations of (7)–(9) that are suitable for the analysis. The approach is motivated by Kružík and Prohl [25], where only particle-particle interactions are taken into account.

First of all, we observe that multiplying (7) with $\mathbf{m}$ on both sides yields

$$\frac{1}{2} \cdot \frac{\mathrm{d}}{\mathrm{d}t} |\mathbf{m}(x, t)|^2 = \mathbf{m}(x, t) \cdot \mathbf{m}_t(x, t) = 0, \tag{14}$$

which shows that the absolute value of $\mathbf{m}$ does not change in time. Since $|\mathbf{m}_0| = m_S$, we have $\mathbf{m}(x, t) \in m_S \cdot \mathcal{S}^2$, where $\mathcal{S}^2 := \{\mathbf{v} \in \mathbb{R}^3 : |\mathbf{v}| = 1\}$ is the unit sphere in $\mathbb{R}^3$. As a consequence, we have $0 = \nabla |\mathbf{m}|^2 = 2\nabla\mathbf{m} \cdot \mathbf{m}$ in Ω, so that, by taking the divergence we get

$$\langle \mathbf{m}, \Delta\mathbf{m} \rangle = -\langle \nabla\mathbf{m}, \nabla\mathbf{m} \rangle. \tag{15}$$

Now we make use of the identity

$$\mathbf{a} \times (\mathbf{b} \times \mathbf{c}) = \langle \mathbf{a}, \mathbf{c} \rangle \mathbf{b} - \langle \mathbf{a}, \mathbf{b} \rangle \mathbf{c}$$

for $\mathbf{a}, \mathbf{b}, \mathbf{c} \in \mathbb{R}^3$ to derive

$$\mathbf{m} \times (\mathbf{m} \times \Delta\mathbf{m}) = \langle \mathbf{m}, \Delta\mathbf{m} \rangle \mathbf{m} - |\mathbf{m}|^2 \Delta\mathbf{m} = -|\nabla\mathbf{m}|^2\mathbf{m} - m_S^2 \Delta\mathbf{m}, \tag{16}$$

$$\mathbf{m} \times (\mathbf{m} \times \mathbf{h}_{\mathrm{ext}}) = \langle \mathbf{m}, \mathbf{h}_{\mathrm{ext}} \rangle \mathbf{m} - |\mathbf{m}|^2\mathbf{h}_{\mathrm{ext}} = \langle \mathbf{m}, \mathbf{h}_{\mathrm{ext}} \rangle \mathbf{m} - m_S^2\mathbf{h}_{\mathrm{ext}}. \tag{17}$$

Using (15) together with (16), (17) and $|\mathbf{m}| = m_S$, we obtain from (7)–(9)

$$\begin{aligned} \mathbf{m}_t - \alpha_1 m_S^2 \Delta\mathbf{m} &= \alpha_1 |\nabla\mathbf{m}|^2\mathbf{m} + \alpha_2\mathbf{m} \times \Delta\mathbf{m} \\ &\quad - \alpha_1 \langle \mathbf{m}, \mathbf{h}_{\mathrm{ext}} \rangle \mathbf{m} + \alpha_1 m_S^2 \mathbf{h}_{\mathrm{ext}} + \alpha_2\mathbf{m} \times \mathbf{h}_{\mathrm{ext}} \end{aligned} \quad \text{in } [0, T] \times \Omega, \tag{18}$$

$$0 = \partial_\nu\mathbf{m} \qquad \text{on } [0, T] \times \partial\Omega, \tag{19}$$

$$\mathbf{m}_0 = \mathbf{m}(t = 0), \quad |\mathbf{m}_0| = m_S \qquad \text{in } \Omega, \tag{20}$$

Taking the cross product of $\mathbf{m}$ with (18) and multiplying with $-\hat{\alpha}_2$, where $\hat{\alpha}_1 = \frac{\alpha_1}{m_S^2\alpha_1^2+\alpha_2^2}$, $\hat{\alpha}_2 = \frac{\alpha_2}{m_S^2\alpha_1^2+\alpha_2^2}$, by (16), (17) and cancellation of the first and third term on the right hand side we get

$$- \hat{\alpha}_2 \mathbf{m} \times \mathbf{m}_t + \alpha_1 \hat{\alpha}_2 m_S^2 \, \mathbf{m} \times \Delta \mathbf{m}$$

$$= \frac{\alpha_2^2}{m_S^2 \alpha_1^2 + \alpha_2^2} \left(|\nabla \mathbf{m}|^2 \mathbf{m} + m_S^2 \Delta \mathbf{m} \right)$$

$$- \alpha_1 \hat{\alpha}_2 m_S^2 \, \mathbf{m} \times \mathbf{h}_{\text{ext}} + \frac{\alpha_2^2}{m_S^2 \alpha_1^2 + \alpha_2^2} \left(m_S^2 \mathbf{h}_{\text{ext}} - \langle \mathbf{m}, \mathbf{h}_{\text{ext}} \rangle \mathbf{m} \right) ,$$

where the second term on the left hand side can be expressed via (18) as

$$\alpha_1 \hat{\alpha}_2 \mathbf{m} \times \Delta \mathbf{m} = \hat{\alpha}_1 \mathbf{m}_t + \frac{\alpha_1^2}{m_S^2 \alpha_1^2 + \alpha_2^2} \left(-m_S^2 \Delta \mathbf{m} - |\nabla \mathbf{m}|^2 \mathbf{m} \right.$$

$$\left. + \langle \mathbf{m}, \mathbf{h}_{\text{ext}} \rangle \mathbf{m} - m_S^2 \mathbf{h}_{\text{ext}} \right) - \alpha_1 \hat{\alpha}_2 \mathbf{m} \times \mathbf{h}_{\text{ext}} .$$

This yields the alternative formulation

$$\hat{\alpha}_1 m_S^2 \mathbf{m}_t - \hat{\alpha}_2 \mathbf{m} \times \mathbf{m}_t - m_S^2 \Delta \mathbf{m} = |\nabla \mathbf{m}|^2 \mathbf{m} + m_S^2 \mathbf{h}_{\text{ext}} - \langle \mathbf{m}, \mathbf{h}_{\text{ext}} \rangle \mathbf{m} \quad \text{in } [0, T] \times \Omega, \tag{21}$$

$$0 = \partial_\nu \mathbf{m} \quad \text{on } [0, T] \times \partial\Omega, \tag{22}$$

$$\mathbf{m}_0 = \mathbf{m}(t = 0), \ |\mathbf{m}_0| = m_S \quad \text{in } \Omega . \tag{23}$$

3 An Inverse Problem for the Calibration Process in MPI

Apart from the obvious inverse problem of determining the concentration c of magnetic particles inside a body from the measurements v_ℓ, $\ell = 1, \ldots, L$, MPI gives rise to a range of further parameter identification problems of entirely different nature. In this work, we are not addressing the imaging process itself, but consider an inverse problem that is essential for the calibration process. Here, calibration refers to determining the system function s_ℓ, which serves as an integral kernel in the imaging process. The system function includes all system parameters of the tomograph and encodes the physical behaviour of the magnetic material in the cores of the magnetic particles inside a temporally changing external magnetic field. Experiments show that a simple model for the magnetization, based on the assumption that the particles are in their equilibrium state at all times, is insufficient for the imaging, see, e.g., [22]. A model-based approach with an enhanced physical model has so far been omitted due to the complexity of the involved physics and the system function is usually measured in a time-consuming calibration process [23, 24].

In this work, we address the inverse problem of calibrating an MPI system for a given set of standard calibration concentrations c_k, $k = 1, \ldots, K$, for which we measure the corresponding signals and obtain the data $v_{k\ell}(t)$, $k = 1, \ldots, K$, $\ell = 1, \ldots, L$. Here we assume that the coil sensitivity $\mathbf{p}_\ell^R$ as well as the transfer function $\tilde{a}_\ell$ are known.

This, together with the fact that $\mathbf{m}$ is supposed to satisfy the LLG equation (21)–(23), is used to determine the system function (4). Actually, since $\mathbf{p}^R$ is known, the inverse problem under consideration here consists of reconstructing $\mathbf{m}$ from (13), (21)–(23). As the initial boundary value problem (21)–(23) has a unique solution $\mathbf{m}$ for given $\hat{\alpha}_1$, $\hat{\alpha}_2$, it actually suffices to determine these two parameters. This is the point of view that we take when using a classical reduced formulation of the calibration problem

$$F(\hat{\alpha}) = y \tag{24}$$

with the data $y_{k\ell} = v_{k\ell}$ and the forward operator

$$F : \mathcal{D}(F)(\subseteq \mathcal{X}) \to \mathcal{Y}, \qquad \hat{\alpha} = (\hat{\alpha}_1, \hat{\alpha}_2) \mapsto \mathcal{K}\frac{\partial}{\partial t}S(\hat{\alpha}) \tag{25}$$

containing the parameter-to-state map

$$S : \mathcal{X} \to \tilde{\mathcal{U}} \tag{26}$$

that maps the parameters $\hat{\alpha}$ into the solution $\mathbf{m} := S(\hat{\alpha})$ of the LLG initial boundary value problem (21)–(23). The linear operator $\mathcal{K}$ is the integral operator defined by the kernels $\mathbf{K}_{k\ell}$, $k = 1, \ldots, K$, $\ell = 1, \ldots, L$, i.e.,

$$\mathcal{K}_{k\ell}\mathbf{u} = \int_0^T \int_\Omega \mathbf{K}_{k\ell}(t, \tau, \mathbf{x}) \cdot \mathbf{u}(\mathbf{x}, \tau)\, d\tau\, d\mathbf{x}. \tag{27}$$

Here, the preimage and image spaces are defined by

$$\mathcal{X} = \mathbb{R}^2, \qquad \mathcal{Y} = L^2(0, T)^{KL} \tag{28}$$

and the state space $\tilde{\mathcal{U}}$ will be chosen appropriately below, see Sect. 4.2.

Alternatively, we also consider the all-at-once formulation of the inverse problem as a simultaneous system

$$\mathbb{F}(\mathbf{m}, \hat{\alpha}) = \mathbf{y} := (0, y)^T \tag{29}$$

for the state $\mathbf{m}$ and the parameters $\hat{\alpha}$, with the forward operator

$$\mathbb{F}(\mathbf{m}, \hat{\alpha}) = \begin{pmatrix} \mathbb{F}_0(\mathbf{m}, \hat{\alpha}) \\ \left(\mathbb{F}_{k\ell}(\mathbf{m}, \hat{\alpha})\right)_{k=1,\ldots,K\,,\ \ell=1,\ldots,L} \end{pmatrix},$$

where

$$\mathbb{F}_0(\mathbf{m}, \hat{\alpha}_1, \hat{\alpha}_2) =: \hat{\alpha}_1 \mathbf{m}_t - \Delta \mathbf{m} - \hat{\alpha}_2 \mathbf{m} \times \mathbf{m}_t - |\nabla \mathbf{m}|^2 \mathbf{m} - \mathbf{h}_{\text{ext}} + (\mathbf{m} \cdot \mathbf{h}_{\text{ext}}) \mathbf{m}$$

and

$$\mathbb{F}_{k\ell}(\mathbf{m}, \hat{\alpha}_1, \hat{\alpha}_2) = \mathcal{K}_{k,\ell} \mathbf{m}_t$$

with $\mathcal{K}_{k,\ell}$ as in (27). Here $\mathbb{F}$ maps between $\mathcal{U} \times X$ and $\mathcal{W} \times \mathcal{Y}$ with X, $\mathcal{Y}$ as in (28), and $\mathcal{U}$, $\mathcal{W}$ appropriately chosen function spaces, see Sect. 4.1.

Iterative methods for solving inverse problems usually require the linearization $F'(\hat{\alpha})$ of the forward operator F and its adjoint $F'(\hat{\alpha})^*$ (and likewise for $\mathbb{F}$) in the given Hilbert space setting.

For example, consider Landweber's iteration cf., e.g., [14, 27] defined by a gradient decent method for the least squares functional $\|F(\hat{\alpha}) - y\|_{\mathcal{Y}}^2$ as

$$\hat{\alpha}_{n+1} = \hat{\alpha}_n - \mu_n F'(\hat{\alpha}_n)^*(F(\hat{\alpha}_n) - y)$$

with an appropriately chosen step size μ_n. Alternatively, one can split the forward operator into a system by considering it row wise $F_k(\hat{\alpha}) = y_k$ with $F_k = (F_{kl})_{\ell=1...L}$ or column wise $F_\ell(\hat{\alpha}) = y_\ell$ with $F_\ell = (F_{kl})_{k=1,...,K}$, or even element wise $F_{kl}(\hat{\alpha}) = y_{kl}$, and cyclically iterating over these equations with gradient descent steps in a Kaczmarz version of the Landweber iteration cf., e.g., [12, 13]. The same can be done with the respective all-at-once versions [16]. These methods extend to Banach spaces as well by using duality mappings, cf., e.g., [35], however, for the sake of simplicity of exposition and implementation, we will concentrate on a Hilbert space setting here; in particular, all adjoints will be Hilbert space adjoints.

4 Derivatives and Adjoints

Motivated by their need in iterative reconstruction methods, we now derive and rigorously justify derivatives of the forward operators as well as their adjoints, both in an all-at-once and in a reduced setting.

To simplify notation for the following analysis sections, the subscript "ext" in the external magnetic field will be skipped. Moreover, to avoid confusion with the dual pairing, we will use the dot notation for the Euclidean inner product.

4.1 All-at-Once Formulation

We split the magnetization additively into its given initial value $\mathbf{m}_0$ and the unknown rest $\hat{\mathbf{m}}$, so that the forward operator reads

$$\mathbb{F}(\hat{\mathbf{m}}, \hat{\alpha}_1, \hat{\alpha}_2) = \begin{pmatrix} \mathbb{F}_0(\hat{\mathbf{m}}, \hat{\alpha}_1, \hat{\alpha}_2) \\ \left(\mathbb{F}_{k\ell}(\hat{\mathbf{m}}, \hat{\alpha}_1, \hat{\alpha}_2) \right)_{k=1,\dots,K,\ \ell=1,\dots,L} \end{pmatrix}$$

$$:= \begin{pmatrix} \hat{\alpha}_1 \hat{\mathbf{m}}_t - \Delta_N(\mathbf{m}_0 + \hat{\mathbf{m}}) - \hat{\alpha}_2(\mathbf{m}_0 + \hat{\mathbf{m}}) \times \hat{\mathbf{m}}_t \\ -|\nabla(\mathbf{m}_0 + \hat{\mathbf{m}})|^2(\mathbf{m}_0 + \hat{\mathbf{m}}) - \mathbf{h} + ((\mathbf{m}_0 + \hat{\mathbf{m}}) \cdot \mathbf{h})(\mathbf{m}_0 + \hat{\mathbf{m}}) \\ \left(\int_0^T \int_\Omega \mathbf{K}_{k\ell}(t, \tau, x) \cdot \mathbf{m}_t(x, \tau) \, dx \, d\tau \right)_{k=1,\dots,K,\ \ell=1,\dots,L} \end{pmatrix},$$

for given $\mathbf{h} \in L^2(0, T; L^P(\Omega; \mathbb{R}^3))$, $p \geq 2$, where $\Delta_N : H^1(\Omega) \to H^1(\Omega)^*$ and, using the same notation, $\Delta_N : H_N^2(\Omega) \to L^2(\Omega)(\subseteq H^1(\Omega)^*)$ with $H_N^2(\Omega) = \{u \in H^2(\Omega) : \partial_\nu u = 0 \text{ on } \partial\Omega\}^1$ is equipped with homogeneous Neumann boundary conditions, i.e, it is defined by

$$\langle -\Delta_N u, v \rangle_{H^1(\Omega)^*, H^1(\Omega)} = (\nabla u, \nabla v)_{L^2(\Omega)} \quad \forall u, v \in H^1(\Omega)$$

and thus satisfies

$$(-\Delta_N u, v)_{L^2(\Omega)} = \int_\Omega \nabla u \cdot \nabla v \, dx \quad \forall u \in H_N^2(\Omega), \ v \in H^1(\Omega). \tag{30}$$

The forward operator is supposed to act between Hilbert spaces

$$\mathbb{F} : \mathcal{U} \times \mathbb{R}^2 \to \mathcal{W} \times L^2(0, T)^{KL}$$

with the linear space

$$\mathcal{U} = \{ \mathbf{u} \in L^2(0, T; H_N^2(\Omega; \mathbb{R}^3)) \cap H^1(0, T; L^2(\Omega; \mathbb{R}^3)) : \mathbf{u}(0) = 0 \}$$
$$\subseteq C(0, T; H^1(\Omega)) \cap H^s(0, T; H^{2-2s}(\Omega)), \tag{31}$$

for $s \in [0, 1]$, where the latter embedding is continuous by, e.g., [34, Lemma 7.3], applied to $\frac{\partial u_i}{\partial x_j}$, and interpolation, as well as

$$\mathcal{W} = H^1(0, T; H^1(\Omega; \mathbb{R}^3))^* \text{ or, in case } p > 2, \mathcal{W} = H^1(0, T; L^2(\Omega; \mathbb{R}^3))^*. \tag{32}$$

We equip $\mathcal{U}$ with the inner product

$$(\mathbf{u}_1, \mathbf{u}_2)_{\mathcal{U}} := \int_0^T \int_\Omega \left((-\Delta_N \mathbf{u}_1) \cdot (-\Delta_N \mathbf{u}_2) + \mathbf{u}_{1t} \cdot \mathbf{u}_{2t} \right) dx \, dt$$
$$+ \int_\Omega \nabla \mathbf{u}_1(T) : \nabla \mathbf{u}_2(T) \, dx,$$

---

[1] Note that as opposed to $H^1(\Omega)$ functions, $H^2(\Omega)$ functions do have a Neumann boundary trace.

which, in spite of the nontrivial nullspace of the Neumann Laplacian $-\Delta_N$, defines a norm equivalent to the usual norm on $L^2(0, T; H^2(\Omega; \mathbb{R}^3)) \cap H^1(0, T; L^2(\Omega; \mathbb{R}^3))$, due to the estimates

$$\|\mathbf{u}\|^2_{L^2(0,T;L^2(\Omega))} = -\int_0^T \int_\Omega \int_0^t \mathbf{u}(s)\, ds\, \mathbf{u}_t(t)\, dx\, dt + \int_\Omega \int_0^t \mathbf{u}(s)\, ds\, \mathbf{u}(T)\, dx$$

$$\leq \left(T \|\mathbf{u}_t\|_{L^2(0,T;L^2(\Omega))} + \sqrt{T} \|\mathbf{u}(T)\|_{L^2(\Omega)}\right) \|\mathbf{u}\|_{L^2(0,T;L^2(\Omega))}$$

$$\|\mathbf{u}(T)\|_{L^2(\Omega)} = \| \int_0^T \mathbf{u}_t(t)\, dt\|_{L^2(\Omega)} \leq \sqrt{T} \|\mathbf{u}_t\|_{L^2(0,T;L^2(\Omega))}.$$

This, together with the definition of the Neumann Laplacian (30), and the use of solutions $\mathbf{z}, \mathbf{v}$ to the auxiliary problems

$$\begin{cases} \mathbf{z}_t - \Delta\mathbf{z} = \mathbf{v} \text{ in } (0, T) \times \Omega \\ \partial_\nu\mathbf{z} = 0 \text{ on } (0, T) \times \partial\Omega \;, \\ \mathbf{z}(0) = 0 \text{ in } \Omega \end{cases} \qquad \begin{cases} -\mathbf{v}_t - \Delta\mathbf{v} = \mathbf{f} \text{ in } (0, T) \times \Omega \\ \partial_\nu\mathbf{v} = 0 \text{ on } (0, T) \times \partial\Omega \;, \\ \mathbf{v}(T) = \mathbf{g} \text{ in } \Omega \end{cases} \qquad (33)$$

allows to derive the identity

$$\begin{aligned}
(\mathbf{u}, \mathbf{z})_{\mathcal{U}} &= \int_0^T \int_\Omega \left(\nabla\mathbf{u} : \nabla(-\Delta_N\mathbf{z}) - \mathbf{u}\cdot\mathbf{z}_{tt}\right) dx\, dt + \int_\Omega \mathbf{u}(T)\cdot\left(\mathbf{z}_t(T) - \Delta_N\mathbf{z}(T)\right) dx \\
&= \int_0^T \int_\Omega \left(\nabla\mathbf{u} : \nabla(\mathbf{v} - \mathbf{z}_t) - \mathbf{u}\cdot(\mathbf{v}_t + \Delta_N\mathbf{z}_t)\right) dx\, dt + \int_\Omega \mathbf{u}(T)\cdot\mathbf{v}(T)\, dx \\
&= \int_0^T \int_\Omega \mathbf{u}\cdot\left(-\Delta_N\mathbf{v} - \mathbf{v}_t\right) dx\, dt + \int_\Omega \mathbf{u}(T)\cdot\mathbf{v}(T)\, dx \\
&= \int_0^T \int_\Omega \mathbf{u}\cdot\mathbf{f}\, dx\, dt + \int_\Omega \mathbf{u}(T)\cdot\mathbf{g}\, dx\,,
\end{aligned}$$

$$(34)$$

which will be needed later on for deriving the adjoint.

On $\mathcal{W} = H^1(0, T; H^1(\Omega; \mathbb{R}^3))^*$ we use the inner product

$$(\mathbf{w}_1, \mathbf{w}_2)_{\mathcal{W}} := \int_0^T \int_\Omega \left(I_1[\nabla(-\Delta_N + \mathrm{id})^{-1}\mathbf{w}_1](t) : I_1[\nabla(-\Delta_N + \mathrm{id})^{-1}\mathbf{w}_2](t)\right.$$

$$\left. + I_1[(-\Delta_N + \mathrm{id})^{-1}\mathbf{w}_1](t) \cdot I_1[(-\Delta_N + \mathrm{id})^{-1}\mathbf{w}_2](t)\, dx\, dt\,,\right.$$

with the isomorphism $-\Delta_N + \mathrm{id} : H^1(\Omega) \to (H^1(\Omega))^*$ and the time integral operators

$$I_1[w](t) := \int_0^t w(s)\,ds - \frac{1}{T}\int_0^T (T-s)w(s)\,ds\,,$$

$$I_2[w](t) := -\int_0^t (t-s)w(s)\,ds + \frac{t}{T}\int_0^T (T-s)w(s)\,ds\,,$$

so that $I_2[w]_t(t) = -I_1[w](t)$, $I_1[w]_t(t) = -I_2[w]_{tt}(t) = w(t)$ and $I_2[w](0) = I_2[w](T) = 0$, hence

$$\int_0^T I_1[w_1](t)\,I_1[w_2](t)\,dt = \int_0^T I_2[w_1](t)\,w_2(t)\,dt,$$

so that in case $\mathbf{w}_2 \in L^2(0,T; L^2(\Omega; \mathbb{R}^3))$,

$$(\mathbf{w}_1, \mathbf{w}_2)_{\mathcal{W}} = \int_0^T \int_\Omega \Big(I_2[\nabla(-\Delta_N + \mathrm{id})^{-1}\mathbf{w}_1](t) : [\nabla(-\Delta_N + \mathrm{id})^{-1}\mathbf{w}_2](t)$$

$$+ I_2[(-\Delta_N + \mathrm{id})^{-1}\mathbf{w}_1](t) \cdot [(-\Delta_N + \mathrm{id})^{-1}\mathbf{w}_2](t)\,dx\,dt$$

$$= \int_0^T \int_\Omega I_2[(-\Delta_N + \mathrm{id})^{-1}\mathbf{w}_1](t) \cdot \mathbf{w}_2(t)\,dx\,dt\,.$$

(35)

In case $p > 2$ in the assumption on $\mathbf{h}$, we can set $\mathcal{W} = H^1(0,T; L^2(\Omega; \mathbb{R}^3))^*$ and use the simpler inner product

$$(\mathbf{w}_1, \mathbf{w}_2)_{\mathcal{W}} := \int_0^T \int_\Omega I_1[\mathbf{w}_1](t) \cdot I_1[\mathbf{w}_2](t)\,dx\,dt\,,$$

which in case $\mathbf{w}_2 \in L^2(0,T; L^2(\Omega; \mathbb{R}^3))$ satisfies

$$(\mathbf{w}_1, \mathbf{w}_2)_{\mathcal{W}} = \int_0^T \int_\Omega I_2[\mathbf{w}_1](t) \cdot \mathbf{w}_2(t)\,dx\,dt\,.$$

4.1.1 Well-Definedness of the Forward Operator

Indeed it can be verified that $\mathbb{F}$ maps between the function spaces introduced above, cf. (31), (32). For the linear (with respect to $\hat{\mathbf{m}}$) parts $\hat{\alpha}_1\hat{\mathbf{m}}_t$, $-\Delta_N\hat{\mathbf{m}}$, and $\int_0^T \int_\Omega \mathbf{K}_{k\ell}(t,\tau,x) \cdot \mathbf{m}_t(x,\tau)\,dx\,d\tau$ of $\mathbb{F}$, this is obvious and for the nonlinear terms $\hat{\alpha}_2(\mathbf{m}_0 + \hat{\mathbf{m}}) \times \hat{\mathbf{m}}_t$, $|\nabla(\mathbf{m}_0 + \hat{\mathbf{m}})|^2(\mathbf{m}_0 + \hat{\mathbf{m}})$, $((\mathbf{m}_0 + \hat{\mathbf{m}}) \cdot \mathbf{h})(\mathbf{m}_0 + \hat{\mathbf{m}})$ we use the following estimates (36), (37), (38), (39), (40), (41), holding for any $\mathbf{u}, \mathbf{w}, \mathbf{z} \in \mathcal{U}$. For the term $\hat{\alpha}_2(\mathbf{m}_0 + \hat{\mathbf{m}}) \times \hat{\mathbf{m}}_t$, we estimate

$$\|\mathbf{u} \times \mathbf{w}_t\|_{H^1(0,T;H^1(\Omega;\mathbb{R}^3))^*}$$

$$\leq \|\mathbf{u} \times \mathbf{w}_t\|_{L^2(0,T;(H^1(\Omega;\mathbb{R}^3))^*)}$$

$$\leq C^\Omega_{H^1 \to L^3} \|\mathbf{u} \times \mathbf{w}_t\|_{L^2(0,T;L^{3/2}(\Omega;\mathbb{R}^3))} \tag{36}$$

$$\leq C^\Omega_{H^1 \to L^3} \|\mathbf{u}\|_{C(0,T;L^6(\Omega;\mathbb{R}^3))} \|\mathbf{w}_t\|_{L^2(0,T;L^2(\Omega;\mathbb{R}^3))}$$

$$\leq C^\Omega_{H^1 \to L^3} C^\Omega_{H^1 \to L^6} \|\mathbf{u}\|_{C(0,T;H^1(\Omega;\mathbb{R}^3))} \|\mathbf{w}_t\|_{L^2(0,T;L^2(\Omega;\mathbb{R}^3))} \,,$$

where we have used duality and continuity of the embeddings $H^1(0,T;H^1(\Omega;\mathbb{R}^3)) \hookrightarrow L^2(0,T;H^1(\Omega;\mathbb{R}^3)) \hookrightarrow L^2(0,T;L^3(\Omega))$ in the first and second estimate, and Hölder's inequality with exponent 4 in the third estimate; For the term $|\nabla(\mathbf{m}_0 + \hat{\mathbf{m}})|^2(\mathbf{m}_0 + \hat{\mathbf{m}})$, we use

$$\|(\nabla\mathbf{u} : \nabla\mathbf{w})\mathbf{z}\|_{H^1(0,T;H^1(\Omega;\mathbb{R}^3))^*}$$

$$\leq C^{(0,T)}_{H^1 \to L^\infty} \|(\nabla\mathbf{u} : \nabla\mathbf{w})\mathbf{z}\|_{L^1(0,T;(H^1(\Omega;\mathbb{R}^3))^*)}$$

$$\leq C^{(0,T)}_{H^1 \to L^\infty} C^\Omega_{H^1 \to L^6} \|(\nabla\mathbf{u} : \nabla\mathbf{w})\mathbf{z}\|_{L^1(0,T;L^{6/5}(\Omega;\mathbb{R}^3))}$$

$$\leq C^{(0,T)}_{H^1 \to L^\infty} C^\Omega_{H^1 \to L^6} \tag{37}$$

$$\|\nabla\mathbf{u}\|_{L^2(0,T;L^6(\Omega;\mathbb{R}^3))} \|\nabla\mathbf{w}\|_{L^2(0,T;L^6(\Omega;\mathbb{R}^3))} \|\mathbf{z}\|_{C(0,T;L^2(\Omega;\mathbb{R}^3))}$$

$$\leq C^{(0,T)}_{H^1 \to L^\infty} C^\Omega_{H^1 \to L^6}$$

$$\|\mathbf{u}\|_{L^2(0,T;H^2(\Omega;\mathbb{R}^3))} \|\mathbf{w}\|_{L^2(0,T;H^2(\Omega;\mathbb{R}^3))} \|\mathbf{z}\|_{C(0,T;H^1(\Omega;\mathbb{R}^3))} \,,$$

again using duality and the embeddings $H^1(0,T;H^1(\Omega;\mathbb{R}^3)) \hookrightarrow L^\infty(0,T;H^1(\Omega)) \hookrightarrow L^\infty(0,T;L^6(\Omega))$;
For the term $((\mathbf{m}_0 + \hat{\mathbf{m}}) \cdot \mathbf{h})(\mathbf{m}_0 + \hat{\mathbf{m}})$, we estimate

$$\|(\mathbf{u} \cdot \mathbf{h})\mathbf{z}\|_{H^1(0,T;H^1(\Omega;\mathbb{R}^3))^*}$$

$$\leq C^\Omega_{H^1 \to L^6} \|(\mathbf{u} \cdot \mathbf{h})\mathbf{z}\|_{L^2(0,T;L^{6/5}(\Omega;\mathbb{R}^3))}$$

$$\leq C^\Omega_{H^1 \to L^6} \|\mathbf{u}\|_{C(0,T;L^6(\Omega;\mathbb{R}^3))} \|\mathbf{z}\|_{C(0,T;L^6(\Omega;\mathbb{R}^3))} \|\mathbf{h}\|_{L^2(0,T;L^2(\Omega;\mathbb{R}^3))}$$

$$\leq (C^\Omega_{H^1 \to L^6}; \mathbb{R}^3) \|\mathbf{u}\|_{C(0,T;H^1(\Omega;\mathbb{R}^3))} \|\mathbf{z}\|_{C(0,T;H^1(\Omega;\mathbb{R}^3))} \|\mathbf{h}\|_{L^2(0,T;L^2(\Omega;\mathbb{R}^3))}$$

$$\tag{38}$$

by duality and the embedding $H^1(0,T;H^1(\Omega;\mathbb{R}^3)) \hookrightarrow L^2(0,T;L^6(\Omega))$, as well as Hölder's inequality.

In case $p > 2$, $\mathbb{F}$ maps into the somewhat stronger space $\mathcal{W} = H^1(0, T; L^2(\Omega; \mathbb{R}^3))^*$, due to the estimates

$$\|\mathbf{u} \times \mathbf{w}_t\|_{H^1(0,T;L^2(\Omega;\mathbb{R}^3))^*}$$
$$\leq C^{(0,T)}_{H^1 \to L^\infty} \|\mathbf{u} \times \mathbf{w}_t\|_{L^1(0,T;L^2(\Omega;\mathbb{R}^3))}$$
$$\leq C^{(0,T)}_{H^1 \to L^\infty} \|\mathbf{u}\|_{L^2(0,T;L^\infty(\Omega;\mathbb{R}^3))} \|\mathbf{w}_t\|_{L^2(0,T;L^2(\Omega;\mathbb{R}^3))} \tag{39}$$
$$\leq C^{(0,T)}_{H^1 \to L^\infty} C^\Omega_{H^2 \to L^\infty} \|\mathbf{u}\|_{L^2(0,T;H^2(\Omega;\mathbb{R}^3))} \|\mathbf{w}_t\|_{L^2(0,T;L^2(\Omega;\mathbb{R}^3))},$$

as well as

$$\|(\nabla\mathbf{u} : \nabla\mathbf{w})\mathbf{z}\|_{H^1(0,T;L^2(\Omega;\mathbb{R}^3))^*}$$
$$\leq C^{(0,T)}_{H^1 \to L^\infty} \|(\nabla\mathbf{u} : \nabla\mathbf{w})\mathbf{z}\|_{L^1(0,T;L^2(\Omega;\mathbb{R}^3))}$$
$$\leq C^{(0,T)}_{H^1 \to L^\infty} \|\nabla\mathbf{u}\|_{L^2(0,T;L^6(\Omega;\mathbb{R}^3))} \|\nabla\mathbf{w}\|_{L^2(0,T;L^6(\Omega;\mathbb{R}^3))} \|\mathbf{z}\|_{C(0,T;L^6(\Omega;\mathbb{R}^3))}$$
$$\leq C^{(0,T)}_{H^1 \to L^\infty} (C^\Omega_{H^1 \to L^6}; \mathbb{R}^3) \|\mathbf{u}\|_{L^2(0,T;H^2(\Omega;\mathbb{R}^3))}$$
$$\|\mathbf{w}\|_{L^2(0,T;H^2(\Omega;\mathbb{R}^3))} \|\mathbf{z}\|_{C(0,T;H^1(\Omega;\mathbb{R}^3))},$$
$$\tag{40}$$

and

$$\|(\mathbf{u} \cdot \mathbf{h})\mathbf{z}\|_{H^1(0,T;L^2(\Omega;\mathbb{R}^3))^*}$$
$$\leq C^{(0,T)}_{H^1 \to L^\infty} \|(\mathbf{u} \cdot \mathbf{h})\mathbf{z}\|_{L^1(0,T;L^2(\Omega;\mathbb{R}^3))}$$
$$\leq C^{(0,T)}_{H^1 \to L^\infty} \|\mathbf{u}\|_{L^4(0,T;L^{p^{**}}(\Omega;\mathbb{R}^3))} \|\mathbf{z}\|_{L^4(0,T;L^{p^{**}}(\Omega;\mathbb{R}^3))} \|\mathbf{h}\|_{L^2(0,T;L^p(\Omega;\mathbb{R}^3))}$$
$$\leq C^{(0,T)}_{H^1 \to L^\infty} (C^{(0,T)}_{H^{1/4},L^4})^2 (C^\Omega_{H^{3/2},L^{p^{**}}})^2$$
$$\|\mathbf{u}\|_{H^{1/4}(0,T;H^{3/2}(\Omega;\mathbb{R}^3))} \|\mathbf{z}\|_{H^{1/4}(0,T;H^{3/2}(\Omega;\mathbb{R}^3))} \|\mathbf{h}\|_{L^2(0,T;L^p(\Omega;\mathbb{R}^3))},$$
$$\tag{41}$$

for $p^{**} = \frac{2p}{p-2} < \infty$, which can be bounded by the $\mathcal{U}$ norm of $\mathbf{u}$ and $\mathbf{z}$, using interpolation with $s = \frac{1}{4}$ in (31).

4.1.2 Differentiability of the Forward Operator

Formally, the derivative of $\mathbb{F}$ is given by

$$\mathbb{F}'(\hat{\mathbf{m}}, \hat{\alpha}_1, \hat{\alpha}_2)(\mathbf{u}, \beta_1, \beta_2)$$

$$= \begin{pmatrix} \begin{array}{c} \beta_1\hat{\mathbf{m}}_t - \beta_2(\mathbf{m}_0 + \hat{\mathbf{m}}) \times \hat{\mathbf{m}}_t \\ +\hat{\alpha}_1\mathbf{u}_t - \Delta_N\mathbf{u} - \hat{\alpha}_2\mathbf{u} \times \hat{\mathbf{m}}_t - \hat{\alpha}_2(\mathbf{m}_0 + \hat{\mathbf{m}}) \times \mathbf{u}_t \\ -2(\nabla(\mathbf{m}_0 + \hat{\mathbf{m}}) : \nabla\mathbf{u})(\mathbf{m}_0 + \hat{\mathbf{m}}) - |\nabla(\mathbf{m}_0 + \hat{\mathbf{m}})|^2\mathbf{u} \\ +((\mathbf{m}_0 + \hat{\mathbf{m}}) \cdot \mathbf{h})\mathbf{u} + (\mathbf{u} \cdot \mathbf{h})(\mathbf{m}_0 + \hat{\mathbf{m}}) \end{array} \\ \left(\int_0^T \int_\Omega \mathbf{K}_{k\ell}(t, \tau, x) \cdot \mathbf{u}_t(x, \tau)\, dx\, d\tau \right)_{k=1,...,K,\ \ell=1,...,L} \end{pmatrix}$$

$$= \begin{pmatrix} \frac{\partial\mathbb{F}_0}{\partial\hat{\mathbf{m}}}(\hat{\mathbf{m}}, \hat{\alpha}) & \frac{\partial\mathbb{F}_0}{\partial\hat{\alpha}_1}(\hat{\mathbf{m}}, \hat{\alpha}) & \frac{\partial\mathbb{F}_0}{\partial\hat{\alpha}_2}(\hat{\mathbf{m}}, \hat{\alpha}) \\ (\frac{\partial\mathbb{F}_{k\ell}}{\partial\hat{\mathbf{m}}}(\hat{\mathbf{m}}, \hat{\alpha}))_{k=1,...,K,\ell=1,...,L} & 0 & 0 \end{pmatrix} \begin{pmatrix} \mathbf{u} \\ \beta_1 \\ \beta_2 \end{pmatrix}$$

where $\frac{\partial\mathbb{F}_0}{\partial\hat{\mathbf{m}}}(\hat{\mathbf{m}}, \hat{\alpha}) : \mathcal{U} \to \mathcal{W}$, $\frac{\partial\mathbb{F}_0}{\partial\hat{\alpha}_1}(\hat{\mathbf{m}}, \hat{\alpha}) : \mathbb{R} \to \mathcal{W}$, $\frac{\partial\mathbb{F}_0}{\partial\hat{\alpha}_2}(\hat{\mathbf{m}}, \hat{\alpha}) : \mathbb{R} \to \mathcal{W}$, $(\frac{\partial\mathbb{F}_{k\ell}}{\partial\hat{\mathbf{m}}}(\hat{\mathbf{m}}, \hat{\alpha}))_{k=1,...,K,\ell=1,...,L} : \mathcal{U} \to L^2(0, T)^{KL}$. Fréchet differentiability follows from the fact that in

$$\mathbb{F}(\hat{\mathbf{m}} + \mathbf{u}, \hat{\alpha}_1 + \beta_1, \hat{\alpha}_2 + \beta_2) - \mathbb{F}(\hat{\mathbf{m}}, \hat{\alpha}_1, \hat{\alpha}_2) - \mathbb{F}'(\hat{\mathbf{m}}, \hat{\alpha}_1, \hat{\alpha}_2)(\mathbf{u}, \beta_1, \beta_2)$$

all linear terms cancel out and the nonlinear ones are given by (abbreviating $\mathbf{m} = \mathbf{m}_0 + \hat{\mathbf{m}}$)

$$(\hat{\alpha}_1 + \beta_1)(\mathbf{m}_t + \mathbf{u}_t) - \hat{\alpha}_1\mathbf{m}_t - \hat{\alpha}_1\mathbf{u}_t - \beta_1\mathbf{m}_t$$
$$= \beta_1\mathbf{u}_t$$
$$(\hat{\alpha}_2 + \beta_2)(\mathbf{m} + \mathbf{u}) \times (\mathbf{m}_t + \mathbf{u}_t) - \hat{\alpha}_2\mathbf{m} \times \mathbf{m}_t - \beta_2\mathbf{m} \times \mathbf{m}_t - \hat{\alpha}_2\mathbf{u} \times \mathbf{m}_t - \hat{\alpha}_2\mathbf{m} \times \mathbf{u}_t$$
$$= \hat{\alpha}_2\mathbf{u} \times \mathbf{u}_t + \beta_2\mathbf{m} \times \mathbf{u}_t + \beta_2\mathbf{u} \times \mathbf{m}_t + \beta_2\mathbf{u} \times \mathbf{u}_t$$

$$|\nabla\mathbf{m} + \nabla\mathbf{u}|^2(\mathbf{m} + \mathbf{u}) - |\nabla\mathbf{m}|^2\mathbf{m} - 2(\nabla\mathbf{m} : \nabla\mathbf{u})\mathbf{m} - |\nabla\mathbf{m}|^2\mathbf{u}$$
$$= |\nabla\mathbf{u}|^2(\mathbf{m} + \mathbf{u}) + 2(\nabla\mathbf{m} : \nabla\mathbf{u})\mathbf{u}$$
$$((\mathbf{m} + \mathbf{u}) \cdot \mathbf{h})(\mathbf{m} + \mathbf{u}) - (\mathbf{m} \cdot \mathbf{h})\mathbf{m} - (\mathbf{u} \cdot \mathbf{h})\mathbf{m} - (\mathbf{m} \cdot \mathbf{h})\mathbf{u}$$
$$= (\mathbf{u} \cdot \mathbf{h})\mathbf{u},$$

hence, using again (36)–(38), they can be estimated by some constant multiplied by $\|\mathbf{u}\|_{\mathcal{U}}^2 + \beta_1^2 + \beta_2^2$.

4.1.3 Adjoints

We start with the adjoint of $\frac{\partial\mathbb{F}_0}{\partial\hat{\mathbf{m}}}(\hat{\mathbf{m}}, \hat{\alpha})$. For any $\mathbf{u} \in \mathcal{U}$, $\mathbf{y} \in L^2(0, T; L^2(\Omega))$, we have, using the definition of $-\Delta_N$, i.e., (30),

$$\int_0^T \int_\Omega (\frac{\partial \mathbb{F}_0}{\partial \hat{\mathbf{m}}}(\hat{\mathbf{m}}, \hat{\alpha})\mathbf{u}) \cdot \mathbf{y} \, dx \, dt$$

$$= \int_0^T \int_\Omega \Big(\hat{\alpha}_1 \mathbf{u}_t \cdot \mathbf{y} + \nabla \mathbf{u} : \nabla \mathbf{y} - \hat{\alpha}_2 (\mathbf{u} \times \hat{\mathbf{m}}_t) \cdot \mathbf{y} - \hat{\alpha}_2 ((\mathbf{m}_0 + \hat{\mathbf{m}}) \times \mathbf{u}_t) \cdot \mathbf{y}$$

$$- 2(\nabla(\mathbf{m}_0 + \hat{\mathbf{m}}) : \nabla \mathbf{u})((\mathbf{m}_0 + \hat{\mathbf{m}}) \cdot \mathbf{y}) - |\nabla(\mathbf{m}_0 + \hat{\mathbf{m}})|^2 (\mathbf{u} \cdot \mathbf{y})$$

$$+ ((\mathbf{m}_0 + \hat{\mathbf{m}}) \cdot \mathbf{h})(\mathbf{u} \cdot \mathbf{y}) + (\mathbf{u} \cdot \mathbf{h})((\mathbf{m}_0 + \hat{\mathbf{m}}) \cdot \mathbf{y}) \Big) dx \, dt$$

$$= \int_0^T \int_\Omega \mathbf{u} \cdot \Big(-\hat{\alpha}_1 \mathbf{y}_t + (-\Delta \mathbf{y}) - \hat{\alpha}_2 \hat{\mathbf{m}}_t \times \mathbf{y} + \hat{\alpha}_2 \mathbf{y}_t \times (\mathbf{m}_0 + \hat{\mathbf{m}}) + \hat{\alpha}_2 \mathbf{y} \times \hat{\mathbf{m}}_t$$

$$- 2((\mathbf{m}_0 + \hat{\mathbf{m}}) \cdot \mathbf{y})(-\Delta_N(\mathbf{m}_0 + \hat{\mathbf{m}})) + 2((\nabla(\mathbf{m}_0 + \hat{\mathbf{m}}))^T (\nabla \mathbf{y}))(\mathbf{m}_0 + \hat{\mathbf{m}})$$

$$+ 2((\nabla(\mathbf{m}_0 + \hat{\mathbf{m}}))^T (\nabla(\mathbf{m}_0 + \hat{\mathbf{m}}))) \mathbf{y} - |\nabla(\mathbf{m}_0 + \hat{\mathbf{m}})|^2 \mathbf{y}$$

$$+ ((\mathbf{m}_0 + \hat{\mathbf{m}}) \cdot \mathbf{h}) \mathbf{y} + ((\mathbf{m}_0 + \hat{\mathbf{m}}) \cdot \mathbf{y}) \mathbf{h} \Big) dx \, dt$$

$$+ \int_\Omega \mathbf{u}(T) \cdot \Big(\hat{\alpha}_1 \mathbf{y}(T) - \hat{\alpha}_2 \mathbf{y}(T) \times (\mathbf{m}_0 + \hat{\mathbf{m}}(T)) \Big) dx$$

$$=: \int_0^T \int_\Omega \mathbf{u} \cdot \mathbf{f}^{\mathbf{y}} \, dx \, dt + \int_\Omega \mathbf{u}(T) \cdot \mathbf{g}_T^{\mathbf{y}} \, dx \, ,$$

where we have integrated by parts with respect to time and used the vector identities

$$\mathbf{a} \cdot (\mathbf{b} \times \mathbf{c}) = \mathbf{b} \cdot (\mathbf{c} \times \mathbf{a}) = \mathbf{c} \cdot (\mathbf{a} \times \mathbf{b}) \, .$$

Matching the integrals over $\Omega \times (0, T)$ and $\Omega \times \{T\}$, respectively, and taking into account the homogeneous Neumann boundary conditions implied by the definition of $-\Delta_N$, (30), as well as the identities (34), (35), we find that $\frac{\partial \mathbb{F}_0}{\partial \hat{\mathbf{m}}}(\hat{\mathbf{m}}, \hat{\alpha})^* \mathbf{y} =: \mathbf{z}$ is the solution of (33) with $\mathbf{f} = \mathbf{f}^{\mathbf{y}}$, $\mathbf{g} = \mathbf{g}_T^{\mathbf{y}}$, where in case $\mathcal{W} = H^1(0, T; H^1(\Omega; \mathbb{R}^3))^*$, $\mathbf{y} = I_2[\tilde{y}]$, with $\tilde{y}(t)$ solving

$$\begin{cases} -\Delta \tilde{y}(t) + \tilde{y}(t) = \mathbf{w}(t) \text{ in } \Omega \\ \qquad\qquad \partial_\nu \tilde{y} = 0 \text{ on } \partial\Omega \end{cases}$$

for each $t \in (0, T)$, or in case $\mathcal{W} = H^1(0, T; L^2(\Omega; \mathbb{R}^3))^*$, just $\mathbf{y} = I_2[\mathbf{w}]$.

With the same $\mathbf{y}$, after pointwise projection onto the mutually orthogonal vectors $\hat{\mathbf{m}}_t(x, t)$ and $(\mathbf{m}_0(x) + \hat{\mathbf{m}}(x, t)) \times \hat{\mathbf{m}}_t(x, t)$ and integration over space and time, we also get the adjoints of $\frac{\partial \mathbb{F}_0}{\partial \hat{\alpha}_1}(\hat{\mathbf{m}}, \hat{\alpha})$, $\frac{\partial \mathbb{F}_0}{\partial \hat{\alpha}_2}(\hat{\mathbf{m}}, \hat{\alpha})$

$$\frac{\partial \mathbb{F}_0}{\partial \hat{\alpha}_1}(\hat{\mathbf{m}}, \hat{\alpha})^* \mathbf{w} = \int_0^T \int_\Omega \hat{\mathbf{m}}_t \cdot \mathbf{y} \, dx \, dt \, ,$$

$$\frac{\partial \mathbb{F}_0}{\partial \hat{\alpha}_2}(\hat{\mathbf{m}}, \hat{\alpha})^* \mathbf{w} = - \int_0^T \int_\Omega ((\mathbf{m}_0 + \hat{\mathbf{m}}) \times \hat{\mathbf{m}}_t) \cdot \mathbf{y} \, dx \, dt \, .$$

Finally, the fact that for $\mathbf{u} \in \mathcal{U}$, $y \in L^2(0, T)^{KL}$

$$\left(\left(\frac{\partial \mathbb{F}_{k\ell}}{\partial \hat{\mathbf{m}}} (\hat{\mathbf{m}}, \hat{\alpha}) \right)_{k=1,\dots,K, \ell=1,\dots,L} \mathbf{u}, y \right)_{L^2(0,T)^{KL}}$$

$$= \sum_{k=1}^{K} \sum_{\ell=1}^{L} \int_0^T \left(\left(\frac{\partial \mathbb{F}_{k\ell}}{\partial \hat{\mathbf{m}}} (\hat{\mathbf{m}}, \hat{\alpha}) \right)_{k=1,\dots,K, \ell=1,\dots,L} \mathbf{u} \right)_{k\ell} (t) \, y_{k\ell}(t) \, dt$$

$$= \sum_{k=1}^{K} \sum_{\ell=1}^{L} \int_0^T \int_0^T \int_\Omega \mathbf{K}_{k\ell}(t, \tau, x) \cdot \mathbf{u}_t(x, \tau) \, dx \, d\tau \, y_{k\ell}(t) \, dt \tag{42}$$

$$= \sum_{k=1}^{K} \sum_{\ell=1}^{L} \int_0^T \left(- \int_0^T \int_\Omega \frac{\partial}{\partial \tau} \mathbf{K}_{k\ell}(t, \tau, x) \cdot \mathbf{u}(x, \tau) \, dx \, d\tau \right.$$

$$\left. + \int_\Omega \mathbf{K}_{k\ell}(t, T, x) \cdot \mathbf{u}(x, T) \, dx \right) y_{k\ell}(t) \, dt \, ,$$

where we have integrated by parts with respect to time, implies that due to (34), $(\frac{\partial \mathbb{F}_{k\ell}}{\partial \hat{\mathbf{m}}} (\hat{\mathbf{m}}, \hat{\alpha}))^*_{k=1,\dots,K, \ell=1,\dots,L} y = \mathbf{z}$ is obtained by solving another auxiliary problem (33) with

$$\mathbf{f}(x, \tau) = - \int_0^T \sum_{k=1}^{K} \sum_{\ell=1}^{L} \frac{\partial}{\partial \tau} \mathbf{K}_{k\ell}(t, \tau, x) y_{k\ell}(t) \, dt,$$

$$\mathbf{g}(x) = \int_0^T \sum_{k=1}^{K} \sum_{\ell=1}^{L} \mathbf{K}_{k\ell}(t, T, x) y_{k\ell}(t) \, dt \, . \tag{43}$$

Remark 2 In case of a Landweber-Kaczmarz method iterating cyclically over the equations defined by $\mathbb{F}_0$, $\mathbb{F}_{k\ell}$, $k = 1, \dots, K$, $\ell = 1, \dots, L$, adjoints of derivatives of $\mathbb{F}_0$ remain unchanged while adjoints of $\frac{\partial \mathbb{F}_{k\ell}}{\partial \hat{\mathbf{m}}} (\hat{\mathbf{m}}, \hat{\alpha}))_{k=1,\dots,K, \ell=1,\dots,L}$ are defined as in (42), (43) by just skipping the sums over k and ℓ there.

4.2 Reduced Formulation

We now consider the formulation (24) with F defined by (25), (26), and (27). Due to the estimate

$$\|\mathcal{K}_{k\ell} \mathbf{m}_t\|^2_{L^2(0,T)} \leq T \|\tilde{a}_\ell\|^2_{L^2(0,T)} \|c_k \mathbf{p}^R_\ell\|^2_{L^2(\Omega, \mathbb{R}^3)} \|\mathbf{m}\|^2_{H^1(0,T;L^2(\Omega, \mathbb{R}^3))} \, ,$$

if $\tilde{a}_\ell \in L^2(0, T)$, $c_k \mathbf{p}^R_\ell \in L^2(\Omega, \mathbb{R}^3)$ we can choose the state space in the reduced setting as

$$\tilde{\mathcal{U}} = H^1(0, T; L^2(\Omega, \mathbb{R}^3)), \tag{44}$$

which is different from the one in the all-at-once setting.

4.2.1 Adjoint Equation

From (25) the derivative of the forward operation takes the form

$$F'(\hat{\alpha})\beta = \mathcal{K}\mathbf{u}_t, \tag{45}$$

where $\mathbf{u}$ solves the linearized LLG equation

$$\begin{aligned}
\hat{\alpha}_1\mathbf{u}_t - \hat{\alpha}_2\mathbf{m} \times \mathbf{u}_t - \hat{\alpha}_2\mathbf{u} \times \mathbf{m}_t - \Delta\mathbf{u} - 2(\nabla\mathbf{u} : \nabla\mathbf{m})\mathbf{m} & \\
+ \mathbf{u}(-|\nabla\mathbf{m}|^2 + (\mathbf{m} \cdot \mathbf{h})) + (\mathbf{u} \cdot \mathbf{h})\mathbf{m} & \\
= -\beta_1\mathbf{m}_t + \beta_2\mathbf{m} \times \mathbf{m}_t & \quad \text{in } (0, T) \times \Omega \\
\partial_\nu\mathbf{u} = 0 & \quad \text{on } (0, T) \times \partial\Omega \\
\mathbf{u}(0) = 0 & \quad \text{in } \Omega,
\end{aligned}$$

and $\mathbf{m}$ is the solution to (21)–(23). This equation can be obtained by formally taking directional derivatives (in the direction of $\mathbf{u}$) in all terms of the LLG equation (21)–(23), or alternatively by subtracting the defining boundary value problems for $S(\mathbf{m} + \epsilon\mathbf{u})$ and $S(\mathbf{m})$, dividing by ϵ and then letting ϵ tend to zero.

The Hilbert space adjoint

$$F'(\hat{\alpha})^* : L^2(0, T)^{KL} \to \mathbb{R}^2$$

of $F'(\hat{\alpha})$ satisfies, for each $z \in L^2(0, T)^{KL}$,

$$(F'(\hat{\alpha})^*z, \beta)_{\mathbb{R}^2}$$

$$= (z, F'(\hat{\alpha})\beta)_{L^2(0,T)^{KL}}$$

$$= \sum_{k=1}^{K}\sum_{\ell=1}^{L}\int_0^T z_{k\ell}(t)\int_0^T\int_\Omega (-\mu_0)\tilde{a}_\ell(t - \tau)c_k(x)\mathbf{p}_\ell^R(x) \cdot \mathbf{u}_\tau(\tau, x)dx\, d\tau\, dt$$

$$= \sum_{k=1}^{K}\sum_{\ell=1}^{L}\int_0^T z_{k\ell}(t)\left(-\int_0^T\int_\Omega (-\mu_0) \cdot (-1)\tilde{a}_{\ell\,t}(t-\tau)c_k(x)\mathbf{p}_\ell^R(x) \cdot \mathbf{u}(\tau, x)\, dx\, d\tau \right.$$

$$\left. + \int_\Omega (-\mu_0)\tilde{a}_\ell(t - T)c_k(x)\mathbf{p}_\ell^R(x) \cdot \mathbf{u}(T, x)\, dx\right)dt$$

$$= \int_0^T \int_\Omega \mathbf{u}(\tau, x) \cdot \sum_{k=1}^K \sum_{\ell=1}^L \left(\int_0^T (-\mu_0) \tilde{a}_{\ell\, t}(t - \tau) z_{k\ell}(t)\, dt \right) c_k(x) \mathbf{p}_\ell^R(x)\, dx\, d\tau$$

$$+ \int_\Omega \mathbf{u}(T, x) \cdot \sum_{k=1}^K \sum_{\ell=1}^L \left(\int_0^T (-\mu_0) \tilde{a}_\ell(t) z_{k\ell}(t)\, dt \right) c_k(x) \mathbf{p}_\ell^R(x)\, dx$$

$$=: (\mathbf{u}, \tilde{K} z)_{L^2(0,T;L^2(\Omega,\mathbb{R}^3))} + (\mathbf{u}(T), \tilde{K}_T z)_{L^2(\Omega,\mathbb{R}^3)} \tag{46}$$

as the transfer function $\tilde{a}$ is periodic with period T, and the continuous embedding $H(0, T) \hookrightarrow C[0, T]$ allows us to evaluate $\mathbf{u}(t = T)$.

Observing

$$\int_0^T \int_\Omega -\hat{\alpha}_1 \mathbf{q}_t^z \cdot \mathbf{u}\, dx\, dt$$

$$= \int_0^T \int_\Omega \hat{\alpha}_1 \mathbf{u}_t \cdot \mathbf{q}^z\, dx - \int_\Omega \hat{\alpha}_1 \mathbf{q}^z(T) \cdot \mathbf{u}(T)\, dx,$$

$$\int_0^T \int_\Omega -\hat{\alpha}_2 (\mathbf{m} \times \mathbf{q}^z)_t \cdot \mathbf{u}\, dx\, dt$$

$$= \int_0^T \int_\Omega -\hat{\alpha}_2 (\mathbf{m} \times \mathbf{u}_t) \cdot \mathbf{q}^z\, dx\, dt - \int_\Omega \hat{\alpha}_2 (\mathbf{m} \times \mathbf{q}^z)(T) \cdot \mathbf{u}(T)\, dx,$$

$$\int_0^T \int_\Omega \hat{\alpha}_2 (\mathbf{q}^z \times \mathbf{m}_t) \cdot \mathbf{u}\, dx\, dt$$

$$= \int_0^T \int_\Omega -\hat{\alpha}_2 (\mathbf{u} \times \mathbf{m}_t) \cdot \mathbf{q}^z\, dx\, dt,$$

$$\int_0^T \int_\Omega -\Delta \mathbf{q}^z \cdot \mathbf{u}\, dx\, dt$$

$$= \int_0^T \int_\Omega -\mathbf{q}^z \cdot \Delta \mathbf{u}\, dx\, dt - \int_0^T \int_{\partial\Omega} \partial_\nu \mathbf{q}^z \cdot \mathbf{u}\, dx\, dt,$$

$$\int_0^T \int_\Omega \mathbf{q}^z(-|\nabla \mathbf{m}|^2 + (\mathbf{m} \cdot \mathbf{h})) \cdot \mathbf{u}\, dx\, dt$$

$$= \int_0^T \int_\Omega \left(\mathbf{u}(-|\nabla \mathbf{m}|^2 + (\mathbf{m} \cdot \mathbf{h})) \right) \cdot \mathbf{q}^z\, dx\, dt,$$

$$\int_0^T \int_\Omega (\mathbf{q}^z \cdot \mathbf{m}) \mathbf{h} \cdot \mathbf{u}\, dx\, dt$$

$$= \int_0^T \int_\Omega (\mathbf{u} \cdot \mathbf{h}) \mathbf{m} \cdot \mathbf{q}^z\, dx\, dt,$$

$$\int_0^T \int_\Omega 2(\mathbf{m} \cdot \mathbf{q}^z) \Delta \mathbf{m} \cdot \mathbf{u} \, dx \, dt$$

$$= -\int_0^T \int_\Omega 2(\nabla \mathbf{m} : \nabla \mathbf{u})(\mathbf{m} \cdot \mathbf{q}^z) \, dx \, dt$$

$$+ 2 \int_0^T \int_\Omega -\mathbf{u} \cdot ((\nabla \mathbf{m})^\top \nabla \mathbf{m}) \mathbf{q}^z - \mathbf{u} \cdot ((\nabla \mathbf{m})^\top \nabla \mathbf{q}^z) \mathbf{m} \, dx \, dt \,,$$

we see that, if $\mathbf{q}^z$ solves the adjoint equation

$$-\hat{\alpha}_1 \mathbf{q}_t^z - \hat{\alpha}_2 \mathbf{m} \times \mathbf{q}_t^z - 2\hat{\alpha}_2 \mathbf{m}_t \times \mathbf{q}^z - \Delta \mathbf{q}^z$$

$$+ 2 \left((\nabla \mathbf{m})^\top \nabla \mathbf{m} \right) \mathbf{q}^z + 2 \left((\nabla \mathbf{m})^\top \nabla \mathbf{q}^z \right) \mathbf{m}$$

$$+ (-|\nabla \mathbf{m}|^2 + (\mathbf{m} \cdot \mathbf{h})) \mathbf{q}^z + (\mathbf{m} \cdot \mathbf{q}^z)(\mathbf{h} + 2\Delta \mathbf{m}) = \tilde{K}z \qquad \text{in } (0, T) \times \Omega \tag{47}$$

$$\partial_\nu \mathbf{q}^z = 0 \qquad \qquad \text{on } (0, T) \times \partial\Omega \tag{48}$$

$$\hat{\alpha}_1 \mathbf{q}^z(T) + \hat{\alpha}_2 (\mathbf{m} \times \mathbf{q}^z)(T) = \tilde{K}_T z \qquad \text{in } \Omega \tag{49}$$

then with (46), we have

$$(F'(\hat{\alpha})^* z, \beta)_{\mathbb{R}^2} = (\mathbf{u}, \tilde{K}z)_{L^2(0,T;L^2(\Omega,\mathbb{R}^3))} + (\mathbf{u}(T), \tilde{K}_T z)_{L(\Omega,\mathbb{R}^3)}$$

$$= \int_0^T \int_\Omega (-\beta_1 \mathbf{m}_t + \beta_2 \mathbf{m} \times \mathbf{m}_t) \cdot \mathbf{q}^z \, dx \, dt$$

$$= (\beta_1, \beta_2) \cdot \left(\int_0^T \int_\Omega -\mathbf{m}_t \cdot \mathbf{q}^z \, dx \, dt, \int_0^T \int_\Omega (\mathbf{m} \times \mathbf{m}_t) \cdot \mathbf{q}^z \, dx \, dt \right),$$

which implies the Hilbert space adjoint $F'(\hat{\alpha})^* : \mathcal{Y} \to \mathbb{R}^2$

$$F'(\hat{\alpha})^* z = \left(\int_0^T \int_\Omega -\mathbf{m}_t \cdot \mathbf{q}^z \, dx \, dt, \int_0^T \int_\Omega (\mathbf{m} \times \mathbf{m}_t) \cdot \mathbf{q}^z \, dx \, dt \right), \tag{50}$$

provided that the adjoint state $\mathbf{q}^z$ exists and belongs to a sufficiently smooth space (see Sect. 4.2.2 below).

The final condition (49) is equivalent to

$$\begin{pmatrix} \hat{\alpha}_1 & -\hat{\alpha}_2 \mathbf{m}_3(T) & \hat{\alpha}_2 \mathbf{m}_2(T) \\ \hat{\alpha}_2 \mathbf{m}_3(T) & \hat{\alpha}_1 & -\hat{\alpha}_2 \mathbf{m}_1(T) \\ -\hat{\alpha}_2 \mathbf{m}_2(T) & \hat{\alpha}_2 \mathbf{m}_1(T) & \hat{\alpha}_1 \end{pmatrix} \mathbf{q}^z(T) =: M_T^{\hat{\alpha}} \mathbf{q}^z(T) = \tilde{K}_T z,$$

where $\mathbf{m}_i(T), i = 1, 2, 3$, denotes the i-th component of $\mathbf{m}(T)$. The matrix $M_T^{\hat{\alpha}}$ with $\det(M_T^{\hat{\alpha}}) = |\hat{\alpha}_1(\hat{\alpha}_1^2 + \hat{\alpha}_2^2)|$ is invertible if $\hat{\alpha}_1 > 0$, which matches the condition for existence of the solution to the LLG equation. Hence, we are able to rewrite the adjoint equation in the form

$$-\hat{\alpha}_1 \mathbf{q}_t^z - \hat{\alpha}_2 \mathbf{m} \times \mathbf{q}_t^z - 2\hat{\alpha}_2 \mathbf{m}_t \times \mathbf{q}^z - \Delta \mathbf{q}^z$$

$$+ 2\left((\nabla \mathbf{m})^\top \nabla \mathbf{m}\right)\mathbf{q}^z + 2\left((\nabla \mathbf{m})^\top \nabla \mathbf{q}^z\right)\mathbf{m}$$

$$+ (-|\nabla \mathbf{m}|^2 + (\mathbf{m} \cdot \mathbf{h}))\mathbf{q}^z + (\mathbf{m} \cdot \mathbf{q}^z)(\mathbf{h} + 2\Delta \mathbf{m}) = \tilde{K} z \qquad \text{in } (0, T) \times \Omega$$
$$(51)$$

$$\partial_\nu \mathbf{q}^z = 0 \qquad\qquad\qquad\qquad\qquad\qquad\qquad\qquad \text{on } (0, T) \times \partial\Omega$$
$$(52)$$

$$\mathbf{q}^z(T) = (M_T^{\hat{\alpha}})^{-1} \tilde{K}_T z \qquad\qquad\qquad\qquad\qquad\qquad \text{in } \Omega.\qquad (53)$$

Remark 3 Formula (50) inspires a Kaczmarz scheme relying on restricting the observation operator to time subintervals for every fixed k, ℓ, namely, we segment $(0, T)$ into several subintervals (t^j, t^{j+1}) with the break points $0 = t^0 < \ldots < t^{n-1} = T$ and

$$F_{k\ell}^j : \mathcal{D}(F)(\subseteq \mathcal{X}) \to \mathcal{Y}^j, \qquad \hat{\alpha} \mapsto y^j := \mathcal{K}_{k\ell} \frac{\partial}{\partial t} S(\hat{\alpha})|_{(t^j, t^{j+1})} \qquad (54)$$

with

$$\mathcal{Y}^j = L^2(t^j, t^{j+1})^{KL} \qquad j = 0 \ldots n - 1, \qquad (55)$$

hence

$$y_{k\ell}^j(t) = \int_{t^j}^{t^{j+1}} \int_\Omega -\mu_0 \tilde{a}_\ell(t - \tau) c_k(x) \mathbf{p}_\ell^R(x) \cdot \mathbf{m}_\tau(x, \tau) dx d\tau. \qquad (56)$$

Here we distinguish between the superscript j for the time subinterval index and subscripts k, ℓ for the index of different receive coils and concentrations.

For $z^j \in \mathcal{Y}^j$,

$$(\tilde{K} z^j)(x, t) = \sum_{k=1}^K \sum_{\ell=1}^L -\mu_0 c_k(x) \mathbf{p}_\ell^R(x) \int_{t^j}^{t^{j+1}} \tilde{a}_{\ell\tau}(\tau - t) z_{k\ell}^j(\tau) d\tau \qquad t \in (0, T),$$

$$(\tilde{K}_T z^j)(x) = \sum_{k=1}^K \sum_{\ell=1}^L -\mu_0 c_k(x) \mathbf{p}_\ell^R(x) \int_{t^j}^{t^{j+1}} \tilde{a}_\ell(\tau) z_{k\ell}^j(\tau) d\tau$$

yield the same Hilbert space adjoint $F^{j'}(\hat{\alpha})^* : \mathcal{Y}^j \to \mathbb{R}^2$ as in (50), and the adjoint state $\mathbf{q}^{z^j}$ still needs to be solved on the whole time line $[0, T]$ with

$$
\begin{aligned}
&- \hat{\alpha}_1 \mathbf{q}_t^{z^j} - \hat{\alpha}_2 \mathbf{m} \times \mathbf{q}_t^{z^j} - 2\hat{\alpha}_2 \mathbf{m}_t \times \mathbf{q}^{z^j} - \Delta \mathbf{q}^{z^j} \\
&+ 2 \left((\nabla \mathbf{m})^\top \nabla \mathbf{m} \right) \mathbf{q}^{z^j} + 2 \left((\nabla \mathbf{m})^\top \nabla \mathbf{q}^{z^j} \right) \mathbf{m} \\
&+ (-|\nabla \mathbf{m}|^2 + (\mathbf{m} \cdot \mathbf{h})) \mathbf{q}^{z^j} + (\mathbf{m} \cdot \mathbf{q}^{z^j})(\mathbf{h} + 2\Delta \mathbf{m}) = \tilde{K} z^j \quad \text{in } (0, T) \times \Omega
\end{aligned}
$$
(57)

$$
\partial_\nu \mathbf{q}^{z^j} = 0 \qquad\qquad\qquad\qquad\qquad\qquad\qquad\qquad \text{on } (0, T) \times \partial\Omega
$$
(58)

$$
\mathbf{q}^{z^j}(T) = (M_T^{\hat{\alpha}})^{-1} \tilde{K}_T z^j \qquad\qquad\qquad\qquad\qquad \text{in } \Omega.
$$
(59)

Besides this, the conventional Kaczmarz method resulting from the collection of observation operators $\mathcal{K}_{k\ell}$ with $k = 1 \ldots K, \ell = 1 \ldots L$ as in (13) is always applicable, where

$$
F_{k\ell} : \mathcal{D}(F)(\subseteq X) \to \mathcal{Y}_{k\ell}, \qquad \hat{\alpha} \mapsto y_{k\ell} := \mathcal{K}_{k\ell} \frac{\partial}{\partial t}(S(\hat{\alpha}))
$$
(60)

with

$$
\mathcal{Y}_{k\ell} = L^2(0, T) \qquad k = 1 \ldots K, \ \ell = 1 \ldots L
$$
(61)

Thus $F_{k\ell}'(\hat{\alpha})^*$ can be seen as (50), where the adjoint state $\mathbf{q}_{k\ell}^z$ solves (51)–(53) with corresponding data

$$
\tilde{K}_{k\ell} z(x, t) = -\mu_0 c_k(x) \mathbf{p}_\ell^R(x) \int_0^T \tilde{a}_{\ell \tau}(\tau - t) z(\tau) \, d\tau \qquad t \in (0, T),
$$

$$
\tilde{K}_{T\,k\ell} z(x) = -\mu_0 c_k(x) \mathbf{p}_\ell^R(x) \int_0^T \tilde{a}_\ell(\tau) z(\tau) \, d\tau
$$

for each $z \in \mathcal{Y}_{k\ell}$.

4.2.2 Solvability of the Adjoint Equation

First of all, we derive a bound for $\mathbf{q}^z$. To begin with, we set $\tau = T - t$ to convert (51)–(53) into an initial boundary value problem. Then we test (51) with $\mathbf{q}_t^z$ and obtain the identities and estimates

$$\int_\Omega \hat{\alpha}_1 \mathbf{q}_t^z(t) \cdot \mathbf{q}_t^z(t)\, dx$$

$$= \hat{\alpha}_1 \|\mathbf{q}_t^z(t)\|_{L^2(\Omega,\mathbb{R}^3)}^2\,,$$

$$\int_\Omega \hat{\alpha}_2(\mathbf{m}(t) \times \mathbf{q}_t^z(t)) \cdot \mathbf{q}_t^z(t)\, dx$$

$$= 0\,,$$

$$\int_\Omega \hat{\alpha}_2(\mathbf{m}_t(t) \times \mathbf{q}^z(t)) \cdot \mathbf{q}_t^z(t)\, dx$$

$$\leq |\hat{\alpha}_2| \|\mathbf{m}_t(t)\|_{L^3(\Omega,\mathbb{R}^3)} \|\mathbf{q}^z(t)\|_{L^6(\Omega,\mathbb{R}^3)} \|\mathbf{q}_t^z(t)\|_{L^2(\Omega,\mathbb{R}^3)}\,,$$

$$\int_\Omega -\Delta \mathbf{q}^z(t) \cdot \mathbf{q}_t^z(t)\, dx$$

$$= \frac{1}{2}\frac{d}{dt} \|\nabla \mathbf{q}^z(t)\|_{L^2(\Omega,\mathbb{R}^3)}^2\,,$$

$$\int_\Omega \left(((\nabla \mathbf{m}(t))^\top \nabla \mathbf{m}(t)) \mathbf{q}^z(t) \right) \cdot \mathbf{q}_t^z(t)\, dx$$

$$\leq (C_{H^1 \to L^6}^\Omega)^2 \|\nabla \mathbf{m}\|_{L^\infty(0,T;H^1(\Omega,\mathbb{R}^3))}^2 \|\mathbf{q}^z(t)\|_{L^6(\Omega,\mathbb{R}^3)} \|\mathbf{q}_t^z(t)\|_{L^2(\Omega,\mathbb{R}^3)}\,,$$

$$\int_\Omega \left(((\nabla \mathbf{m}(t))^\top \nabla \mathbf{q}^z(t)) \mathbf{m}(t) \right) \cdot \mathbf{q}_t^z(t)\, dx$$

$$\leq C_{H^2 \to L^\infty}^\Omega \|\nabla \mathbf{m}(t)\|_{H^2(\Omega,\mathbb{R}^3)} \|\nabla \mathbf{q}^z(t)\|_{L^2(\Omega,\mathbb{R}^3)} \|\mathbf{q}_t^z(t)\|_{L^2(\Omega,\mathbb{R}^3)}\,,$$

$$\int_\Omega (-|\nabla \mathbf{m}(t)|^2 + (\mathbf{m}(t) \cdot \mathbf{h})) \mathbf{q}^z(t) \cdot \mathbf{q}_t^z(t)\, dx$$

$$\leq \left((C_{H^1 \to L^6}^\Omega)^2 \|\nabla \mathbf{m}\|_{L^\infty(0,T;H^1(\Omega,\mathbb{R}^3))}^2 + \|\mathbf{h}(t)\|_{L^3(\Omega,\mathbb{R}^3)} \right)$$

$$\|\mathbf{q}^z(t)\|_{L^6(\Omega,\mathbb{R}^3)} \|\mathbf{q}_t^z(t)\|_{L^2(\Omega,\mathbb{R}^3)}\,,$$

$$\int_\Omega \left(\mathbf{m}(t) \cdot \mathbf{q}^z(t) \right) \mathbf{h}(t) \cdot \mathbf{q}_t^z(t)\, dx$$

$$\leq \|\mathbf{h}(t)\|_{L^3(\Omega,\mathbb{R}^3)} \|\mathbf{q}^z(t)\|_{L^6(\Omega,\mathbb{R}^3)} \|\mathbf{q}_t^z(t)\|_{L^2(\Omega,\mathbb{R}^3)}\,,$$

$$\int_\Omega (\mathbf{m}(t) \cdot \mathbf{q}^z(t)) \Delta \mathbf{m}(t) \cdot \mathbf{q}_t^z(t)\, dx$$

$$\leq C_{H^1 \to L^3}^\Omega \|\Delta \mathbf{m}(t)\|_{H^1(\Omega,\mathbb{R}^3)} \|\mathbf{q}^z(t)\|_{L^6(\Omega,\mathbb{R}^3)} \|v q_t^z(t)\|_{L^2(\Omega,\mathbb{R}^3)}\,,$$

$$\int_\Omega \tilde{K} z(t) \cdot \mathbf{q}_t^z(t)\, dx$$

$$\leq \|\tilde{K} z(t)\|_{L^2(\Omega,\mathbb{R}^3)} \|\mathbf{q}_t^z(t)\|_{L^2(\Omega,\mathbb{R}^3)}\,.$$

Above, we employ the fact that the solution $\mathbf{m}$ to the LLG equation has $|\mathbf{m}| = 1$ and the continuity of the embeddings $H^1(\Omega, \mathbb{R}^3) \hookrightarrow L^6(\Omega, \mathbb{R}^3) \hookrightarrow L^3(\Omega, \mathbb{R}^3)$, $H^2(\Omega, \mathbb{R}^3) \hookrightarrow L^\infty(\Omega, \mathbb{R}^3)$ through the constants $C^\Omega_{H^1 \to L^6}$, $C^\Omega_{H^1 \to L^3}$ and $C^\Omega_{H^2 \to L^\infty}$, respectively.

Employing Young's inequality we deduce, for each $t \leq T$ and $\epsilon > 0$ sufficiently small,

$$\frac{1}{2}\frac{d}{dt}\|\nabla \mathbf{q}^z(t)\|^2_{L^2(\Omega,\mathbb{R}^3)} + (\hat{\alpha}_1 - \epsilon)\|\mathbf{q}^z_t(t)\|^2_{L^2(\Omega,\mathbb{R}^3)}$$

$$\leq \left[\left(\|\nabla\mathbf{m}\|^4_{L^\infty(0,T;H^1(\Omega,\mathbb{R}^3))} + \|\nabla\mathbf{m}(t)\|^2_{H^2(\Omega,\mathbb{R}^3)} + \|\mathbf{m}_t(t)\|^2_{L^3(\Omega,\mathbb{R}^3)} + \|\mathbf{h}(t)\|^2_{L^3(\Omega,\mathbb{R}^3)} \right) \right.$$

$$\left. \cdot \|\mathbf{q}^z(t)\|^2_{H^1(\Omega,\mathbb{R}^3)} + \|\tilde{K}z(t)\|^2_{L^2(\Omega,\mathbb{R}^3)} \right]\frac{C}{4\epsilon}. \tag{62}$$

The generic constant C might take different values whenever it appears.

To have the full H^1-norm on the left hand side of this estimate, we apply the transformation $\tilde{\mathbf{q}}^z(t) = e^t \mathbf{q}^z(t)$, which yields $\tilde{\mathbf{q}}^z_t(t) = e^t(\mathbf{q}^z(t) + \mathbf{q}^z_t(t))$. After testing by $\mathbf{q}^z_t$, the term $\int_\Omega \mathbf{q}^z(t) \cdot \mathbf{q}^z_t(t)\,dx = \frac{1}{2}\frac{d}{dt}\|\mathbf{q}^z(t)\|^2_{L^2(\Omega,\mathbb{R}^3)}$ will contribute to $\frac{1}{2}\frac{d}{dt}\|\nabla\mathbf{q}^z(t)\|^2_{L^2(\Omega,\mathbb{R}^3)}$ forming the full H^1-norm on the left hand side. Alternatively, one can add $\mathbf{q}^z$ to both sides of (51) and evaluate the right hand side with $\int_\Omega \mathbf{q}^z(t) \cdot \mathbf{q}^z_t(t)\,dx \leq \frac{1}{4\epsilon}\|\mathbf{q}^z(t)\|^2_{H^1(\Omega,\mathbb{R}^3)} + \epsilon\|\mathbf{q}^z_t(t)\|^2_{L^2(\Omega,\mathbb{R}^3)}$.

Integrating over $(0, t)$, we get

$$\frac{1}{2}\|\mathbf{q}^z(t)\|^2_{H^1(\Omega,\mathbb{R}^3)} + (\hat{\alpha}_1 - \epsilon)\|\mathbf{q}^z_t\|^2_{L^2(0,t;L^2(\Omega,\mathbb{R}^3))}$$

$$\leq \frac{C}{4\epsilon}\left[\int_0^t \left(\|\nabla\mathbf{m}\|^4_{L^\infty(0,T;H^1(\Omega,\mathbb{R}^3))} + \|\nabla\mathbf{m}(\tau)\|^2_{H^2(\Omega,\mathbb{R}^3)} + \|\mathbf{m}_t(\tau)\|^2_{L^3(\Omega,\mathbb{R}^3)} \right.\right.$$

$$\left. + \|\mathbf{h}(\tau)\|^2_{L^3(\Omega,\mathbb{R}^3)} \right)\cdot\|\mathbf{q}^z(\tau)\|^2_{H^1(\Omega,\mathbb{R}^3)}\,d\tau$$

$$\left. + \|\tilde{K}z\|^2_{L^2(0,T;L^2(\Omega,\mathbb{R}^3))} + \|(M^{\hat{\alpha}}_T)^{-1}\tilde{K}_T z\|^2_{H^1(\Omega,\mathbb{R}^3)} \right]$$

with the evaluation for the terms $\|\tilde{K}z\|_{L^2(0,T;L^2(\Omega,\mathbb{R}^3))}$ and $\|(M^{\hat{\alpha}}_T)^{-1}\tilde{K}_T z\|^2_{H^1(\Omega,\mathbb{R}^3)}$ (not causing any misunderstanding, we omit here the subscripts k, ℓ for indices of concentrations and coil sensitivities)

$$\|\tilde{K}z(t)\|^2_{L^2(\Omega,\mathbb{R}^3)} \leq C\|c\mathbf{p}^R\|^2_{L^2(\Omega,\mathbb{R}^3)}\|\tilde{a}\|^2_{H^1(0,T)}\|z\|^2_{L^2(0,T)} \leq C^{\tilde{a},c,\mathbf{p}^R}\|z\|^2_{L^2(0,T)},$$

$$\|(M^{\hat{\alpha}}_T)^{-1}\tilde{K}_T z\|^2_{H^1(\Omega,\mathbb{R}^3)}$$

$$\leq C^{\hat{\alpha}}\|z\|^2_{L^2(0,T)}\|\tilde{a}\|^2_{L^2(0,T)}$$

$$
\cdot \left(\| c\mathbf{p}^R \|^2_{H^1(\Omega,\mathbb{R}^3)} + \| c\mathbf{p}\mathbf{m}_i(T) \|^2_{H^1(\Omega,\mathbb{R}^3)} + \| c\mathbf{p}^R \mathbf{m}_j(T)\mathbf{m}_k(T) \|^2_{H^1(\Omega,\mathbb{R}^3)} \right)
$$

$$
\leq C^{\hat{\alpha}_0,\rho,\tilde{a}} \| z \|^2_{L^2(0,T)} \left(\| c\mathbf{p}^R \|^2_{H^1(\Omega,\mathbb{R}^3)} + \| c\mathbf{p}^R \|^2_{L^6(\Omega,\mathbb{R}^3)} \| \nabla \mathbf{m}(T) \|^2_{L^3(\Omega,\mathbb{R}^3)} \right)
$$

$$
\leq C^{\tilde{a}} \| z \|^2_{L^2(0,T)}
$$

$$
\cdot \left(\| c\mathbf{p}^R \|^2_{H^1(\Omega,\mathbb{R}^3)} + (C^\Omega_{H^1 \to L^6} C^\Omega_{H^1 \to L^3})^2 \| c\mathbf{p}^R \|^2_{H^1(\Omega,\mathbb{R}^3)} \| \nabla \mathbf{m} \|^2_{L^\infty(0,T;H^1(\Omega,\mathbb{R}^3))} \right)
$$

$$
\leq C^{\tilde{a},c,\mathbf{p}^R} \| z \|^2_{L^2(0,T)} \| \nabla \mathbf{m} \|^2_{L^\infty(0,T;H^1(\Omega,\mathbb{R}^3))}
$$

with some $i,j,k = 1,2,3$. This estimate holds for $c\mathbf{p}^R \in H^1(\Omega,\mathbb{R}^3)$ and thus requires some smoothness of the concentration c, while the coil sensitivity $\mathbf{p}^R$ is usually smooth in practice.

Then applying Grönwall's inequality yields

$$
\| \mathbf{q}^z \|_{L^\infty(0,T;H^1(\Omega,\mathbb{R}^3))}
$$

$$
\leq C \exp \left(\| \nabla \mathbf{m} \|^2_{L^\infty(0,T;H^1(\Omega,\mathbb{R}^3))} + \| \nabla \mathbf{m} \|_{L^2(0,T;H^2(\Omega,\mathbb{R}^3))} + \| \mathbf{m}_t \|_{L^2(0,T;L^3(\Omega,\mathbb{R}^3))} \right.
$$

$$
\left. + \| \mathbf{h} \|_{L^2(0,T;L^3(\Omega,\mathbb{R}^3))} \right) \cdot \left(\| \tilde{K} z \|_{L^2(0,T;L^2(\Omega,\mathbb{R}^3))} + \| (M^{\hat{\alpha}}_T)^{-1} \tilde{K}_T z \|_{H^1(\Omega,\mathbb{R}^3)} \right)
$$

$$
\leq C^{\tilde{a},c,\mathbf{p}^R} \left(\| \nabla \mathbf{m} \|_{L^\infty(0,T;H^1(\Omega,\mathbb{R}^3)) \cap L^2(0,T;H^2(\Omega,\mathbb{R}^3))}, \| \mathbf{m}_t \|_{L^2(0,T;L^3(\Omega,\mathbb{R}^3))} \right.
$$

$$
\left. , \| \mathbf{h} \|_{L^2(0,T;L^3(\Omega,\mathbb{R}^3))} \right) \cdot \| z \|_{L^2(0,T)}.
$$

Integrating (62) on $(0,T)$, we also get

$$
\| \mathbf{q}^z_t \|_{L^2(0,T;L^2(\Omega,\mathbb{R}^3))}
$$

$$
\leq C^{\tilde{a},c,\mathbf{p}^R} \left(\| \nabla \mathbf{m} \|_{L^\infty(0,T;H^1(\Omega,\mathbb{R}^3)) \cap L^2(0,T;H^2(\Omega,\mathbb{R}^3))}, \| \mathbf{m}_t \|_{L^2(0,T;L^3(\Omega,\mathbb{R}^3))} \right.
$$

$$
\left. , \| \mathbf{h} \|_{L^2(0,T;L^3(\Omega,\mathbb{R}^3))} \right) \cdot \| z \|_{L^2(0,T)}.
$$

Altogether, we obtain

$$
\| \mathbf{q}^z \|_{L^\infty(0,T;H^1(\Omega,\mathbb{R}^3))} + \| \mathbf{q}^z_t \|_{L^2(0,T;L^2(\Omega,\mathbb{R}^3))}
$$

$$
\leq C^{\tilde{a},c,\mathbf{p}^R} \left(\| \nabla \mathbf{m} \|_{L^\infty(0,T;H^1(\Omega,\mathbb{R}^3)) \cap L^2(0,T;H^2(\Omega,\mathbb{R}^3))}, \| \mathbf{m}_t \|_{L^2(0,T;L^3(\Omega,\mathbb{R}^3))} \right.
$$

$$
\left. , \| \mathbf{h} \|_{L^2(0,T;L^3(\Omega,\mathbb{R}^3))} \right) \cdot \| z \|_{L^2(0,T)}. \tag{63}
$$

This result applied to the Galerkin approximation implies existence of the solution to the adjoint equation. Uniqueness also follows from (63).

4.2.3 Regularity of the Solution to the LLG Equation

In (63), first of all we need the solution $\mathbf{m} \in L^\infty(0, T; H^2(\Omega, \mathbb{R}^3))$ $\cap L^2(0, T; H^3(\Omega, \mathbb{R}^3))$ to the LLG equation. This can be obtained from the regularity result in [11, Lemma 2.3] for $\mathbf{m}_0 \in H^2(\Omega, \mathbb{R}^3)$ with small $\|\nabla \mathbf{m}_0\|_{L^2(\Omega, \mathbb{R}^3)}$. The remaining task is verifying that the estimate still holds in case the external field $\mathbf{h}$ is present, i.e., the right hand side of (21) contains the additional term $\mathrm{Proj}_{\mathbf{m}\perp} \mathbf{h}$.

Following the lines of the proof in [11, Lemma 2.3], we take the second spatial derivative of $\mathrm{Proj}_{\mathbf{m}\perp} \mathbf{h}$, then test it by $\Delta \mathbf{m}$ such that

$$\int_\Omega \Delta \mathbf{h}(t) \cdot \Delta \mathbf{m}(t)\, dx$$

$$\leq \begin{cases} \|\Delta \mathbf{h}(t)\|_{L^2(\Omega,\mathbb{R}^3)} \|\Delta \mathbf{m}(t)\|_{L^2(\Omega,\mathbb{R}^3)} & \text{if } \mathbf{h} \in L^2(0, T; H^2(\Omega, \mathbb{R}^3)) \\ \|\nabla \mathbf{h}(t)\|_{L^2(\Omega,\mathbb{R}^3)} \|\nabla^3 \mathbf{m}(t)\|_{L^2(\Omega,\mathbb{R}^3)} & \text{if } \mathbf{h} \in L^2(0, T; H^1(\Omega, \mathbb{R}^3)), \ \partial_\nu \mathbf{h} = 0 \text{ on } \partial\Omega \end{cases},$$

$$\int_\Omega \Delta((\mathbf{m}(t) \cdot \mathbf{h}(t))\mathbf{m}(t)) \cdot \Delta \mathbf{m}(t)\, dx$$

$$\leq \begin{cases} C\|\mathbf{h}(t)\|_{H^2(\Omega,\mathbb{R}^3)} \left(1 + 6\|\nabla \mathbf{m}(t)\|_{H^1(\Omega,\mathbb{R}^3)} + 2\|\nabla \mathbf{m}(t)\|_{H^2(\Omega,\mathbb{R}^3)} \|\nabla \mathbf{m}\|_{L^\infty(0,T;L^2(\Omega,\mathbb{R}^3))}\right) \\ \qquad \cdot \|\Delta \mathbf{m}(t)\|_{L^2(\Omega,\mathbb{R}^3)} & \text{if } \mathbf{h} \in L^2(0, T; H^2(\Omega, \mathbb{R}^3)) \\ C\|\mathbf{h}(t)\|_{H^1(\Omega,\mathbb{R}^3)} \left(1 + 2\|\nabla \mathbf{m}(t)\|_{L^3(\Omega,\mathbb{R}^3)}\right) \|\nabla^3 \mathbf{m}(t)\|_{L^2(\Omega,\mathbb{R}^3)} \\ \qquad\qquad\qquad\qquad\qquad\qquad \text{if } \mathbf{h} \in L^2(0, T; H^1(\Omega, \mathbb{R}^3)), \ \partial_\nu \mathbf{h} = 0 \text{ on } \partial\Omega \end{cases}$$

with C just depending on the constants in the embeddings $H^1(\Omega, \mathbb{R}^3) \hookrightarrow L^6(\Omega, \mathbb{R}^3) \hookrightarrow L^3(\Omega, \mathbb{R}^3)$. Then we can proceed similarly to the proof of [11, Lemma 2.3] by applying Young's inequality, Gronwall's inequality and time integration to arrive at

$$\|\nabla \mathbf{m}\|_{L^\infty(0,T;H^1(\Omega,\mathbb{R}^3)) \cap L^2(0,T;H^2(\Omega,\mathbb{R}^3))}$$
$$\leq \left(\|\nabla \mathbf{m}_0\|_{H^1(\Omega,\mathbb{R}^3)} + \|\mathbf{h}\|\right) C(\|\nabla \mathbf{m}_0\|_{H^1(\Omega,\mathbb{R}^3)}, \|\mathbf{h}\|), \tag{64}$$

where $\|\mathbf{h}\|$ is evaluated in $L^2(0, T; H^1(\Omega, \mathbb{R}^3))$ or $L^2(0, T; H^2(\Omega, \mathbb{R}^3))$ as in the two cases mentioned above.

It remains to prove $\mathbf{m}_t \in L^2(0, T; H^1(\Omega, \mathbb{R}^3)) \hookrightarrow L^2(0, T; L^3(\Omega, \mathbb{R}^3))$ to validate (63). For this purpose, instead of working with (21) we test (18) by $-\Delta \mathbf{m}_t$ and obtain

$$\int_\Omega \mathbf{m}_t(t) \cdot (-\Delta \mathbf{m}_t(t))\, dx$$

$$= \|\nabla \mathbf{m}_t(t)\|^2_{L^2(\Omega,\mathbb{R}^3)},$$

$$\int_\Omega -\alpha_1 \Delta \mathbf{m}(t) \cdot (-\Delta \mathbf{m}_t(t))\, dx$$

$$= \frac{\alpha_1}{2} \frac{d}{dt} \|\Delta \mathbf{m}(t)\|^2_{L^2(\Omega,\mathbb{R}^3)},$$

$$\int_\Omega -\alpha_1 |\nabla \mathbf{m}(t)|^2 \mathbf{m}(t) \cdot (-\Delta \mathbf{m}_t(t))\, dx$$

$$= -\alpha_1 \int_\Omega \nabla \left(|\nabla \mathbf{m}(t)|^2 \mathbf{m}(t) \right) : \nabla \mathbf{m}_t(t)\, dx$$

$$\leq \alpha_1 \Big(2 C^\Omega_{H^1 \to L^6} C^\Omega_{H^1 \to L^3} \|\nabla \mathbf{m}\|_{L^\infty(0,T;H^1(\Omega,\mathbb{R}^3))} \|\Delta \mathbf{m}(t)\|_{H^1(\Omega,\mathbb{R}^3)}$$

$$+ (C^\Omega_{H^1 \to L^6})^3 \|\nabla \mathbf{m}\|^3_{L^\infty(0,T;H^1(\Omega,\mathbb{R}^3))} \Big) . \|\nabla \mathbf{m}_t(t)\|_{L^2(\Omega,\mathbb{R}^3)},$$

$$\int_\Omega -\alpha_1 (\mathbf{h}(t) - (\mathbf{m}(t) \cdot \mathbf{h}(t))\mathbf{m}(t)) \cdot (-\Delta \mathbf{m}_t(t))\, dx$$

$$= -\alpha_1 \int_\Omega \nabla (\mathbf{h}(t) - (\mathbf{m}(t) \cdot \mathbf{h}(t))\mathbf{m}(t)) : \nabla \mathbf{m}_t(t)\, dx$$

$$\leq 2\alpha_1 \Big(\|\nabla \mathbf{h}(t)\|_{L^2(\Omega,\mathbb{R}^3)}$$

$$+ C^\Omega_{H^1 \to L^6} \|\mathbf{h}(t)\|_{L^3(\Omega,\mathbb{R}^3)} \|\nabla \mathbf{m}\|_{L^\infty(0,T;H^1(\Omega,\mathbb{R}^3))} \Big) . \|\nabla \mathbf{m}_t(t)\|_{L^2(\Omega,\mathbb{R}^3)},$$

$$\int_\Omega -\alpha_2 (\mathbf{m}(t) \times \Delta \mathbf{m}(t)) \cdot (-\Delta \mathbf{m}_t(t))\, dx$$

$$= \int_\Omega -\alpha_2 \nabla (\mathbf{m}(t) \times \Delta \mathbf{m}(t)) : \nabla \mathbf{m}_t(t)\, dx$$

$$\leq |\alpha_2| \Big(C^\Omega_{H^1 \to L^6} C^\Omega_{H^1 \to L^3} \|\nabla \mathbf{m}\|_{L^\infty(0,T;H^1(\Omega,\mathbb{R}^3))} \|\Delta \mathbf{m}(t)\|_{H^1(\Omega,\mathbb{R}^3)}$$

$$+ \|\nabla^3 \mathbf{m}(t)\|_{L^2(\Omega,\mathbb{R}^3)} \Big) . \|\nabla \mathbf{m}_t(t)\|_{L^2(\Omega,\mathbb{R}^3)},$$

$$\int_\Omega -\alpha_2 (\mathbf{m}(t) \times \mathbf{h}(t)) \cdot (-\Delta \mathbf{m}_t(t))\, dx$$

$$= \int_\Omega -\alpha_2 \nabla (\mathbf{m}(t) \times \mathbf{h}(t)) : (\nabla \mathbf{m}_t(t))\, dx$$

$$\leq |\alpha_2| \Big(C^\Omega_{H^1 \to L^6} \|\mathbf{h}(t)\|_{L^3(\Omega,\mathbb{R}^3)} \|\nabla \mathbf{m}\|_{L^\infty(0,T;H^1(\Omega,\mathbb{R}^3))}$$

$$+ \|\nabla \mathbf{h}(t)\|_{L^2(\Omega,\mathbb{R}^3)} \Big) . \|\nabla \mathbf{m}_t(t)\|_{L^2(\Omega,\mathbb{R}^3)}.$$

Integrating over $(0, T)$ then employing Hölder's inequality, Young's inequality and (64), it follows that

$$(1 - \epsilon)\|\nabla \mathbf{m}_t\|_{L^2(0,T;L^2(\Omega,\mathbb{R}^3))}$$

$$\leq \frac{C}{4\epsilon}\Big(\|\nabla \mathbf{m}\|_{L^\infty(0,T;H^1(\Omega,\mathbb{R}^3))}\|\nabla \mathbf{m}\|_{L^2(0,T;H^2(\Omega,\mathbb{R}^3))} + \|\nabla \mathbf{m}\|^3_{L^\infty(0,T;H^1(\Omega,\mathbb{R}^3))}$$

$$+ \|\nabla \mathbf{m}\|_{L^2(0,T;H^2(\Omega,\mathbb{R}^3))} + \|\mathbf{h}\|_{L^2(0,T;H^1(\Omega,\mathbb{R}^3))}\|\nabla \mathbf{m}\|_{L^\infty(0,T;H^1(\Omega,\mathbb{R}^3))}$$

$$+ \|\mathbf{h}\|_{L^2(0,T;H^1(\Omega,\mathbb{R}^3))}\Big)$$

$$\leq \big(\|\nabla \mathbf{m}_0\|_{H^1(\Omega,\mathbb{R}^3)} + \|\mathbf{h}\|\big) C(\|\nabla \mathbf{m}_0\|_{H^1(\Omega,\mathbb{R}^3)}, \|\mathbf{h}\|). \tag{65}$$

Also $\|\mathbf{m}_t\|_{L^2(0,T;L^2(\Omega,\mathbb{R}^3))} < C\left(\|\nabla \mathbf{m}_0\|_{L^2(\Omega,\mathbb{R}^3)} + \|\mathbf{h}\|_{L^2(0,T;L^2(\Omega,\mathbb{R}^3))}\right)$ according to [25] with taking into account the presence of $\mathbf{h}$, we arrive at

$$\|\mathbf{m}_t\|_{L^2(0,T;H^1(\Omega,\mathbb{R}^3))} \leq \big(\|\nabla \mathbf{m}_0\|_{H^1(\Omega,\mathbb{R}^3)} + \|\mathbf{h}\|\big) C(\|\nabla \mathbf{m}_0\|_{H^1(\Omega,\mathbb{R}^3)}, \|\mathbf{h}\|), \tag{66}$$

where $\|\mathbf{h}\|$ is evaluated in $L^2(0, T; H^1(\Omega, \mathbb{R}^3))$ or $L^2(0, T; H^2(\Omega, \mathbb{R}^3))$.

In conclusion, the fact that $\mathbf{m} \in L^\infty(0, T; H^2(\Omega, \mathbb{R}^3)) \cap L^2(0, T; H^3(\Omega, \mathbb{R}^3)) \cap H^1(0, T; H^1(\Omega, \mathbb{R}^3))$ for $\mathbf{m}_0 \in H^2(\Omega, \mathbb{R}^3)$ with small $\|\nabla \mathbf{m}_0\|_{L^2(\Omega,\mathbb{R}^3)}$, and $\mathbf{h} \in L^2(0, T; H^1(\Omega, \mathbb{R}^3))$, $\partial_\nu \mathbf{h} = 0$ on $\partial\Omega$ or $\mathbf{h} \in L^2(0, T; H^2(\Omega, \mathbb{R}^3))$ guarantee unique existence of the adjoint state $\mathbf{q}^z \in L^\infty(0, T; H^1(\Omega, \mathbb{R}^3)) \cap H^1(0, T; L^2(\Omega, \mathbb{R}^3))$. And this regularity of $\mathbf{q}^z$ ensures the adjoint $F'(\hat{\alpha})^*$ in (50) to be well-defined.

Remark 4

- The LLG equation (21)–(23) is uniquely solvable for $\hat{\alpha}_1 > 0$ and arbitrary $\hat{\alpha}_2$. Therefore, the regularization problem should be locally solved within the ball $\mathcal{B}_\rho(\hat{\alpha}^0)$ of center $\hat{\alpha}^0$ with $\hat{\alpha}_1^0 > 0$ and radius $\rho < \hat{\alpha}_1^0$.
- [11, Lemma 2.3] requires smallness $\|\nabla \mathbf{m}_0\|_{L^2(\Omega,\mathbb{R}^3)} \leq \lambda$, and this smallness depends on $\hat{\alpha}$ through the relation $C^I\left(\lambda^2 + 2\lambda + \frac{\hat{\alpha}_2}{\hat{\alpha}_1}\lambda\right) < 1$ with C^I depending on the constants in the interpolation inequalities.

Altogether, we arrive at

$$\mathcal{D}(F) = \left\{\hat{\alpha} = (\hat{\alpha}_1, \hat{\alpha}_2) \in \mathcal{B}_\rho(\hat{\alpha}^0) : 0 < \hat{\alpha}_1^0, \rho < \hat{\alpha}_1^0, C^I\left(\lambda^2 + 2\lambda + \frac{\hat{\alpha}_2}{\hat{\alpha}_1}\lambda\right) < 1\right\}. \tag{67}$$

4.2.4 Differentiability of the Forward Operator

Since the observation operator $\mathcal{K}$ is linear, differentiability of F is just the question of differentiability of S.

Let us rewrite the LLG equation (21) in the following form

$$\tilde{g}(\hat{\alpha}, \mathbf{m}) - \Delta\mathbf{m} = \tilde{f}(\mathbf{m})$$

and denote

$$\tilde{\mathbf{v}}^\epsilon := \frac{S(\hat{\alpha} + \epsilon\beta) - S(\hat{\alpha})}{\epsilon} - \mathbf{u} =: \frac{\mathbf{n} - \mathbf{m}}{\epsilon} - \mathbf{u} =: \mathbf{v}^\epsilon - \mathbf{u}.$$

Considering the system of equations

$$\tilde{g}(\hat{\alpha} + \epsilon\beta, \mathbf{n}) \qquad\qquad - \Delta\mathbf{n} = \tilde{f}(\mathbf{n}),$$

$$\tilde{g}(\hat{\alpha}, \mathbf{m}) \qquad\qquad - \Delta\mathbf{m} = \tilde{f}(\mathbf{m}),$$

$$\tilde{g}'_{\mathbf{m}}(\hat{\alpha}, \mathbf{m})\mathbf{u} + \tilde{g}'_{\hat{\alpha}}(\hat{\alpha}, \mathbf{m})\beta - \Delta\mathbf{u} = \tilde{f}'_{\mathbf{m}}(\mathbf{m})\mathbf{u},$$

with the same boundary and initial data for each, we see that $\tilde{\mathbf{v}}^\epsilon$ solves

$$\tilde{g}'_{\mathbf{m}}(\hat{\alpha}, \mathbf{m})\tilde{\mathbf{v}}^\epsilon - \Delta\tilde{\mathbf{v}}^\epsilon - \tilde{f}'_{\mathbf{m}}(\mathbf{m})\tilde{\mathbf{v}}^\epsilon$$

$$= \frac{\tilde{f}(\mathbf{n}) - \tilde{f}(\mathbf{m})}{\epsilon} - \tilde{f}'_{\mathbf{m}}(\mathbf{m})\mathbf{v}^\epsilon - \frac{\tilde{g}(\hat{\alpha} + \epsilon\beta, \mathbf{n}) - \tilde{g}(\hat{\alpha}, \mathbf{m})}{\epsilon} \qquad (68)$$

$$+ \tilde{g}'_{\mathbf{m}}(\hat{\alpha}, \mathbf{m})\mathbf{v}^\epsilon + \tilde{g}'_{\hat{\alpha}}(\hat{\alpha}, \mathbf{m})\beta \qquad\qquad\qquad \text{in } (0, T) \times \Omega$$

$$\partial_\nu\tilde{\mathbf{v}}^\epsilon = 0 \qquad\qquad\qquad\qquad\qquad\qquad \text{on } [0, T] \times \partial\Omega \qquad (69)$$

$$\tilde{\mathbf{v}}^\epsilon(0) = 0 \qquad\qquad\qquad\qquad\qquad\qquad \text{in } \Omega, \qquad (70)$$

explicitly

$$\hat{\alpha}_1\tilde{\mathbf{v}}^\epsilon_t - \hat{\alpha}_2\mathbf{m} \times \tilde{\mathbf{v}}^\epsilon_t - \hat{\alpha}_2\tilde{\mathbf{v}}^\epsilon \times \mathbf{m}_t - \Delta\tilde{\mathbf{v}}^\epsilon$$

$$- 2(\nabla\tilde{\mathbf{v}}^\epsilon : \nabla\mathbf{m})\mathbf{m} + \tilde{\mathbf{v}}^\epsilon(-|\nabla\mathbf{m}|^2 + (\mathbf{m} \cdot \mathbf{h})) + (\tilde{\mathbf{v}}^\epsilon \cdot \mathbf{h})\mathbf{m}$$

$$= \frac{1}{\epsilon}\left(|\nabla\mathbf{n}|^2\mathbf{n} + \text{Proj}_{\mathbf{n}\perp}\mathbf{h} - |\nabla\mathbf{m}|^2\mathbf{m} - \text{Proj}_{\mathbf{m}\perp}\mathbf{h}\right) \qquad (71)$$

$$- 2(\nabla\mathbf{v}^\epsilon : \nabla\mathbf{m})\mathbf{m} + \mathbf{v}^\epsilon(-|\nabla\mathbf{m}|^2 + (\mathbf{m} \cdot \mathbf{h})) + (\mathbf{v}^\epsilon \cdot \mathbf{h})\mathbf{m}$$

$$- \frac{1}{\epsilon}\left((\hat{\alpha}_1 + \epsilon\beta_1)\mathbf{n}_t - (\hat{\alpha}_2 + \epsilon\beta_2)\mathbf{n} \times \mathbf{n}_t - \hat{\alpha}_1\mathbf{m}_t + \hat{\alpha}_2\mathbf{m}\times\mathbf{m}_t\right)$$

$$+ \hat{\alpha}_1\mathbf{v}^\epsilon_t - \hat{\alpha}_2\mathbf{m} \times \mathbf{v}^\epsilon_t - \hat{\alpha}_2\mathbf{v}^\epsilon \times \mathbf{m}_t$$

$$+ \beta_1\mathbf{m}_t - \beta_2\mathbf{m} \times \mathbf{m}_t \qquad\qquad\qquad\qquad \text{in } (0, T) \times \Omega$$

$$\partial_\nu\tilde{\mathbf{v}}^\epsilon = 0 \qquad\qquad\qquad\qquad\qquad\qquad \text{on } [0, T] \times \partial\Omega$$
$$\qquad\qquad\qquad\qquad\qquad\qquad\qquad\qquad\qquad\qquad (72)$$

$$\tilde{\mathbf{v}}^\epsilon(0) = 0 \qquad\qquad\qquad\qquad\qquad\qquad \text{in } \Omega. \qquad (73)$$

Observing the similarity of (71)–(73) to the adjoint equation (51)–(53) with $\tilde{\mathbf{v}}^\epsilon$ in place of $\mathbf{q}^z$ and denoting by $\mathbf{b}^\epsilon$ the right-hand side of (68) or (71), one can evaluate $\|\tilde{\mathbf{v}}^\epsilon\|$ using the same technique as in Sect. 4.2.2. By this way, one achieves, for each $\epsilon \in [0, \bar{\epsilon}]$,

$$\|\tilde{\mathbf{v}}^\epsilon\|_{L^\infty(0,T;H^1(\Omega,\mathbb{R}^3))\cap H^1(0,T;L^2(\Omega,\mathbb{R}^3))} \leq C\|\mathbf{b}^\epsilon\|_{L^2(0,T;L^2(\Omega,\mathbb{R}^3))}$$

with $\mathbf{b}^\epsilon \in L^2(0, T; L^2(\Omega, \mathbb{R}^3))$ also by analogously estimating and employing $\mathbf{m}, \mathbf{n} \in L^\infty(0, T; H^2(\Omega, \mathbb{R}^3)) \cap L^2(0, T; H^3(\Omega, \mathbb{R}^3)) \cap H^1(0, T; H^1(\Omega, \mathbb{R}^3))$. We note that the constant C here is independent of ϵ.

Next letting $\mathcal{V} := L^\infty(0, T; H^1(\Omega, \mathbb{R}^3)) \cap H^1(0, T; L^2(\Omega, \mathbb{R}^3))$, we have

$$\|\mathbf{b}^\epsilon\|_{L^2(0,T;L^2(\Omega,\mathbb{R}^3))} = \left\| \frac{\tilde{f}(\mathbf{n}) - \tilde{f}(\mathbf{m})}{\epsilon} - \tilde{f}'_{\mathbf{m}}(\mathbf{m})\mathbf{v}^\epsilon - \frac{\tilde{g}(\hat{\alpha} + \epsilon\beta, \mathbf{n}) - \tilde{g}(\hat{\alpha}, \mathbf{m})}{\epsilon} \right.$$
$$\left. + \tilde{g}'_{\mathbf{m}}(\hat{\alpha}, \mathbf{m})\mathbf{v}^\epsilon + \tilde{g}'_{\hat{\alpha}}(\hat{\alpha}, \mathbf{m})\beta \right\|_{L^2(0,T;L^2(\Omega,\mathbb{R}^3))}$$

$$\leq \left\| \int_0^1 \left((\tilde{f}'_{\mathbf{m}}(\mathbf{m} + \lambda\epsilon\mathbf{v}^\epsilon) - \tilde{f}'_{\mathbf{m}}(\mathbf{m}))\mathbf{v}^\epsilon - (\tilde{g}'_{\mathbf{m}}(\hat{\alpha} + \lambda\epsilon\beta, \mathbf{m} + \lambda\epsilon\mathbf{v}^\epsilon) - \tilde{g}'_{\mathbf{m}}(\hat{\alpha}, \mathbf{m}))\mathbf{v}^\epsilon \right.\right.$$
$$\left.\left. - (\tilde{g}'_{\hat{\alpha}}(\hat{\alpha} + \lambda\epsilon\beta, \mathbf{m} + \lambda\epsilon\mathbf{v}^\epsilon) - \tilde{g}'_{\hat{\alpha}}(\hat{\alpha}, \mathbf{m}))\beta \right) d\lambda \right\|_{L^2(0,T;L^2(\Omega,\mathbb{R}^3))}$$

$$\leq 2 \sup_{\substack{\lambda\in[0,1]\\ \epsilon\in[0,\bar{\epsilon}]}} \left(\|\tilde{f}'_{\mathbf{m}}(\mathbf{m} + \lambda\epsilon\mathbf{v}^\epsilon)\|_{\mathcal{V}\to L^2(0,T;L^2(\Omega,\mathbb{R}^3))} \|\mathbf{v}^\epsilon\|_{\mathcal{V}} \right.$$
$$+ \|\tilde{g}'_{\mathbf{m}}(\hat{\alpha} + \lambda\epsilon\beta, \mathbf{m} + \lambda\epsilon\mathbf{v}^\epsilon)\|_{\mathcal{V}\to L^2(0,T;L^2(\Omega,\mathbb{R}^3))} \|\mathbf{v}^\epsilon\|_{\mathcal{V}}$$
$$\left. + \|\tilde{g}'_{\hat{\alpha}}(\hat{\alpha} + \lambda\epsilon\beta, \mathbf{m} + \lambda\epsilon\mathbf{v}^\epsilon)\|_{\mathbb{R}^2\to L^2(0,T;L^2(\Omega,\mathbb{R}^3))} |\beta| \right).$$

In order to prove uniform boundedness of the derivatives of $\tilde{f}, \tilde{g}$ w.r.t λ, ϵ in the above estimate, we again proceed in a similar manner as in Sect. 4.2.2 since the space for $\mathbf{q}^z$ in Sect. 4.2.2 (c.f. (64)) coincides with $\mathcal{V}$ here and by the fact that

$$\max\{\|\mathbf{m}\|, \|\mathbf{n}\|\} \leq \max\left\{ \frac{1}{\hat{\alpha}_1}, \frac{1}{\hat{\alpha}_1 + \epsilon\beta_1} \right\} C \left(\|\mathbf{m}_0\|_{H^2(\Omega,\mathbb{R}^3)}, \|\mathbf{h}\|_{L^2(0,T;H^2(\Omega,\mathbb{R}^3))} \right)$$
$$\leq \frac{C}{\hat{\alpha}_1^0 - \rho} \tag{74}$$

for $\mathbf{m}, \mathbf{n} \in L^\infty(0, T; H^2(\Omega, \mathbb{R}^3)) \cap L^2(0, T; H^3(\Omega, \mathbb{R}^3)) \cap H^1(0, T; H^1(\Omega, \mathbb{R}^3))$. If $\partial_\nu \mathbf{h} = 0$ on $\partial\Omega$, we just need the $\|.\|_{L^2(0,T;H^1(\Omega,\mathbb{R}^3))}$-norm for $\mathbf{h}$ as claimed in (64). This estimate holds for any $\epsilon \in [0, \bar{\epsilon}]$, and the constant C is independent of ϵ.

To accomplish uniform boundedness for $\|\mathbf{b}^\epsilon\|_{L^2(0,T;L^2(\Omega,\mathbb{R}^3))}$, we need to show that $\|\mathbf{v}^\epsilon\|_{\mathcal{V}}$ is also uniformly bounded w.r.t ϵ. It is seen from

$$\tilde{g}(\hat{\alpha} + \epsilon\beta, \mathbf{n}) - \Delta\mathbf{n} = \tilde{f}(\mathbf{n}),$$

$$\tilde{g}(\hat{\alpha}, \mathbf{m}) \qquad - \Delta\mathbf{m} = \tilde{f}(\mathbf{m})$$

that $\mathbf{v}^\epsilon$ solves

$$\int_0^1 \tilde{g}'_{\mathbf{m}}(\hat{\alpha} + \lambda\epsilon\beta, \mathbf{m} + \lambda\epsilon\mathbf{v}^\epsilon)\mathbf{v}^\epsilon + \tilde{g}'_{\hat{\alpha}}(\hat{\alpha} + \lambda\epsilon\beta, \mathbf{m} + \lambda\epsilon\mathbf{v}^\epsilon)\beta \, d\lambda - \Delta\mathbf{v}^\epsilon$$

$$= \int_0^1 \tilde{f}'_{\mathbf{m}}(\mathbf{m} + \lambda\epsilon\mathbf{v}^\epsilon)\mathbf{v}^\epsilon \, d\lambda \qquad\qquad \text{in } (0, T) \times \Omega \tag{75}$$

$$\partial_\nu\mathbf{v}^\epsilon = 0 \qquad\qquad\qquad\qquad\qquad \text{on } [0, T] \times \partial\Omega \tag{76}$$

$$\mathbf{v}^\epsilon(0) = 0 \qquad\qquad\qquad\qquad\qquad \text{in } \Omega. \tag{77}$$

Noting that $\mathbf{M} := \mathbf{m} + \lambda\epsilon\mathbf{v}^\epsilon = \lambda\mathbf{n} + (1 - \lambda)\mathbf{m}$ has $\|\mathbf{M}\| \leq \frac{C}{\hat{\alpha}_1^0 - \rho}$ for all $\lambda \in [0, 1]$ with C being independent of ϵ, and $\tilde{g}$ is first order in $\hat{\alpha}$, we can rewrite (75) into the linear equation

$$\tilde{G}(\hat{\alpha} + \lambda\epsilon\beta, \mathbf{M})\mathbf{v}^\epsilon - \Delta\mathbf{v}^\epsilon + \tilde{F}(\mathbf{M})\mathbf{v}^\epsilon = \tilde{B}(\mathbf{M})\beta. \tag{78}$$

Following the lines of the proof in Sect. 4.2.2, boundedness of the terms $-\Delta$, $\tilde{F}(\mathbf{M})$, $\tilde{B}(\mathbf{M})$ are straightforward, while the main term in $\tilde{G}(\hat{\alpha} + \lambda\epsilon\beta, \mathbf{M})$ producing the single square norm of $\mathbf{v}^\epsilon_{\mathbf{t}}$, after being tested by $\mathbf{v}^\epsilon_{\mathbf{t}}$ is

$$\int_0^1 (\hat{\alpha}_1 + \lambda\epsilon\beta_1) \int_\Omega \mathbf{v}^\epsilon_{\mathbf{t}}(t) \cdot \mathbf{v}^\epsilon_{\mathbf{t}}(t) \, dx \, d\lambda = \|\mathbf{v}^\epsilon_{\mathbf{t}}(t)\|^2_{L^2(\Omega,\mathbb{R}^3)} \left(\hat{\alpha}_1 + \frac{\epsilon\beta_1}{2}\right)$$

$$\geq \|\mathbf{v}^\epsilon_{\mathbf{t}}(t)\|^2_{L^2(\Omega,\mathbb{R}^3)}(\hat{\alpha}_1^0 - \rho).$$

According to this, one gets, for all $\epsilon \in [0, \bar{\epsilon}]$,

$$\|\mathbf{v}^\epsilon\|_{\mathcal{V}} \leq C|\beta| \|\tilde{B}(\mathbf{M})\|_{\mathbb{R}^2 \to L^2(0,T;L(\Omega,\mathbb{R}^3))} \leq |\beta|C \tag{79}$$

with C depending only on $\mathbf{m}_0$, $\mathbf{h}$, $\hat{\alpha}^0$, ρ.

Since $\mathbf{b}^\epsilon \to 0$ pointwise and $\|\mathbf{b}^\epsilon\|_{L^2(0,T;L^2(\Omega,\mathbb{R}^3))} \leq C$ for all $\epsilon \in [0, \bar{\epsilon}]$, applying Lebesgue's Dominated Convergence Theorem yields convergence of $\|\mathbf{b}^\epsilon\|_{L^2(0,T;L^2(\Omega,\mathbb{R}^3))}$, thus of $\|\tilde{\mathbf{v}}^\epsilon\|_{\mathcal{V}}$, to zero. Fréchet differentiability of the forward operator in the reduced setting is therefore proved.

5   Conclusion

In this contribution we outlined a mathematical model of MPI taking into account relaxation effects, which led us to the LLG equation describing the behavior of the magnetic material inside the particles on a microscale level. For calibrating the MPI device it is necessary to compute the system function, which mathematically can be interpreted as an inverse parameter identification problem for an initial boundary value problem based on the LLG equation. To this end we deduced a detailed analysis of the forward model, i.e., the operator mapping the coefficients to the solution of the PDE as well as of the underlying inverse problem. The inverse problem itself was investigated in an all-at-once and a reduced approach. The analysis includes representations of the respective adjoint operators and Fréchet derivatives. These results are necessary for a subsequent numerical computation of the system function in a robust manner, which will be subject of future research. Even beyond this, the analysis might be useful for the development of solution methods for other inverse problems that are connected to the LLG equation.

Acknowledgments The work of Anne Wald and Thomas Schuster was partly funded by Hermann und Dr. Charlotte Deutsch–Stiftung and by the German Federal Ministry of Education and Research (Bundesministerium für Bildung und Forschung, BMBF) under 05M16TSA. This article was written during Tram Nguyen's employment at Alpen-Adria-Universität Klagenfurt.

References

1. L. Borcea, Electrical impedance tomography. Inverse Prob. **18**, R99–R136 (2002)
2. F. Bruckner, D. Suess, M. Feischl, T. Führer, P. Goldenits, M. Page, D. Praetorius, M. Ruggeri, Multiscale modeling in micromagnetics: existence of solutions and numerical integration. Math. Models Methods Appl. Sci. **24**, 2627–2662 (2014)
3. D. Colton, R. Kress, *Inverse Acoustic and Electromagnetic Scattering Theory* (Springer, New York, 2013)
4. L.R. Croft, P.W. Goodwill, S.M. Conolly, Relaxation in x-space magnetic particle imaging. IEEE Trans. Med. Imaging **31**, 2335–2342 (2012)
5. B.D. Cullity, C.D. Graham, *Introduction to Magnetic Materials* (Wiley, New York, 2011)
6. W. Demtroeder, *Experimentalphysik* (Springer, Berlin, 2013)
7. T. Dunst, M. Klein, A. Prohl, A. Schäfer, Optimal control in evolutionary micromagnetism. IMA J. Numer. Anal. **35**, 1342–1380 (2015)
8. A.S. Fokas, G.A. Kastis, *Mathematical Methods in PET and SPECT Imaging* (Springer, New York, 2015), pp. 903–936
9. T.L. Gilbert, A phenomenological theory of damping in ferromagnetic materials. IEEE Trans. Magn. **40**, 3443–3449 (2004)
10. B. Gleich, J. Weizenecker, Tomographic imaging using the nonlinear response of magnetic particles. Nature **435**, 1214–1217 (2005)
11. B. Guo, M.-C. Hong, The Landau-Lifshitz equation of the ferromagnetic spin chain and harmonic maps. Calculus Variations Partial Differ. Equ. **1**, 311–334 (1993)
12. M. Haltmeier, R. Kowar, A. Leitao, O. Scherzer, Kaczmarz methods for regularizing nonlinear ill-posed equations II: applications. Inverse Prob. Imaging **1**, 507–523 (2007)
13. M. Haltmeier, A. Leitao, O. Scherzer, Kaczmarz methods for regularizing nonlinear ill-posed equations I: convergence analysis. Inverse Prob. Imaging **1**, 289–298 (2007)

14. M. Hanke, A. Neubauer, O. Scherzer, A convergence analysis of the Landweber iteration for nonlinear ill-posed problems. Numer. Math. **72**, 21–37 (1995)
15. W. Hinshaw, A. Lent, An introduction to NMR imaging: From the Bloch equation to the imaging equation. Proc. IEEE **71**, 338–350 (1983)
16. B. Kaltenbacher, Regularization based on all-at-once formulations for inverse problems. SIAM J. Numer. Anal. **54**, 2594–2618 (2016)
17. B. Kaltenbacher, All-at-once versus reduced iterative methods for time dependent inverse problems. Inverse Prob. **33**, 064002 (2017)
18. B. Kaltenbacher, A. Neubauer, O. Scherzer, *Iterative Regularization Methods for Nonlinear Ill-Posed Problems* (De Gruyter, Berlin, 2008)
19. B. Kaltenbacher, F. Schöpfer, T. Schuster, Iterative methods for nonlinear ill-posed problems in Banach spaces: convergence and applications to parameter identification problems. Inverse Prob. **25**, 065003 (2009)
20. A. Kirsch, A. Rieder, Inverse problems for abstract evolution equations with applications in electrodynamics and elasticity. Inverse Prob. **32**, 085001 (2016)
21. T. Kluth, Mathematical models for magnetic particle imaging. Inverse Prob. **34**, 083001 (2018)
22. T. Kluth, P. Maass, Model uncertainty in magnetic particle imaging: nonlinear problem formulation and model-based sparse reconstruction. Int. J. Magn. Part. Imaging **3**, 1707004 (2017)
23. T. Knopp, T.M. Buzug, *Magnetic Particle Imaging: an Introduction to Imaging Principles and Scanner Instrumentation* (Springer, Berlin, 2012)
24. T. Knopp, N. Gdaniec, M. Möddel, Magnetic particle imaging: from proof of principle to preclinical applications. Phys. Med. Biol. **62**, R124 (2017)
25. M. Kruzík, A. Prohl, Recent Developments in the Modeling, Analysis, and Numerics of Ferromagnetism. SIAM Rev. **48**, 439–483 (2006)
26. L. Landau, E. Lifshitz, 3—On the theory of the dispersion of magnetic permeability in ferromagnetic bodies Reprinted from Physikalische Zeitschrift der Sowjetunion 8, Part 2, 153, 1935, in *Perspectives in Theoretical Physics*, ed. by L. Pitaevski (Pergamon, Amsterdam, 1992), pp. 51–65
27. L. Landweber, An iteration formula for Fredholm integral equations of the first kind. Am. J. Math. **73**, 615–624 (1951)
28. T. März, A. Weinmann, Model-based reconstruction for magnetic particle imaging in 2D and 3D. Inverse Prob. Imaging **10**, 1087–1110 (2016)
29. F. Natterer, *The Mathematics of Computerized Tomography* (Vieweg+Teubner, Wiesbaden, 1986)
30. F. Natterer, F. Wübbeling, *Mathematical Methods in Image Reconstruction* (SIAM, Philadelphia, 2001)
31. T.T.N. Nguyen, Landweber–Kaczmarz for parameter identification in time-dependent inverse problems: all-at-once versus reduced version. Inverse Prob. **35**, 035009 (2019)
32. D.B. Reeves, J.B. Weaver, Approaches for modeling magnetic nanoparticle dynamics. Critical Rev. Biomed. Eng. **42**, 85–93 (2014)
33. A. Rieder, On the regularization of nonlinear ill-posed problems via inexact Newton iterations. Inverse Prob. **15**, 309–327 (1999)
34. T. Roubíček, in *Nonlinear Partial Differential Equations with Applications*. International Series of Numerical Mathematics (Springer, Basel, 2013)
35. T. Schuster, B. Kaltenbacher, B. Hofmann, K. Kazimierski, *Regularization Methods in Banach Spaces* (de Gruyter, Berlin, 2012). Radon Series on Computational and Applied Mathematics
36. L. Shepp, Y. Vardi, Maximum likelihood reconstruction for emission tomography. IEEE Trans. Med. Imag. **1**, 113–122 (1982)
37. A. Wald, T. Schuster, Sequential subspace optimization for nonlinear inverse problems. J. Inverse Ill-posed Prob. **25**, 99–117 (2016)
38. A. Wald, T. Schuster, Tomographic terahertz imaging using sequential subspace optimization, in *New Trends in Parameter Identification for Mathematical Models*, ed. by B. Hofmann, A. Leitao, J. Zubelli (Birkhäuser, Basel, 2018)

An Inverse Source Problem Related to Acoustic Nonlinearity Parameter Imaging

Masahiro Yamamoto and Barbara Kaltenbacher

Abstract In this article, we discuss an inverse source problem of determining a spatially varying factor of a source term in a linearized higher order model of nonlinear acoustics, which is a partial differential equation of the third order in time and the fourth order in space. We establish two kinds of stability for the inverse source problems: (1) space-local stability of Hölder type and (2) space-global stability of Lipschitz type. Our key is two types of Carleman estimates with a regular and a singular weight functions.

1 Introduction and Main Results

We consider the general higher order model of nonlinear acoustics proposed by Brunnhuber and Jordan [5, Equation (4)] and see also Kaltenbacher [25], Kaltenbacher and Thalhammer [26]:

$$
\partial_t^3 \psi(x,t) + A_1 \Delta^2 \psi(x,t) + A_2 \Delta^2 \partial_t \psi(x,t) - A_3 \Delta \partial_t^2 \psi(x,t) - A_4 \Delta \partial_t \psi(x,t)
$$
$$
= -\partial_t^2 (\kappa(x)(\partial_t \psi(x,t))^2 + |\nabla \psi(x,t)|^2) \quad x \in \Omega, \, 0 < t < T. \tag{1}
$$

Here $\Omega \subset \mathbb{R}^n$ denotes a bounded domain with sufficiently smooth boundary $\partial\Omega$ and $x = (x_1, \ldots, x_n) \in \mathbb{R}^n$, $\partial_j = \frac{\partial}{\partial x_j}$, $\partial_i \partial_j = \frac{\partial^2}{\partial x_i \partial x_j}$, $\partial_j^2 = \frac{\partial^2}{\partial x_j^2}$, $1 \leq i, j \leq n$,

M. Yamamoto (✉)
Graduate School of Mathematical Sciences, The University of Tokyo, Komaba, Meguro, Tokyo, Japan

Academy of Romanian Scientists, Bucharest, Romania

Peoples' Friendship University of Russia (RUDN University), Moscow, Russian Federation
e-mail: myama@ms.u-tokyo.ac.jp

B. Kaltenbacher
Department of Mathematics, Alpen-Adria-Universität Klagenfurt, Klagenfurt, Austria
e-mail: barbara.kaltenbacher@aau.at

© Springer Nature Switzerland AG 2021
B. Kaltenbacher et al. (eds.), *Time-dependent Problems in Imaging and Parameter Identification*, https://doi.org/10.1007/978-3-030-57784-1_14

$\partial_0 = \partial_t$, $\Delta = \sum_{j=1}^n \partial_j^2$, $\nabla = (\partial_1, \ldots, \partial_n)$, $\nabla_{x,t} = (\nabla, \partial_t)$. Throughout this article, we further use the following notations: $\partial_x^\gamma = \partial_1^{\gamma_1} \cdots \partial_n^{\gamma_n}$ for $\gamma = (\gamma_1, \ldots, \gamma_n) \in (\mathbb{N} \cup \{0\})^n$, where we set $|\gamma| = \sum_{i=1}^n \gamma_j$. We set

$$Q = \Omega \times (0, T).$$

Recently the field of nonlinear acoustics received great attention not only from the physical but also from the mathematical viewpoint and we refer to, e.g., Kaltenbacher [24] for a recent review and further references on the mathematical analysis such as the well-posedness for initial boundary value problems, qualitative properties of solutions such as long time behavior, and some optimization problems. However, to the best knowledge of the authors, there are no works on the mathematical analysis of inverse problems for (1), although results on quantitative identification of parameters in (1) are crucial, e.g., for the below mentioned application of nonlinearity imaging.

We describe more physical backgrounds for (1). In (1), ψ is the acoustic velocity potential,

$$A_1 = \frac{v c_0^2}{\mathrm{Pr}}, \ A_2 = \frac{b(1 + B/A)v^2}{\mathrm{Pr}}, \ A_3 = v\left(b + \frac{1 + B/A}{\mathrm{Pr}}\right), \ A_4 = c_0^2, \ \kappa = \frac{B/A}{2c_0^2}$$

and the quantities have the physical meaning indicated in Table 1. It has been observed that during ultrasound propagation through biological tissue, the parameter of nonlinearity B/A exhibits a dependence on the type of tissue and therefore determining κ as a spatially variable coefficient allows to image such media [4, 7, 8, 15, 31, 34, 35]. Therefore we will consider κ as a function of x, whereas (neglecting the mild dependence of A_2, A_3 on B/A as it is also done in the above references), we assume that A_1, A_2, A_3, A_4 in (1) are all positive constants.

We assume that all the coefficients are sufficiently smooth, and we do not pursue the optimal regularity conditions for the concise exposition.

We discuss an inverse source problem for a linearization of Eq. (1):

$$\partial_t^3 u + A_1 \Delta^2 u + A_2 \Delta^2 \partial_t u - A_3 \Delta \partial_t^2 u \tag{2}$$

Table 1 Physical parameters

$b = \frac{4}{3} + \frac{\mu_B}{\mu}$	…viscosity number
μ_B	…bulk viscosity
μ	…shear viscosity
v	…kinematic viscosity
Pr	…Prandtl number
c_0	…speed of sound
B/A	…nonlinearity parameter

$$= - p_0 \partial_t^3 u - p_1 \partial_t u - p_2 \partial_t^2 u - p_3 \cdot \nabla \partial_t u$$
$$- p_4 \cdot \nabla \partial_t^2 u - p_5 \cdot \nabla u + R(x,t) f(x), \quad (x,t) \in Q.$$

Indeed (2) follows from the linearization of the right-hand side of (1): for two coefficients κ_1 and κ_2 and corresponding solutions ψ_1 and ψ_2, setting $f = \kappa_1 - \kappa_2$, $u = \psi_1 - \psi_2$, we have

$$- \partial_t^2 \left[\kappa_1 (\partial_t \psi_1)^2 - \kappa_2 (\partial_t \psi_2)^2 + (|\nabla \psi_1|^2 - |\nabla \psi_2|^2) \right]$$

$$= - f \partial_t^2 ((\partial_t \psi_1)^2) - \kappa_2 \partial_t^2 ((\partial_t \psi_1 + \partial_t \psi_2)\partial_t u) - \partial_t^2 \left(\sum_{j=1}^{n} (\partial_j \psi_1 + \partial_j \psi_2)\partial_j u \right)$$

$$= - f \partial_t^2 ((\partial_t \psi_1)^2) - \kappa_2 (\partial_t \psi_1 + \partial_t \psi_2)\partial_t^3 u - 2\kappa_2 \partial_t ((\partial_t \psi_1 + \partial_t \psi_2))\partial_t^2 u$$

$$- \kappa_2 (\partial_t^2 (\partial_t \psi_1 + \partial_t \psi_2))\partial_t u - \sum_{j=1}^{n} (\partial_j \psi_1 + \partial_j \psi_2)\partial_t^2 \partial_j u) - 2\sum_{j=1}^{n} (\partial_t (\partial_j \psi_1 + \partial_j \psi_2))\partial_t \partial_j u$$

$$- \partial_t^2 \left(\sum_{j=1}^{n} (\partial_j \psi_1 + \partial_j \psi_2) \right) \partial_j u.$$

Therefore, setting $p_0, p_1, p_2, p_3, p_4, p_5$ suitably and $R(x,t) = -\partial_t^2 (\partial_t \psi_1)^2(x,t)$, we reach Eq. (2).

Our main purpose here is to discuss the inverse source problem arising from this linearization of (1) and establish the uniqueness and the conditional stability for our inverse problem. Uniqueness and stability are the primary theoretical issues for the inverse problem, and we expect that the current work should provide a theoretical basement for further related research, such as numerical reconstruction of the source term, and, via a differential approach, also approximation of the nonlinearity parameter. Moreover our method here for (1) can be widely applied to other types of model equations of nonlinear acoustics.

Our main subject is

Inverse Source Problem
Let $\Gamma \subset \partial \Omega$ be a non-empty relatively open subset of $\partial \Omega$ and ω be a non-empty subdomain of Ω, and let $t_0 \in (0, T)$ be arbitrarily given. Then determine $f(x)$ in Ω or on some subdomain of Ω by $u(\cdot, t_0)$ and data of u on $\Gamma \times (0, T)$ or $\omega \times (0, T)$.

Since we can consider $f(x) = \kappa_1(x) - \kappa_2(x)$ in the reduced Eq. (2), we see that the inverse source problem is a linearization of the parameter identification of determining $\kappa(x)$.

From the above definition of A_1–A_4 and realistic values of the physical parameters in Table 1, we can conclude that the positive constants A_3 and A_4 satisfy

$$A_3^2 - 4A_2 \geq 0.$$

Moreover we assume that $p_j, 0 \leq j \leq 5$ are sufficiently smooth, and

$$\begin{cases} 1 + p_0(x,t) > 0, & (x,t) \in \overline{Q}, \\ A_3^2 - 4A_2(1 + p_0(x,t)) > 0, & (x,t) \in \overline{Q}. \end{cases} \quad (3)$$

Condition (3) is satisfied if

$$-1 < p_0(x,t) < \frac{A_3^2 - 4A_2}{4A_2}, \quad (x,t) \in \overline{Q}.$$

We rewrite (2) as

$$(1 + p_0)\partial_t^3 u - A_3 \Delta \partial_t^2 u + A_1 \Delta^2 u + A_2 \Delta^2 \partial_t u + G(u) = R(x,t)f(x), \quad (x,t) \in Q. \quad (4)$$

Here we set

$$G(u) = \sum_{k=1}^{2} \sum_{\ell=0}^{1} q_{k\ell}(x,t) \cdot \partial_t^k \nabla^\ell u + q_{01}(x,t) \cdot \nabla u, \quad (5)$$

where $q_{k\ell} \in L^\infty(Q)$ are some sufficiently smooth functions.

We set

$$H^{2,1}(Q) = \left\{ u = u(x,t); \; u, \partial_t u, \sum_{|\gamma| \leq 2} \partial_x^\gamma u \in L^2(Q) \right\}.$$

For simplicity, we always assume that solutions u to (4) under consideration is sufficiently smooth, for example,

$$\partial_t^j u \in H^{2,1}(Q), \quad j = 0, 1, 2, 3, \qquad \partial_t^k \Delta u \in H^{2,1}(Q), \quad k = 0, 1, 2.$$

Also for the coefficients, we can discuss under more relaxed regularity conditions for u, but we omit.

We will establish two kinds of stability results.

To state the first result, we assume that $\Gamma \subset \partial\Omega$ is an arbitrarily fixed non-empty relatively open subset. We arbitrarily choose a subdomain $\Omega_0 \subset \Omega$ such that $\overline{\Omega_0} \subset \Omega \cup \Gamma$, $\partial\Omega_0 \cap \partial\Omega$ is a non-empty relatively open subset of $\partial\Omega$ and $\overline{\partial\Omega_0 \cap \partial\Omega} \subset \Gamma$. We note that the intersection of $\partial\Omega$ and some small neighborhood of $\partial\Omega_0 \cap \partial\Omega$ is included in Γ. Moreover, we fix $t_0 \in (0, T)$ and $\delta > 0$ such that $0 < t_0 - \delta < t_0 + \delta < T$ and set $I = (t_0 - \delta, t_0 + \delta)$.

Theorem 1 (Local Hölder Stability) *We assume* (3) *and the existence of a constant $r_0 > 0$ such that*

$$R \in H^1(0, T; L^\infty(\Omega)), \quad |R(x, t_0)| \geq r_0 > 0, \quad x \in \overline{\Omega} \tag{6}$$

and an a priori bound

$$\sum_{j+k\leq 5, j,k\in\mathbb{N}\cup\{0\}} (\|u\|_{H^j(I; H^{k+\frac{1}{2}+\varepsilon_0}(\Omega))} + \|u\|_{H^{j+\frac{1}{2}+\varepsilon_0}(I; H^k(\Omega))}) \leq M \tag{7}$$

with some constant $M > 0$ and $\varepsilon_0 > 0$. Then there exist constants $C > 0$ and $\chi \in (0, 1)$ depending on M, Γ, Ω_0 such that

$$\|f\|_{L^2(\Omega_0)} \leq C \left(\sum_{j+|\gamma|\leq 5} \|\partial_t^j \partial_x^\gamma u\|_{L^2(I; L^2(\Gamma))} \right.$$

$$\left. + \|u(\cdot, t_0)\|_{H^4(\Omega)} + \|\partial_t u(\cdot, t_0)\|_{H^4(\Omega)} + \|\partial_t^2 u(\cdot, t_0)\|_{H^2(\Omega)} \right)^\chi$$

for each $f \in L^2(\Omega)$.

The conclusion is a stability estimate in determining f under the a priori bound condition (7), and is called conditional stability. We note that Eq. (4) is essentially of parabolic type as is seen in the proof in Sects. 2–3, and for the determination of f, we can choose any small $T > 0$. We do not know the stability in the cases of $t_0 = 0$ and $t_0 = T$.

Since $\Omega_0 \subset \Omega$ is arbitrary such that $\overline{\Omega_0} \subset \Omega \cup \Gamma$, as the proof in Sect. 3 shows, the uniqueness in the inverse problem holds:

Corollary 1 *In Theorem 1, if*

$$\partial_x^\gamma u = 0 \quad \text{on } \Gamma \times I \text{ with } |\gamma| \leq 5 \text{ and } \partial_t^j u(\cdot, t_0) = 0 \text{ in } \Omega \text{ with } j = 0, 1, 2,$$

then $f = 0$ in Ω.

Next we state the second stability result. Assuming the boundary condition on the whole $\partial\Omega \times (0, T)$, we derive the Lipschitz stability which is global over Ω.

Here we choose interior measurements. More precisely, let $\omega \subset \Omega$ be an arbitrarily fixed subdomain such that $\overline{\omega} \subset \Omega$ and let $I := (t_0 - \delta, t_0 + \delta) \subset (0, T)$.

Theorem 2 (Global Lipschitz Stability) *Let $u \in H^4(0, T; L^2(\Omega)) \cap H^2(0, T; H^4(\Omega))$ satisfy* (4) *and*

$$u = \Delta u = 0 \quad \text{on } \partial\Omega \times (0, T). \tag{8}$$

We assume (3) *and* (6). *Then there exists a constant* $C > 0$ *such that*

$$\|f\|_{L^2(\Omega)} \leq C(\|u(\cdot, t_0)\|_{H^4(\Omega)} + \|\partial_t u(\cdot, t_0)\|_{H^4(\Omega)} + \|\partial_t^2 u(\cdot, t_0)\|_{H^2(\Omega)}$$

$$+ \|u\|_{H^2(I;H^2(\omega))} + \|u\|_{H^3(I;L^2(\omega))})$$

for each $f \in L^2(\Omega)$.

Theorem 2 immediately produces the uniqueness in determining f in Ω: $u = 0$ in $\omega \times I$ and $\partial_t^j u(\cdot, t_0) = 0$ in Ω with $j = 0, 1, 2$, yield $f = 0$ in Ω.

In Theorems 1 and 2, we can weaken the norms of data of u, but we omit the details. Correspondingly to Theorems 1 and 2, we can similarly prove the local stability with data on $\Gamma \times I$ and the global stability with data in $\omega \times I$ respectively, but we omit the details.

Our results assert stability estimates for the inverse source problem by a single measurement of data of solution to an initial boundary value problem, and in Theorem 1, the stability is conditional under a priori bound assumption (7). For this kind of inverse problems for initial boundary value problems, a method by Carleman estimates is very effective and Bukhgeim and Klibanov [6] first proved the uniqueness for partial differential equations of the second order. Also see Klibanov [27, 28]. After [6], the works Imanuvilov and Yamamoto [17–20] established the stability mainly for the hyperbolic and parabolic equations by modifying the method by Bukhgeim and Klibanov [6]. There have been many publications and here as very limited articles, we refer to Beilina et al. [2], Cannarsa et al. [9, 10], Gölgeleyen and Yamamoto [13], and also to monographs Beilina and Klibanov [1], Bellassoued and Yamamoto [3], Klibanov and Timonov [29], and to a survey article Yamamoto [33].

This paper is composed of five sections. In Sect. 2, we establish two key Carleman estimates. Sections 3 and 4 are devoted to the proofs of Theorems 1 and 2 respectively. In Sect. 5 we give remarks on the necessity of spatial data for the uniqueness.

2 Key Carleman Estimates

The proofs of Theorems 1 and 2 originate from Bukhgeim and Klibanov [6] and we will use a method from Huang et al. [14] and Imanuvilov and Yamamoto [17]. The key is Carleman estimates and we need to establish Carleman estimates for our Eq. (4). Equation (4) is of higher order and it is more feasible to factorize into two partial differential operators of lower orders. In this section, we first carry out such a reduction.

Henceforth $C > 0$ denotes generic constants which are independent of the parameters $s, \lambda > 0$ later introduced, and $C_0(\lambda) > 0$ means generic constants which are dependent on λ but independent of s.

2.1   Reduction of (4)

Let

$$P_0 u := \partial_t^3 u - \frac{A_3}{1 + p_0} \Delta \partial_t^2 u + \frac{A_2}{1 + p_0} \Delta^2 \partial_t u + \frac{A_1}{1 + p_0} \Delta^2 u.$$

For R and $q_{k\ell}$ in (4) and (5), we set

$$\begin{cases} \widetilde{R}(x, t) = \frac{R(x,t)}{1+p_0(x,t)}, \quad \widetilde{q_{k\ell}}(x, t) = \frac{q_{k\ell}(x,t)}{1+p_0(x,t)}, \\ u^0(x) = u(x, t_0), \quad u^1(x) = \partial_t u(x, t_0), \quad u^2(x) = \partial_t^2 u(x, t_0), \\ D_1 = \|u^0\|_{H^3(\Omega)}, \\ D_2 = \|u^0\|_{H^4(\Omega)} + \|u^1\|_{H^4(\Omega)} + \|u^2\|_{H^2(\Omega)}. \end{cases} \tag{9}$$

We set

$$\begin{cases} a_1(x, t) = \frac{1}{2(1+p_0(x,t))}\left(A_3 - \sqrt{A_3^2 - 4A_2(1 + p_0(x, t))}\right), \\[2mm] a_2(x, t) = \frac{1}{2(1+p_0(x,t))}\left(A_3 + \sqrt{A_3^2 - 4A_2(1 + p_0(x, t))}\right), \\[2mm] a_3(x, t) = \frac{A_1}{a_1(1+p_0)}. \end{cases} \tag{10}$$

By (3) we note that a_1, a_2 are real-valued and and by the assumed smoothness of p_0, the derivatives of a_1, a_2, a_3 appearing below exist and are bounded, and we note that $a_2(x, t) > 0$ for $(x, t) \in \overline{Q}$.

Then

$$P_0 u = \partial_t^3 u - \frac{A_3}{1 + p_0} \Delta \partial_t^2 u + \frac{A_2}{1 + p_0} \Delta^2 \partial_t u + \frac{A_1}{1 + p_0} \Delta^2 u$$

$$= (\partial_t - a_1 \Delta)(\partial_t^2 u - a_3 \Delta u - a_2 \Delta \partial_t u)$$

$$+ (\partial_t a_3) \Delta u + (\partial_t a_2) \Delta \partial_t u - 2a_1 \nabla a_3 \cdot \nabla(\Delta u) - a_1 (\Delta a_3) \Delta u$$

$$- 2a_1 \nabla a_2 \cdot \nabla(\Delta \partial_t u) - (a_1 \Delta a_2 - a_3) \Delta \partial_t u$$

$$= \partial_t^3 u - (a_1 + a_2) \Delta \partial_t^2 u + a_1 a_3 \Delta^2 u + a_1 a_2 \Delta^2 \partial_t u$$

$$+ (\partial_t a_3) \Delta u + (\partial_t a_2) \Delta \partial_t u - 2a_1 \nabla a_3 \cdot \nabla(\Delta u) - a_1 (\Delta a_3) \Delta u$$

$$- 2a_1 \nabla a_2 \cdot \nabla(\Delta \partial_t u) - (a_1 \Delta a_2 - a_3) \Delta \partial_t u.$$

Hence we can rewrite (4) as

$$(\partial_t - a_1 \Delta)(\partial_t^2 u - a_3 \Delta u - a_2 \Delta \partial_t u)$$

$$+ \sum_{j,k=0}^{1} a_{jk}(x,t) \partial_t^j \nabla^k \Delta u + \sum_{k=0}^{1} \sum_{j=0}^{2} b_{jk}(x,t) \partial_t^j \nabla^k u = \widetilde{R}(x,t) f(x) \quad \text{in } Q,$$
(11)

where a_{jk} and b_{jk} are also sufficiently smooth.

Moreover we introduce a function by

$$w = \partial_t^2 u - a_2 \Delta \partial_t u - a_3 \Delta u. \tag{12}$$

Therefore each of (4) and (11) is equivalent to

$$\begin{cases} v = \partial_t u, \\ \partial_t v - a_2(x,t) \Delta v = a_3(x,t) \int_{t_0}^t \Delta v(x,\xi) d\xi - a_3(x,t) \Delta u^0(x) + w(x,t). \end{cases}$$
(13)

Since

$$u(x,t) = \int_{t_0}^t v(x,\xi) d\xi + u^0(x), \quad (x,t) \in Q,$$

we can represent

$$- \sum_{j,k=0}^{1} a_{jk}(x,t) \partial_t^j \nabla^k \Delta u - \sum_{k=0}^{1} \sum_{j=0}^{2} b_{jk}(x,t) \partial_t^j \nabla^k u$$

in terms of $v = \partial_t u$, we have

$$\partial_t w - a_1 \Delta w = - \sum_{k=0}^{1} a_{1k} \nabla^k \Delta v - \sum_{k=0}^{1} \sum_{j=0}^{1} b_{j+1,k} \partial_t^j \nabla^k v$$

$$- \sum_{k=0}^{1} a_{0k} \int_{t_0}^t \nabla^k \Delta v(x,\xi) d\xi - \sum_{k=0}^{1} b_{0k} \int_{t_0}^t \nabla^k v(x,\xi) d\xi + b_0(x,t) + \widetilde{R}(x,t) f(x) \text{ in } Q,$$
(14)

where

$$b_0(x,t) = \sum_{k=0}^{1} (a_{0k}(x,t) \nabla^k \Delta u^0 + b_{0k}(x,t) \nabla^k u^0).$$

In particular,

$$|b^0(x,t)| \leq C \sum_{|\gamma| \leq 3} |\partial_x^\gamma u^0(x)|, \quad (x,t) \in Q. \tag{15}$$

Henceforth we mainly discuss Carleman estimates for system (14).

2.2   The First Carleman Estimate: Proposition 1

Let $d \in C^2(\overline{\Omega})$ and $|\nabla d| \neq 0$ on $\overline{\Omega}$. We arbitrarily fix $t_0 \in (0, T)$ and $\delta > 0$ such that $0 < t_0 - \delta < t_0 + \delta < T$. In addition to $Q := \Omega \times (0, T)$, we further set

$$I = (t_0 - \delta, t_0 + \delta), \qquad Q_I = \Omega \times I,$$

and

$$\varphi(x, t) = e^{\lambda \psi(x,t)}, \quad \psi(x, t) = d(x) - \beta |t - t_0|^2,$$

where $\lambda, \beta > 0$ are parameters chosen later. Moreover for convenience, we set

$$\Phi(v) := \int_{\partial\Omega \times I} s^3 \lambda^3 \varphi^3 (|v|^2 + |\nabla_{x,t} v|^2 + |\nabla^2 v|^2) e^{2s\varphi} dSdt$$

$$+ \int_\Omega s^5 \lambda^6 \varphi^5 (|v(x, t_0 - \delta)|^2 + |\nabla v(x, t_0 - \delta)|^2 + |v(x, t_0 + \delta)|^2$$

$$+ |\nabla v(x, t_0 + \delta)|^2) e^{2s\varphi(x, t_0 + \delta)} dx. \tag{16}$$

Here we notice that $\varphi(x, t_0 - \delta) = \varphi(x, t_0 + \delta)$ for $x \in \Omega$.

Now we are ready to state our first Carleman estimate for system (14) with the weight function $\varphi(x, t)$.

Proposition 1 *There exists a constant $\lambda_0 > 0$ such that for each $\lambda \geq \lambda_0$, we can choose constants $s_0 = s_0(\lambda) > 0, C = C(\lambda) > 0$, and $C_0 = C_0(\lambda) > 0$ satisfying*

(i)

$$\int_{Q_I} (s^{-1}\varphi^{-1}|\partial_t^2 w|^2 + s^3 \lambda^4 \varphi^3 |\partial_t w|^2) e^{2s\varphi} dxdt$$

$$\leq C \int_{Q_I} s\varphi(|\partial_t \widetilde{R}|^2 + |\widetilde{R}|^2)|f(x)|^2 e^{2s\varphi} dxdt$$

$$+ Ce^{C_0(\lambda)s} D_1^2 + C(\Phi(v) + \Phi(\nabla_{x,t} v) + \Phi(\partial_t \nabla v) + \Phi(w) + \Phi(\partial_t w))$$

for all $s \geq s_0$.

(ii)

$$\int_{Q_I} (s^{-1}\varphi^{-1}(|\partial_t w|^2 + |\Delta w|^2) + s\lambda^2\varphi|\nabla w|^2 + s^3\lambda^4\varphi^3|w|^2)e^{2s\varphi}dxdt$$

$$\leq C\int_{Q_I} |\tilde{R}f|^2 e^{2s\varphi}dxdt + Ce^{C_0(\lambda)s}D_1^2 + C(\Phi(\nabla v) + \Phi(v) + \Phi(w)).$$

for all $s \geq s_0$.

Here w *solves* (14), Φ *is defined in* (16) *and* D_1 *in* (9).

We can further estimate

$$\int_{Q_I} (s^{-1}\varphi^{-1}(|\partial_t^2 w|^2 + |\partial_t\Delta w|^2) + s\lambda^2\varphi|\nabla\partial_t w|^2 + s^2\lambda^2\varphi^2|\nabla w|^2$$

$$+s^3\lambda^4\varphi^3|\partial_t w|^2 + s^4\lambda^4\varphi^4|w|^2)e^{2s\varphi}dxdt,$$

but for the proof of Theorem 1, the estimates of $|\partial_t^2 w|^2$ and $|\partial_t w|^2$ in (i) are sufficient.

We can describe an estimate in terms of the original solution u to (4), but for the proof, the estimate of w is convenient. Proposition 1 asserts a weighted L^2-estimate which is uniform in all large $s > 0$: more precisely, the constants $C > 0$ and $C_0(\lambda) > 0$ are uniform for all large $s > 0$. As for the general theory for Carleman estimates, see Isakov [23], but the general theory does not work for our equation. Moreover in Carleman estimates proved in [23], it is assumed that boundary data of w should vanish especially at the final and the initial times $t_0\pm\delta$. Such an assumption makes the application of the Carleman estimate to the inverse problem more complicated. A recent work [14] simplifies the argument, which we follow in Sect. 3.

Proof of Proposition 1

First Step
The proof is based on the Carleman estimate for a parabolic equation.

Lemma 1 *Let* $m = 0, 1, 2$ *and* $\kappa \in C^1(\overline{Q_I})$, > 0 *on* $\overline{Q_I}$. *There exists a constant* $\lambda_0 > 0$ *such that for each* $\lambda \geq \lambda_0$, *we can choose constants* $s_0 = s_0(\lambda) > 0$ *and* $C = C(\lambda) > 0$ *such that*

$$\int_{Q_I} \left\{ s^{m-1}\lambda^m\varphi^{m-1}\left(|\partial_t z|^2 + \sum_{|\gamma|\leq 2}|\partial_x^\gamma z|^2\right) \right.$$

$$\left. +s^{m+1}\lambda^{m+2}\varphi^{m+1}|\nabla z|^2 + s^{m+3}\lambda^{m+4}\varphi^{m+3}|z|^2 \right\}e^{2s\varphi}dxdt$$

$$\leq C\int_{Q_I} s^m\lambda^m\varphi^m|\partial_t z(x,t) - \kappa(x,t)\Delta z(x,t)|^2 e^{2s\varphi}dxdt + C\Phi(z) \qquad (17)$$

for all $s \geq s_0$ *and all* $z \in H^1(0, T; L^2(\Omega)) \cap L^2(0, T; H^3(\Omega))$ *satisfying* $\partial_t z \in L^2(\partial\Omega \times (0, T))$, $z(\cdot, t_0 \pm \delta) \in H^1(\Omega)$.

Lemma 1 is a classical Carleman estimate for $m = 0$ especially in the case of $u(\cdot, t_0 \pm \delta) = 0$ in Ω (e.g., Bellassoued and Yamamoto [3], Yamamoto [33]). We note that without such a vanishing assumption, the proof is the same as in [33] by keeping all the boundary terms in $\Omega \times \{t_0 \pm \delta\}$ which are created by the integration by parts during the proof. We can prove similar Carleman estimates for general $m > 0$. Indeed, for $m \neq 0$, setting $v_1 = \varphi^{\frac{m}{2}} v$ and applying (17) with $m = 0$ to v_1, we can directly complete the proof (e.g., Lemma 7.2 (p. 195) in [3]).

Furthermore we need a weighted integral inequality.

Lemma 2 *Let* $\ell \in \mathbb{N} \cup \{0\}$. *Then there exists a constant* $C > 0$ *depending on* ℓ *such that*

$$
\int_{Q_I} s^\ell \lambda^\ell \varphi^\ell \left| \int_{t_0}^t |g(x, \xi)| d\xi \right|^2 e^{2s\varphi(x,t)} dx dt
$$

$$
\leq C \int_{Q_I} s^{\ell-1} \lambda^{\ell-1} \varphi^{\ell-1} |g(x, t)|^2 e^{2s\varphi(x,t)} dx dt
$$

for all $s > 0$ *and all* $g \in L^2(Q_I)$.

The lemma relies on the fact that that φ gains the maximum at $t = t_0$ where the definite integral of $|v(x, \xi)|$ on the left-hand side is considered. This type of inequality is essential for the application of Carleman estimates to inverse problems (Bukhgeim and Klibanov [6], Klibanov [27]) especially in the case of $\ell = 0$. For general ℓ, we can prove it in the same way (e.g., Loreti et al. [30]), and so the proof is omitted. An inequality of the type of Lemma 2 is essential for Carleman estimates and inverse problems for integro-differential equations (e.g., Cavaterra et al. [11], Imanuvilov and Yamamoto [21, 22], and Loreti et al. [30]).

The proof of Proposition 1 is done by applications of Lemma 1. It is an essence in the proof that we absorb integrating terms with lower powers of s and λ into the terms with the highest powers. In particular, thanks to Lemma 2 with $g = v$, we can regard a term $\left| \int_{t_0}^t |v(x, \xi)| d\xi \right|^2$, etc., as a term $|v(x, t)|^2$ with lower power which is reduced by $s^{-1} \lambda^{-1} \varphi^{-1}$. For such reduction of the integral terms, we have to be given data $u(\cdot, t_0)$ in Ω.

Remark: Main Idea of the Proofs of the Key Carleman Estimates for Our System (11) The proof is based on Carleman estimates Lemmata 1 and 3 below and these Carleman estimates can estimate derivatives of at most first order but in our system (11) we have to estimate derivatives of higher orders. Therefore we will apply the basic Carleman estimates for $v := \partial_t u$ and its derivatives by Eq. (13), and then for $w := \partial_t^2 u - a_2 \Delta \partial_t u - a_3 \Delta u$ by Eq. (14). For raising the orders of the derivatives to be estimated for v, we take t- and x-derivatives of (13) at the expense of extra higher-order derivatives of w, while for w we can take only t-derivatives of (14) because (14) is attached with $\widetilde{R}f$, and so the x-derivatives would produce ∇f, which cannot be controlled by $\|f\|_{L^2(\Omega)}$. For completing the proofs, we finally synthesize the gained Carleman estimates for v and w, and we absorb minor terms on the right-hand sides into the left-hand sides by choosing $s, \lambda > 0$ sufficiently large. The proof of the second key Carleman estimate Proposition 2 stated below is based on the same strategy. However the underlying Lemma 3 below requires the zero Dirichlet boundary condition, so that we cannot take arbitrary x-derivatives of v while keeping this boundary condition. Hence we will only do so by taking Δ for gaining estimates of higher spatial derivatives.

Second Step
We recall $v = \partial_t u$ from (13). Setting

$$w_0 = \partial_t w,$$

by the equation in w given by (14), we have

$$\partial_t w_0 - a_1 \Delta w_0 = (\partial_t \widetilde{R}) f(x) + \partial_t b_0$$

$$- \sum_{k=0}^{1} a_{1k} \nabla^k \Delta \partial_t v - \sum_{k=0}^{1} \sum_{j=0}^{1} b_{j+1,k} \partial_t^{j+1} \nabla^k v - \sum_{k=0}^{1} a_{0k} \nabla^k \Delta v - \sum_{k=0}^{1} b_{0k} \nabla^k v$$

$$+ (\partial_t a_1) \Delta w - \sum_{k=0}^{1} \sum_{j=0}^{1} (\partial_t b_{j+1,k}) \partial_t^{j} \nabla^k v - \sum_{k=0}^{1} \partial_t a_{0k} \int_{t_0}^{t} \nabla^k \Delta v(x, \xi) d\xi$$

$$- \sum_{k=0}^{1} \partial_t b_{0k} \int_{t_0}^{t} \nabla^k v(x, \xi) d\xi \quad \text{in } Q,$$

and so

$$|\partial_t w_0 - a_1 \Delta w_0| \le C |(\partial_t \widetilde{R}) f(x)| + C \sum_{|\gamma| \le 3} |\partial_x^\gamma u^0|$$

$$+ C \sum_{j=0}^{1} (|\partial_t^j v| + |\partial_t^j \nabla v| + |\partial_t^j \Delta v| + |\partial_t^j \nabla \Delta v|) + C(|\partial_t^2 v| + |\partial_t^2 \nabla v|) + C |\Delta w|$$

$$+ C \int_{t_0}^{t} (|\Delta v| + |\nabla \Delta v| + |\nabla v| + |v|)(x, \xi) d\xi \quad \text{in } Q. \tag{18}$$

We set

$$v_0 = \partial_t v, \quad v_j = \partial_j v, \quad v_{0j} = \partial_t \partial_j v, \quad 1 \le j \le n.$$

Then (13) yields

$$\partial_t v_0 - a_2 \Delta v_0 = \partial_t w + (\partial_t a_2 + a_3) \Delta v + (\partial_t a_3) \int_{t_0}^{t} \Delta v(x, \xi) d\xi - (\partial_t a_3) \Delta u^0, \tag{19}$$

$$\partial_t v_j - a_2 \Delta v_j = \partial_j w - \partial_j (a_3 \Delta u^0) + (\partial_j a_2) \Delta v$$

$$+ a_3 \int_{t_0}^t \Delta v_j(x, \xi) d\xi + (\partial_j a_3) \int_{t_0}^t \Delta v(x, \xi) d\xi \tag{20}$$

and

$$\partial_t v_{0j} - a_2 \Delta v_{0j} = \partial_t \partial_j w + (\partial_j a_2) \Delta v_0 + (\partial_t a_2 + a_3) \Delta v_j + (\partial_j \partial_t a_2 + \partial_j a_3) \Delta v$$

$$+ (\partial_t a_3) \int_{t_0}^t \Delta v_j(x, \xi) d\xi + (\partial_t \partial_j a_3) \int_{t_0}^t \Delta v d\xi - \partial_j \partial_t (a_3 \Delta u^0) \tag{21}$$

for $(x, t) \in Q$. First we will derive a Carleman estimate for v and its derivatives.

Now by applying Lemma 2, we see

$$\int_{Q_I} s^m \lambda^m \varphi^m \left| \int_{t_0}^t |\Delta v(x, \xi)| d\xi \right|^2 e^{2s\varphi(x,t)} dx dt$$

$$\leq C \int_{Q_I} s^{m-1} \lambda^{m-1} \varphi^{m-1} |\Delta v(x, t)|^2 e^{2s\varphi(x,t)} dx dt \tag{22}$$

for all $s > 0$. Applying Lemma 1 to (13), we have

$$\int_{Q_I} (s^{m-1} \lambda^m \varphi^{m-1} (|\partial_t v|^2 + |\Delta v|^2) + s^{m+1} \lambda^{m+2} \varphi^{m+1} |\nabla v|^2$$

$$+ s^{m+3} \lambda^{m+4} \varphi^{m+3} |v|^2) e^{2s\varphi} dx dt$$

$$\leq C \int_{Q_I} s^m \lambda^m \varphi^m \left| \int_{t_0}^t |\Delta v(x, \xi)| d\xi \right|^2 e^{2s\varphi} dx dt + C \int_{Q_I} s^m \lambda^m \varphi^m |w|^2 e^{2s\varphi} dx dt$$

$$+ C e^{C_0(\lambda)s} D_1^2 + C \Phi(v)$$

$$\leq C \int_{Q_I} s^{m-1} \lambda^{m-1} \varphi^{m-1} |\Delta v|^2 e^{2s\varphi} dx dt + C \int_{Q_I} s^m \lambda^m \varphi^m |w|^2 e^{2s\varphi} dx dt$$

$$+ C e^{C_0(\lambda)s} D_1^2 + C \Phi(v).$$

Therefore, choosing $\lambda > 0$ large, we can absorb the first term on the right-hand side into the first term on the left-hand side, and we have

$$\int_{Q_I} (s^{m-1} \lambda^m \varphi^{m-1} (|\partial_t v|^2 + |\Delta v|^2) + s^{m+1} \lambda^{m+2} \varphi^{m+1} |\nabla v|^2 + s^{m+3} \lambda^{m+4} \varphi^{m+3} |v|^2) e^{2s\varphi} dx dt$$

$$\leq C \int_{Q_I} s^m \lambda^m \varphi^m |w|^2 e^{2s\varphi} dx dt + C e^{C_0(\lambda)s} D_1^2 + C \Phi(v). \tag{23}$$

Next we apply Lemma 1 to (19):

$$
\int_{Q_I} (s^{m-1}\lambda^m\varphi^{m-1}(|\partial_t^2 v|^2 + |\Delta\partial_t v|^2) + s^{m+1}\lambda^{m+2}\varphi^{m+1}|\partial_t\nabla v|^2
$$

$$
+ s^{m+3}\lambda^{m+4}\varphi^{m+3}|\partial_t v|^2)e^{2s\varphi}dxdt \le C\int_{Q_I} s^m\lambda^m\varphi^m|\partial_t w|^2 e^{2s\varphi}dxdt
$$

$$
+ C\int_{Q_I} s^m\lambda^m\varphi^m\left(|\Delta v|^2 + \left|\int_{t_0}^t |\Delta v(x,\xi)|d\xi\right|^2\right)e^{2s\varphi}dxdt
$$

$$
+ Ce^{C_0(\lambda)s}D_1^2 + C\Phi(\partial_t v). \tag{24}
$$

Replacing m by $m+1$ in (23), we have

$$
\int_{Q_I} s^m\lambda^m\varphi^m|\Delta v|^2 e^{2s\varphi}dxdt
$$
$$
\le C\int_{Q_I} s^{m+1}\lambda^m\varphi^{m+1}|w|^2 e^{2s\varphi}dxdt + Ce^{C_0(\lambda)s}D_1^2 + C\Phi(v). \tag{25}
$$

Hence Lemma 2 yields

$$
\int_{Q_I} s^m\lambda^m\varphi^m\left(|\Delta v|^2 + \left|\int_{t_0}^t |\Delta v(x,\xi)|d\xi\right|^2\right)e^{2s\varphi}dxdt
$$

$$
\le C\int_{Q_I} (s^m\lambda^m\varphi^m|\Delta v|^2 + s^{m-1}\lambda^{m-1}\varphi^{m-1}|\Delta v|^2)e^{2s\varphi}dxdt
$$

$$
\le C\int_{Q_I} s^m\lambda^m\varphi^m|\Delta v|^2 e^{2s\varphi}dxdt
$$

$$
\le C\int_{Q_I} s^{m+1}\lambda^m\varphi^{m+1}|w|^2 e^{2s\varphi}dxdt + Ce^{C_0(\lambda)s}D_1^2 + C\Phi(v).
$$

Therefore with (24) we reach

$$
\int_{Q_I} (s^{m-1}\lambda^m\varphi^{m-1}(|\partial_t^2 v|^2 + |\Delta\partial_t v|^2) + s^{m+3}\lambda^{m+4}\varphi^{m+3}|\partial_t v|^2)e^{2s\varphi}dxdt
$$

$$
\le C\int_{Q_I} (s^m\lambda^m\varphi^m|\partial_t w|^2 + s^{m+1}\lambda^m\varphi^{m+1}|w|^2)e^{2s\varphi}dxdt \tag{26}
$$

$$
+ Ce^{C_0(\lambda)s}D_1^2 + C(\Phi(v) + \Phi(\partial_t v)).
$$

Similarly from (20), applying Lemma 1 and (25), we can obtain

$$
\int_{Q_I} \left\{ s^{m-1}\lambda^m\varphi^{m-1} \left(|\partial_t\nabla v|^2 + \sum_{|\gamma|\le 3} |\partial_x^\gamma v|^2 \right) \right.
$$
$$
\left. + s^{m+1}\lambda^{m+2}\varphi^{m+1} \sum_{i,j=1}^n |\partial_i\partial_j v|^2 + s^{m+3}\lambda^{m+4}\varphi^{m+3}|\nabla v|^2 \right\} e^{2s\varphi} dxdt \qquad (27)
$$
$$
\le C \int_{Q_I} (s^m\lambda^m\varphi^m|\nabla w|^2 + s^{m+1}\lambda^m\varphi^{m+1}|w|^2)e^{2s\varphi} dxdt
$$
$$
+ Ce^{C_0(\lambda)s}D_1^2 + C(\Phi(\nabla v) + \Phi(v)).
$$

Here we used an estimate

$$
\int_{Q_I} s\lambda\varphi \left| \int_{t_0}^t |\Delta v(x,\xi)|d\xi \right|^2 e^{2s\varphi} dxdt \le C \int_{Q_I} |\Delta v(x,t)|^2 e^{2s\varphi} dxdt
$$

by Lemma 2.

Finally from (21), applying Lemma 2 to the integral terms $\left| \int_{t_0}^t \cdots d\xi \right|^2$, by (25) we can argue in the same way to have

$$
\sum_{j=1}^n \int_{Q_I} \left\{ s^{m-1}\lambda^m\varphi^{m-1} \left(|\partial_t v_{0j}|^2 + \sum_{i,k=1}^n |\partial_i\partial_k v_{0j}|^2 \right) \right.
$$
$$
\left. + s^{m+1}\lambda^{m+2}\varphi^{m+1}|\nabla v_{0j}|^2 + s^{m+3}\lambda^{m+4}\varphi^{m+3}|v_{0j}|^2 \right\} e^{2s\varphi} dxdt
$$
$$
\le C \int_{Q_I} s^m\lambda^m\varphi^m|\nabla\partial_t w|^2 e^{2s\varphi} dxdt
$$
$$
+ C \int_{Q_I} s^m\lambda^m\varphi^m \left(|\Delta v_0|^2 + |\Delta v|^2 + |\Delta(\partial_j v)|^2 \right.
$$
$$
\left. + \left| \int_{t_0}^t |\Delta v(x,\xi)|d\xi \right|^2 + \left| \int_{t_0}^t |\Delta(\partial_j v)(x,\xi)|d\xi \right|^2 \right) e^{2s\varphi} dxdt
$$
$$
+ Ce^{C_0(\lambda)s}D_1^2 + C\Phi(\partial_t\nabla v)
$$
$$
\le C \int_{Q_I} (s^m\lambda^m\varphi^m|\nabla\partial_t w|^2 + s^{m+1}\lambda^m\varphi^{m+1}|w|^2)e^{2s\varphi} dxdt
$$
$$
+ C \int_{Q_I} s^m\lambda^m\varphi^m \left(|\Delta v_0|^2 + \sum_{|\gamma|\le 3} |\partial_x^\gamma v|^2 \right) e^{2s\varphi} dxdt
$$
$$
+ Ce^{C_0(\lambda)s}D_1^2 + C(\Phi(\partial_t\nabla v) + \Phi(v)). \qquad (28)
$$

Applying (26)–(27) where m is replaced by $m + 1$, we have

[the second term on the right-hand side of (28)]

$$\leq C \int_{Q_I} (s^{m+1} \lambda^m \varphi^{m+1} |\nabla_{x,t} w|^2 + s^{m+2} \lambda^m \varphi^{m+2} |w|^2) e^{2s\varphi} dx dt$$

$$+ C e^{C_0(\lambda)s} D_1^2 + C(\Phi(\nabla_{x,t} v) + \Phi(v)).$$

Therefore (28) yields

$$\int_{Q_I} \left\{ s^{m-1} \lambda^m \varphi^{m-1} \left(|\nabla \partial_t^2 v|^2 + \sum_{|\gamma| \leq 3} |\partial_t \partial_x^\gamma v|^2 \right) \right.$$

$$\left. + s^{m+1} \lambda^{m+2} \varphi^{m+1} \sum_{|\gamma| \leq 2} |\partial_x^\gamma \partial_t v|^2 + s^{m+3} \lambda^{m+4} \varphi^{m+3} |\nabla \partial_t v|^2 \right\} e^{2s\varphi} dx dt$$

$$\leq C \int_{Q_I} (s^m \lambda^m \varphi^m |\partial_t \nabla w|^2 + s^{m+1} \lambda^m \varphi^{m+1} |\nabla_{x,t} w|^2 + s^{m+2} \lambda^m \varphi^{m+2} |w|^2) e^{2s\varphi} dx dt$$

$$+ C e^{C_0(\lambda)s} D_1^2 + C(\Phi(v) + \Phi(\nabla_{x,t} v) + \Phi(\partial_t \nabla v)).$$

(29)

Third Step
We will derive a Carleman estimate for w. Applying Lemma 1 to (14) by using Lemma 2 for estimating

$$\int_{t_0}^t |\nabla^k \Delta v(x, \xi)| d\xi, \quad \int_{t_0}^t |\nabla^k v(x, \xi)| d\xi, \quad k = 0, 1,$$

we obtain

$$\int_{Q_I} (s^{m-1} \lambda^m \varphi^{m-1} (|\partial_t w|^2 + |\Delta w|^2) + s^{m+1} \lambda^{m+2} \varphi^{m+1} |\nabla w|^2$$

$$+ s^{m+3} \lambda^{m+4} \varphi^{m+3} |w|^2) e^{2s\varphi} dx dt \leq C \int_{Q_I} s^m \lambda^m \varphi^m |\tilde{R} f|^2 e^{2s\varphi} dx dt$$

$$+ C \int_{Q_I} s^m \lambda^m \varphi^m \left(\sum_{|\gamma| \leq 3} |\partial_x^\gamma v|^2 + \sum_{k=0}^1 |\partial_t \nabla^k v|^2 \right) e^{2s\varphi} dx dt$$

$$+ C e^{C_0(\lambda)s} D_1^2 + C \Phi(w).$$

Replacing m by $m + 1$ in (27) and substituting it into the second term on the right-hand side, we have

$$\int_{Q_I} (s^{m-1}\lambda^m\varphi^{m-1}(|\partial_t w|^2 + |\Delta w|^2) + s^{m+1}\lambda^{m+2}\varphi^{m+1}|\nabla w|^2$$

$$+ s^{m+3}\lambda^{m+4}\varphi^{m+3}|w|^2)e^{2s\varphi}dxdt \leq C\int_{Q_I} s^m\lambda^m\varphi^m|\tilde{R}f|^2e^{2s\varphi}dxdt$$

$$+C\int_{Q_I}(s^{m+1}\lambda^m\varphi^{m+1}|\nabla w|^2 + s^{m+2}\lambda^m\varphi^{m+2}|w|^2)e^{2s\varphi}dxdt$$

$$+Ce^{C_0(\lambda)s}D_1^2 + C(\Phi(\nabla v) + \Phi(v) + \Phi(w)).$$

Choosing $s, \lambda > 0$ sufficiently large, we can absorb the second term on the right-hand side into the left-hand side, so that

$$\int_{Q_I}(s^{m-1}\lambda^m\varphi^{m-1}(|\partial_t w|^2 + |\Delta w|^2)$$

$$+ s^{m+1}\lambda^{m+2}\varphi^{m+1}|\nabla w|^2 + s^{m+3}\lambda^{m+4}\varphi^{m+3}|w|^2)e^{2s\varphi}dxdt$$

$$\leq C\int_{Q_I} s^m\lambda^m\varphi^m|\tilde{R}f|^2e^{2s\varphi}dxdt + Ce^{C_0(\lambda)s}D_1^2 + C(\Phi(\nabla v) + \Phi(v) + \Phi(w)).$$

$$(30)$$

Thus by setting $m = 0$, this proves Proposition 1 (ii). Finally, applying Lemma 1 to (18), we obtain

$$\int_{Q_I}(s^{m-1}\lambda^m\varphi^{m-1}(|\partial_t^2 w|^2 + |\partial_t\Delta w|^2) + s^{m+1}\lambda^{m+2}\varphi^{m+1}|\nabla\partial_t w|^2)e^{2s\varphi}dxdt$$

$$\leq C\int_{Q_I} s^m\lambda^m\varphi^m|(\partial_t\tilde{R})f|^2e^{2s\varphi}dxdt + C\int_{Q_I} s^m\lambda^m\varphi^m|\Delta w|^2e^{2s\varphi}dxdt$$

$$+\int_{Q_I} s^m\lambda^m\varphi^m\left(\sum_{|\gamma|\leq 3}(|\partial_x^\gamma\partial_t v|^2 + |\partial_x^\gamma v|^2) + \sum_{|\gamma|\leq 1}|\partial_x^\gamma\partial_t^2 v|^2\right)e^{2s\varphi}dxdt$$

$$+C\int_{Q_I} s^m\lambda^m\varphi^m\sum_{|\gamma|\leq 3}\left|\int_{t_0}^t|\partial_x^\gamma v(x,\xi)|d\xi\right|^2 e^{2s\varphi}dxdt + Ce^{C_0(\lambda)s}D_1^2 + C\Phi(\partial_t w).$$

In (29) and (30), we replace m by $m + 1$. Then, applying Lemma 2 to the fourth term on the right-hand side and (30) and (29) respectively to the second and the third terms on the right-hand side, we can obtain

$$\int_{Q_I} (s^{m-1}\lambda^m \varphi^{m-1}(|\partial_t^2 w|^2 + |\partial_t \Delta w|^2)$$

$$+ s^{m+1}\lambda^{m+2}\varphi^{m+1}|\nabla \partial_t w|^2 + s^{m+3}\lambda^{m+4}\varphi^{m+3}|\partial_t w|^2)e^{2s\varphi}dxdt$$

$$\leq C \int_{Q_I} s^m \lambda^m \varphi^m |\partial_t \widetilde{R}|^2 |f|^2 e^{2s\varphi}dxdt$$

$$+ \left[C \int_{Q_I} s^{m+1}\lambda^m \varphi^{m+1}|\widetilde{R}|^2 |f|^2 e^{2s\varphi}dxdt + C(e^{C_0(\lambda)s} D_1^2 + \Phi(v) + \Phi(\nabla v) + \Phi(w)) \right]$$

$$+ \left[C \int_{Q_I} (s^{m+1}\lambda^m \varphi^{m+1}|\partial_t \nabla w|^2 + s^{m+2}\lambda^m \varphi^{m+2}|\nabla_{x,t} w|^2 + s^{m+3}\lambda^m \varphi^{m+3}|w|^2)e^{2s\varphi}dxdt \right.$$

$$\left. + C(e^{C_0(\lambda)s} D_1^2 + \Phi(v) + \Phi(\nabla_{x,t}v) + \Phi(\nabla \partial_t v)) \right] + [C(e^{C_0(\lambda)s} D_1^2 + \Phi(\partial_t w))].$$

Choosing $s, \lambda > 0$ large, we can absorb the terms $s^{m+1}\lambda^m \varphi^{m+1}|\partial_t \nabla w|^2$ and $s^{m+2}\lambda^m \varphi^{m+2}|\partial_t w|^2$ into the left-hand side, and so we have

$$\int_{Q_I} (s^{m-1}\lambda^m \varphi^{m-1}(|\partial_t^2 w|^2 + |\partial_t \Delta w|^2)$$

$$+ s^{m+1}\lambda^{m+2}\varphi^{m+1}|\nabla \partial_t w|^2 + s^{m+3}\lambda^{m+4}\varphi^{m+3}|\partial_t w|^2)e^{2s\varphi}dxdt$$

$$\leq C \int_{Q_I} s^{m+1}\lambda^m \varphi^{m+1}(|\partial_t \widetilde{R}|^2 + |\widetilde{R}|^2)|f|^2 e^{2s\varphi}dxdt$$

$$+ C \int_{Q_I} (s^{m+2}\lambda^m \varphi^{m+2}|\nabla w|^2 + s^{m+3}\lambda^m \varphi^{m+3}|w|^2)e^{2s\varphi}dxdt$$

$$+ C(e^{C_0(\lambda)s} D_1^2 + \Phi(v) + \Phi(\nabla_{x,t}v) + \Phi(\nabla \partial_t v) + \Phi(w) + \Phi(\partial_t w)). \tag{31}$$

Again replacing m by $m + 1$ in (30), we have

$$\int_{Q_I} (s^{m+2}\lambda^m \varphi^{m+2}|\nabla w|^2 + s^{m+4}\lambda^{m+2}\varphi^{m+4}|w|^2)e^{2s\varphi}dxdt$$

$$\leq C \int_{Q_I} s^{m+1}\lambda^{m-2}\varphi^{m+1}|\widetilde{R}|^2 |f|^2 e^{2s\varphi}dxdt$$

$$+ C(e^{C_0(\lambda)s} D_1^2 + \Phi(v) + \Phi(\nabla v) + \Phi(w)).$$

Substituting this into the second term on the right-hand side of (31), we reach

$$\int_{Q_I} (s^{m-1}\lambda^m \varphi^{m-1}(|\partial_t^2 w|^2 + |\partial_t \Delta w|^2) + s^{m+1}\lambda^{m+2}\varphi^{m+1}|\nabla \partial_t w|^2$$

$$+ s^{m+3}\lambda^{m+4}\varphi^{m+3}|\partial_t w|^2)e^{2s\varphi}dxdt$$

$$\leq C \int_{Q_l} s^{m+1} \lambda^m \varphi^{m+1} (|\partial_t \widetilde{R}|^2 + |\widetilde{R}|^2) |f|^2 e^{2s\varphi} dx dt$$

$$+ C(e^{C_0(\lambda)s} D_1^2 + \Phi(v) + \Phi(\nabla_{x,t} v) + \Phi(\nabla \partial_t v) + \Phi(w) + \Phi(\partial_t w)).$$

Setting $m = 0$, we complete the proof of Proposition 1.

2.3 The Second Carleman Estimate: Proposition 2

We will establish another underlying Carleman estimate. For $\omega \subset \Omega$, we arbitrarily choose a subdomain ω_0 such that $\overline{\omega_0} \subset \omega$. Then, it is known (e.g., Imanuvilov [16]) that there exists $d_0 \in C^2(\overline{\Omega})$ such that

$$d_0 > 0 \quad \text{in } \Omega, \quad d_0 = 0 \quad \text{on } \partial\Omega, \quad |\nabla d_0| > 0 \quad \text{on } \overline{\Omega \setminus \omega_0}. \tag{32}$$

Later we fix a constant $\lambda > 0$ sufficiently large. We set

$$\begin{cases} \eta(x) = e^{\lambda d_0(x)} - e^{2\lambda \|d_0\|_{C(\overline{\Omega})}} < 0, \\ \alpha(x,t) = \frac{\eta(x)}{t(T-t)}, \quad \theta(x,t) = \frac{e^{\lambda d_0(x)}}{t(T-t)}, \quad t_0 = \frac{T}{2}. \end{cases} \tag{33}$$

We note

$$t(T-t) = e^{\lambda d_0(x)} \theta^{-1}(x,t), \quad \partial_t \theta(x,t) = \frac{2t-T}{t^2(T-t)^2} e^{\lambda d_0(x)}, \quad t - t_0 = \frac{t^2(T-t)^2}{2\eta(x)} \partial_t \alpha, \tag{34}$$

and

$$\begin{cases} \frac{1}{t^2(T-t)^2} = e^{-2\lambda d_0(x)} \theta(x,t)^2 \leq \theta(x,t)^2, \\ |\partial_t \theta(x,t)| = \frac{|2t-T|}{t^2(T-t)^2} e^{\lambda d_0(x)} = |2t - T| e^{-\lambda d_0(x)} \theta(x,t)^2 \leq C\theta(x,t)^2 \\ \qquad \text{in } \Omega \times (0,T). \end{cases} \tag{35}$$

Here $C > 0$ is independent of $\lambda > 0$.
   We recall (9):

$$D_1 := \|u^0\|_{H^3(\Omega)}, \quad D_2 = \|u^0\|_{H^4(\Omega)} + \|u^1\|_{H^4(\Omega)} + \|u^2\|_{H^2(\Omega)},$$

and that v, w are defined by (12) and (13) for a solution u to (4).
   We state our second key Carleman estimate for a solution w to (14).

Proposition 2 *There exists a constant $\lambda_0 > 0$ such that for each $\lambda \geq \lambda_0$ we can choose constants $s_0 = s_0(\lambda) > 0$, $C = C(\lambda) > 0$, and $C_0 = C_0(\lambda) > 0$ such that*

$$\int_Q (s^{-1}\theta^{-1}|\partial_t^2 w|^2 + s^3\lambda^4\theta^3|\partial_t w|^2)e^{2s\alpha}dxdt$$

$$\leq C\int_Q |\partial_t \widetilde{R}|^2|f|^2 e^{2s\alpha}dxdt + C_0(\lambda)(D_2^2 + \|u\|_{H^3(0,T;L^2(\omega))}^2 + \|u\|_{H^2(0,T;H^2(\omega))}^2)$$

for all $s > s_0$.

Proof of Proposition 2 We set

$$E(v) = \int_{\omega\times(0,T)} s^4\lambda^5\theta^4|v|^2 e^{2s\alpha}dxdt.$$

First Step

The proof is similar to Proposition 1, but we need some other estimation. First we show

Lemma 3 *Let $m = -1, 0, 1$. There exists a constant $\lambda_0 > 0$ such that for each $\lambda \geq \lambda_0$ we can choose constants $s_0 = s_0(\lambda) > 0$ and $C = C(\lambda) > 0$ such that*

$$\int_Q \left\{ s^{m-1}\lambda^m\theta^{m-1}\left(|\partial_t z|^2 + \sum_{|\gamma|\leq 2}|\partial_x^\gamma z|^2\right)\right.$$

$$\left. + s^{m+1}\lambda^{m+2}\theta^{m+1}|\nabla z|^2 + s^{m+3}\lambda^{m+4}\theta^{m+3}|z|^2 \right\}e^{2s\alpha}dxdt$$

$$\leq C\int_Q s^m\lambda^m\theta^m|F|^2 e^{2s\alpha}dxdt + CE(z)$$

for all $s \geq s_0$ and all z satisfying

$$\partial_t z - \kappa(x,t)\Delta z = F \quad in\ Q, \quad z|_{\partial\Omega} = 0$$

with $\kappa \in C^1(\overline{Q})$, > 0 on $\overline{Q}$.

This is a Carleman estimate with the singular weight function $\alpha(x,t)$ and the proof can be found for example in Fursikov and Imanuvilov [12], Imanuvilov [16].

Second we show an inequality on a definite integral starting at t_0 where the weight function $\alpha(x,t)$ gains the maximum value for each x. This lemma corresponds to Lemma 2.

Lemma 4 *There exist constants $C > 0$ and $\delta_0 > 0$ such that*

$$\int_Q s^\ell\theta^\ell\left|\int_{t_0}^t |g(x,\xi)|d\xi\right|^2 e^{2s\alpha(x,t)}dxdt \leq C\int_Q s^{\ell-1}\theta^{\ell-1}e^{-\lambda\delta_0}|g|^2 e^{2s\alpha(x,t)}dxdt$$

for all $s \geq 0$, $\ell \in \mathbb{N}$ and all $g \in L^2(Q)$.

Proof The proof is similar to Lemma 2 and we provide it for completeness. We can estimate $\int_\Omega \int_{t_0}^T s^\ell \theta^\ell \left| \int_{t_0}^t |g(x,\xi)| d\xi \right|^2 e^{2s\alpha(x,t)} dx dt$ because we can estimate for the integral over $\Omega \times (0, t_0)$ in the same way. Noting (34), $\alpha(x,T) = -\infty$ and $\partial_t e^{2s\alpha} = 2s(\partial_t \alpha)e^{2s\alpha}$, we apply the Cauchy-Schwarz inequality and the integration by parts, so that

$$\int_\Omega \int_{t_0}^T (s\theta)^\ell \left| \int_{t_0}^t |g(x,\xi)| d\xi \right|^2 e^{2s\alpha(x,t)} dt dx$$

$$\leq \int_\Omega \int_{t_0}^T (s\theta)^\ell (t-t_0) \left(\int_{t_0}^t |g(x,\xi)|^2 d\xi \right) e^{2s\alpha(x,t)} dt dx$$

$$= \int_\Omega \int_{t_0}^T (s\theta)^\ell \frac{t^2(T-t)^2}{2\eta(x)} (\partial_t \alpha) e^{2s\alpha(x,t)} \left(\int_{t_0}^t |g(x,\xi)|^2 d\xi \right) dt dx$$

$$= \int_\Omega \int_{t_0}^T s^{\ell-1} \theta^\ell \frac{t^2(T-t)^2}{4\eta(x)} \partial_t (e^{2s\alpha(x,t)}) \left(\int_{t_0}^t |g(x,\xi)|^2 d\xi \right) dt dx$$

$$= \int_\Omega \left[s^{\ell-1} \theta^\ell \frac{t^2(T-t)^2}{4\eta(x)} e^{2s\alpha(x,t)} \left(\int_{t_0}^t |g(x,\xi)|^2 d\xi \right) \right]_{t=t_0}^{t=T}$$

$$- \int_\Omega \int_{t_0}^T s^{\ell-1} \partial_t (\theta^\ell t^2 (T-t)^2) \frac{1}{4\eta(x)} \left(\int_{t_0}^t |g(x,\xi)|^2 d\xi \right) e^{2s\alpha(x,t)} dt dx$$

$$- \int_\Omega \int_{t_0}^T s^{\ell-1} \theta^\ell \frac{t^2(T-t)^2}{4\eta(x)} |g(x,t)|^2 e^{2s\alpha(x,t)} dt dx$$

$$= - \int_\Omega \int_{t_0}^T \left(\ell \theta^{\ell-1} (\partial_t \theta) t^2 (T-t)^2 + \theta^\ell 2t(T-t)(T-2t) \right)$$

$$\frac{s^{\ell-1}}{4\eta(x)} \left(\int_{t_0}^t |g(x,\xi)|^2 d\xi \right) e^{2s\alpha(x,t)} dt dx$$

$$+ \int_\Omega \int_{t_0}^T s^{\ell-1} \theta^\ell \left(\frac{e^{\lambda d_0(x)} \theta^{-1} t(T-t)}{-4\eta(x)} \right) |g(x,t)|^2 e^{2s\alpha(x,t)} dt dx. \tag{36}$$

For the last term, we used $t^2(T-t)^2 = t(T-t)(e^{\lambda d_0(x)}\theta^{-1})$ by (34).

On the other hand, we set $\delta_0 = \min_{x\in\overline{\Omega}} (2\|d_0\|_{C(\overline{\Omega})} - d_0(x)) > 0$. There exists a constant $C_1 = C_1(\delta_0) > 0$ such that

$$\theta^{-1} \frac{e^{\lambda d_0(x)} t(T-t)}{-4\eta(x)} = \theta^{-1} t(T-t) \frac{e^{\lambda d_0(x)}}{4(e^{2\lambda\|d_0\|_{C(\overline{\Omega})}} - e^{\lambda d_0(x)})}$$

$$= \frac{\theta^{-1} t(T-t)}{4} \frac{e^{-\lambda(2\|d_0\|_{C(\overline{\Omega})} - d_0(x))}}{1 - e^{-\lambda(2\|d_0\|_{C(\overline{\Omega})} - d_0(x))}} \leq \frac{\theta^{-1} t(T-t)}{4} \frac{e^{-\lambda\delta_0}}{1 - e^{-\lambda\delta_0}}$$

$$\leq C_1 \frac{\theta^{-1} t(T-t)}{4} e^{-\lambda \delta_0} \leq C_2 \theta^{-1} e^{-\lambda \delta_0}, \quad (x,t) \in \overline{Q}$$

for all $\lambda \geq 1$. Indeed, by setting $h(\zeta) = \frac{\zeta}{1-\zeta}$, the function $h(\zeta)$ is monotone increasing for $0 < \zeta < 1$, and

$$\frac{e^{-\lambda(2\|d_0\|_{C(\overline{\Omega})} - d_0(x))}}{1 - e^{-\lambda(2\|d_0\|_{C(\overline{\Omega})} - d_0(x))}} = h(e^{-\lambda(2\|d_0\|_{C(\overline{\Omega})} - d_0(x))}) \leq h(e^{-\lambda \delta_0})$$

by $\lambda(2\|d_0\|_{C(\overline{\Omega})} - d_0(x)) \geq \lambda \delta_0$.

Therefore

$$\int_\Omega \int_{t_0}^T s^{\ell-1} \theta^\ell \left(\frac{e^{\lambda d_0(x)} \theta^{-1} t(T-t)}{-4\eta(x)} \right) |g(x,t)|^2 e^{2s\alpha(x,t)} dt dx$$

$$\leq C \int_\Omega \int_{t_0}^T s^{\ell-1} \theta^{\ell-1} e^{-\lambda \delta_0} |g(x,t)|^2 e^{2s\alpha(x,t)} dt dx.$$

On the other hand, by (34) we see

$$|\ell \theta^{\ell-1}(\partial_t \theta) t^2 (T-t)^2| = \left| \ell \theta^{\ell-1} \frac{2t-T}{t^2(T-t)^2} e^{\lambda d_0(x)} t^2 (T-t)^2 \right|$$

$$= |2\ell \theta^{\ell-1}(t-t_0) e^{\lambda d_0(x)}| = |2\ell \theta^{\ell-1}(t-t_0)(\theta(x,t) t(T-t))| \leq C\theta^\ell |t-t_0|$$

and

$$-\eta(x) = e^{2\lambda \|d_0\|_{C(\overline{\Omega})}} - e^{\lambda d_0(x)} \geq e^{2\lambda \|d_0\|_{C(\overline{\Omega})}} - e^{\lambda \|d_0\|_{C(\overline{\Omega})}}$$

$$= e^{\lambda \|d_0\|_{C(\overline{\Omega})}} (e^{\lambda \|d_0\|_{C(\overline{\Omega})}} - 1) \geq e^{\|d_0\|_{C(\overline{\Omega})}} (e^{\|d_0\|_{C(\overline{\Omega})}} - 1) > 0,$$

that is,

$$\left| -\frac{1}{\eta(x)} \right| \leq \frac{1}{e^{\|d_0\|_{C(\overline{\Omega})}} (e^{\|d_0\|_{C(\overline{\Omega})}} - 1)}$$

if $\lambda \geq 1$. Applying these to the first term on the right-side of (36), we can obtain

$$\int_\Omega \int_{t_0}^T (s\theta)^\ell (t-t_0) \left| \int_{t_0}^t |g(x,\xi)| d\xi \right|^2 e^{2s\alpha(x,t)} dt dx$$

$$\leq C \int_\Omega \int_{t_0}^T s^{\ell-1} \theta^\ell |t-t_0| \left(\int_{t_0}^t |g|^2 d\xi \right) e^{2s\alpha(x,t)} dt dx$$

$$+ \int_\Omega \int_{t_0}^T (s\theta)^{\ell-1} e^{-\lambda \delta_0} |g| e^{2s\alpha(x,t)} dt dx.$$

The first term on the right-hand side can be absorbed into the left-hand side by choosing $s > 0$ large, and the proof of Lemma 4 is complete.

In particular, Lemma 4 implies

Lemma 5

$$\int_Q (s\theta)^\ell e^{\lambda\delta_0}|w|^2 e^{2s\alpha}\,dxdt \leq C \int_Q (s\theta)^{\ell-1}|\partial_t w|^2 e^{2s\alpha}\,dxdt + Ce^{\lambda\delta_0}\|w(\cdot,t_0)\|^2_{L^2(\Omega)}.$$

Proof We have

$$w(x,t) = \int_{t_0}^t \partial_t w(x,\xi)\,d\xi + w(x,t_0), \quad (x,t) \in Q.$$

Therefore Lemma 4 yields

$$\int_Q e^{\lambda\delta_0}s^\ell\theta^\ell|w|^2 e^{2s\alpha}\,dxdt$$

$$\leq C\int_Q s^\ell\theta^\ell e^{\lambda\delta_0}\left|\int_{t_0}^t |\partial_t w(x,\xi)|d\xi\right|^2 e^{2s\alpha}t\,dxdt + C\int_Q |w(x,t_0)|^2 e^{\lambda\delta_0}s^\ell\theta^\ell e^{2s\alpha}\,dxdt$$

$$\leq C\int_Q s^{\ell-1}\theta^{\ell-1}|\partial_t w|^2 e^{2s\alpha}\,dxdt + C\int_Q |w(x,t_0)|^2 e^{\lambda\delta_0}s^\ell\theta^\ell e^{2s\alpha}\,dxdt.$$

Here

$$s^\ell\theta^\ell e^{2s\alpha} = \left(s\frac{e^{\lambda d_0(x)}}{t(T-t)}\right)^\ell \exp\left(-2s\frac{e^{2\lambda\|d_0\|_{C(\overline{\Omega})}} - e^{\lambda d_0(x)}}{t(T-t)}\right)$$

$$\leq \left(s\frac{e^{\lambda d_0(x)}}{t(T-t)}\right)^\ell \exp\left(-2s\frac{e^{2\lambda\|d_0\|_{C(\overline{\Omega})}} - e^{\lambda\|d_0\|_{C(\overline{\Omega})}}}{t(T-t)}\right).$$

We choose $\lambda > 0$ large such that $e^{\lambda\|d_0\|_{C(\overline{\Omega})}} \geq 2$. Therefore

$$e^{2\lambda\|d_0\|_{C(\overline{\Omega})}} - e^{\lambda\|d_0\|_{C(\overline{\Omega})}} = e^{\lambda\|d_0\|_{C(\overline{\Omega})}}(e^{\lambda\|d_0\|_{C(\overline{\Omega})}} - 1) \geq e^{\lambda\|d_0\|_{C(\overline{\Omega})}}.$$

Hence

$$s^\ell\theta^\ell e^{2s\alpha} \leq \left(s\frac{e^{\lambda\|d_0\|_{C(\overline{\Omega})}}}{t(T-t)}\right)^\ell \exp\left(-2s\frac{e^{\lambda\|d_0\|_{C(\overline{\Omega})}}}{t(T-t)}\right) \leq \sup_{\eta\geq 0}\eta^\ell e^{-2\eta} = \left(\frac{\ell}{2}\right)^\ell e^{-\ell}.$$

$$(37)$$

Thus the proof of Lemma 5 is complete.

Second Step

The main idea for the proof of Proposition 2 is similar to Proposition 1, and additionally, in terms of Lemma 5, we regard the terms $\partial_x^\gamma v(x, t)$ as a term with lower powers in s, λ and θ than $\partial_t \partial_x^\gamma v(x, t)$, not only by comparing $\left| \int_{t_0}^t |v(x, \xi)| d\xi \right|^2$ with $|v(x, t)|^2$. For this, we further have to assume data $\partial_t u(\cdot, t_0)$ and $\partial_t^2 u(\cdot, t_0)$ as well as $u(\cdot, t_0)$ in Ω. We can remove data $\partial_t u(\cdot, t_0)$ and $\partial_t^2 u(\cdot, t_0)$ but the proof is more complicated.

In this step, we complete the proof of Proposition 2. We apply Lemma 3 to (19):

$$\int_Q s^{m+3} \lambda^{m+4} \theta^{m+3} |\partial_t v|^2 e^{2s\alpha} dx dt + \int_Q s^{m+1} \lambda^{m+2} \theta^{m+1} |\partial_t \nabla v|^2 e^{2s\alpha} dx dt$$

$$+ \int_Q s^{m-1} \lambda^m \theta^{m-1} (|\partial_t \Delta v|^2 + |\partial_t^2 v|^2) e^{2s\alpha} dx dt$$

$$\leq C \int_Q s^m \lambda^m \theta^m |\partial_t w|^2 e^{2s\alpha} dx dt$$

$$+ C \int_Q s^m \lambda^m \theta^m \left(|\Delta v|^2 + \left| \int_{t_0}^t |\Delta v(x, \xi)| d\xi \right|^2 \right)$$

$$e^{2s\alpha} dx dt + C_0(\lambda) D_2^2 + C E(\partial_t v)$$

for all $s \geq s_0$. Here we used (37) for estimating $\int_Q s^m \lambda^m \theta^m |\partial_t (a_3 \Delta u^0)|^2 e^{2s\alpha} dx dt$.

Lemmata 4 and 5 yield

$$\int_Q s^m \lambda^m \theta^m |\Delta v|^2 e^{2s\alpha} dx dt \leq C \int_Q s^{m-1} \lambda^m \theta^{m-1} e^{-\lambda \delta_0} |\partial_t \Delta v|^2 e^{2s\alpha} dx dt + C_0(\lambda) D_2^2$$

and

$$\int_Q s^m \lambda^m \theta^m \left| \int_{t_0}^t |\Delta v| d\xi \right|^2 e^{2s\alpha} dx dt \leq C \int_Q s^{m-1} \lambda^m \theta^{m-1} e^{-\lambda \delta_0} |\Delta v|^2 e^{2s\alpha} dx dt$$

$$\leq C \int_Q s^{m-2} \lambda^m \theta^{m-2} e^{-\lambda \delta_0} |\partial_t \Delta v|^2 e^{2s\alpha} dx dt + C_0(\lambda) D_2^2.$$

Consequently

$$\int_Q s^m \lambda^m \theta^m \left(|\Delta v|^2 + \left| \int_{t_0}^t |\Delta v(x, \xi)| d\xi \right|^2 \right) e^{2s\alpha} dx dt$$

$$\leq C \int_Q s^{m-1} \lambda^m \theta^{m-1} e^{-\lambda \delta_0} |\partial_t \Delta v|^2 e^{2s\alpha} dx dt + C_0(\lambda) D_2^2.$$

Hence

$$
\int_Q s^{m+3}\lambda^{m+4}\theta^{m+3}|\partial_t v|^2 e^{2s\alpha}dxdt + \int_Q s^{m+1}\lambda^{m+2}\theta^{m+1}|\partial_t \nabla v|^2 e^{2s\alpha}dxdt
$$

$$
+ \int_Q s^{m-1}\lambda^m\theta^{m-1}(|\partial_t \Delta v|^2 + |\partial_t^2 v|^2)e^{2s\alpha}dxdt
$$

$$
\le C \int_Q s^m \lambda^m \theta^m |\partial_t w|^2 e^{2s\alpha}dxdt + C \int_Q s^{m-1}\lambda^m\theta^{m-1}e^{-\lambda\delta_0}|\partial_t \Delta v|^2 e^{2s\alpha}dxdt
$$

$$
+ C_0(\lambda) D_2^2 + C E(\partial_t v).
$$

Choosing $\lambda > 0$ in $e^{-\lambda\delta_0}$ sufficiently large, we can absorb the second term on the right-hand side into the second term on the left-hand side, we have

$$
\int_Q (s^{m+3}\lambda^{m+4}\theta^{m+3}|\partial_t v|^2 + s^{m+1}\lambda^{m+2}\theta^{m+1}|\partial_t \nabla v|^2
$$

$$
+ s^{m-1}\lambda^m\theta^{m-1}(|\partial_t \Delta v|^2 + |\partial_t^2 v|^2)e^{2s\alpha}dxdt
$$

$$
\le C \int_Q s^m \lambda^m \theta^m |\partial_t w|^2 e^{2s\alpha}dxdt + C_0(\lambda)D_2^2 + C E(\partial_t v).
$$

We apply Lemma 5 to the left-hand side and estimate $|v|^2$ and $|\Delta v|^2$ respectively by $|\partial_t v|^2$ and $|\partial_t \Delta v|^2$, so that we obtain

$$
\int_Q (s^{m+4}\lambda^{m+4}\theta^{m+4}|v|^2 + s^{m+3}\lambda^{m+4}\theta^{m+3}|\partial_t v|^2
$$

$$
+ s^{m+1}\lambda^{m+2}\theta^{m+1}|\partial_t \nabla v|^2 + s^m \lambda^m \theta^m |\Delta v|^2 + s^{m-1}\lambda^m\theta^{m-1}(|\partial_t \Delta v|^2
$$

$$
+ |\partial_t^2 v|)e^{2s\alpha}dxdt \le C \int_Q s^m \lambda^m \theta^m |\partial_t w|^2 e^{2s\alpha}dxdt + C_0(\lambda)D_2^2 + C E(\partial_t^2 u).
$$

$$(38)$$

We recall $v_0 = \partial_t v$ and set $V := \partial_t \Delta v = \partial_t^2 \Delta u$. We further take Δ in (19) to have

$$
\partial_t V - a_2 \Delta V
$$

$$
= (\partial_t a_2 + a_3)\Delta^2 v + (\partial_t a_3)\int_{t_0}^t \Delta^2 v(x, \xi)d\xi
$$

$$
+ 2\nabla a_2 \cdot \nabla(\Delta v_0) + (\Delta a_2)\Delta v_0 + \partial_t \Delta w + 2\nabla(\partial_t a_2 + a_3) \cdot \nabla(\Delta v)
$$

$$
+ \Delta(\partial_t a_2 + a_3)\Delta v + 2\nabla(\partial_t a_2)\int_{t_0}^t \nabla(\Delta v)(x, \xi)d\xi
$$

$$
+ \Delta(\partial_t a_3)\int_{t_0}^t \Delta v(x, \xi)d\xi - \Delta((\partial_t a_3)\Delta u^0).
$$

By (8), we note that $V = \partial_t^2 \Delta u = 0$ on $\partial\Omega \times (0, T)$. Consequently we can apply Lemma 3 to V and obtain

$$\int_Q (s^{m-1}\lambda^m\theta^{m-1}(|\partial_t^2\Delta v|^2 + |\partial_t\Delta^2 v|^2) + s^{m+1}\lambda^{m+2}\theta^{m+1}|\partial_t\nabla\Delta v|^2$$

$$+ s^{m+3}\lambda^{m+4}\theta^{m+3}|\partial_t\Delta v|^2)e^{2s\alpha}\,dxdt$$

$$= \int_Q (s^{m-1}\lambda^m\theta^{m-1}(|\partial_t V|^2 + |\Delta V|^2) + s^{m+1}\lambda^{m+2}\theta^{m+1}|\nabla V|^2$$

$$+ s^{m+3}\lambda^{m+4}\theta^{m+3}|V|^2)e^{2s\alpha}\,dxdt$$

$$\leq C\int_Q s^m\lambda^m\theta^m\left\{|\Delta^2 v|^2 + \left|\int_{t_0}^t |\Delta^2 v(x,\xi)|d\xi\right|^2 + |\nabla\Delta v|^2 + \left|\int_{t_0}^t |\nabla\Delta v(x,\xi)|d\xi\right|^2\right.$$

$$\left. + |\Delta v|^2 + \left|\int_{t_0}^t |\Delta v(x,\xi)|d\xi\right|^2\right\}e^{2s\alpha}\,dxdt$$

$$+ C\int_Q s^m\lambda^m\theta^m(|\partial_t\nabla\Delta v|^2 + |\partial_t\Delta v|^2)e^{2s\alpha}\,dxdt + C\int_Q s^m\lambda^m\theta^m|\partial_t\Delta w|^2 e^{2s\alpha}\,dxdt$$

$$+ C\int_Q s^m\lambda^m\theta^m|\Delta((\partial_t a_3)\Delta u^0)|^2 e^{2s\alpha}\,dxdt + CE(V). \tag{39}$$

By Lemma 4 and (37), choosing large λ and $s > 0$ in the first term on the right-hand side, we can absorb the integral terms $\left|\int_{t_0}^t \cdots d\xi\right|$ into the terms which are the same as the integrands, and so

[the right-hand side of (39)]

$$\leq C\int_Q s^m\lambda^m\theta^m(|\Delta^2 v|^2 + |\nabla\Delta v|^2 + |\Delta v|^2)e^{2s\alpha}\,dxdt$$

$$+ C\int_Q s^m\lambda^m\theta^m(|\partial_t\nabla\Delta v|^2 + |\partial_t\Delta v|^2)e^{2s\alpha}\,dxdt$$

$$+ C\int_Q s^m\lambda^m\theta^m|\partial_t\Delta w|^2 e^{2s\alpha}\,dxdt + C_0(\lambda)D_2^2 + CE(\partial_t^2\Delta u).$$

Applying Lemma 5 to the first term here, we have

[the right-hand side of (39)]

$$\leq \left[C \int_Q s^{m-1}\lambda^m\theta^{m-1}e^{-\lambda\delta_0}(|\partial_t\Delta^2 v|^2 + |\partial_t\nabla\Delta v|^2 + |\partial_t\Delta v|^2)e^{2s\alpha}dxdt + C_0(\lambda)D_2^2 \right]$$

$$+C \int_Q s^m\lambda^m\theta^m(|\partial_t\nabla\Delta v|^2 + |\partial_t\Delta v|^2)e^{2s\alpha}dxdt$$

$$+C \int_Q s^m\lambda^m\theta^m|\partial_t\Delta w|^2e^{2s\alpha}dxdt + C_0(\lambda)D_2^2 + CE(\partial_t^2\Delta u).$$

We note that in applying Lemma 5, we need the norm $\|u^1\|_{H^4(\Omega)}$. Therefore

$$\int_Q (s^{m-1}\lambda^m\theta^{m-1}(|\partial_t^2\Delta v|^2 + |\partial_t\Delta^2 v|^2) + s^{m+1}\lambda^{m+2}\theta^{m+1}|\partial_t\nabla\Delta v|^2$$

$$+s^{m+3}\lambda^{m+4}\theta^{m+3}|\partial_t\Delta v|^2)e^{2s\alpha}dxdt$$

$$\leq C \int_Q s^{m-1}\lambda^m\theta^{m-1}e^{-\lambda\delta_0}(|\partial_t\Delta^2 v|^2 + |\partial_t\nabla\Delta v|^2 + |\partial_t\Delta v|^2)e^{2s\alpha}dxdt$$

$$+C \int_Q s^m\lambda^m\theta^m(|\partial_t\nabla\Delta v|^2 + |\partial_t\Delta v|^2)e^{2s\alpha}dxdt$$

$$+C \int_Q s^m\lambda^m\theta^m|\partial_t\Delta w|^2e^{2s\alpha}dxdt + C_0(\lambda)D_2^2 + CE(\partial_t^2\Delta u).$$

Thanks to the factor $e^{-\lambda\delta_0}$, choosing $s > 0$ and $\lambda > 0$ sufficiently large, we can absorb the first and the second terms on the right-hand side into the left-hand side, so that

$$\int_Q (s^{m-1}\lambda^m\theta^{m-1}(|\partial_t^2\Delta v|^2 + |\partial_t\Delta^2 v|^2) + s^{m+1}\lambda^{m+2}\theta^{m+1}|\partial_t\nabla\Delta v|^2$$

$$+s^{m+3}\lambda^{m+4}\theta^{m+3}|\partial_t\Delta v|^2)e^{2s\alpha}dxdt$$

$$\leq C \int_Q s^m\lambda^m\theta^m|\partial_t\Delta w|^2e^{2s\alpha}dxdt + C_0(\lambda)D_2^2 + CE(\partial_t^2\Delta u). \tag{40}$$

Next, differentiating (19) with respect to t and setting $v_{00} := \partial_t v_0 = \partial_t^2 v$, we have

$$\partial_t v_{00} - a_2\Delta v_{00} = \partial_t^2 w + (\partial_t a_2 + a_3)\Delta\partial_t v$$

$$+(\partial_t a_3)\Delta v - (\partial_t^2 a_3)\Delta u^0 + (\partial_t a_2)\Delta\partial_t v + \partial_t(\partial_t a_2 + a_3)\Delta v + \partial_t^2 a_3 \int_{t_0}^t \Delta v(x, \xi)d\xi$$

in Q. Since $v_{00} = \partial_t^3 u = 0$ on $\partial\Omega \times (0, T)$ by (8), we can apply Lemma 3 to v_{00} and use (37) to obtain

$$\int_Q s^m \lambda^{m+1} \theta^m |\nabla v_{00}|^2 e^{2s\alpha} dxdt = \int_Q s^m \lambda^{m+1} \theta^m |\nabla \partial_t^2 v|^2 e^{2s\alpha} dxdt$$

$$\leq C \int_Q s^{m-1} \lambda^{m-1} \theta^{m-1} |\partial_t^2 w|^2 e^{2s\alpha} dxdt$$

$$+ C \int_Q s^{m-1} \lambda^{m-1} \theta^{m-1} |\partial_t \Delta v|^2 e^{2s\alpha} dxdt$$

$$+ C \int_Q s^{m-1} \lambda^{m-1} \theta^{m-1} \left(|\Delta v|^2 + \left| \int_{t_0}^t |\Delta v(x, \xi)| d\xi \right|^2 \right) e^{2s\alpha} dxdt$$

$$+ C_0(\lambda) D_2^2 + CE(\partial_t^3 u).$$

Then, in terms of Lemmata 4 and 5 to the third term on the right-hand side, we absorb it into the second term, and we reach

$$\int_Q s^m \lambda^{m+1} \theta^m |\partial_t^2 \nabla v|^2 e^{2s\alpha} dxdt$$

$$\leq C \int_Q s^{m-1} \lambda^{m-1} \theta^{m-1} |\partial_t^2 w|^2 e^{2s\alpha} dxdt \qquad (41)$$

$$+ C \int_Q s^{m-1} \lambda^{m-1} \theta^{m-1} |\partial_t \Delta v|^2 e^{2s\alpha} dxdt$$

$$+ C_0(\lambda) D_2^2 + CE(\partial_t^3 u).$$

On the other hand, we see

$$w_0 = \partial_t (\partial_t^2 u - a_2 \Delta \partial_t u - a_3 \Delta u)$$

$$= \partial_t^3 u - (\partial_t a_2) \partial_t (\Delta u) - a_2 \partial_t^2 (\Delta u) - (\partial_t a_3) \Delta u - a_3 \partial_t (\Delta u) = 0 \quad \text{on } \partial\Omega$$

by $u = \Delta u = 0$ on $\partial\Omega$. Therefore we can apply Lemma 3 with $m = 0$ to (18), and in terms of Lemma 4, we have

$$\int_Q (s^{-1} \theta^{-1} (|\partial_t^2 w|^2 + |\Delta \partial_t w|^2) + s\lambda^2 \theta |\partial_t \nabla w|^2 + s^3 \lambda^4 \theta^3 |\partial_t w|^2) e^{2s\alpha} dxdt$$

$$\leq C \int_Q |\partial_t \widetilde{R}|^2 |f(x)|^2 e^{2s\alpha} dxdt + C \int_Q |\Delta w|^2 e^{2s\alpha} dxdt$$

$$+ C \int_Q \left\{ \sum_{j=0}^1 (|\partial_t^j v|^2 + |\partial_t^j \nabla v|^2 + |\partial_t^j \Delta v|^2 + |\partial_t^j \nabla \Delta v|^2) + |\partial_t^2 v|^2 + |\partial_t^2 \nabla v|^2 \right\}$$

$$e^{2s\alpha} dxdt + C_0(\lambda) D_2^2 + CE(\partial_t w).$$

Applying Lemma 5 to the third term on the right-hand side, similarly to (41), we obtain

$$\int_Q (s^{-1}\theta^{-1}(|\partial_t^2 w|^2 + |\Delta\partial_t w|^2) + s\lambda^2\theta|\partial_t\nabla w|^2 + s^3\lambda^4\theta^3|\partial_t w|^2)e^{2s\alpha}dxdt$$

$$\leq C\int_Q |\partial_t\widetilde{R}|^2|f(x)|^2 e^{2s\alpha}dxdt + C\int_Q |\Delta w|^2 e^{2s\alpha}dxdt$$

$$+C\int_Q (|\partial_t\Delta v|^2 + |\partial_t\nabla\Delta v|^2 + |\partial_t^2 v|^2 + |\partial_t^2\nabla v|^2)e^{2s\alpha}dxdt$$

$$+ C_0(\lambda)D_2^2 + CE(\partial_t w). \tag{42}$$

Moreover (38) with $m = 1$ yields

$$\int_Q (|\partial_t\Delta v|^2 + |\partial_t^2 v|^2)e^{2s\alpha}dxdt \leq C\int_Q s\theta|\partial_t w|^2 e^{2s\alpha}dxdt + C_0(\lambda)D_2^2 + CE(\partial_t^2 u). \tag{43}$$

We apply (41) with $m = 0$ and (38) with $m = 0$ successively, and we can reach

$$\int_Q |\partial_t^2\nabla v|^2 e^{2s\alpha}dxdt \leq C\int_Q s^{-1}\lambda^{-2}\theta^{-1}|\partial_t^2 w|^2 e^{2s\alpha}dxdt$$

$$+ C\int_Q s^{-1}\lambda^{-2}\theta^{-1}|\partial_t\Delta v|^2 e^{2s\alpha}dxdt + C(D_2^2 + E(\partial_t^3 u))$$

$$\leq C\int_Q s^{-1}\lambda^{-2}\theta^{-1}|\partial_t^2 w|^2 e^{2s\alpha}dxdt$$

$$+ C\int_Q \lambda^{-2}|\partial_t w|^2 e^{2s\alpha}dxdt + C_0(\lambda)D_2^2 + CE(\partial_t^3 u). \tag{44}$$

Next (40) with $m = -1$ yields

$$\int_Q |\partial_t\nabla\Delta v|^2 e^{2s\alpha}dxdt$$

$$\leq C\int_Q s^{-1}\lambda^{-2}\theta^{-1}|\partial_t\Delta w|^2 e^{2s\alpha}dxdt \tag{45}$$

$$+ C_0(\lambda)D_2^2 + CE(\partial_t^2\Delta u).$$

Now we will improve (42). First, applying Lemma 5 to the second term on the right-hand side of (42), we can absorb it into the left-hand side. Moreover, substituting (43)–(45) into the third term on the right-hand side of (42), we obtain

$$\int_Q (s^{-1}\theta^{-1}(|\partial_t^2 w|^2 + |\Delta \partial_t w|^2) + s^3\lambda^4\theta^3|\partial_t w|^2)e^{2s\alpha}dxdt$$

$$\leq C \int_Q |\partial_t \widetilde{R}|^2 |f(x)|^2 e^{2s\alpha}dxdt$$

$$+C \int_Q (s\theta|\partial_t w|^2 + \lambda^{-2}|\partial_t w|^2 + s^{-1}\lambda^{-2}\theta^{-1}(|\partial_t^2 w|^2 + |\partial_t \Delta w|^2)e^{2s\alpha}dxdt$$

$$+C_0(\lambda)D_2^2 + C(E(\partial_t^2 u) + E(\partial_t^3 u) + E(\partial_t^2 \Delta u) + E(\partial_t w)).$$

Choosing $s > 0$ and $\lambda > 0$ sufficiently large and comparing the powers s, λ, θ, we can absorb the second integral term on the right-hand side into the left-hand side, and we finally reach

$$\int_Q (s^{-1}\theta^{-1}(|\partial_t^2 w|^2 + |\Delta \partial_t w|^2) + s^3\lambda^4\theta^3|\partial_t w|^2)e^{2s\alpha}dxdt$$

$$\leq C \int_Q |\partial_t \widetilde{R}|^2 |f(x)|^2 e^{2s\alpha}dxdt$$

$$+C_0(\lambda)D_2^2 + C(E(\partial_t^2 u) + E(\partial_t^3 u) + E(\partial_t^2 \Delta u) + E(\partial_t w)).$$

By the definition of $E(v)$, (12) and (37), we estimate

$$E(\partial_t^2 u) + E(\partial_t^3 u) + E(\partial_t^2 \Delta u) + E(\partial_t w) \leq C_0(\lambda)(\|u\|_{H^3(0,T;L^2(\omega))}^2 + \|u\|_{H^2(0,T;H^2(\omega))}^2).$$

Thus the proof of Proposition 2 is complete.

3   Proof of Theorem 1

The proof of the local stability in Ω for the inverse source problem is based on the method by Bukhgeim and Klibanov [6] using a cut-off argument. Except for a recent work by Huang et al. [14], all the existing works use a cut-off argument. However, the cut-off argument makes the proof lengthy and untransparent. Huang et al. [14] provides a simpler proof without any cut-off arguments. Here we follow the way in [14].

First Step
We choose $d(x)$ in the weight function in Proposition 1 in Sect. 2. First we construct some domain Ω_1. For given $\Gamma \subset \partial\Omega$, we choose a bounded domain Ω_1 with smooth boundary such that

$$\Omega \subsetneq \Omega_1, \quad \Gamma = \overline{\partial\Omega \cap \Omega_1}, \quad \partial\Omega \setminus \Gamma \subset \partial\Omega_1. \tag{46}$$

In particular, $\Omega_1 \setminus \overline{\Omega}$ contains some non-empty open subset. We note that Ω_1 can be constructed as the interior of a union of $\overline{\Omega}$ and the closure of a non-empty domain $\widehat{\Omega}$ satisfying $\widehat{\Omega} \subset \mathbb{R}^3 \setminus \overline{\Omega}$ and $\partial\widehat{\Omega} \cap \partial\Omega = \Gamma$.

We choose a domain ω such that $\overline{\omega} \subset \Omega_1 \setminus \overline{\Omega}$. Then, by Imanuvilov [16] for example, we can find $d \in C^2(\overline{\Omega_1})$ such that

$$d > 0 \quad \text{in } \Omega_1, \quad |\nabla d| > 0 \quad \text{on } \overline{\Omega_1 \setminus \omega}, \quad d = 0 \quad \text{on } \partial\Omega_1.$$

In particular,

$$d \geq 0 \quad \text{on } \overline{\Omega}, \quad d > 0 \quad \text{on } \overline{\Omega_0}, \quad d = 0 \quad \text{on } \partial\Omega \setminus \Gamma, \quad |\nabla d| > 0 \quad \text{on } \overline{\Omega}. \tag{47}$$

We recall that for relatively open subset Γ of $\partial\Omega$, we choose a domain $\Omega_0 \subset \Omega$ satisfying $\overline{\partial\Omega_0 \cap \partial\Omega} \subset \Gamma$ and $\overline{\Omega_0} \subset \Omega \cup \Gamma$.

In the weight function $\varphi(x, t) = e^{\lambda(d(x) - \beta|t - t_0|^2)}$ of the Carleman estimate, we choose a constant $\beta > 0$ sufficiently large such that

$$\|d\|_{C(\overline{\Omega})} \leq \beta\delta^2. \tag{48}$$

We fix $\lambda > 0$ sufficiently large, and we recall $I = (t_0 - \delta, t_0 + \delta)$. Then we can neglect the dependency on λ and φ in Proposition 1. Before applying Proposition 1, we estimate

$$\frac{1}{s}\Big(\Phi(v) + \Phi(\nabla_{x,t}v) + \Phi(\partial_t\nabla v) + \Phi(w) + \Phi(\partial_t w)\Big),$$

where we recall (16). First representing v and w in terms of u, by $v = \partial_t u$ and (12) we have

$$J_1 := \int_{\partial\Omega \times I} s^2\big(|v|^2 + |\nabla_{x,t}v|^2 + |\nabla_{x,t}\nabla_{x,t}v|^2 + |\nabla_{x,t}\partial_t\nabla v|^2$$

$$+ |w|^2 + |\nabla_{x,t}w|^2 + |\nabla_{x,t}\partial_t w|^2 + |\nabla^3 v|^2 + |\nabla^2\partial_t v|^2 + |\nabla^3\partial_t v|^2$$

$$+ |\nabla^2 w|^2 + |\nabla^2\partial_t w|^2\big)e^{2s\varphi}\,dS dt$$

$$\leq Cs^2 \int_{\partial\Omega \times I} \sum_{|\gamma| + j \leq 5} |\partial_x^\gamma \partial_t^j u|^2 e^{2s\varphi}\,dS dt$$

$$= Cs^2 \left(\int_{\Gamma \times I} + \int_{(\partial\Omega \setminus \Gamma) \times I}\right) \sum_{|\gamma| + j \leq 5} |\partial_x^\gamma \partial_t^j u|^2 e^{2s\varphi}\,dS dt$$

$$=: J_{11} + J_{12}.$$

We set

$$D_3 := \sum_{|\gamma|+j \leq 5} \|\partial_x^\gamma \partial_t^j u\|_{L^2(I;L^2(\Gamma))}. \tag{49}$$

We recall that D_1 and D_2 are defined by (9). Then

$$|J_{11}| \leq Cs^2 D_3^2 \exp(2s \max_{x \in \overline{\Gamma}, t \in \overline{I}} \varphi(x, t)). \tag{50}$$

By the trace theorem and (7), we see

$$|J_{12}| \leq Cs^2 \sum_{|\gamma|+j \leq 5} \|\partial_x^\gamma \partial_t^j u\|^2_{L^2(I;L^2(\partial\Omega\backslash\Gamma))} \exp(2s \max_{x \in \overline{\partial\Omega\backslash\Gamma}, t \in \overline{I}} \varphi(x, t))$$

$$\leq Cs^2 M^2 \exp(2s \max_{x \in \overline{\partial\Omega\backslash\Gamma}, t \in \overline{I}} \varphi(x, t)).$$

Hence, with (50), we see

$$|J| \leq Cs^2 D_3^2 \exp(2s \max_{x \in \overline{\Gamma}, t \in \overline{I}} \varphi(x, t)) + Cs^2 M^2 \exp(2s \max_{x \in \overline{\partial\Omega\backslash\Gamma}, t \in \overline{I}} \varphi(x, t)). \tag{51}$$

Next by (7) and the Sobolev embedding, we have

$$s^4 \int_\Omega \{|v(x, t_0 - \delta)|^2 + |\nabla_{x,t} v(x, t_0 - \delta)|^2 + |\nabla(\nabla_{x,t} v)(x, t_0 - \delta)|^2$$

$$+ |\nabla(\nabla \partial_t v)(x, t_0 - \delta)|^2 + |w(x, t_0 - \delta)|^2$$

$$+ |\nabla w(x, t_0 - \delta)|^2 + |\partial_t w(x, t_0 - \delta)|^2 + |\nabla \partial_t w(x, t_0 - \delta)|^2$$

$$+ |v(x, t_0 + \delta)|^2 + |\nabla_{x,t} v(x, t_0 + \delta)|^2 + |\nabla(\nabla_{x,t} v)(x, t_0 + \delta)|^2$$

$$+ |\nabla(\nabla \partial_t v)(x, t_0 + \delta)|^2 + |w(x, t_0 + \delta)|^2$$

$$+ |\nabla w(x, t_0 + \delta)|^2 + |\partial_t w(x, t_0 + \delta)|^2 + |\nabla \partial_t w(x, t_0 + \delta)|^2\} e^{2s\varphi(x, t_0 + \delta)} dx$$

$$\leq Cs^4 M^2 \exp(2s \max_{x \in \overline{\Omega}} \varphi(x, t_0 + \delta)). \tag{52}$$

In terms of (51) and (52), we reach

$$\frac{1}{s}\left(\Phi(v) + \Phi(\nabla_{x,t} v) + \Phi(\partial_t \nabla v) + \Phi(w) + \Phi(\partial_t w)\right)$$

$$\leq Cs^2 D_3^2 e^{Cs} + Cs^2 M^2 \exp(2s \max_{x \in \overline{\partial\Omega\backslash\Gamma}, t \in \overline{I}} \varphi(x, t))$$

$$+ Cs^4 M^2 \exp(2s \max_{x \in \overline{\Omega}} \varphi(x, t_0 + \delta)).$$

On the other hand, since $d = 0$ on $\partial\Omega \setminus \Gamma$ by (47), we have

$$\max_{x\in\partial\Omega\setminus\Gamma,t\in\overline{I}} \varphi(x, t) = \exp(\lambda \max_{x\in\partial\Omega\setminus\Gamma,t\in\overline{I}}(d(x) - \beta(t - t_0)^2)) \le 1.$$

Moreover, by (48) we can see that $\|d\|_{C(\overline{\Omega})} - \beta\delta^2 \le 0$, and so

$$\max_{x\in\Omega} \varphi(x, t_0 + \delta) = \exp(\lambda(\max_{x\in\Omega}(d(x) - \beta\delta^2))) = \exp(\lambda(\|d\|_{C(\overline{\Omega})} - \beta\delta^2)) \le 1.$$

$$(53)$$

Hence

$$\frac{1}{s}(\Phi(v) + \Phi(\nabla_{x,t}v) + \Phi(\partial_t \nabla v) + \Phi(w) + \Phi(\partial_t w))$$

$$\le Cs^2 D_3^2 e^{Cs} + Cs^4 M^2 e^{2s}.$$

Therefore by Proposition 1, we obtain

$$\int_{Q_I} (s^{-2}|\partial_t^2 w|^2 + s^2|\partial_t w|^2)e^{2s\varphi}dxdt$$

$$\le C \int_{Q_I} (|\partial_t \widetilde{R}|^2 + |\widetilde{R}|^2)|f(x)|^2 e^{2s\varphi}dxdt + Cs^2 e^{Cs}(D_1^2 + D_3^2) + Cs^4 M^2 e^{2s} \quad (54)$$

for all large $s > 0$.

Furthermore (47) implies

$$\delta_1 := \min_{x\in\overline{\Omega_0}} d(x) > 0. \quad (55)$$

Therefore

$$\min_{x\in\overline{\Omega_0}} \varphi(x, t_0) = \exp(\lambda \min_{x\in\overline{\Omega_0}} d(x)) = e^{\lambda\delta_1}. \quad (56)$$

Second Step
We recall $u^0 = u(\cdot, t_0)$, $u^1 = \partial_t u(\cdot, t_0)$, $u^2 = \partial_t^2 u(\cdot, t_0)$, and set

$$H(u)(x) = -\sum_{k=0}^{1} a_{1k}(x, t_0)\nabla^k \Delta u^1(x, t_0)$$

$$-\sum_{k=0}^{1}\sum_{j=0}^{1} b_{k+1,j}(x, t_0)\nabla^k u^{j+1}(x, t_0), \quad x \in \Omega.$$

Then we see

$$|H(u)(x)| \le C \left(\sum_{|\gamma| \le 3} |\partial_x^\gamma u^1(x)| + \sum_{|\gamma| \le 1} |\partial_x^\gamma u^2(x)| \right), \quad x \in \Omega.$$

It follows from (14) that

$$\widetilde{R}(x, t_0) f(x) = \partial_t w(x, t_0) - a_1(x, t_0) \Delta w(x, t_0) + H(u)(x) + b_0(x, t_0), \quad x \in \Omega.$$

Since

$$\Delta w(x, t_0) = \Delta(\partial_t^2 u(x, t_0) - a_2(x, t_0) \Delta \partial_t u(x, t_0) - a_3(x, t_0) \Delta u(x, t_0))$$

by (12), noting the second condition of (6) and $\partial_t u(\cdot, t_0) = u^1$ and $\partial_t^2 u(\cdot, t_0) = u^2$, we obtain

$$|f(x)|^2 \le C \left(|\partial_t w(x, t_0)|^2 + \sum_{|\gamma| \le 4} (|\partial_x^\gamma u^0(x)|^2 + |\partial_x^\gamma u^1(x)|^2) + \sum_{|\gamma| \le 2} |\partial_x^\gamma u^2(x)|^2 \right), \tag{57}$$

$$x \in \Omega_0.$$

Now we estimate

$$\int_\Omega |\partial_t w(x, t_0)|^2 e^{2s\varphi(x, t_0)} dx$$

as follows:

$$\int_\Omega |\partial_t w(x, t_0)| e^{2s\varphi(x, t_0)} dx$$

$$= \int_{t_0-\delta}^{t_0} \partial_t \left(\int_\Omega |\partial_t w(x, t)|^2 e^{2s\varphi(x, t)} dx \right) dt + \int_\Omega |\partial_t w(x, t_0 - \delta)|^2 e^{2s\varphi(x, t_0-\delta)} dx$$

$$= \int_{t_0-\delta}^{t_0} \int_\Omega (2\partial_t w(x, t) \partial_t^2 w(x, t) + 2s(\partial_t\varphi)|\partial_t w(x, t)|^2) e^{2s\varphi(x, t)} dx dt$$

$$+ \int_\Omega |\partial_t w(x, t_0 - \delta)|^2 e^{2s\varphi(x, t_0-\delta)} dx$$

and so (53) yields

$$\int_\Omega |\partial_t w(x, t_0)| e^{2s\varphi(x, t_0)} dx \le C \int_{Q_l} (|\partial_t w||\partial_t^2 w| + s|\partial_t w|^2) e^{2s\varphi(x, t)} dx dt + CM^2 e^{2s}.$$

For the last inequality, by the Sobolev embedding, (7) and (12), we used

$$\|\partial_t w(\cdot, t_0 - \delta)\|_{L^2(\Omega)} \leq C\|w\|_{H^2(I;L^2(\Omega))} \leq CM.$$

We estimate the first term on the right-hand side:

$$|\partial_t w||\partial_t^2 w| = (s^{-1}|\partial_t^2 w|)(s|\partial_t w|) \leq \frac{1}{2}(s^{-2}|\partial_t^2 w|^2 + s^2|\partial_t w|^2),$$

and so Proposition 1 yields

$$\int_{Q_I} (|\partial_t w||\partial_t^2 w| + s|\partial_t w|^2)e^{2s\varphi} dx dt \leq C \int_{Q_I} (s^{-2}|\partial_t^2 w|^2 + s^2|\partial_t w|^2)e^{2s\varphi} dx dt$$

$$\leq C \int_{Q_I} (|\partial_t \widetilde{R}|^2 + |\widetilde{R}|^2)|f(x)|^2 e^{2s\varphi} dx dt + Cs^2 e^{Cs} D_1^2 + Cs^2 e^{Cs} D_3^2 + Cs^4 M^2 e^{2s}$$

for $s \geq 1$. Here we applied also (54) and recall that D_1, D_2 and D_3 are defined by (9) and (49). Consequently

$$\int_{\Omega} |\partial_t w(x, t_0)|^2 e^{2s\varphi(x,t_0)} dx$$

$$\leq C \int_{Q_I} (|\partial_t \widetilde{R}|^2 + |\widetilde{R}|^2)|f(x)|^2 e^{2s\varphi} dx dt + Ce^{Cs} s^2(D_1^2 + D_3^2) + Cs^4 M^2 e^{2s}.$$

Therefore with (57), we obtain

$$\int_{\Omega} |f(x)|^2 e^{2s\varphi(x,t_0)} dx \leq C \int_{Q_I} (|\partial_t \widetilde{R}|^2 + |\widetilde{R}|^2)|f(x)|^2 e^{2s\varphi} dx dt$$

$$+ Cs^2 e^{Cs}(D_2^2 + D_3^2) + Cs^4 M^2 e^{2s}. \tag{58}$$

Since $\partial_t \widetilde{R}, \widetilde{R} \in L^2(I; L^\infty(\Omega))$ by (6) and $d \geq 0$ on $\overline{\Omega}$ by (47), we have

$$\int_{Q_I} (|\partial_t \widetilde{R}|^2 + |\widetilde{R}|^2)|f(x)|^2 e^{2s\varphi} dx dt$$

$$= \int_{\Omega} |f(x)|^2 e^{2s\varphi(x,t_0)} \left(\int_{t_0-\delta}^{t_0+\delta} (|\partial_t \widetilde{R}(x, t)|^2 + |\widetilde{R}(x, t)|^2)e^{-2s(\varphi(x,t_0)-\varphi(x,t))} dt \right) dx$$

$$\leq \int_{\Omega} |f(x)|^2 e^{2s\varphi(x,t_0)} \left(\int_{t_0-\delta}^{t_0+\delta} (\|\partial_t \widetilde{R}(\cdot, t)\|_{L^\infty(\Omega)}^2 + \|\widetilde{R}(\cdot, t)\|_{L^\infty(\Omega)}^2) \right.$$

$$\times \exp(-2s \min_{x\in\overline{\Omega}} e^{\lambda d(x)}(1 - e^{-\beta\lambda(t-t_0)^2})) dt \bigg) dx$$

$$\leq \int_{\Omega} |f(x)|^2 e^{2s\varphi(x,t_0)}$$

$$\times \left(\int_{t_0-\delta}^{t_0+\delta} (\|\partial_t \widetilde{R}(\cdot, t)\|_{L^\infty(\Omega)}^2 + \|\widetilde{R}(\cdot, t)\|_{L^\infty(\Omega)}^2) \exp(-2s(1 - e^{-\beta\lambda(t-t_0)^2}))) dt \right) dx.$$

Since

$$\lim_{s\to\infty} \exp(-2s(1 - e^{-\beta\lambda(t-t_0)^2})) = 0$$

if $t \neq t_0$ and

$$\|\partial_t \widetilde{R}(\cdot, t)\|_{L^\infty(\Omega)}^2 + \|\widetilde{R}(\cdot, t)\|_{L^\infty(\Omega)}^2 \in L^1(t_0 - \delta, t_0 + \delta),$$

we apply the Lebesgue convergence theorem, so that

$$\int_{Q_l} (|\partial_t \widetilde{R}|^2 + |\widetilde{R}|^2)|f(x)|^2 e^{2s\varphi} dxdt \leq o(1) \int_\Omega |f(x)|^2 e^{2s\varphi(x,t_0)} dx$$

as $s \to \infty$. Substituting this into the first term on the right-hand side of (58) and choosing $s > 0$ sufficiently large to absorb it into the left-hand side, we can reach

$$\int_\Omega |f(x)|^2 e^{2s\varphi(x,t_0)} dx \leq Cs^2 e^{Cs}(D_2^2 + D_3^2) + Cs^4 M^2 e^{2s}$$

for all $s \geq s_1$, where the constant $s_1 > 0$ is sufficiently large. Shrinking the integration domain Ω on the left-hand side to Ω_0, by (55), we have

$$e^{2se^{\lambda\delta_1}} \int_{\Omega_0} |f(x)|^2 dx \leq Cs^2 e^{Cs}(D_2^2 + D_3^2) + Cs^4 M^2 e^{2s},$$

that is,

$$\|f\|_{L^2(\Omega_0)}^2 \leq Cs^2 e^{Cs}(D_2^2 + D_3^2) + Cs^4 M^2 e^{-2s\mu}.$$

Here we have

$$\mu := e^{\lambda\delta_1} - 1 > 0$$

by $\delta_1 > 0$. Since $s^2 e^{Cs} \leq C_1 e^{C_1 s}$ for all $s > 0$ and $\sup_{s>0} s^4 e^{-s\mu} < \infty$, we obtain

$$\|f\|_{L^2(\Omega_0)}^2 \leq C_1 M^2 e^{-s\mu} + C_1 e^{C_1 s}(D_2^2 + D_3^2)$$

for $s \geq s_1$. Replacing $C_1 > 0$ by $C_1 e^{C_1 s_1}$ and changing s into $s + s_1$ with $s \geq 0$, we obtain

$$\|f\|_{L^2(\Omega_0)}^2 \leq C_1 M^2 e^{-s\mu} + C_1 e^{C_1 s}(D_2^2 + D_3^2) \tag{59}$$

for each $s \geq 0$. We minimize the right-hand side by choosing an appropriate value of parameter $s \geq 0$.

Case 1: $M^2 > D_2^2 + D_3^2$ Then we can solve

$$M^2 e^{-s\mu} = e^{C_1 s}(D_2^2 + D_3^2), \quad \text{that is,} \quad s = \frac{1}{C_1 + \mu} \log \frac{M^2}{D_2^2 + D_3^2} > 0,$$

so that

$$\|f\|_{L^2(\Omega_0)}^2 \leq 2C_1 M^{2(1-\chi)}(D_2^2 + D_3^2)^\chi,$$

where $\chi = \frac{\mu}{C_1 + \mu} \in (0, 1)$.

Case 2: $M^2 \leq D_2^2 + D_3^2$ Then $\|f\|_{L^2(\Omega_0)}^2 \leq C_1(1 + e^{C_1 s})(D_2^2 + D_3^2)$. By the trace theorem and the Sobolev embedding, we readily see that $D_2 + D_3 \leq CM$, and

$$D_2^2 + D_3^2 = (D_2^2 + D_3^2)^\chi (D_2^2 + D_3^2)^{1-\chi} \leq (CM)^{2(1-\chi)}(D_2^2 + D_3^2)^\chi.$$

Therefore, in both Cases 1 and 2, we can obtain

$$\|f\|_{L^2(\Omega_0)}^2 \leq C(M)(D_2^2 + D_3^2)^\chi.$$

Thus the proof of Theorem 1 is completed. ∎

4 Proof of Theorem 2

Now in Proposition 2, we fix sufficiently large $\lambda > 0$. Then (35) yields $|\partial_t \alpha| \leq C\theta^2$ in Q. By Eq. (14) and $\widetilde{R}(x, t_0) \neq 0$ for $x \in \overline{\Omega}$, noting that $v(\cdot, t_0) = \partial_t u(\cdot, t_0) = u^1(x)$ and $\partial_t v(\cdot, t_0) = \partial_t^2 u(\cdot, t_0) = u^2(x)$, we have (57). Therefore, since $w = \partial_t^2 u - a_2 \Delta \partial_t u - a_3 \Delta u$ by (12), we see

$$\int_\Omega |f(x)|^2 e^{2s\alpha(x,t_0)} dx \leq C \int_\Omega |\partial_t w(x, t_0)|^2 e^{2s\alpha(x,t_0)} dx + C D_2^2. \quad (60)$$

Now we estimate $\int_\Omega |\partial_t w(x, t_0)|^2 e^{2s\alpha(x,t_0)} dx$ as follows. Since $\lim_{t \downarrow 0} e^{2s\alpha(x,t)} = 0$ by (33), using (35), we see

$$\int_\Omega |\partial_t w(x, t_0)|^2 e^{2s\alpha(x,t_0)} dx = \int_0^{t_0} \partial_t \left(\int_\Omega |\partial_t w(x, t)|^2 e^{2s\alpha(x,t)} dx \right) dt$$

$$= \int_0^{t_0} \int_\Omega (2(\partial_t^2 w)\partial_t w + 2s(\partial_t \alpha)|\partial_t w|^2) e^{2s\alpha} dx dt$$

$$\leq C \int_Q (|\partial_t^2 w||\partial_t w| + s\theta^2|\partial_t w|^2) e^{2s\alpha} dx dt.$$

Since

$$|\partial_t w||\partial_t^2 w| = s^{-1}\theta^{-\frac{1}{2}}|\partial_t^2 w|s\theta^{\frac{1}{2}}|\partial_t w| \le \frac{1}{2}(s^{-2}\theta^{-1}|\partial_t^2 w|^2 + s^2\theta|\partial_t w|^2),$$

by Proposition 2 we obtain

$$\int_\Omega |\partial_t w(x,t_0)|^2 e^{2s\alpha(x,t_0)} dx \le Cs^{-1}\int_Q |\partial_t \widetilde{R}|^2 |f(x)|^2 e^{2s\alpha} dxdt$$

$$+CD_2^2 + C(\|u\|_{H^3(I;L^2(\omega))}^2 + \|u\|_{H^2(I;H^2(\omega))}^2).$$

Hence with (60) we see

$$\int_\Omega |f(x)|^2 e^{2s\alpha(x,t_0)} dx \le Cs^{-1}\int_Q |\partial_t \widetilde{R}|^2 |f(x)|^2 e^{2s\alpha} dxdt$$

$$+CD_2^2 + C(\|u\|_{H^3(I;L^2(\omega))}^2 + \|u\|_{H^2(I;H^2(\omega))}^2).$$

for all $s \ge s_0$.

Since $\alpha(x,t_0) \ge \alpha(x,t)$ for $(x,t) \in Q$, we have

$$s^{-1}\int_Q |\partial_t \widetilde{R}|^2 |f(x)|^2 e^{2s\alpha} dxdt$$

$$\le s^{-1}\int_\Omega |f(x)|^2 e^{2s\alpha(x,t_0)} \left(\int_0^T |\partial_t \widetilde{R}(x,t)|^2 dt\right) dx$$

$$\le s^{-1}\int_\Omega |f(x)|^2 e^{2s\alpha(x,t_0)} \left(\int_0^T \|\partial_t \widetilde{R}(\cdot,t)\|_{L^\infty(\Omega)}^2 dt\right) dx$$

$$\le s^{-1}\|\partial_t \widetilde{R}\|_{L^2(0,T;L^\infty(\Omega))}^2 \int_\Omega |f(x)|^2 e^{2s\alpha(x,t_0)} dx.$$

Therefore

$$(1 - Cs^{-1}\|\partial_t \widetilde{R}\|_{L^2(0,T;L^\infty(\Omega))}^2)\int_\Omega |f(x)|^2 e^{2s\alpha(x,t_0)} dx$$

$$\le C(D_2^2 + \|u\|_{H^3(I;L^2(\omega))}^2 + \|u\|_{H^2(I;H^2(\omega))}^2)$$

for all $s \ge s_0$. Thus we complete the proof of Theorem 2.

5 Concluding Remarks

5.1 Unique Continuation with Initial Condition

We have established Carleman estimates for our system (4) by the factorization in (11). In the first Carleman estimate Proposition 1, we have to be given $u(\cdot, t_0)$ in Ω. In the second one, we assume that $u(\cdot, t_0)$, $\partial_t u(\cdot, t_0)$, and $\partial_t^2 u(\cdot, t_0)$ in Ω are given as data, although we can prove a Carleman estimate only with additional data $u(\cdot, t_0)$. For the inverse problem, we need spatial data such as $u(\cdot, t_0)$ in Ω. We note that for the inverse problem, we do not assume to know initial values, but we need such spatial data at $t = t_0 \in (0, T)$. On the other hand, the unique continuation for partial differential equations should not require such spatial data. More precisely, the unique continuation for our system means that if u satisfies (4) with $Rf \equiv 0$ and suitable boundary values on a subboundary of $\partial\Omega \times (0, T)$ vanish, then u vanishes in some subdomain in $\Omega \times (0, T)$ without further spatial data. It is well-known that a relevant Carleman estimate yields such a unique continuation property (e.g., Isakov [23]). However our Carleman estimate Proposition 1 cannot produce the unique continuation without extra spatial data owing to the extra term containing $u(\cdot, t_0)$ on the right-hand side. However Proposition 1 implies

Proposition 3 *Let $\Gamma \subset \partial\Omega$ be an arbitrarily fixed subboundary. Let smooth u satisfy*

$$(1+p_0(x, t))\partial_t^3 u(x, t) - A_3\Delta\partial_t^2 u + A_1\Delta^2 u + A_2\Delta^2\partial_t u + G(u) = 0 \quad in \ Q. \quad (61)$$

If $\partial_x^\gamma u = 0$ on $\Gamma \times (0, T)$ for all $|\gamma| \le 3$, and

$$u(\cdot, t_0) = 0 \quad in \ \Omega,$$

then $u(x, t) = 0$ for $(x, t) \in Q$.

We can also prove a conditional stability estimate similarly to Theorem 1, and here we discuss only the uniqueness.

Proof We fix $\lambda > 0$ sufficiently large and in Proposition 1, we can neglect the dependency on λ and φ. We recall that the constant $M > 0$ satisfies (7). We choose $t_0 \in (0, T)$ arbitrarily. Then choose $\varepsilon > 0$ sufficiently small such that $0 < t_0 - \varepsilon < t_0 + \varepsilon < T$. We set $Q_\varepsilon = \Omega \times (t_0 - \varepsilon, t_0 + \varepsilon)$. We choose the same $d \in C^2(\overline{\Omega_1})$ satisfying (47). In Q_ε, we apply Proposition 1 (ii). We choose $N > 1$ large such that

$$N \min_{x \in \overline{\Omega_0}} d(x) > \|d\|_{C(\overline{\Omega})}.$$

This is possible because $\min_{x \in \overline{\Omega_0}} d(x) > 0$ by (47). Since

$$\frac{\|d\|_{C(\overline{\Omega})}}{\varepsilon^2} < \frac{N \min_{x \in \overline{\Omega_0}} d(x)}{\varepsilon^2},$$

we can choose $\beta > 0$ such that

$$\frac{\|d\|_{C(\overline{\Omega})}}{\varepsilon^2} < \beta < \frac{N \min_{x \in \overline{\Omega_0}} d(x)}{\varepsilon^2}.$$

Then we see

$$\|d\|_{C(\overline{\Omega})} - \beta \varepsilon^2 < 0, \quad \min_{x \in \overline{\Omega_0}} d(x) - \frac{\beta}{N} \varepsilon^2 > 0,$$

which imply

$$\max_{x \in \overline{\Omega}} \varphi(x, t_0 + \varepsilon) = \max_{x \in \overline{\Omega}} \exp(\lambda(d(x) - \beta \varepsilon^2)) < 1 \tag{62}$$

and

$$\min \left\{ \varphi(x, t); \ x \in \overline{\Omega_0}, \ t_0 - \frac{\varepsilon}{\sqrt{N}} \le t \le t_0 + \frac{\varepsilon}{\sqrt{N}} \right\}$$

$$= \exp \left(\lambda \left(\min_{x \in \overline{\Omega_0}} d(x) - \beta \frac{\varepsilon^2}{N} \right) \right) =: \mu_0 > 1. \tag{63}$$

By (47), we have $d|_{\partial \Omega \setminus \Gamma} = 0$, so that

$$\max\{\varphi(x, t); \ x \in \overline{\partial \Omega \setminus \Gamma}, \ t_0 - \varepsilon \le t \le t_0 + \varepsilon\}$$

$$= \max\{\exp(\lambda(d(x) - \beta(t - t_0)^2)); \ x \in \overline{\partial \Omega \setminus \Gamma}, \ t_0 - \varepsilon \le t \le t_0 + \varepsilon\} \le 1. \tag{64}$$

Hence, since $\partial_x^\gamma u = 0$ on $\Gamma \times (0, T)$ for all $|\gamma| \le 3$, and

$$\|\partial_t u(\cdot, t)\|_{H^1(\Omega)} + \|\nabla \partial_t u(\cdot, t)\|_{H^1(\Omega)} + \|(\partial_t^2 u - a_2 \Delta \partial_t u - a_3 \Delta u)(\cdot, t)\|_{H^1(\Omega)} \le M$$

by the Sobolev embedding, recalling (7), (12) and (16), we have

$$\Phi(v) + \Phi(\nabla v) + \Phi(w)$$

$$= s^3 \left(\int_{(\partial \Omega \setminus \Gamma) \times (t_0 - \varepsilon, t_0 + \varepsilon)} + \int_{\Gamma \times (t_0 - \varepsilon, t_0 + \varepsilon)} \right) (|\partial_t u|^2 + |\nabla_{x,t} \partial_t u|^2 + |\nabla_{x,t} \nabla \partial_t u|^2$$

$$+ |\partial_t^2 u - a_2 \Delta \partial_t u - a_3 \Delta u|^2 + |(\nabla_{x,t} + \nabla^2)(\partial_t^2 u - a_2 \Delta \partial_t u - a_3 \Delta u)|^2 + |\nabla^3 \partial_t u|^2) e^{2s\varphi} \, dS \, dt$$

$$+s^5 \int_\Omega \{(|\partial_t u|^2 + |\nabla \partial_t u|^2 + |\nabla^2 \partial_t u|^2$$

$$+|\partial_t^2 u - a_2 \Delta \partial_t u - a_3 \Delta u|^2 + |\nabla(\partial_t^2 u - a_2 \Delta \partial_t u - a_3 \Delta u)|^2)(x, t_0 - \varepsilon)$$

$$+(|\partial_t u|^2 + |\nabla \partial_t u|^2 + |\nabla^2 \partial_t u|^2$$

$$+|\partial_t^2 u - a_2 \Delta \partial_t u - a_3 \Delta u|^2 + |\nabla(\partial_t^2 u - a_2 \Delta \partial_t u - a_3 \Delta u)|^2)(x, t_0 + \varepsilon)\}e^{2s\varphi(x, t_0 + \varepsilon)} dx$$

$$\leq Cs^3 M^2 \exp(2s \max_{x \in \overline{\partial\Omega \setminus \Gamma}, t_0 - \varepsilon \leq t \leq t_0 + \varepsilon} \varphi(x, t)) + Cs^5 M^2 \exp(2s \max_{x \in \overline{\Omega}} \varphi(x, t_0 + \varepsilon)).$$

Applying (62) and (64), we obtain

$$\Phi(v) + \Phi(\nabla v) + \Phi(w) \leq Cs^3 M^2 e^{2s} + Cs^5 M^2 e^{2s} \tag{65}$$

for all $s \geq s_0$.

Applying Proposition 1 (ii) and using $D_1 := \|u(\cdot, t_0)\|_{H^3(\Omega)} = 0$, we see

$$\int_{\Omega \times (t_0 - \varepsilon, t_0 + \varepsilon)} s^3 |w|^2 e^{2s\varphi} dx dt \leq Cs^5 M^2 e^{2s}$$

for all $s \geq s_0$. Shrinking $\Omega \times (t_0 - \varepsilon, t_0 + \varepsilon)$ to $\Omega_0 \times \left(t_0 - \frac{\varepsilon}{\sqrt{N}}, t_0 + \frac{\varepsilon}{\sqrt{N}}\right)$ on the left-hand side, we have

$$e^{2s\mu_0} \int_{t_0 - \frac{\varepsilon}{\sqrt{N}}}^{t_0 + \frac{\varepsilon}{\sqrt{N}}} \int_{\Omega_0} |w|^2 dx dt \leq Cs^2 M^2 e^{2s}$$

for all $s \geq s_0$. Dividing by $e^{2s\mu_0}$ and using $\mu_0 > 1$ by (63), we let $s \to \infty$ to obtain $w(x, t) = 0$ for $x \in \Omega_0$ and $t_0 - \frac{\varepsilon}{\sqrt{N}} < t < t_0 + \frac{\varepsilon}{\sqrt{N}}$.

Since $\Omega_0 \subset \Omega$ can be arbitrary provided that $\overline{\Omega_0} \subset \Omega \cup \Gamma$ and $\partial\Omega_0 \cap \partial\Omega$ is a nonempty relatively open subset of $\partial\Omega$ and $\overline{\partial\Omega_0 \cap \partial\Omega} \subset \Gamma$, changing all the possible t_0, ε, we reach

$$w = 0 \quad \text{in } \Omega \times (0, T). \tag{66}$$

Therefore in terms of $D_1 = 0$ and (65), estimate (23) with $m = 0$ yields

$$\int_{\Omega \times (t_0 - \varepsilon, t_0 + \varepsilon)} s^3 |\partial_t u|^2 e^{2s\varphi} dx dt \leq Cs^5 M^2 e^{2s}$$

for $s \geq s_0$. Similarly to (66), we can obtain $\partial_t u = 0$ in $\Omega \times (0, T)$. Applying $u(\cdot, t_0) = 0$ in Ω, we reach $u = 0$ in $\Omega \times (0, T)$. Thus the proof of Proposition 3 is complete.

5.2   Unique Continuation without Initial Condition

If in Proposition 3 we want to remove the condition $u(\cdot, t_0) = 0$ in Ω, then in terms
of the factorization (11), we should discuss at least the following:
if $u \in C_0^\infty(Q)$ satisfies

$$\partial_t^2 u - a_2(x, t)\Delta\partial_t u(x, t) - a_3(x, t)\Delta u = 0 \quad \text{in } Q, \tag{67}$$

then $u = 0$ in Q holds?

In a special case of $a_3 = 0$, we see that $u(x, t) := u_0(x)$ with arbitrary $u_0 \in C_0^\infty(\Omega)$ satisfies (67), then we cannot expect that $u \in C_0^\infty(Q)$ and (67) yield $u \equiv 0$ in Q. On the other hand, this counter-example does not work for the presence of the lower-order term $a_3\Delta$, and so for (67) with $a_3 \not\equiv 0$, we do not know whether the unique continuation holds or not. Since any lower-order terms do not matter for a Carleman estimate and there is a possibility that the lower-order term may recover the unique continuation, we are suggested to have to prepare a different method not based on the Carleman estimate in discussing the unique continuation for system (67). As a similar problem, we refer to Yamamoto [32] who considers a linearized Benjamin-Bona-Mahony equation:

$$\partial_t u(x, t) - \frac{\partial^2}{\partial x^2}\partial_t u(x, t) = p(x, t)\frac{\partial u}{\partial x}(x, t) + q(x, t)u(x, t)$$

for $x \in (0, \ell)$ and $t > 0$.

Acknowledgments The first author was supported by Grant-in-Aid for Scientific Research (S) 15H05740 of Japan Society for the Promotion of Science and by The National Natural Science Foundation of China (no. 11771270, 91730303). This work was supported by A3 Foresight Program "Modeling and Computation of Applied Inverse Problems" of Japan Society for the Promotion of Science and prepared with the support of the "RUDN University Program 5–100".

References

1. L. Beilina, M.V. Klibanov, *Approximate Global Convergence and Adaptivity for Coefficient Inverse Problems* (Springer, Berlin, 2012)
2. L. Beilina, M. Cristofol, S. Li, M. Yamamoto, Lipschitz stability for an inverse hyperbolic problem of determining two coefficients by a finite number of observations. Inverse Probl. **34**, 015001 (2018)
3. M. Bellassoued, M. Yamamoto, *Carleman Estimates and Applications to Inverse Problems for Hyperbolic Systems* (Springer, Tokyo, 2017)
4. L. Bjørnø, Characterization of biological media by means of their non-linearity. Ultrasonics **24**, 254–259 (1986)
5. R. Brunnhuber, P. Jordan, On the reduction of Blackstock's model of thermoviscous compressible flow via Becker's assumption. Int. J. Non-Linear Mech. **78**, 131–132 (2016)

6. A.L. Bukhgeim, M.V. Klibanov, Global uniqueness of class of multidimensional inverse problems. Sov. Math. Dokl. **24**, 244–247 (1981)
7. V. Burov, I. Gurinovich, O. Rudenko, E. Tagunov, Reconstruction of the spatial distribution of the nonlinearity parameter and sound velocity in acoustic nonlinear tomography. Acoust. Phys. **40**, 816–823 (1994)
8. C.A. Cain, Ultrasonic reflection mode imaging of the nonlinear parameter b/a: I. A theoretical basis. J. Acoust. Soc. Am. **80**, 28–32 (1986)
9. P. Cannarsa, G. Floridia, F. Gölgeleyen, M. Yamamoto, Inverse coefficient problems for a transport equation by local Carleman estimate. Inverse Probl. **35**, 105013 (2019)
10. P. Cannarsa, G. Floridia, M. Yamamoto, *Observability Inequalities for Transport Equations Through Carleman Estimates*. in "Trends in Control Theory and Partial Differential Equations", Springer INdAM Series, Springer, Cham. **32**, 69–87 (2019)
11. C. Cavaterra, A. Lorenzi, M. Yamamoto, A stability result via Carleman estimates for an inverse source problem related to a hyperbolic integro-differential equation. Comput. Appl. Math. **25**, 229–250 (2006)
12. A.V. Fursikov, O.Yu. Imanuvilov, *Controllability of Evolution Equations* (Seoul National University, Seoul, 1996)
13. F. Gölgeleyen, M. Yamamoto, Stability for some inverse problems for transport equations. SIAM J. Math. Anal. **48**, 2319–2344 (2016)
14. X. Huang, O.Yu. Imanuvilov, M. Yamamoto, Stability for inverse source problems by Carleman estimates. Inverse Problems **36**, 125006 (2020)
15. N. Ichida, T. Sato, M. Linzer, Imaging the nonlinear ultrasonic parameter of a medium. Ultrason. Imaging **5**, 295–299 (1983). PMID: 6686896
16. O.Yu. Imanuvilov, Controllability of parabolic equations. Math. Sb. **186**, 879–900 (1995)
17. O.Yu. Imanuvilov, M. Yamamoto, Lipschitz stability in inverse parabolic problems by the Carleman estimate. Inverse Probl. **14**, 1229–1245 (1998)
18. O.Yu. Imanuvilov, M. Yamamoto, Global Lipschitz stability in an inverse hyperbolic problem by interior observations. Inverse Probl. **17**, 717–728 (2001)
19. O.Yu. Imanuvilov, M. Yamamoto, Global uniqueness and stability in determining coefficients of wave equations. Commun. Partial Differ. Equ. **26**, 1409–1425 (2001)
20. O.Yu. Imanuvilov, M. Yamamoto, Determination of a coefficient in an acoustic equation with a single measurement. Inverse Probl. **19**, 157–171 (2003)
21. O.Yu. Imanuvilov, M. Yamamoto, Carleman estimate and an inverse source problem for the Kelvin-Voigt model for viscoelasticity. Inverse Probl. **35**, 125001 (2019)
22. O.Yu. Imanuvilov, M. Yamamoto, Carleman estimate for linear viscoelasticity equations and an inverse source problem. SIAM J. Math. Anal. **52**, 718–791 (2020)
23. V. Isakov, *Inverse Problems for Partial Differential Equations* (Springer, Berlin, 2006)
24. B. Kaltenbacher, Mathematics of nonlinear acoustics. Evol. Equ. Control Theory **4**, 447–491 (2015)
25. B. Kaltenbacher, Well-posedness of a general higher order model in nonlinear acoustics. Appl. Math. Lett. **63**, 21–27 (2017)
26. B. Kaltenbacher, M. Thalhammer, Fundamental models in nonlinear acoustics - Part I. analytical comparison. Math. Models Methods Appl. Sci. **28**, 2403–2455 (2018)
27. M.V. Klibanov, Inverse problems and Carleman estimates. Inverse Probl. **8**, 575–596 (1992)
28. M.V. Klibanov, Carleman estimates for global uniqueness, stability and numerical methods for coefficient inverse problems. J. Inverse Ill-Posed Probl. **21**, 477–560 (2013)
29. M.V. Klibanov, A. Timonov, *Carleman Estimates for Coefficient Inverse Problems and Numerical Applications* (VSP, Utrecht, 2004)
30. P. Loreti, D. Sforza, M. Yamamoto, Carleman estimate and application to an inverse source problem for a viscoelasticity model in anisotropic case. Inverse Probl. **33**, 125014 (2017)
31. F. Varray, O. Basset, P. Tortoli, C. Cachard, Extensions of nonlinear b/a parameter imaging methods for echo mode. IEEE Trans. Ultrason. Ferroelectr. Freq. Control **58**, 1232–1244 (2011)

32. M. Yamamoto, One unique continuation for a linearized Benjamin-Bona-Mahony equation. J. Inverse Ill-Posed Probl. **11**, 537–543 (2003)
33. M. Yamamoto, Carleman estimates for parabolic equations and applications. Inverse Probl. **25**, 123013 (2009)
34. D. Zhang, X. Gong, S. Ye, Acoustic nonlinearity parameter tomography for biological specimens via measurements of the second harmonics. J. Acoust. Soc. Am. **99**, 2397–2402 (1996)
35. D. Zhang, X. Chen, X. Gong, Acoustic nonlinearity parameter tomography for biological tissues via parametric array from a circular piston source theoretical analysis and computer simulations. J. Acoust. Soc. Am. **109**, 1219–1225 (2001)

Printed in the United States
by Baker & Taylor Publisher Services